A New Chemistry

A USED BOOK FROM
UNIVERSITY BOOK STORE
North Gate - Culpepper Plaza

The New School Series
General Editor
R. Stone, M.A., A.Inst.P.

A New Biology
K. G. Brocklehurst, M.A., M.I.Biol.
H. Ward, B.Sc., M.I.Biol.

A New Chemistry
S. Clynes, B.Sc., C.Chem., F.R.I.C.
D. J. W. Williams, M.A.
J. S. Clarke, B.Sc., M.A., C.Chem., F.R.I.C.

General School Physics
(in three volumes)
R. Stone, M.A., A.Inst.P.
N. Bronner, M.A.

New School Geography
(in five volumes)
F. R. Dobson, B.A.
H. E. Virgo, M.A.

A New Geology
M. J. Bradshaw, M.A.

A New Chemistry

S. CLYNES, B.Sc., C.Chem., F.R.I.C.
Head of Science, The Manchester Grammar School

D. J. W. WILLIAMS, M.A.
Headmaster, Trinity School, Carlisle

J. S. CLARKE, B.Sc., M.A., C.Chem., F.R.I.C.
Head of Science, Alleyn's School, London

HODDER AND STOUGHTON
LONDON SYDNEY AUCKLAND TORONTO

ISBN 0 340 20092 8

First printed 1971. Reprinted (twice) 1972
Second edition 1972. Reprinted 1973
Third edition 1975. Reprinted 1976

Printed in Great Britain for
Hodder and Stoughton Educational,
a division of Hodder and Stoughton Ltd.,
Mill Road, Dunton Green, Sevenoaks, Kent, TW13 2YD.
by Hazell Watson and Viney Ltd, Aylesbury.

Editor's Foreword

We live in a rapidly changing world; some might say in a too rapidly changing one. Today ideas can alter as much in a decade as in the whole of the preceding century. As the kaleidoscopic patterns of syllabus and method pass before us, it is not always easy to distinguish the important and permanent from the ephemeral. It is our hope that, in the New School Series, we shall be able to produce a group of books which will help colleagues in the classroom, and their pupils, to meet the challenge of the next ten years.

Our main purpose is to provide a wide range of books, covering both the Arts and Sciences at a number of levels, which will allow the teacher increased latitude in his approach, and offer him full scope to develop the creative aspects of his work. It must not be forgotten that the suitability of a particular text for a given form depends largely on the way in which the teacher uses it: the speed and depth of the work will be dictated by the abilities and interests of the pupils rather than by the wishes of the teacher or the nature of the text. Many of the new books will be suitable for all but the lower streams of the comprehensive schools; others will express the newer conceptual, as opposed to factual approach to teaching, but may be contained within somewhat narrower academic boundaries. We shall be greatly in debt to the many teachers, professional bodies, and others whose untiring efforts have done so much to change the pattern of teaching in recent years. Nor must we forget those pupils who, whether consciously or not, have been the guinea pigs in the experiments which were necessary to prove the new ideas.

We hope to arrange for books to be written by teams of two or more experienced teachers who have tried out the new methods and syllabuses in the classroom, and who will be able to engender in their readers the same enthusiasm which they have instilled into their pupils. It is this dual interest of those who teach and those who are taught which is the key to all successful learning, a process in which we hope to play our part.

R. STONE

Preface

The fifteen years following the publication of General School Chemistry saw many changes in syllabus and approach to the study of chemistry. This book, like its predecessor, aims to cover fully the requirements of the Ordinary Level Examination for the General Certificate of all the Examining Boards in the British Isles. The modern chemist must be interested in the materials involved from the time that they are found, in their processing and in the study of their properties and uses, to the time that they are discarded: it is not for him to live in splendid isolation in a laboratory. He is also a geographer and an economist, considers the human aspects of scientific endeavour and industry, and is aware of the possible dangers of the materials studied.

The chapters have been re-ordered but users of the previous book should soon be able to find their way around this one, in which basic chemistry is followed by a study of atoms and molecules, then the important systems of classification, and some miscellaneous topics. The final section contains the chemistry of the elements and their compounds in periodic table order; it is therefore a section to be consulted at many levels in the pursuit of knowledge.

There is a tendency in academic circles to forget the contribution that industry makes to our wealth and comfort, as well as, occasionally, to our lack of health and discomfort: in modern life there are advantages and disadvantages at every turn, and quality always competes with quantity. Biographies of famous scientists have been included to emphasize that science is one aspect of human endeavour.

The authors are grateful to many people, to the Examining Boards and to many firms and public bodies for their assistance in the preparation of this book. Particular thanks for their assistance with the revision are due to Messrs. M. N. Carver, M. P. Dodds, G. Howard and J. D. Ladd (all of Alleyn's School), Mrs. P. Porter (of Bury Convent School) and Mr. C. T. Robertson (of Archbishop Tenison's Grammar School); and, for typing the entire manuscript, to Mrs. J. S. Clarke. Mr. R. Stone, Mr. P. R. Turk (of Manchester Grammar School) and Professor M. L. McGlashan (of Exeter University) also gave helpful advice. The responsibility for any errors of omission or of commission remains with the authors and they would be grateful for their attention being drawn to any that exist.

S.C. D.J.W.W. J.S.C.

Acknowledgements

We are indebted to many firms and public bodies for supplying information, advice and photographs (*) in the compilation of this book:

*A.E.I. Scientific Apparatus Ltd.
Albright & Wilson Ltd.
The Aluminium Federation
*The Associated Octel Co. Ltd.
Association of Jute Spinners & Manufacturers
*British Aircraft Corporation Ltd.
*British Aluminium Co. Ltd.
The British Gas Corporation
British Gypsum Ltd.
*British Insulated Callender's Cables Ltd.
British Man-Made Fibres Federation
The British Oxygen Co. Ltd.
The British Petroleum Co. Ltd.
British Standards Institute
*British Steel Corporation
British Sugar Bureau
*British Sulphur Corporation Ltd.
*Cambridge Scientific Instruments Ltd. (and Dr. R. Blaschke)
*Camera Press (camera portrait of Sir Lawrence Bragg by Tom Blan)
Cement Makers' Federation
Central Statistical Office (Monthly Digest of Statistics)
*Cheddar Caves
Commonwealth Secretariat, Commodities Division
*Copper Development Association
Cotton Research Corporation
*Cunard Steam Ship Co. Ltd.
Department of Industry & Commerce, Eire
Engelhard Industries Ltd.
English China Clays Sales Co. Ltd.
Europlastics Monthly
The Ever Ready Co. (G.B.) Ltd.
Glass Textile Association
*High Duty Alloys Ltd.
Imperial Chemical Industries Ltd., Agriculture, *Dyestuffs, Fibres, Mond (especially Buxton Lime Group) and *Plastics Divisions
The Institute of Geological Sciences (U.K. Mineral Statistics)
The Institute of Petroleum
International Union of Pure and Applied Chemistry and Butterworths Scientific Publications for permission to reprint the Table of Relative Atomic Masses
International Wool Secretariat
Laporte Industries Ltd.
*Lead Development Association
*Magnesium Elektron Ltd.
*Mansell Collection (camera portrait of Sir J. J. Thompson by E. O. Hoppé)
The Mining Journal (Annual Review)
Ministry of Commerce, Northern Ireland
National Coal Board
The National Sulphuric Acid Association Ltd.
Nitrate Corporation of Chile Ltd.
The Permutit Co. Ltd.
The Plastics Institute
*Radio Times Hulton Picture Library
*The Royal Society
R.T.Z. Services Ltd.
*Shell International Petroleum Ltd.
*Sterling Metals Ltd.
Tin Research Institute
World Bureau of Metal Statistics
*Zinc Development Association

The statistics of production quoted are the latest available and are for the whole world.

We are also indebted to the Examining Boards for permission to reproduce questions from their examination papers; the responsibility of rewording many of them using modern nomenclature and units is ours:

[A] Associated Examining Board for the G.C.E.
[C] University of Cambridge Local Examinations Syndicate
[CO] University of Cambridge Local Examinations Syndicate (Oversea Centres)
[E] Eire, The Department of Education, Secondary Education Branch
[J] Joint Matriculation Board
[L] University of London School Examinations Council
[Nf] Nuffield course papers as published by the University of London
[NI] Northern Ireland G.C.E. Committee
[O] Oxford Delegacy of Local Examinations
[OC] Oxford and Cambridge Schools Examination Board
[Sc] Scottish Certificate of Education Examination Board
[S] Southern Universities' Joint Board for School Examinations
[W] Welsh Joint Education Committee

Finally we must mention the series of books published by I.C.I. and the O.U.P.:

—	Sulphuric Acid
D. W. F. Hardie:	The Electrolytic Manufacture of Chemicals from Salt
A. J. Harding:	Ammonia: Manufacture and Uses
F. D. Miles:	Nitric Acid
F. P. Stowell:	Limestone as a Raw Material in Industry

describe carfully how you would separate a sample of carbon from a mixture of carbon and Salt

Contents

A Modern Form of the Periodic Table

Period \ Group	I	II											III	IV	V	VI	VII	0
1								1 H	2 He									
2	3 Li	4 Be											5 B	6 C	7 N	8 O	9 F	10 Ne
3	11 Na	12 Mg	Transition (d-block) elements ⟷										13 Al	14 Si	15 P	16 S	17 Cl	18 Ar
4	19 K	20 Ca	21 Sc	22 Ti	23 V	24 Cr	25 Mn	26 Fe	27 Co	28 Ni	29 Cu	30 Zn	31 Ga	32 Ge	33 As	34 Se	35 Br	36 Kr
5	37 Rb	38 Sr	39 Y	40 Zr	41 Nb	42 Mo	43 Tc	44 Ru	45 Rh	46 Pd	47 Ag	48 Cd	49 In	50 Sn	51 Sb	52 Te	53 I	54 Xe
6	55 Cs	56 Ba	57 La*	72 Hf	73 Ta	74 W	75 Re	76 Os	77 Ir	78 Pt	79 Au	80 Hg	81 Tl	82 Pb	83 Bi	84 Po	85 At	86 Rn
7	87 Fr	88 Ra	89 Ac**	104 Rf?	105 Ha?	106 ?						112 ?						

*	58 Ce	59 Pr	60 Nd	61 Pm	62 Sm	63 Eu	64 Gd	65 Tb	66 Dy	67 Ho	68 Er	69 Tm	70 Yb	71 Lu
**	90 Th	91 Pa	92 U	93 Np	94 Pu	95 Am	96 Cm	97 Bk	98 Cf	99 Es	100 Fm	101 Md	102 No	103 Lr

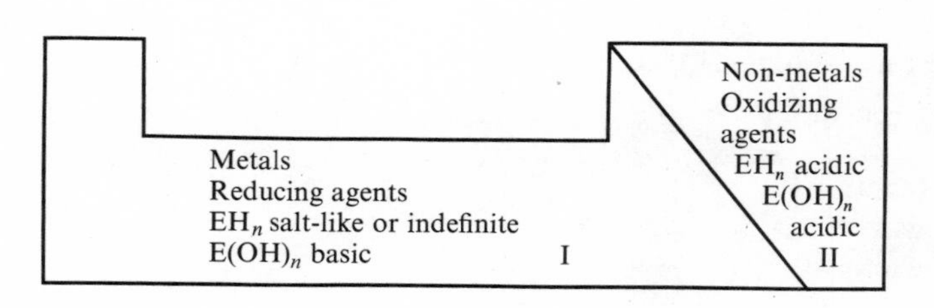

Binary compounds between elements in area I and in area II: electrovalent
Binary compounds of elements both in area II: covalent
Compounds and mixtures of elements all in area I: alloys

Table of Relative Atomic Masses

Based on the Assigned Relative Atomic Mass of $^{12}C = 12$, the following values apply to elements as they exist in materials of terrestrial origin and to certain artificial elements.

Atomic Number	*Name*	*Symbol*	A_r
89	Actinium	Ac	(227)
13	Aluminium	Al	26·98
95	Americium	Am	(243)
51	Antimony	Sb	121·8
18	Argon	Ar	39·95
33	Arsenic	As	74·92
85	Astatine	At	(210)
56	Barium	Ba	137·3
97	Berkelium	Bk	(247)
4	Beryllium	Be	9·012
83	Bismuth	Bi	209·0
5	Boron	B	10·81
35	Bromine	Br	79·90
48	Cadmium	Cd	112·4
55	Caesium	Cs	132·9
20	Calcium	Ca	40·08
98	Californium	Cf	(251)
6	Carbon	C	12·01
58	Cerium	Ce	140·1
17	Chlorine	Cl	35·45
24	Chromium	Cr	52·00
27	Cobalt	Co	58·93
29	Copper	Cu	63·55
96	Curium	Cm	(247)
66	Dysprosium	Dy	162·5
99	Einsteinium	Es	(254)
68	Erbium	Er	167·3
63	Europium	Eu	152·0
100	Fermium	Fm	(257)
9	Fluorine	F	19·00
87	Francium	Fr	(223)
64	Gadolinium	Gd	157·3
31	Gallium	Ga	69·72
32	Germanium	Ge	72·6
79	Gold	Au	197·0
72	Hafnium	Hf	178·5
2	Helium	He	4·003
67	Holmium	Ho	164·9
1	Hydrogen	H	1·008
49	Indium	In	114·8
53	Iodine	I	126·9
77	Iridium	Ir	192·2
26	Iron	Fe	55·85
36	Krypton	Kr	83·80
57	Lanthanum	La	138·9
103	Lawrencium	Lr	(260)
82	Lead	Pb	207·2
3	Lithium	Li	6·94
71	Lutetium	Lu	175·0
12	Magnesium	Mg	24·30
25	Manganese	Mn	54·94
101	Mendelevium	Md	(258)
80	Mercury	Hg	200·6
42	Molybdenum	Mo	95·9
60	Neodymium	Nd	144·2
10	Neon	Ne	20·18
93	Neptunium	Np	(237)
28	Nickel	Ni	58·70
41	Niobium	Nb	92·91
7	Nitrogen	N	14·01
102	Nobelium	No	(259)
76	Osmium	Os	190·2
8	Oxygen	O	16·00
46	Palladium	Pd	106·4
15	Phosphorus	P	30·97
78	Platinum	Pt	195·1
94	Plutonium	Pu	(244)
84	Polonium	Po	(209)
19	Potassium	K	39·10
59	Praseodymium	Pr	140·9
61	Promethium	Pm	(145)
91	Protactinium	Pa	231·0
88	Radium	Ra	226·0
86	Radon	Rn	(222)
75	Rhenium	Re	186·2
45	Rhodium	Rh	102·9
37	Rubidium	Rb	85·47
44	Ruthenium	Ru	101·1
62	Samarium	Sm	150·4
21	Scandium	Sc	44·96
34	Selenium	Se	78·9
14	Silicon	Si	28·09
47	Silver	Ag	107·9
11	Sodium	Na	22·99
38	Strontium	Sr	87·62
16	Sulphur	S	32·06
73	Tantalum	Ta	180·9
43	Technetium	Tc	(97)
52	Tellurium	Te	127·6
65	Terbium	Tb	158·9
81	Thallium	Tl	204·4
90	Thorium	Th	232·0
69	Thulium	Tm	168·9
50	Tin	Sn	118·7
22	Titanium	Ti	47·9
74	Tungsten	W	183·8
92	Uranium	U	238·0
23	Vanadium	V	50·94
54	Xenon	Xe	131·3
70	Ytterbium	Yb	173·0
39	Yttrium	Y	88·91
30	Zinc	Zn	65·38
40	Zirconium	Zr	91·22

Basic Chemistry

1 Introduction

1.1 Historical Aspects

The earth, the air and the sea provide the materials we need for our lives. Sometimes they are put to good use, sometimes not. The age of the earth has been estimated as several thousand million years but man has only inhabited it for the last hundred thousand years. The continents reached their present shape in comparatively modern times—the sea at one time covered most of the British Isles, and even today mountains in the form of volcanoes occasionally appear, e.g. Surtsey near Iceland. Man's mastery over his environment has been attained slowly: in the past there have been stone, bronze and iron ages. The pace of development started to accelerate noticeably in the eighteenth century when the Industrial Revolution swept across Britain: it has not relaxed since. Technology and pure science go hand in hand; advances in one usually lead—though in Britain often painfully slowly—to advances in the other. The Revolution has been accompanied by the congregation of people in towns and a decrease in the population of the countryside; the total population has increased immensely. All these developments bring attendant problems.

1.2 Raw Materials and Factories

The earliest rocks are of the igneous type: they were formed by the crystallization of magma (the initial fluid mass of elements and compounds). These igneous rocks have been subjected to the action of wind, rain, sea and heat: they weather slowly to give most of the sedimentary rocks. Some sedimentary rocks have been formed by precipitation (e.g. chalk, which also comes from sea shells), and evaporation and crystallization, e.g. sodium chloride and calcium sulphate. Metamorphic rocks are rocks that have been formed from igneous or sedimentary rocks by intense heat or pressure (or both). Most of our mineral resources are obtained from the sedimentary rocks. See figure 1.1.

96·5% by mass of the ocean consists of water, leaving 3·5% of solids which can be obtained by evaporating off the water. The abundances of substances in sea water are given in figure 1.2.

The air often contains up to 1% water vapour and because this is a variable quantity dried air is usually analyzed. See figure 1.3.

The relative importance of the various energy sources in the world is shown in figure 1.4.

The import–export balance sheet of the United Kingdom is shown opposite. The United Kingdom mines about 1486 million pounds worth of minerals every year (see figure 1.5) but imports of crude materials and minerals cost more than twice that value (see figure 1.6). Whilst iron ore is obtained on a large scale at home and from abroad the foreign ores are much richer in iron. The home production of tin and lead is very small and most of these metals are obtained from abroad. Also, most of our petroleum is imported and hence arises the importance of developing our resources of natural gas and petroleum.

There are many problems in the siting of an industry: supply of materials, availability of labour, supply of fuel and cooling water, distribution of products, etc. The sites of some mines and factories are shown in figures 1.8 and 1.9 and for further information see figures 24.1, 25.3, 26.3, 29.3, 29.5 and 29.7. The industries considered are those in which elements and compounds of interest at this stage in the study of chemistry are involved. They do not show the sites of factories where those substances are processed further, e.g. they show where the aluminium ores are converted into aluminium but not where the metal is fabricated into sheets, rods and tubes for use in engineering.

Not all materials manufactured in the British Isles are consumed there; the export trade must thrive or it is impossible to import the food that we need and like. British farmers grow 60% of the food consumed.

1.3 People

As the world's population approaches 4000

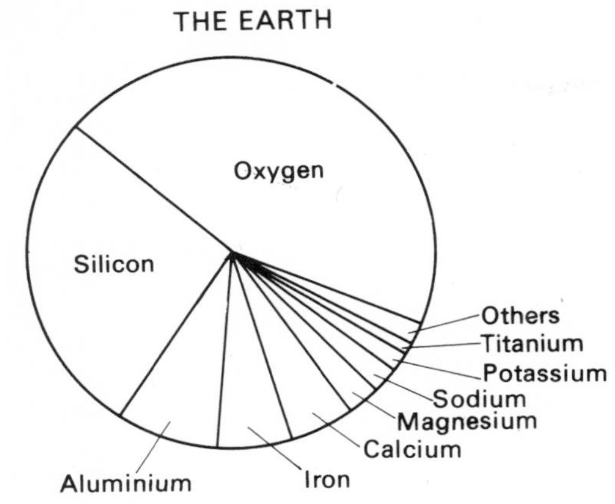

	% by mass		% by mass
Oxygen	45·0	Magnesium	2·8
Silicon	27·0	Sodium	2·3
Aluminium	8·0	Potassium	1·7
Iron	5·8	Titanium	0·86
Calcium	5·1	All others, total	1·4

Figure 1.1
Abundances of substances in the crustal rocks (whether present as elements or compounds)

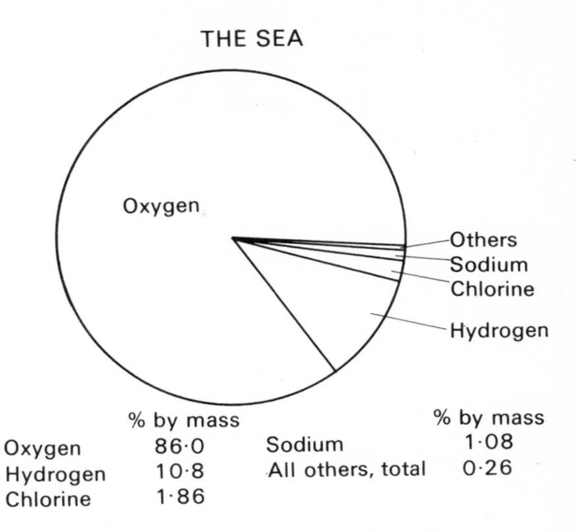

	% by mass		% by mass
Oxygen	86·0	Sodium	1·08
Hydrogen	10·8	All others, total	0·26
Chlorine	1·86		

Although present in a very small proportion, magnesium (0·13%) and bromine (0·0067%) are obtained from sea water

Figure 1.2
Abundances of substances in sea water (whether present as elements or compounds)

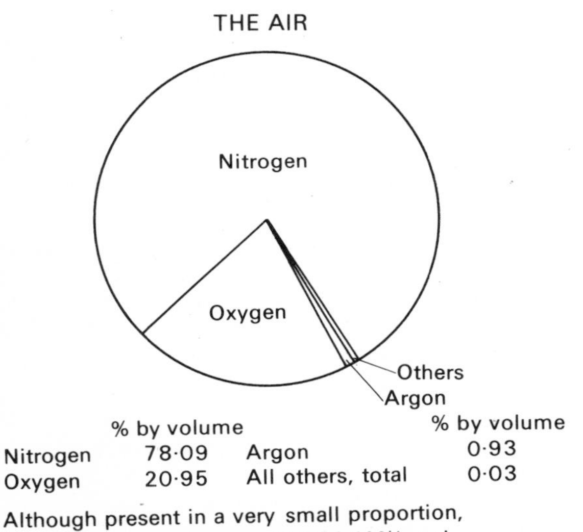

	% by volume		% by volume
Nitrogen	78·09	Argon	0·93
Oxygen	20·95	All others, total	0·03

Although present in a very small proportion, neon (0·0018%), krypton (0·000 11%) and xenon (0·000 01%) are obtained from air. Carbon dioxide 0·03%

Figure 1.3
Abundances of substances in dried air

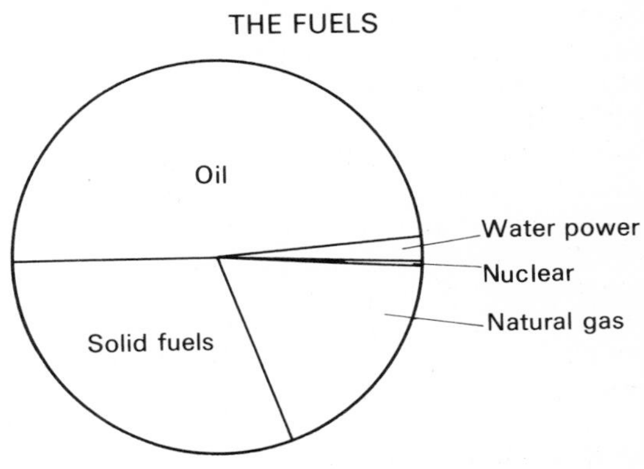

	Percentage		
	U.K.	U.S.A.	WORLD
Oil	52·0	47·2	48·5
Solid fuels	33·6	17·8	30·6
Natural gas	13·2	32·8	18·6
Water power	0·2	1·4	1·9
Nuclear	1·0	0·8	0·4

Figure 1.4
Primary energy consumption in 1973

U.K. Imports and Exports, 1973, in millions of £'s

	Import	Export
Food and live animals	2714	512
Beverages and tobacco	383	363
Crude materials, inedible	1835	416
Minerals, fuels, lubricants	1727	370
Animal and vegetable oil, fats	130	17
Chemicals	897	1272
Manufactured goods	3 383	3 258
Machines and transport equipment	3 293	4 775
Miscellaneous manufactured goods	1 340	1 143
Miscellaneous	151	320
Total	15 853	12 446

Mineral	Chemical Compound	Thousands of tonnes
Coal	Carbon (complex)	132 000
Natural gas and natural gas liquids	Methane	39 400
Petroleum	Carbon (complex)	88
Tin ores	Tin(IV) oxide	10
Lead ores	Lead(II) sulphide, etc.	8
Zinc ores	Zinc(II) sulphide, etc.	8
Iron ore and ironstone	Iron(II) carbonate	7 105
Sands, gravel, sandstone	Silicon dioxide	198 000
Limestone, dolomite, chalk, chert and flint	Calcium carbonate	127 000
Clay, shale, slate and fireclay		37 000
Igneous rock		49 000
Salt	Sodium chloride	8 400
Gypsum and anhydrite	Calcium sulphate	4 200
Fluorspar	Calcium fluoride	233
China clay, china stone, and potters' clay		4 500
Miscellaneous		138

Total mass about 507 million tonnes
Total value £1486 million

Figure 1.5
The United Kingdom mineral production 1973

Imports: metals in 1973	*Thousands of tonnes of ores and concentrates*	*Thousands of tonnes of metal (pure or in alloys)*
Iron and steel	23 160	3624
Manganese	588	—
Titanium	408	—
Aluminium	299	287
Zinc	140	229
Tin	50	—
Nickel	48	—
Copper	—	467
Lead	—	218

Imports: other non-food in 1973	*Thousands of tonnes of material*
Timber	765
Wood pulp	2 544
Paper and board	3 636
Rubber	326
Petroleum	133 968
Jute	72
Cotton	168
Wool	151

Imports: food in 1973	*Thousands of tonnes of material*
Meat and bacon	1244
Butter, cheese and margarine	674
Wheat and flour	3738
Maize and barley	3623
Fruit	1837
Sugar	1997
Coffee, cocoa and tea	438
Oil cake and meal	859
Edible oils	1872

Figure 1.6
Some United Kingdom mineral and other imports 1973

million that of the United Kingdom is about 56 million. One of the problems which the world faces is that of feeding its people. The reserves of nitrogen in the soil are depleted by continued cultivation and the only rapid way to build these up again is to rest arable soils under grass. The world applies about 60 million tonnes of fertilizer to the soil but one quarter of the land receives three quarters of the fertilizer.

An analysis of the population of the United Kingdom is shown in figure 1.7. A large proportion of people are in industries which make chemicals or fabricate metals into useful articles. A knowledge of chemistry is thus useful to many people, but only a few of them will be intimately concerned with it day by day.

The geologist studies the rocks, the agriculturist studies the land and its crops, the oceanographer studies the seas and the meteorologist studies the atmosphere. Together they tell us about our resources and their possible utilization. The chemist and the physicist study the properties of these materials and of the substances which can be derived from them. The chemist is interested in where the materials come from, their composition, the methods of making pure substances from the natural materials, the reactions of one material with another and the uses to which the natural and derived materials can be put. He seeks to offer an explanation of these properties in terms of the structures of the materials and of the energy relationships in their reactions. The physicist provides the picture of matter as composed of minute particles and studies the properties of materials, seeking eventually to explain them in detail. There are many instances of overlap between chemistry and physics, e.g. study of gases and electricity, but the divergent approaches and interests merit separate studies, e.g. the physicist is interested in obtaining detailed relationships between the pressure, volume and temperature of a gas whereas the chemist is possibly satisfied with an approximate relationship and then proceeds to study the reaction of one gas with another. Bunsen said (in German) 'A chemist who is not a physicist also is no good', to which one might add today that he should be a mathematician as well.

1.4 Nomenclature

People give names to the distinct substances that they discover; sometimes this is done according to the place of origin or to the appearance or to the name of the person, e.g. Chile saltpetre, ruby ore, Glauber's salt.

The chemist then analyzes these materials in order to establish their composition. He prefers to give a name to a substance which indicates its composition, particularly if this is fairly simple. If a substance consists of one element only, then the name of the element is used rather than a trivial or descriptive name, e.g. oxygen, rather than the 'vital' gas, and hydrogen, rather than the 'lightest' gas. If a substance consists of two elements then the metal (or more metallic substance) is named first and the non-metal (or less metallic substance) second

1. Total population	55 933 000
of which male	48·6%
and female	51·4%
2. Working population	25 490 000
Employed labour force	97·7%
Unemployed	2·3%
3. Employed labour force	24 914 000
Employers and self-employed	7·6%
Employees in employment	91·0%
H.M. forces	1·4%
4. Employees in employment	22 662 000
of which male	60·8%
and female	39·2%
5. Employees in employment	22 662 000
Professional and scientific services	14·3%
Distributive trades	12·1
Transport and communication	6·7
Construction	6·1
Insurance, banking, finance and business services	4·7
Mechanical engineering	4·3
Local government services	4·3
Electrical engineering	3·6
Vehicles	3·5
Catering and hotels	3·5
Food, drink and tobacco	3·3
National government service	2·7
Textiles	2·6
Paper, printing and publishing	2·5
Metal goods	2·5
Metal manufacture	2·3
Clothing and footwear	1·9
Agriculture, forestry and fishing	1·9
Chemicals and allied industries	1·9
Mining and quarrying	1·6
Gas, electricity and water	1·5
Bricks, pottery, glass, cement, etc.	1·3
Timber, furniture, etc.	1·3
Shipbuilding and marine engineering	0·8
Instrument engineering	0·7
Leather, leather goods and fur	0·2
Coal and petroleum products	0·2
Other manufacturing industries	1·5
Miscellaneous services	6·0

Figure 1.7
The population of the United Kingdom 1973

Figure 1.8
The extraction and manufacturing centres for elements and inorganic compounds in the British Isles

with the ending -ide, e.g. calcium oxide, aluminium oxide, magnesium oxide.

There are many examples of substances containing three elements one of which is oxygen associated with another non-metal: in this case the ending of the word naming the non-metal is -ate, e.g. sodium carbonate, calcium sulphate and potassium nitrate.

If a metal forms two compounds with a non-metal it has been customary to use the endings -ic and -ous on the name of the metal. Thus copper, which has the symbol Cu from the Latin 'cuprum', forms two chlorides: cuprous chloride and cupric chloride. An obvious difficulty arises when two elements form three compounds; also a source of confusion is found when comparing the compounds of different metals. The result is that nowadays the Stock system is used more and more: the chlorides mentioned are known as copper(I) chloride and copper(II) chloride. For further details see sections 26.9 and 26.12.

The chemist will therefore refer to the ores in terms of their composition in a manner that is as informative as possible. Thus Chile saltpetre is sodium nitrate, ruby ore is copper(I) oxide and Glauber's salt is sodium sulphate-10-water (or decahydrate). Some trivial names such as water (hydrogen oxide) and ammonia (hydrogen nitride) survive but others such as caustic soda (sodium hydroxide), caustic potash (potassium hydroxide), quicklime (calcium oxide), slaked lime (calcium hydroxide) and limestone (calcium carbonate) are being used less and less in chemical literature. A recent example of the possibilities of confusion is the discovery of potassium chloride in Yorkshire—as a mineral it is referred to as potash, but to the chemist this is more likely to mean potassium carbonate or potassium hydroxide.

1.5 Units

The investigation of the properties of materials received a tremendous stimulus with the improvements in weighing which were made about 200 years ago. Thus Black (1755) found

---

The map shows some of the principal centres and a detailed study of an area should soon lead to the discovery of many other important places.

General: Ashford, Ipswich, London (Dagenham and Romford) and Poole.
Hydrogen: usually from petroleum feedstock (see figure 29.3) but some also from sodium chloride solution (see figure 24.1).
Sodium, sodium hydroxide, chlorine and hydrogen; sodium carbonate: see figure 24.1, e.g. Runcorn and Widnes.
Magnesium oxide: Hartlepool.
Calcium compounds: for occurrence see figure 25.1.
Iron and steel: see figure 26.2.
Copper (new): Gortdrum and Tynagh.
Copper (recovery): Birmingham, Prescot and Swansea.
Zinc (new): Avonmouth, Silvermines and Tynagh.
Zinc (recovery): Newcastle upon Tyne and Swansea.
Aluminium: Fort William, Holyhead, Invergordon, Kinlochleven and Lynemouth.
Aluminium oxide: Burntisland.
Carbon (black) and carbon dioxide: from petroleum (see figure 29.3).
China clay: St. Austell and Plymouth.
Glass: Birkenhead, Doncaster, Gateshead, Glasgow, London, Pontypool, St. Asaph and St. Helens.
Silicones: Barry and Stevenston.
Tin: Camborne, Redruth, St. Just and Truro.
Tin plating: see figure 26.2.
Lead (new): Avonmouth, Silvermines and Tynagh.
Lead (recovery): Chester, Glasgow, Leeds, London, Manchester, Matlock (Darley Dale) and Newcastle upon Tyne.
Liquid air: at most large steelworks and at many large towns, e.g. Corby, Glasgow, London, Manchester, Port Talbot, Stockton, Widnes and Wolverhampton.
Ammonia and nitric acid: Arklow, Billingham, Heysham, Immingham and Severnside; usually from air and petroleum (see figure 29.3). Nitric acid is also made at Avonmouth, Belfast, Huddersfield, Leith, Stevenston and Wilton.
Phosphoric acid: Arklow, Severnside and Whitehaven.
Fertilizers: Arklow, Avonmouth, Barton-upon-Humber, Belfast, Billingham, Chedburgh, Dublin, Ince, Immingham, Micheldever, Severnside, Stanford-le-Hope and Widnes.
Hydrogen peroxide: Warrington.
Sulphuric acid: Arklow, Avonmouth, Belfast, Billingham, Bradford, Carrickfergus, Cork, Dublin, Grimsby (Great Coates and Pyewipe), Holywell (Greenfield), Huddersfield, Immingham, Leith, London, Runcorn, St. Helens, Severnside, Stallingborough, Swansea (Llansamlet), Trafford Park, Whitehaven and Widnes.
Bromine: Amlwch.

Figure 1.9
The manufacturing centres for organic compounds in the British Isles

the relationships between calcium hydroxide, calcium carbonate, sodium hydroxide and potassium hydroxide and these results were in the background when Dalton proposed his atomic theory.

The unit of mass is the kilogram **(1 kg = 1000 grams)**: this is a reasonable unit when weighing people, food and many other materials but it is much too small when dealing with the quantities of materials manufactured in a country—**a metric ton (1000 kilograms, or 1 tonne)** is more suitable. At the other extreme in the world of atoms the kilogram is much too large and a relative scale in which the hydrogen atom, the lightest atom, is given a mass of 1 is more useful. The relative atomic mass (weight) scale runs from 1 to about 260, the heaviest atoms known being 260 times as heavy as hydrogen atoms. In terms of mass a hydrogen atom weighs about

0·000 000 000 000 000 000 000 000 001 kg
(or 10^{-27} kg)

(see figure 1.10).

The unit of volume is the cubic metre (correctly abbreviated as m^3); this is a reasonable unit when measuring volumes of air or petroleum or concrete on a commercial scale but the unit more familiar to the average person is one thousandth of this i.e. 1 cubic decimetre **(1 dm^3, popularly known as 1 litre). In the laboratory many experiments are done using a unit of 1 cm^3 which is one thousandth of 1 dm^3 and sometimes known as 1 ml.**

Element	Mass	Element	Mass
Aluminium	27	Magnesium	24
Argon	40	Manganese	55
Barium	137	Mercury	201
Bromine	80	Nickel	59
Calcium	40	Nitrogen	14
Carbon	12	Oxygen	16
Chlorine	35·5	Phosphorus	31
Chromium	52	Potassium	39
Copper	64	Silicon	28
Helium	4	Silver	108
Hydrogen	1	Sodium	23
Iodine	127	Sulphur	32
Iron	56	Tin	119
Lead	207	Titanium	48
		Zinc	65

(The accurate values for all the elements are given in figure 10.2; the values quoted above will suffice for most calculations.)

Figure 1.10
Approximate relative atomic masses

1.6 The Scientific Method

It has been said that science began when the first man asked the first question. The scientist makes use of his senses—principally sight, touch, hearing and smell but not so often taste—to investigate the materials he finds or makes. He attempts to unravel what is happening in the natural world so that he can then repeat the changes at will or prevent unwanted changes taking place. He may devise models to explain the changes and describe those changes in terms of the substances involved and their structures and energy relationships, but he cannot answer the question 'Why?'. He may build upon the knowledge he gains to make better materials for a particular purpose but he never knows what he might discover as he pursues his investigations.

The map shows some of the principal centres and a detailed study of an area should soon lead to the discovery of many other important places.

General: Ashford, Barnard Castle, London (Colnbrook, Dagenham and Romford), Macclesfield, Pontypool, Poole, Speke (Liverpool) and Ulverston.

Petrochemicals: Baglan Bay, Carrington, Fawley, Grangemouth, Killingholme, Milford Haven and Wilton.

Ethyne: Barry, Chester-le-Street, Fleetwood, London and Londonderry (Maydown).

Methanol: Heysham and Billingham.

Ethanol: Baglan Bay, Cork, Dublin, Grangemouth and Kingston-upon-Hull.

Glycerol: Bebington.

Ethanoic acid: Grangemouth and Kingston-upon-Hull.

Margarine: Manchester, Port Sunlight and London (Purfleet).

Soaps and detergents: Belfast, Carrickfergus, Manchester, Newcastle upon Tyne, Portadown, Port Sunlight, Whitehaven and Wilton.

Plastics: Aycliffe, Baglan Bay, Barry, Carrington, Fawley, Grangemouth, Hyde, Immingham, Newport, Runcorn, Staveley, Welwyn Garden City and Wilton.

Terylene: Carrickfergus (Kilroot), Limavady and Wilton.

Nylon: Antrim, Billingham, Coleraine, Doncaster, Dundonald, Gloucester, Harrogate, Pontypool, Sligo, Stevenston (Ardeer) and Wilton.

Rayon: Carrickfergus, Grimsby, Holywell (Greenfield), Liverpool and Wilton.

Starch: Kingston-upon-Hull and Norwich.

Sugar (beet): Colchester (Wissington), Spalding and York.

Sugar (cane): Greenock, Liverpool and London.

Paints: Birmingham, Bristol, Kingston-upon-Hull, Manchester, Newcastle upon Tyne and Slough.

The scientific method is to do experiments and then to look for patterns in the behaviour of substances. The problem is not simply one of doing experiments but of designing experiments that yield results. It is easy to ask questions but the answers may not be understandable! The success of an experiment depends upon asking the right questions and then proceeding to answer them by means of the experiments devised. At an elementary level experiments may be performed to verify what is already known so that one has a check upon the accuracy of one's experiments or alternatively to learn basic information by seeing for oneself.

Having done a series of experiments the scientist may propose a hypothesis—a preliminary explanation of his results. Then, when others repeat and verify his results, the hypothesis may be upgraded to a theory. Sometimes a hypothesis, e.g. that of Avogadro, remains an hypothesis for so long that it is referred to as such long after it has become a theory. Later further experiments may be done which make it necessary to extend the theory: thus Dalton's Atomic Theory of 1808 was, and still is, adequate to explain many experiments but has had to be extended to deal with many other phenomena.

1.7 Uses and Misuses

The scientist in his research has discovered many things which can be used for better or for worse. He has discovered anaesthetics that enable operations to take place while the patient is unaware of what is happening, found materials which are antiseptic so that the wounds will heal and other materials that will ease the pain. An aspirin is a common mild pain killer but too many aspirins on one occasion may cause death.

The changes in the occupations of people as they go from country to town, together with the fact that the population of the world increases, means that the land must provide more and more food. Thus fertilizers must be manufactured to improve the productivity of the land and herbicides spread to prevent weeds growing in competition with the food. If too much fertilizer is used, or it is spread at the wrong time, then the fertilizer will be washed away into rivers and seas. Fertilizers in the rivers not only produce nitrates which may cause internal troubles if the water is drunk by young children, but they also cause the excessive growth of algae which leads to stagnant waters. The herbicides may be poisonous to human beings so care must be exercised in their use.

Radioactivity was discovered about 1900 but it was many years before its attendant dangers were recognized. The controlled use of radioactive substances has been a boon to sufferers from cancer and thyroid troubles and the United Kingdom leads the world in the generation of electrical energy from nuclear reactors. On the other hand the most destructive explosives known to mankind are those dependent upon radioactive substances.

1.8 Pollution or Environmental Control?

The waste products from our factories and our homes have increased as our society becomes more industrial and affluent, thus creating environmental and social problems. The disposal of such unwanted and unusable materials is an increasing problem in a crowded world. If, when the material is thrown away, it offends the senses of others or is detrimental to their well-being then pollution is occurring. The environment has not been the same throughout the history of the earth: if it had been, strange beasts and plants might still inhabit the earth in preference to us. The changes that have taken place have been gradual ones and for many centuries the natural world has been very much like the one that now exists. The Industrial Revolution which started about two hundred years ago has caused far-reaching changes with serious consequences which must be carefully considered.

The **pollution of the atmosphere** can be considered along the following lines:

(*a*) From domestic and industrial chimneys the products of combustion pour forth day by day. In the United Kingdom alone this means every year:

- 2·0 million tonnes of smoke (finely divided carbon etc.)
- 1·5 million tonnes of grit and ash (deposited later, 1 kg/m^2)
- 7·0 million tonnes of sulphur dioxide (oil is worse than coal)

We may be warm, but what of our lungs?

(*b*) Car exhausts cause many substances to be present in the atmosphere:

oxides of nitrogen (in the U.S.A. farmers spread 9 million tonnes of nitrates and 3 million more come from cars),

carbon monoxide (from incomplete combustion of the fuel, causes headaches and lack of concentration),
unburnt hydrocarbons (which will increase if measures to control the oxides of nitrogen are carried out),
lead compounds (lead is a cumulative poison).

It is proposed to eliminate lead compounds from petrol and measures are being taken to improve the combustion of the fuel particularly in the U.S.A.

(*c*) The waste materials and sometimes the escape of wanted materials from factories may pollute the atmosphere:

fluorine compounds from brickworks, pottery kilns, steelplants and aluminium smelters,
cyanides from steelworks and electroplating factories,
oxides of sulphur and nitrogen from factories when sulphuric and nitric acid are made,
asbestos dust in and near factories and places where it is used.

(*d*) Radioactivity: the escape of materials being used constitutes a health hazard if the activity is very high.

(*e*) Loud and unnecessary noise.

The **pollution of water** can be considered similarly. A survey of the rivers in the United Kingdom showed that 12% of their total length contained water which was grossly polluted or of poor quality and 15% which was of doubtful quality. The continued flow of unnatural materials into the sea and direct dumping may not be advantageous when the long term consequences are discovered. At the moment some things may be offensive but may not be detrimental: it is hard to tell. Some of the Great Lakes in the U.S.A. have been described as 'rank, muddy sinks', and just off Long Island, New York there are 50 square kilometres of 'dead sea'. Some factors in this pollution are:

(*a*) Sewage: it is true to say that people could not safely congregate in large towns before the advent of a satisfactory sewage system but the dumping of raw sewage into rivers and seas is indefensible. The sewage can be processed to give useful products.

(*b*) Oil: about a million tonnes of petroleum are lost or dumped at sea every year. The Torrey Canyon disaster led to beaches being spoiled, the death of seabirds etc. but it is now reckoned that the measures taken at that time to destroy the oil slicks did more damage than the oil itself.

(*c*) The dumping of materials in the sea:

coal wastes—Northumberland and Durham collieries throw in 4 million tonnes of waste each year,
china clay—the Cornish quarries discolour the rivers and seas,
poisonous gases—it is 'cheaper' to throw them away than to destroy them by chemical reactions,
radioactive substances—which later may be absorbed by growing crops or be absorbed into animals and people.

(*d*) The dumping of materials in rivers:

detergents—unless these are decomposed they cause foaming. Most of them contain phosphates which are excellent fertilizers for algae and thus the river becomes deoxygenated.

The **pollution of the land** is often more obvious and some of it can be prevented much more easily:

(*a*) Litter—in the New Forest alone 750 tonnes of litter are dropped every year.

(*b*) Factory waste—the coal mines and china clay works have led to large waste heaps: 65% of the dereliction is in Cornwall and Northern England. The offence to the eye is not as deeply felt as the loss of life when a tip disintegrates, e.g. Aberfan.

(*c*) Gruinard Island off the west coast of Scotland has been uninhabitable since 1941 when it was purposely infected by anthrax (a deadly disease, most common in sheep and cattle but communicable to man) as an experiment in biological warfare.

(*d*) Ugly and derelict buildings.

Last but by no means least, what of **animals and human beings**? Many materials are poisonous if taken in excessive quantities even if they are enjoyable or vital in smaller ones: many medicines come into this category.

(*a*) Tobacco—the link between smoking and lung cancer has been proved statistically but the reduction in the proportion of people smoking is small so far (65% are non-smokers).

(*b*) Alcohol—an excessive consumption of alcohol causes internal disease and leads to death.

(*c*) Many drugs, tranquillizers, stimulants etc. may have damaging effects on the chemistry of our bodies if used in excess (even tea and coffee in **very** large quantities) or even if used at all (heroin, L.S.D. etc.).

(*d*) Dangerous gases. In 1968 6400 sheep died in Skull Valley, U.S.A., because some nerve gas escaped 60 km away. The use of C.S. gas in riot control causes horrible burns and affects the lungs.

(*e*) Materials used elsewhere with advantage but harmful to human beings and animals: pesticides, e.g. D.D.T.; fungicides, e.g. mercury compounds (in tuna fish); herbicides, e.g. 2,4,5-T.

Some of the **action** that can be taken against pollution is a matter of educating people, some of it is a matter of spending money. The question is really one of economics in terms of human life and of human enjoyment. An early measure against pollution was a Royal Proclamation against the use of coal in London in 1273 but the Industrial Revolution has brought with it some of the legislation necessary to control our environment.

The first 'Alkali' Act was passed in 1868 to govern the emission of fumes from chemical works. The death in London of 4000 more people than was to be expected in 1952 led to the Clean Air Act of 1956 and another followed in 1968. In 1951 and 1961 measures were passed to control the pollution of rivers. International Conventions for the prevention of pollution of the sea by oil were signed in 1954 and 1962. In 1972 the United Nations held an international conference on environmental planning. The scientist knows something about the consequences of his past and present procedures. He must try to foresee the consequences of future operations: he cannot stand idly by and let an ignorant world commit suicide or condemn all endeavour, good and bad.

1.9 Safety

In view of all that has been said about the wide consequences of scientific and industrial practices it is reasonable to offer some advice to the new chemist.

(*a*) Follow the directions offered for your experiment very carefully; do not do anything else without obtaining permission.

(*b*) Be sure that the chemical substances employed are those which are stated and that if a solution is required it is of the correct concentration.

(*c*) Use the correct, or if in doubt, small quantities of materials: if something goes amiss the consequences should not be too serious.

(*d*) If you have to move around in the laboratory during a practical session do so at a reasonable pace; do not disturb other people.

(*e*) Concentrate on your own experiment and record all the observations you make which you think are significant: smell gases carefully and do not taste any substance (most of them are poisonous).

(*f*) Use clean apparatus and after the session clean and put away your apparatus tidily. Do not return any surplus chemicals directly to the bottle unless told to do so.

(*g*) The quickest and often the best treatment for a burn or scald or accidental contact with a corrosive chemical is washing with plenty of cold water. Report any accident.

(*h*) It is advisable to wear an overall to protect your normal clothing. If you do not wear spectacles, safety spectacles should be worn.

(*i*) When tidying up solid materials (paper, matches, etc.) put them in the waste bins only, not in the sink.

QUESTIONS 1

Introduction

A map showing the distribution of the population in the British Isles may be useful for answering some questions.

1 Why are the petroleum refineries distributed around the coastline?

2 Why is there an oil pipe-line from Finnart to Grangemouth?

3 What are the advantages and disadvantages of the Government sponsoring the setting-up of industry in Northern Scotland and Northern Ireland?

4 Why is there so much industry along the Lancashire-Cheshire border?

5 Write an essay on *pollution* as it affects your town.

6 Justify the removal of the Mint from Tower Hill, London to Llantrisant (Glam.)—most nickel is obtained from Canada, some is refined at Clydach (Glam.).

7 Three refineries and an oil terminal existed at Milford Haven and Pembroke when a fifth installation was proposed. The area is a National Park. Comment.

8 The study of chemistry is said to have benefited mankind. Oppose this view by describing one chemical discovery which, in your view, has not benefited mankind. You should describe the chemical processes involved, the nature of the product, and the use to which the product is put.

2 Physical and Chemical Changes

2.1 States of Matter

Matter can be divided into three main categories: solids, liquids and gases. These are called the three states of matter and are sometimes referred to as phases.

Solids have a definite shape and offer resistance to any attempt to change that shape. Their volumes are hardly affected by changes in temperature and pressure.

Liquids have no definite shape and take the shape of the base of the vessel containing them. However, they have definite surfaces which limit the amount of space they occupy. Liquids can flow, that is they can change their shape under the influence of very small forces. Like solids, their volumes are slightly affected by changes in temperature and pressure.

Gases are like liquids in that they have no definite shape and can flow: they are both fluids. Unlike a liquid a gas has no bounding surface and will fill completely any container in which it is placed. The volumes of gases are considerably affected by even small changes in temperature and pressure (see chapter 7).

Most substances can be fitted into one of these three categories without much difficulty. Pitch and sealing wax appear to be solid but will flow if left for a long time and so are considered to be very viscous (sticky) liquids; glass is also thought to be a viscous liquid which, because of difficulties in forming crystals, has been supercooled: it is a long way below the temperature at which it would be expected to melt. A collection of small, solid particles, for example sand or lead shot, may behave very like a liquid in its ability to flow and take the shape of the container but each individual particle shows the characteristics of a solid.

Many substances in everyday use are referred to as plastics. They fall into two types: thermoplastic and thermosetting. The thermoplastic substances soften on gentle heating and can then be blown, sucked or extruded (forced through a hole or into a mould) into a variety of shapes. On cooling the shape is retained, e.g. polythene can be made into bowls, pipes and bottles. The thermosetting plastics are made by heating moulded materials: they do not soften on heating but become harder and harder. Their shape is thus retained, e.g. bakelite can be made into light switches and many other articles where rigidity is important.

A pure substance can usually be made to exist in any of the three states of matter by altering the conditions of temperature and pressure. When a solid is heated it becomes a liquid at a definite temperature (the melting point) and the liquid on further heating will boil and become a gas (at the boiling point). These changes are reversed by reversing the conditions.

$$\text{Solid} \underset{\text{m.p.}}{\rightleftharpoons} \text{Liquid} \underset{\text{b.p.}}{\rightleftharpoons} \text{Gas}$$

Whereas the melting point of a solid is hardly affected by changes in the atmospheric pressure the boiling point of a liquid is profoundly affected by pressure changes. Ice melts at 0°C to give water (the liquid is usually implied by the term water) and at 100°C, when the atmospheric pressure is equivalent to 760 mmHg, the water boils to form steam. Below 100°C some of the water does exist as a gas, referred to as water vapour, and even below 0°C less, but still some, water vapour is present as well as the ice. Water can only exist as a liquid above 100°C if the pressure is greater than one standard atmosphere: 101 325 Pa (1 pascal = 1 N/m^2) or in terms of a column of mercury, 760 mm. For the effect of impurities see section 6.9.

Occasionally, when a solid is heated, it passes directly to the gaseous phase: this process is called sublimation, e.g. carbon dioxide (dry ice) and iodine if heated rapidly (see section 4.12). Some gases can be liquefied by merely increasing the pressure, while the temperature can remain at its usual value for the room, e.g. chlorine, ammonia and sulphur dioxide.

In chapters 7 to 11 evidence from the properties of substances will be given to show that matter consists of particles. At very low temperatures the particles are almost stationary and as the material is heated up the particles vibrate more and more and may rotate. At the melting point these motions become too violent for the stability of the solid, which breaks down to give a liquid. In the liquid the particles can move around from point to point, (translational motion), although their motion is considerably hindered by their neighbours. In the gaseous state their motion is not hindered to a very great extent and a gas has a high mobility (fluidity), that is to say a low viscosity, compared to a liquid. In a solid and a liquid the particles are touching one another and compressibilities in these two phases of substances are very low. However, in a gas the particles of the material are widely spaced and so a gas is easily compressed, e.g. into a car tyre.

The term **vapour** is often used for a gas close to its boiling point.

The properties of solids, liquids and gases can be summarized in outline:

	Solid	*Liquid*	*Gas*
Shape	Definite—may be characteristic	That of bottom of the container	That of whole container
Volume	Definite	Definite	Indefinite—as large as container
Flow	Resistant, rigid	Fluid—high viscosity	Fluid—low viscosity
Compressibility	Very low	Low	High
Density	High	Medium	Low
Expansion on heating	Low	Medium	High
Motion of particles	Slow—mostly vibration	Medium—vibration, some translation	Fast—translational motion most obvious
Boundary	Melting point (Solid/Liquid)		Boiling point (Liquid/Gas)

2.2 Types of Change

Everything in the world is continually changing, sometimes for the better, sometimes for worse.

The earth rotates on its axis and goes round the sun; the sun in turn is one star in but one galaxy in the universe. The water in the oceans evaporates, the clouds climb into the atmosphere and pass overland to deposit rain on the earth and rivers convey the water back to the oceans. Seeds are planted in the earth, plants grow using water, nutrients from the soil, gases from the air, gaining energy from the light of the sun and the plants then die and rot away. Man is born, grows for several years and then remains at a nearly constant height, if not mass, before he too dies and is buried or burned. If there had been no change since earliest times dinosaurs and pterodactyls would still inhabit the earth.

The changes that occur appear at first sight to be of many different kinds. Even when the object does not move as a whole the particles of which it is composed are in continual motion. The scientists can try to classify the changes; consider the following situations and experiments:

(*a*) A chair which is moved across a room compared with a chair which is broken up and burnt on a fire.

(*b*) An iron nail which is hammered into a piece of wood or is magnetized compared with a nail which is left out in the garden for several months. In the latter case a lot of the iron is converted into red scales of rust.

(*c*) Sulphur is heated gently in a test tube: it melts to a mobile liquid which returns to the yellow solid on cooling. Compare this with a few grains of sulphur which are heated on a spatula: they burn with a blue flame giving off a colourless pungent gas and the sulphur is seen no more.

(*d*) Sugar crystals are poured into water which is then heated; the water evaporates and many sugar crystals, if not all, can be regained. Compare this with sugar that is heated in a test tube to yield caramel and then carbon or with sugar that is eaten and digested.

A distinction can be made between the

changes in which the substances remain with their properties more or less the same, and those which produce new kinds of substances. Thus in the above examples, the movement of the chair, the magnetization of iron, the melting of sulphur and the dissolution of sugar, the materials are left substantially unaltered. The sugar in experiment (*d*) has apparently disappeared but it is quite easily recovered. These changes are **physical** changes.

The alternative experiments, the burning of the chair, the rusting of the nail, the burning of sulphur and the digestion of sugar, all involve the disappearance of the original material and the formation of new substances. These changes are **chemical** changes.

At an elementary level it is often easy to classify changes but as the experiments devised or observed become more subtle and the interpretation more detailed it may be hard or unwise to attempt to make the distinction (see section 6.4).

2.3 Distinction between Physical and Chemical Changes

In a physical change the particles remain in the original arrangements or move to arrangements which can easily be changed back to the original. No new combinations of particles are formed. When a solid melts or boils the particles move further apart but they keep whatever pattern and composition they had at the start. When a substance has two types of crystals the change from one to the other is a physical change (see section 8.4).

In a chemical change the particles undergo internal rearrangement and new combinations are formed with new and often very different properties from the original substance, e.g. hydrogen burns in oxygen and forms steam which condenses to give water. Hydrogen molecules (the particles that freely exist) each consist of two atoms (the smallest particles which retain the identity of the species). Oxygen molecules likewise consist of pairs of atoms. When the hydrogen burns in the oxygen a rearrangement of atoms takes place: two hydrogen atoms combine with an oxygen atom to give a molecule of water. The properties of water (melting point, boiling point, inflammability, effect on metals etc.) are very different from the properties of oxygen and of hydrogen.

The **evidence that enables one to decide** whether a change is physical or chemical can be arrived at by considering the answers to the following questions:

(*a*) Has a new substance been formed? Are the properties very different from those of the starting material?

(*b*) Can the change be reversed easily? A physical change such as melting ice can be reversed by cooling the water. A chemical change such as the combustion of hydrogen to give water is much harder to reverse. This criterion is not as easy to apply as the first one.

(*c*) Is the energy change, often apparent as heat rather than as light or sound or electricity, large or small? In changes that are usually referred to as physical the energy changes are small e.g. on dissolving salt in water it cools slightly, but when petrol is burnt much heat is evolved.

2.4 Examples of Physical Changes

(*a*) The most important physical changes are those which occur when a substance is heated provided that it does not burn nor decompose. Some materials such as the alloy nichrome (nickel and chromium), the wire in electric fires, can be heated to very high temperature without any chemical change occurring: the wire glows and radiates its heat to the surroundings; when the electricity is switched off it returns to its original state. When heated in a Bunsen burner flame the wire behaves similarly. Other materials have very low melting points and boiling points and so they normally occur as liquids or as gases. Only helium has proved impossible to solidify on cooling the liquid at atmospheric pressure.

All the changes of state (melting, boiling, sublimation and the reverse changes freezing and condensation) and the changes of crystal structure are physical changes.

(*b*) All mechanical changes are physical in nature. A chair is the same chemically whether at rest or when being carried about in a room. A shattered windscreen is the same chemically as the original glass: the supercooled liquid has suddenly crystallized (devitrified) and disintegrated.

(*c*) The process of making a solution is a physical change on the grounds that it is usually easy to reverse but when the behaviour of the particles is described on a molecular level it is in many cases better to refer to it as a chemical change because of the changes undergone (see sections 17.2 and 17.3).

(*d*) The magnetization of iron is a physical change. Nothing visible has happened to the iron and hammering the iron demagnetizes it. Its ability to rust is not impaired.

2.5 Examples of Chemical Changes

Chemical changes involve the breaking up of molecules and the rearrangement of the atoms into new molecules. In order to break bonds energy must be supplied and so the first stage may be to heat the reagents or to spark them; when the atoms combine to form the new molecules energy is released and this may mean that the overall reaction will evolve heat. Equations can be written to convey the information (see section 11.9).

(*a*) When mercury(II) oxide, a red powder, is heated in a test tube it turns black, silver beads of mercury collect on the upper part of the tube which is cooler, and a gas, oxygen, is produced which will relight a glowing splint.

$$\text{Mercury oxide} \rightarrow \text{Mercury} + \text{Oxygen}$$

or

$$2HgO \rightarrow 2Hg + O_2\uparrow$$

This type of reaction is called **thermal decomposition**. Another example of this type of reaction is the conversion of lead nitrate to lead oxide etc.

$$\begin{matrix}\text{Lead} \\ \text{nitrate}\end{matrix} \rightarrow \begin{matrix}\text{Lead} \\ \text{oxide}\end{matrix} + \begin{matrix}\text{Nitrogen} \\ \text{dioxide}\end{matrix} + \text{Oxygen}$$

$$2Pb(NO_3)_2 \rightarrow 2PbO + 4NO_2\uparrow + O_2\uparrow$$

(*b*) When a colourless solution of silver nitrate is added to a colourless solution of sodium chloride (common salt) in a test tube a white solid is formed which slowly settles to the bottom.

$$\begin{matrix}\text{Silver} \\ \text{nitrate}\end{matrix} + \begin{matrix}\text{Sodium} \\ \text{chloride}\end{matrix} \rightarrow \begin{matrix}\text{Silver} \\ \text{chloride}\end{matrix} + \begin{matrix}\text{Sodium} \\ \text{nitrate}\end{matrix}$$

or

$$AgNO_3 + NaCl \rightarrow AgCl\downarrow + NaNO_3$$

When a solid is formed by mixing solutions it is called a precipitate; the reaction is a **precipitation** or an example of **ionic association** (see section 17.12).

$$Ag^+ + Cl^- \rightarrow AgCl$$

(*c*) When a piece of iron (grey) is put into a blue solution of copper sulphate the iron becomes coated with a layer of copper (red).

$$\text{Iron} + \begin{matrix}\text{Copper} \\ \text{sulphate}\end{matrix} \rightarrow \text{Copper} + \begin{matrix}\text{Iron} \\ \text{sulphate}\end{matrix}$$

or

$$Fe + CuSO_4 \rightarrow Cu\downarrow + FeSO_4$$

Some of the iron has displaced the copper from the copper sulphate. This type of reaction is called a **displacement** (see section 15.6(d)): if it proceeds to completion the final solution is almost colourless.

(*d*) When small pieces of zinc (grey) are warmed with dilute sulphuric acid (colourless) the zinc dissolves. At the same time bubbles of a gas (hydrogen) are produced and if a burning splint is brought up to the mouth of the test tube the gas ignites with a squeaky 'pop'. The solution left is also colourless and if it is evaporated crystals of a white solid (zinc sulphate) are obtained. The chemical change which takes place when the zinc dissolves in the acid is thus quite different from the physical change occurring when sugar dissolves in water, as new substances are formed.

$$\text{Zinc} + \begin{matrix}\text{Sulphuric} \\ \text{acid}\end{matrix} \rightarrow \begin{matrix}\text{Zinc} \\ \text{sulphate}\end{matrix} + \text{Hydrogen}$$

or

$$Zn + H_2SO_4 \rightarrow ZnSO_4 + H_2\uparrow$$

The reaction is another example of **displacement** but this time the displaced substance is a gas rather than a solid.

(*e*) When the reactive metal magnesium (grey) is heated in air it burns with a blinding light leaving magnesium oxide (a white powder). This is quite a contrast with heating nichrome wire.

$$\text{Magnesium} + \text{Oxygen} \rightarrow \text{Magnesium oxide}$$

or

$$2Mg + O_2 \rightarrow 2MgO$$

A reaction in which a substance combines with oxygen is called an **oxidation** (see section 16.1). This type of reaction, in which a substance is formed from its elements, is also an example of **synthesis**.

Other types of chemical change include the following: reduction (see section 9.3 and chapter 16), neutralization (chapter 17), reversible reactions (section 19.2) and thermal dissociation (section 19.3).

2.6 Conservation of Mass

In a chemical change there may be an apparent change in mass: a gas may be trapped in the reaction when a metal burns or a gas may escape when a metal dissolves in an acid. However, *if all the reagents and all the products are weighed the total mass before the reaction is found to be equal to the total mass after the reaction* (see section 9.2). No atoms are destroyed or created—they are just rearranged.

2.7 Summary

Property	*Physical change*	*Chemical change*
Substances	No new ones	New ones
Reversibility of process	Easy	Hard
Change in mass	None	None, although at first sight may seem so
Heat change	Small	Large
Solvents for substances	Same as before	Usually different
Chemical properties	Same as before	New ones

QUESTIONS 2

Physical and Chemical Changes

1 Give four ways of distinguishing physical changes from chemical changes. State if the following are physical or chemical changes, giving reasons for your answers: (*a*) the souring of wine; (*b*) the action of water on anhydrous cobalt chloride; (*c*) the dissolution of sugar in water; (*d*) the fading of a dyed cloth.

2 Do you think that the following are chemical or physical changes? Give reasons. (*a*) The mixing of a cake; (*b*) the baking of a cake; (*c*) the digestion of a cake; (*d*) the setting of a jelly; (*e*) the melting of cooking fat; (*f*) the eating of butter.

3 When a large log of wood is sawn into small pieces, both the saw and the wood get warm, the change cannot be reversed, and the mass of the sticks is less than that of the original log. Explain why this change is regarded as a physical one.

4 Indicate three ways in which a chemical change differs from a physical change. Describe the changes which occur when each of the following substances is heated in a soft-glass test tube, giving equations for any chemical changes which occur: (*a*) copper sulphate crystals; (*b*) copper carbonate; (*c*) mercury(II) oxide; (*d*) ammonium chloride; (*e*) sugar.

5 Explain what is meant by: (*a*) distillation; (*b*) sublimation; (*c*) precipitation. Give one example of each and state whether it can be described as a chemical or a physical change. Draw a diagram of an apparatus with which a distillation could be performed and give an equation for a reaction in which precipitation occurs, naming the actual precipitate. [J]

3 Elements, Compounds and Mixtures

3.1 Elements

The world we live in is a complicated system composed, apparently, of so many different things that it may be a surprise to learn that there are only about one hundred fundamentally different materials. Buildings of many varieties can be constructed using a very limited number of different kinds of bricks; so too, words of many varieties can be constructed using the 26 letters of the English alphabet.

The fundamental materials are called **elements** and 106 are known at present. All matter, living or dead, is composed of these elements and the chemist is concerned with the changes and transformations that they might undergo. An element cannot be split into anything simpler by chemical methods but the physicist using powerful magnets and electrostatic fields can change one element into another: the process is costly and irrelevant to this book. The elements are all composed of particles smaller than atoms known as electrons, protons and neutrons (see section 12.1).

An element is a substance which cannot be split up into two or more substances by chemical means.

Elements may be solid, like iron and carbon; liquid, like mercury and bromine; or gaseous, like oxygen and hydrogen. The chemist often divides them into two classes: metals and non-metals (see chapter 15). The abundances of the common elements are given in section 1.2.

3.2 Atoms

A piece of an element such as iron cannot be subdivided indefinitely: just as a wall is made up of individual bricks, so a piece of iron is made up of very many tiny particles called **atoms**. The bricks and the atoms cannot be subdivided further without destroying their recognizable properties.

An atom is the smallest possible particle of an element that can exist.

There are 106 kinds of elements because there are 106 kinds of atoms; a statement which will have to be modified later (see section 13.2). Atoms are very small but some idea of their size can be gained from the following:

(*a*) If one drop of water were magnified to the size of the earth the atoms would be the size of table tennis balls.
(*b*) If 1 cm^3 of water were allowed to escape from a pipette at the rate of one million atoms every second it would take 1740 years to empty the pipette.
(*c*) Two million atoms would cover this full stop.
(*d*) If all the atoms in a drop of water were marked and then evenly dispersed throughout all the oceans in the world a glass of water taken at random would contain 2000 of those atoms.

Only with the most powerful electron microscopes have the outlines of some atoms and molecules been photographed.

3.3 Compounds

The four primary substances of the ancient world were earth, air, fire and water, and of these water is a compound, air and earth are mixtures and fire an energy-yielding chemical reaction.

Only a few elements are found free in nature, e.g. carbon, nitrogen, oxygen, sulphur, argon and the other noble gases, copper, gold. Many of these elements are also found as compounds. Compounds are formed by two or more elements combining together: millions of compounds are known, carbon being the element which forms the most. The combination occurs because the atoms of the elements often exert strong forces upon one another: these forces are discussed under the heading of valency (see chapters 11 and 14). The elements in a compound may be bound together by valency forces that are so strong that the compound is very hard to decompose. The molecules of the compound formed do not have the same properties as the particles (which may be atoms or molecules) of the original elements.

Some examples will illustrate this:

(*a*) Sodium is a soft, white metal which reacts violently with water; chlorine is a green gas with a choking smell. Sodium will burn in chlorine to give sodium chloride (common salt), a compound that is seen as white crystals which dissolve slowly in water giving it a characteristic taste. Salt is an essential article of diet but neither of its component elements would be beneficial to a person if taken directly.

(*b*) Carbon is a black solid with no smell which burns when strongly heated; oxygen is a colourless gas with no smell which is essential for the combustion of many materials including carbon. Carbon dioxide, formed by the combustion of carbon in oxygen, is a colourless gas with a slight smell and taste; the liquid, under pressure, is used in fire extinguishers; below $-78°C$, it is a white solid ('dry ice').

A compound is a substance which contains two or more elements combined in such a way that their properties are changed.

3.4 Molecules

The atoms of an element may or may not be able to exist singly: the **atomicity** of an element is the number of atoms in one molecule.

The **molecule** is the smallest particle of an element which can exist in the free state. The atoms of the noble gases (helium, neon, etc.), of mercury and other metals in the vapour state can exist on their own: they are monatomic substances, the atom and the molecule being the same. Oxygen, nitrogen, hydrogen, chlorine, are diatomic and hence the formulae are O_2, N_2, H_2 and Cl_2. There are a few well-known cases of higher atomicities: ozone, O_3; white phosphorus, P_4; and rhombic and monoclinic sulphur, S_8, and there are the giant molecules such as diamond (carbon). The forces holding the atoms together in these molecules are described under the heading valency in chapters 11 and 14.

The molecule is also the smallest particle of a compound which can exist in the free state. If a drop of water is divided more and more finely there comes a stage when it is impossible to continue division any further without separating the two hydrogen atoms and the oxygen atom that have joined to give water. These molecules can be split up, e.g. by an electric current, but the result is that the water no longer exists.

The number of atoms in a molecule of a compound varies considerably from one substance to another. The simplest possible case is a molecule of hydrogen chloride in which one atom of hydrogen has united with one atom of chlorine. Most of the substances considered in detail at an early stage in the study of chemistry contain up to ten atoms in a molecule. Most of the substances in nature and the substances that we eat, wear and consist of are far more complex:

sugar — 12 atoms of carbon, 22 atoms of hydrogen and 11 atoms of oxygen.

vitamin C — 6 atoms of carbon, 8 atoms of hydrogen and 6 atoms of oxygen.

nylon — 12 atoms of carbon, 22 atoms of hydrogen, 2 atoms of nitrogen and 2 atoms of oxygen (all multiplied about 2 000 times).

cotton — 6 atoms of carbon, 10 atoms of hydrogen and 5 atoms of oxygen (all multiplied about 8 000 times).

The atomicity of a substance, whether an element or a compound, is the number of atoms in one molecule.

The molecule is the smallest particle of a substance, whether an element or a compound, which can exist in the free state.

3.5 Mixtures

A pure substance is one in which all the molecules are alike. Elements and compounds are the only pure substances that can exist. If two or more kinds of molecules are present together there is a mixture. Most of the materials encountered are mixtures: air, earth, sea water and plants. One of the chemist's most important and difficult jobs is to sort out the naturally-occurring mixtures into their pure components in order to characterize them.

A mixture contains two or more different substances, either elements or compounds, which are not chemically joined together.

The presence of water of crystallization in a substance, e.g. copper sulphate-5-water, does not make it a mixture because the composition is fixed (see next section).

3.6 Mixtures and Compounds

Mixtures and compounds both contain more than one element and so should be separable into their component elements. This process may be easy or it may be difficult but ease is not the only way of deciding whether a given material is a mixture or a compound.

Consider the two elements iron and sulphur. The iron can be in the form of filings or as steel wool and the sulphur obtained as a powder or in the roll form. Their properties can be summarized as:

Property	*Iron*	*Sulphur*
Colour	Grey	Yellow
Action of magnet	Attracted	Not attracted
Solubility in warm dimethylbenzene	Does not dissolve	Dissolves
Action of dilute hydrochloric acid	Colourless gas evolved which burns with a squeaky 'pop' and blue flame when a light is applied—hydrogen. It usually has an unpleasant smell derived from impurities in the iron	No reaction

The first three properties are usually considered as physical properties and the fourth as a chemical property. Any changes in the first three cases are physical changes, the change in the fourth case being a chemical change. No test indicates iron or sulphur can be broken down to anything simpler: they are elements.

Next the iron filings (or chopped steel wool) can be put with the sulphur in powder form: For the best results the mass of iron should be about twice that of sulphur (56:32) or if measured by volume about twice as much sulphur as iron (16:7). On shaking them together, nothing is observed except an intermingling of the particles and under a microscope they can still be seen clearly. A mixture has been made. It is easy to separate the mixture because a magnet will attract the iron away from the sulphur or warm dimethylbenzene can be used to dissolve the sulphur away. The action of dilute hydrochloric acid on the mixture is the same as on iron alone, the sulphur being left behind.

Then the mixture can be heated in a test tube over a Bunsen burner: after a few moments it will continue to glow even if the tube is removed from the flame; heat is evolved once the transformation of the mixture into a compound is started. The same four lines of investigation this time yield different results and the conclusion is that a compound has been produced,

Property	*Mixture: Iron and Sulphur*	*Compound: Iron(II) sulphide*
Appearance	Grey and yellow—individual particles can be seen with a microscope if not directly. The overall colour varies with the proportions	Black and no differences in texture can be seen
Action of magnet	Iron attracted; some sulphur may also come but this can be removed by shaking	None or very little effect
Solubility in warm dimethylbenzene	Sulphur dissolves leaving iron behind. Sulphur can be recovered from solution by evaporation	No effect
Action of dilute hydrochloric acid	As for iron alone—see above. Sulphur remains	In fume cupboard—colourless gas evolved which burns with blue flame without explosion when ignited. It smells of rotten eggs and is poisonous—hydrogen sulphide

known as iron(II) sulphide. This reaction is another example of synthesis (see section 2.5). It is possible to regain iron and sulphur as elements but this is more difficult than can be tackled at this stage.

A very important difference between a mixture and a compound is not immediately apparent from the above tests: it is that a mixture may contain the substances in any proportion but that in a compound the substances are in a definite proportion by mass. A mixture of iron and sulphur might contain 1 g of iron to 100 g of sulphur or 1 g of sulphur to 100 g of iron but in the compound iron(II) sulphide the proportion is always very close to 1·75 g of iron for 1 g of sulphur (56 iron:32 sulphur). The law of constant composition is discussed further in section 9.3.

An investigation of common salt (sodium chloride) and water would lead to the conclusion that the solution is a mixture.

(*a*) The solution tastes like salt and looks like water.
(*b*) The salt can be recovered by evaporation and the water evaporating can be condensed (distillation)—both processes are physical changes.
(*c*) Although chilling may be apparent when salt is dissolved in water the effect is slight, not great.
(*d*) The amount of salt dissolved in the water can be decreased towards zero or increased up to the limit known as its solubility. Furthermore if the temperature is varied the solubility varies.

A summary is given below.

3.7 The Separation of Mixtures

The formation of a mixture is a physical process and is reversible. In practice the separation of a mixture into its components may not be an easy task. Some simple ways of separating different mixtures are:

(a) Liquid with insoluble solid e.g. sand and water

The solid may be everywhere in the liquid (a suspension) or it may be at the bottom of the liquid (a precipitate). The mixture is filtered: the liquid passes through the paper but the solid does not. For speed this is best done hot, unless on heating the solid dissolves. The solid remaining on the paper can be washed and dried. See section 4.8.

(b) Liquid containing soluble solid e.g. salt or sugar in water, sulphur in dimethylbenzene, naphthalene in ethanol

The solid (known as the solute) can be obtained by careful evaporation of the solution in an evaporating basin over a tripod, gauze and Bunsen burner or on a hot water bath (see section 4.10). Some solids, however, decompose on heating. If the liquid (known as the solvent) is inflammable a water bath, heated beforehand, is the safest way or else the experiment should be conducted in a manner known as distillation.

The liquid can be obtained by distillation using a flask and condenser (see section 4.11). It is collected in a receiver. It is not usually safe to obtain a solid by distilling off the last traces of the liquid: the flask usually cracks.

(c) Two solids, one soluble and one insoluble, e.g. salt and sand, iron and sulphur, chalk and washing soda

It is first necessary to find a solvent that will dissolve one of the substances either at room temperature or on heating. The most common solvent used is water but other solvents include ethanol, propanone, and dimethylbenzene. The mixture is shaken with the most suitable solvent and then filtered. The insoluble component remains on the filter paper and hence should be washed and dried; the soluble component can be

Property	*Mixture*	*Compound*
Formation	Involves a physical change Reversible Usually only small heat changes	Involves a chemical change Not reversible May be large heat change
Composition	Variable	Fixed—usually measured by mass but can be measured by volume
Properties	Of one or other, sometimes the average of the elements	Quite different from the elements

recovered by evaporation of the filtrate.

(d) Two solids both soluble, e.g. in black ink or screened methyl orange

The method known as chromatography is widely used for the separation of coloured materials. Put a drop of the material (Parker's Super Quink Black Ink gives good results) at the centre of a piece of filter paper and cut, but do not detach, a wick as shown in figure 3.1. Then put the paper between glass plates with the wick dipping into a solvent: water or ethanol alone will do but a mixture of water, ethanol, butanol and ammonia gives better results. After a while the separation of the ink into its components is apparent and the paper can be taken out and dried. Chromatography can be used for the separation of colourless materials but with difficulty. See also section 6.17.

Traditionally the separation has been done by fractional crystallization (see section 6.14).

(e) Two solids, one of which sublimes, e.g. ammonium chloride and sodium chloride

On gentle heating ammonium chloride becomes a gas but it is easily reformed on cooling. Sublimation is the process whereby a solid becomes a gas directly on heating, the liquid stage not being evident. Place the mixture in an evaporating basin covered with a pierced filter paper and inverted funnel. Heat the basin gently and the ammonium chloride will sublime and reform on the cold funnel. Sublimation is rarely of use (see section 4.12).

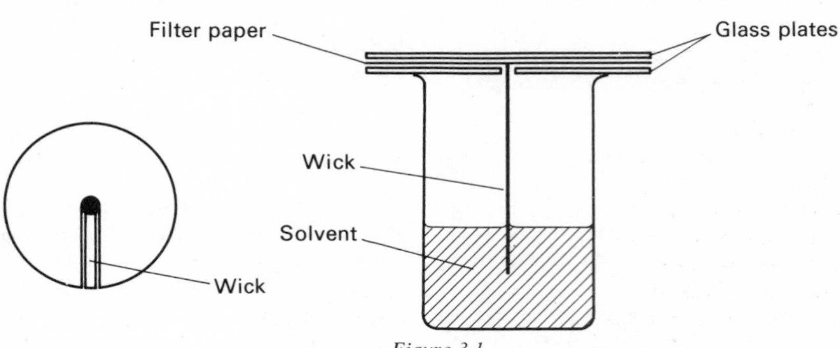

Figure 3.1
Radial chromatography

(f) Two liquids, e.g. liquid nitrogen and liquid oxygen, ethanol and water

If the boiling points of the liquids differ by 10°C, the liquids may be separated by distillation. The liquid with the lower boiling point will distil first and may be condensed. If the heating is continued then the second liquid will also distil.

The method is used on a large scale to separate air and petroleum into their respective components. The separation is not always complete even if the experiment is conducted very carefully by fractional distillation. One reason for the failure is that some pairs of liquids associate to some extent—without forming definite compounds—and, for example, ethanol and water cannot be separated even though methanol and water can.

QUESTIONS 3

Elements, Compounds and Mixtures

1 It is thought that common salt has been upset into a bowl of granulated sugar. By what simple chemical test could you find out if it had? Give the equation for the reaction you use. [S]

2 Classify the following as element, mixture or compound: (*a*) air, (*b*) aluminium, (*c*) brass, (*d*) calcium oxide. [S]

3 (*a*) Which of the following is the most accurate statement about the composition of air? (i) It is a mixture of 50% nitrogen and 50% oxygen. (ii) It is a compound consisting of 80% nitrogen and 20% oxygen. (iii) It is a mixture consisting of about 80% nitrogen, about 20% oxygen and small quantities of other gases. (iv) It is a mixture consisting of about 80% nitrogen and about 20% oxygen (percentages by volume).

(*b*) Which of the following best describes carbohydrates? (i) They are elements. (ii) They are compounds containing carbon and hydrogen. (iii) They are compounds containing carbon, hydrogen and oxygen. (iv) They are composed of carbon and water.

4 State which of the following mixtures can be completely separated by sublimation. For any mixture that cannot be separated by this method, explain why not and describe how separation may be made. (i) Sand and iodine; (ii) ammonium chloride and sodium chloride; (iii) ammonium chloride and copper(II) oxide; (iv) iron filings and flowers of sulphur. [C]

5 When a finely divided mixture of two elements was added to dilute hydrochloric acid, hydrogen was evolved, and a yellow powder remained in suspension. When some of the original mixture was heated in a test tube it began to glow, and the glow spread throughout the whole mixture. After the residue had cooled, a grey mass (P) was seen, and this was found to dissolve in dilute hydrochloric acid forming a solution (Q) and liberating a gas (R). R had an unpleasant smell and turned a lead ethanoate paper black. The solution Q was divided into two portions which gave the following results when tested. (*a*) The addition of an excess of sodium hydroxide solution to the first portion caused the formation of a dirty green precipitate. (*b*) The addition of a few drops of concentrated nitric acid to the second portion of solution Q caused this portion to turn yellow. When an excess of sodium hydroxide solution was added to this yellow solution (S), a brown precipitate was observed. Identify the constituents of the original mixture, name the compounds P, Q, R and S, and explain all the observations, giving equations where possible. [J]

6 State, giving reasons, whether each of the following is an element, compound or mixture: gold, chalk, ink, a Christmas pudding, iron, salt solution, a 'silver' coin.

7 How could you prepare pure specimens of sand, salt and water from a sample of sandy salt water?

8 How would you separate the following mixtures into their compounds: (*a*) sulphur and chalk (calcium carbonate); (*b*) ethanol and ethoxyethane?

9 How would you try to separate a finely-divided mixture of zinc dust and flowers of sulphur? (Zinc is a non-magnetic metal.)

10 State four important differences between mixtures and compounds, illustrating your answer by referring to iron, sulphur and iron(II) sulphide.

11 State the reasons for describing sulphur as an element, air as a mixture and water as a compound. Given a mixture of chalk and common salt, describe how you would find the percentage of each present in the mixture.

12 Explain the difference between an element and a compound, giving two examples of each. How could you obtain reasonably pure samples of (*a*) a blue dye from a mixture of red and blue dyes; (*b*) sulphur from a mixture of zinc filings and sulphur; (*c*) water from sea water?

13 Classify the following substances as elements, compounds or mixtures: water, wine, copper, petrol, air, graphite, calcium oxide, sodium carbonate, diamond, gas. Describe the process of distillation and explain how it can be used to effect a partial separation of a mixture of ethanol and water, their respective boiling points being 78°C and 100°C.

14 State briefly how you would separate from each of the following mixtures a specimen of the substance first named: (*a*) sulphur and iron filings; (*b*) nitrogen and carbon dioxide; (*c*) hydrogen and ammonia. [J]

15 What are the chief differences between mixtures and compounds? Suggest methods for separating the following mixtures into their constituents, obtaining a sample of each component: (*a*) rock salt (sand and sodium chloride); (*b*) oxygen and nitrogen dioxide; (*c*) ammonium chloride and sodium chloride.

16 Hydrides are compounds of hydrogen with another element. Write down a list of six hydrides, giving their names and formulae. Describe a chemical reaction occurring between (*a*) any two of the hydrides in your list (*b*) one of the hydrides and chlorine. [OC]

17 How would you separate the following mixtures so as to get pure samples of each component of each mixture: (*a*) carbon dioxide and hydrogen, (*b*) sodium chloride and ammonium chloride? In the case of the carbon dioxide—hydrogen mixture, how would you prove that the separated hydrogen contained no carbon dioxide and that the carbon dioxide contained no hydrogen? [OC]

18 Give two important differences between a mixture and a compound. Describe how the following can be obtained: (*a*) hydrogen from a mixture of carbon monoxide and hydrogen, (*b*) oxygen from air, (*c*) carbon dioxide from a mixture of carbon monoxide and carbon dioxide. [A]

19 Describe briefly experiments in which specimens of compounds are prepared by (*a*) the action of gases on each other; (*b*) the action of a gas on an aqueous solution of a salt; (*c*) the interaction of ions in solution; (*d*) the action of a solid on an alkaline solution. For each section one preparation only is required. [A]

20 I have a bottle of after-shave lotion which I suspect to be a solution of a solid green dye dissolved in ethanol (b.p. 78°C) or in a mixture of ethanol and water. Describe in detail, giving a diagram of the apparatus, an experiment which would show whether my suspicions were correct and whether the solvent was pure ethanol or a mixture of ethanol and water. [L]

4 Fundamental Processes and Statistics

4.1 Heating

The Bunsen burner is usually used for heating objects in the laboratory (see figure 4.1). Gas enters through a small jet at the base. Air can be mixed with the gas by adjusting the collar and in this manner the temperature of the flame can be altered. The energy output can also be regulated by adjusting the rate of flow of gas from the tap. The mixture of gas and air burns at the top of the tube. Always light the Bunsen with the airhole shut. There are two types of flame:

(*a*) The airhole is shut. The flame is luminous (gives out much light) and smoky. The flame is relatively cool and is unsteady. It is quiet.
(*b*) The airhole is wide open. The flame is non-luminous (by comparison) and not smoky. The flame is hot and steady. It is noisy.

For many purposes the airhole is not fully open nor is the gas tap wide open.

The variations in temperature in a Bunsen flame can be studied by seeing how it affects a piece of nichrome wire held in tongs (*a*) vertically, and (*b*) horizontally at various heights. One of the hottest regions (1200°C) is just above the inner blue cone of the flame when the airhole is wide open. See also section 5.15.

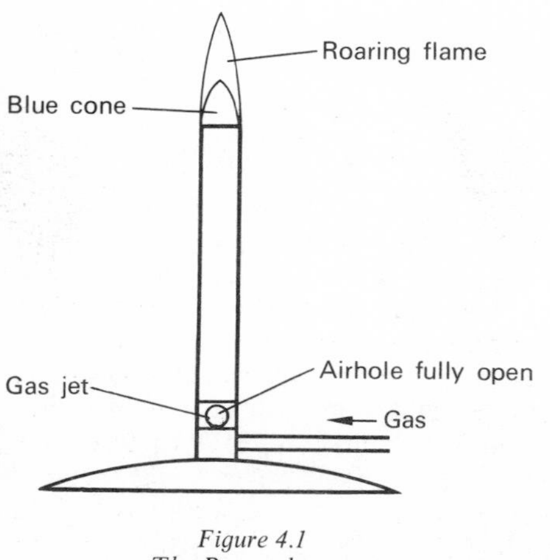

Figure 4.1
The Bunsen burner

If a fixed piece of apparatus is to be heated for a while it is best to move the Bunsen about periodically. If a liquid or solution in a test tube (not more than one third full) is to be heated the tube should be gently oscillated all the time otherwise the formation of vapour may cause the hot liquid suddenly to spurt out.

4.2 Weighing

The balance is the most delicate and expensive instrument used by the beginner in a laboratory and it must always be treated with care.

(*a*) Always weigh materials by difference if that is possible e.g. weigh a crucible empty and then containing the substance and so obtain the mass of the substance by subtraction. This approach eliminates any minor inaccuracy of the balance and prevents materials coming into contact with the pan of the balance which would probably lead to corrosion.
(*b*) Hot objects should not be put on the pan.
(*c*) Do not put objects or weights on or take them off the pans while the pans are released from their rest position. This prohibition does not apply to top pan balances.
(*d*) If there are weights to be put on and off the right hand pan—as in the case of a two pan balance—use tweezers (forceps) not fingers.
(*e*) Record the masses involved immediately.
(*f*) Respect the mass limit of a balance—this may be as low as 120 g.

N.B. The operation of weighing determines the **mass** of the material.

4.3 Pouring from a bottle

(*a*) Always look carefully at the label to check that the substance is precisely the one required not only in kind but also whether it is a concentrated solution, a dilute solution or a solid.
(*b*) Either retain the stopper in the hand or put it down carefully on some inert, clean surface.

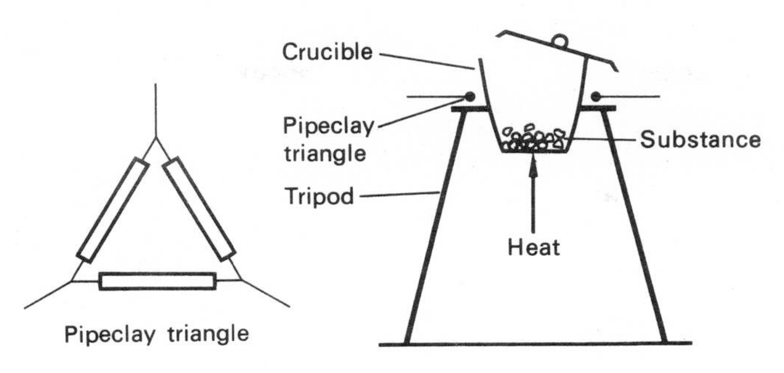

Figure 4.2
Heating a crucible

(*c*) Hold the bottle so that the label faces towards you and tilt sideways: this means that any drips will not deface the label and that any drips left by the previous user will not affect you.
(*d*) Replace the stopper and return the bottle to its correct place as soon as you have finished.

4.4 Using an Ignition Tube

To observe the effect of heat upon a solid put a little solid in a soft glass tube, e.g. to a depth of about 10 mm in a tube 50 × 10 mm. It may be advantageous to load the test tube using a small clean piece of folded paper. The tube is heated gently at first, and finally, if necessary, strongly.

4.5 Using a Crucible

In order to weigh a substance before and after heating it may be preferable to use a clean, dry porcelain crucible rather than a test tube. As shown in figure 4.2, the crucible may be supported in a pipeclay triangle. The crucible should be heated, and then allowed to cool, before being weighed empty in order that the conditions are the same previous to each weighing. The lid of the crucible should be left half-on so that gases can escape but losses by spitting do not occur. After cooling, with the lid on the crucible, it is weighed again. In order to be sure that any change is completed the heating may be repeated; the mass should not have changed again. This is called heating to constant mass.

4.6 Using a Test Tube

The larger versions are usually referred to as boiling tubes.

A common size of test tube is one 100 × 16 mm which, if full, will contain about 12 cm^3 of material. For many tests 2 cm^3 of material are sufficient: using more is wasteful and may be dangerous. The test tube can be held with tongs, special holders or with a folded strip of paper if it is to be heated. If a material is to be heated the material should not occupy more than a third of the tube and the tube should be gently oscillated all the time in a moderate Bunsen flame. Never point the tube at anyone while heating it nor look directly down into the tube to see what is happening.

4.7 Making a Solution

When sugar or salt is shaken with water it seems that it disappears. It is said to dissolve to form a solution: the process is known as dissolution. The solids are known as solutes and the liquid, not always water, is the solvent.

Solute + Solvent → Solution

The rate of obtaining a solution is increased by grinding the solid to a powder in a mortar with a pestle, by warming the solvent and by shaking or stirring the mixture vigorously.

4.8 Filtration

Filtration is used to separate an insoluble solid from a liquid or crystals from the solution remaining. A filter paper is a paper which is not coated or glazed; the fibres in it are not so dense as in writing paper. The filter paper is folded into a cone either with three layers on one half and one layer on the other half or for faster results into a zig-zag form by folding it four times one way, turning it over and folding it four times again (see figure 4.3). An 11 cm paper is used for a 7·5 cm funnel. It may be moistened with solvent before use to hold it in place.

The liquid in the beaker or test tube is then poured, preferably down a glass rod, into the filter paper so that a sudden rush of fluid does not break the paper. Hot liquids filter faster than cold ones. The liquid level should not rise

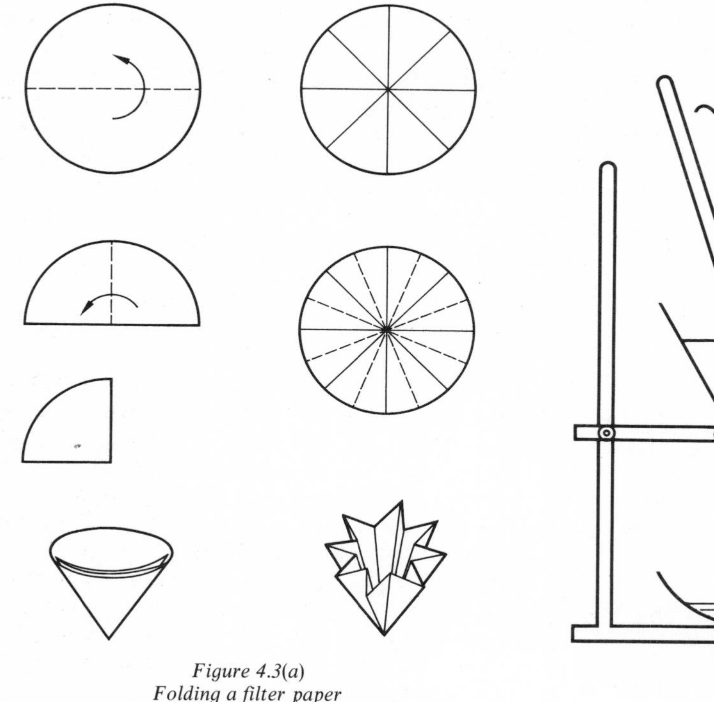

Figure 4.3(a)
Folding a filter paper

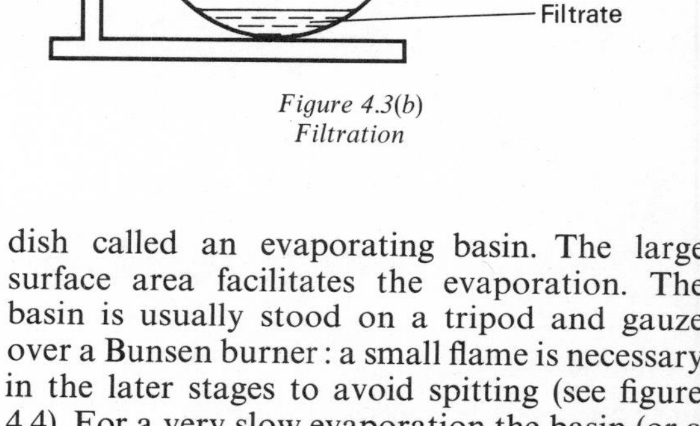

Figure 4.3(b)
Filtration

more than two thirds the way up the paper. The liquid filtrate passes through and is collected. The solid residue (precipitate, suspension) remains on the paper.

4.9 The Centrifuge

An alternative to filtration is to use a centrifuge to separate a solid from a liquid. The principle is applied at home in a spin drier when clothes are rapidly revolved and the water drains away through the sieve-like container. As a test tube is non-porous and no liquid or solid can escape, the solid settles firmly at the bottom of the tube and the liquid above can be decanted, i.e. poured off. The test tube put in a centrifuge should be counterbalanced by a tube containing roughly the same quantity of water to avoid undue strain on the rotor. Whilst the motor is switched on, and until the rotor has come to rest, the centrifuge should be kept shut for safety.

4.10 Evaporation

If a solvent is to be removed from a solution and it is not required for further use (not poisonous or flammable) then evaporation of the solvent is carried out from a porcelain dish called an evaporating basin. The large surface area facilitates the evaporation. The basin is usually stood on a tripod and gauze over a Bunsen burner: a small flame is necessary in the later stages to avoid spitting (see figure 4.4). For a very slow evaporation the basin (or a watch glass) can be stood on top of a beaker of boiling water.

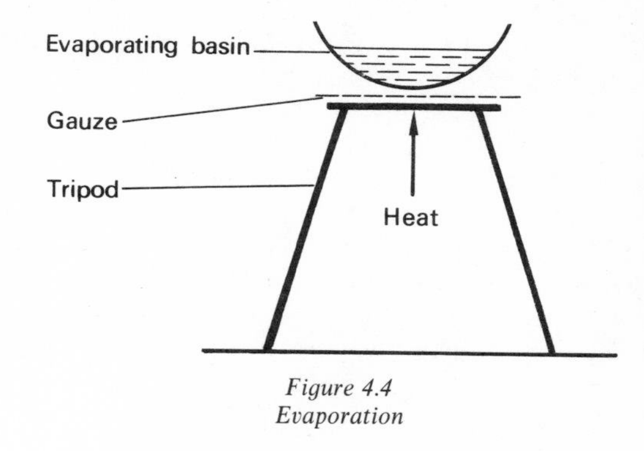

Figure 4.4
Evaporation

4.11 Distillation

Evaporation enables the solute to be recovered from a solution but if the solvent is required distillation must be performed. Thus to obtain, for example, water from sea water the solution is boiled, the solvent becomes a vapour (a gas) and then the vapour must be condensed back to a liquid again (see figure 4.5).

The solution is boiled in a distillation flask (a flask with a side arm) fitted with a cork and thermometer. The side arm leads into a long glass tube surrounded by an outer jacket through which cold water is continually passed from the tap: this form of condenser was invented by Liebig (1803–73). If the liquid has a high boiling point (over 150°C) the water jacket of the condenser should be empty: water not necessary as air cooling will be sufficient and safer. The vapour condenses and the pure liquid, the distillate, is collected in a suitable vessel.

The following points should be noted:

(*a*) The distillation flask should not be more than half full.
(*b*) The flask is heated on a gauze or sand tray so that the heating is not concentrated on a single spot.
(*c*) Small pieces of porous pot or anti-bumping granules (aluminium oxide) are put in the flask to smooth the vaporization of the liquid.
(*d*) The flask is not heated to dryness (i.e. only the solute left), because the flask would probably crack.
(*e*) The thermometer is present to give the temperature at which the distillation takes place: it is in the vapour, not in the liquid, with the bulb opposite to the side arm exit from the flask.
(*f*) The water passes in at the bottom of the condenser and out at the top. Why?
(*g*) The liquid which distils first is discarded, as it is likely to contain gaseous impurities and dirt from the apparatus.
(*h*) The collecting vessel must not be sealed to the condenser: if it is, an explosion will take place. A knee tube may be used to guide the distillate into the collecting vessel. If the distillate is inflammable a Buchner flask can be attached to the condenser and a tube to carry any surplus vapour attached to the side arm and led far away from any flames.
(*i*) At least two clamps must be used to support the apparatus: on the flask, on the condenser and possibly a third one on the receiver.

An older and less efficient method of carrying out a distillation is in a retort. Often used as a symbol for a chemist or alchemist, although a pharmaceutical chemist might choose a mortar and pestle, the retort is an all-glass vessel with a long stem and a small opening with a ground glass stopper. The liquid is heated in the bulb and the stem functions as an air condenser: the stem soon gets hot and then either it or the receiver must be cooled. It is used for preparing nitric acid and bromine in the laboratory (see figures 30.16 and 32.3).

4.12 Sublimation

A few solids e.g. iodine, ammonium chloride, benzoic acid and carbon dioxide on heating turn directly into gases, there being no liquid state at ordinary atmospheric pressure: the transformation is known as sublimation. Sublimation can be used for purification of these substances or substances in which they are present as impurities.

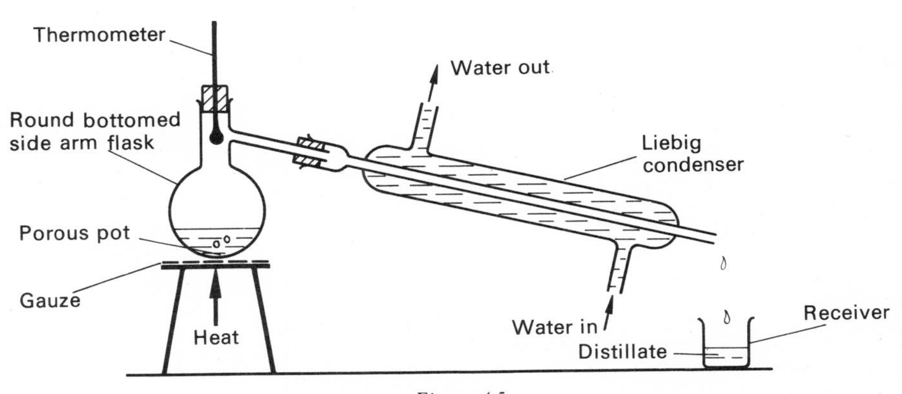

Figure 4.5
Distillation

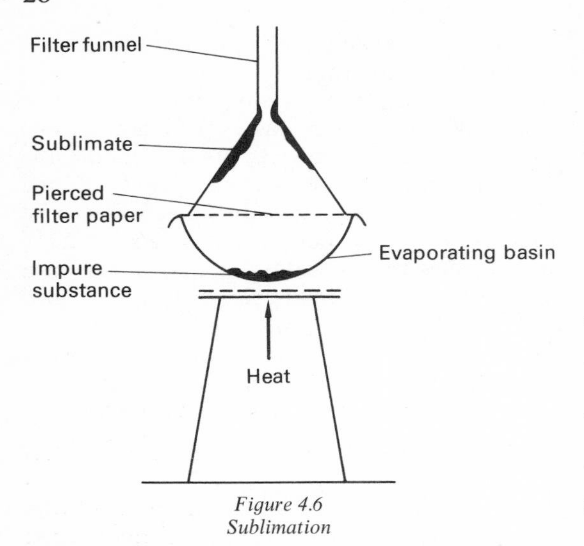

Figure 4.6
Sublimation

The solid is heated and the vapour condensed back to a solid on a cold surface using the apparatus shown in figure 4.6. The solid is placed in an evaporating basin and this is covered with a piece of filter paper in which some small holes have been pierced. An inverted funnel rests on the paper and basin. When the basin is heated gently with a very small Bunsen flame the solid sublimes on to the cold sides of the funnel. A more efficient condenser is a side arm flask through which cold water flows.

4.13 Statistics

Whenever a quantitative experiment is performed the experimental error is an aspect that must be considered for its adequate assessment. Consider the following experiments:
(*a*) an object, e.g. a test tube, is taken round all the balances in a laboratory and weighed on each of them;
(*b*) a person does the same experiment, e.g. the analysis of copper(II) oxide, on numerous samples prepared by the same method;
(*c*) a class does the same experiment, e.g. the analysis of copper(II) oxide, on a sample prepared by one method.

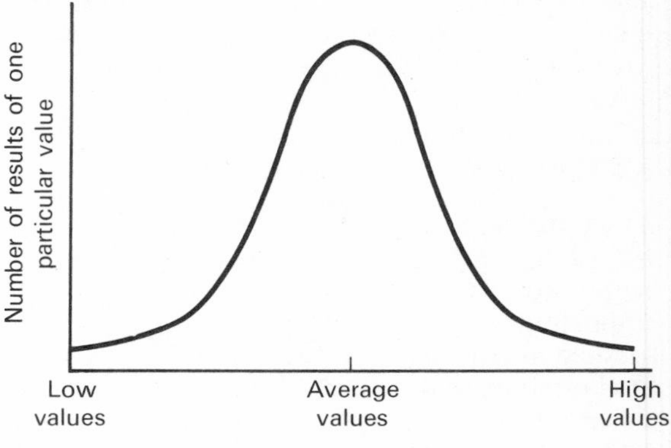

Figure 4.7(a)
The normal distribution curve

In each case either because of errors in the balance or because of errors made by people in carrying out the procedure there are found to be some high results and some low results with a majority of results fairly close to the average. A graph can be plotted of the number of results of one particular value against the value concerned; it should have the form shown in figure 4.7(*a*) The sharper the peak the better the experiment has been carried out: with a large number of results a symmetrical graph should be obtained. Thus in the analysis of copper(II) oxide some results are less than 80% copper, some are more than 80%, whilst the majority are 80%. The vertical axis is usually drawn centrally as shown in figure 4.7(*b*).

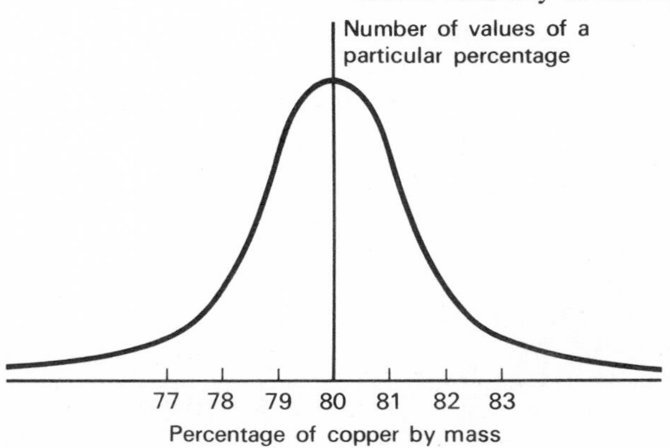

Figure 4.7(b)
The results of analyzing copper(II) oxide

5 Air and Combustion

I BURNING AND RUSTING

5.1 The Change in Mass on Burning

The burning of wood and coal have been known for a long time but that of town gas and petroleum are comparatively recent. These are important chemical changes and have been studied intensively for the last two centuries. It is obvious that new substances are produced when something burns. Wood and coal may leave only a small quantity of an off-white residue; petroleum and town gas leave no residue if there is a plentiful supply of air. The shiny metal magnesium yields a white ash but iron yields a black powder when burnt. The first problem is to explore the relationship of these materials to the original substances: in particular are they the same mass?

Experiment (*a*). A candle standing on a porcelain tile or a large cork is weighed. It is lit and left to burn for a few minutes, care being taken not to lose any of the melted wax. It is extinguished and reweighed: a loss in mass is found.

Experiment (*b*). A small coil of magnesium ribbon is weighed into a porcelain crucible with a lid. It is placed in a pipeclay triangle and heated over a Bunsen burner, gently at first and then strongly. The lid of the crucible is half on. The magnesium burns brightly and care is taken not to let much of the ash escape. After cooling and reweighing, there is found to be an increase in mass.

These results appear to be contradictory. Substances like paper, paraffin, wood and coal behave like the candle but others like iron filings, zinc and mercury behave like magnesium. As the experiment (a) proceeded, however, some smoke may have been noticed or, at least, currents of hot air rising above the flames. Thus it is reasonable to suggest that the product of combustion of the candle is a gas which escapes, whereas that of the magnesium is seen to be a solid. Hence, using calcium chloride to trap water vapour and soda-lime to trap carbon dioxide, experiment (a) can be improved:

Experiment (*c*). A candle is placed on a cork which has several holes bored in it to ensure good ventilation. The cork is put in the bottom of a wide glass tube or lamp chimney about 5 cm in diameter and about 30 cm high (see figure 5.1). A layer of calcium chloride is put on a gauze at the top of the tube and then one of soda-lime. The apparatus is suspended from one arm of a balance and counterpoised. After the candle has burnt for a while the apparatus is found to have increased in mass.

Thus the results are not contradictory: there is an increase in mass whenever a substance is burnt.

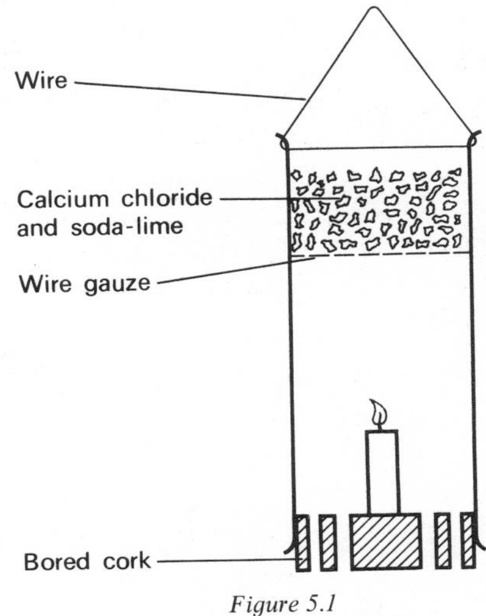

Figure 5.1
The burning of a candle

5.2 The Source of the Increase in Mass

It is well known that a good supply of air is essential if a fire is to burn well (and safely). The majority of combustions are of substances in air. This suggests that a substance from the air may be reacting with the material that is burning. It is fairly easy to weigh air:

Experiment (*a*). A stout round-bottomed flask is fitted with a tight rubber stopper, glass tubing and tap. The flask is weighed full of air and then attached to a vacuum pump and all the air sucked out. On reweighing the flask it is found to be lighter than before.

The next problem is to decide whether it is all or part of the air that is participating in the combustion, thus accounting for the increase in mass.

$$\text{Magnesium} + \text{Air} \rightarrow \text{Magnesium ash}$$

If then a substance is burnt in a closed container so that no air can enter or escape there should be no change in mass. It is easier to burn white phosphorus than magnesium because it will ignite at a very low temperature.

Experiment (*b*). Some sand is placed in the bottom of a stout flask fitted with a rubber stopper. Using tongs (**not** fingers) a small piece of white phosphorus is placed in the centre of the surface of the sand. The flask is weighed and then heated gently until the phosphorus ignites (see figure 5.2). When the reaction is finished, and the flask is cool, the flask and its contents are reweighed. No change in mass is found. Finally, if the stopper is removed when the mouth of the flask is under water, water rushes in and partly fills the flask showing that some of the air has been consumed.

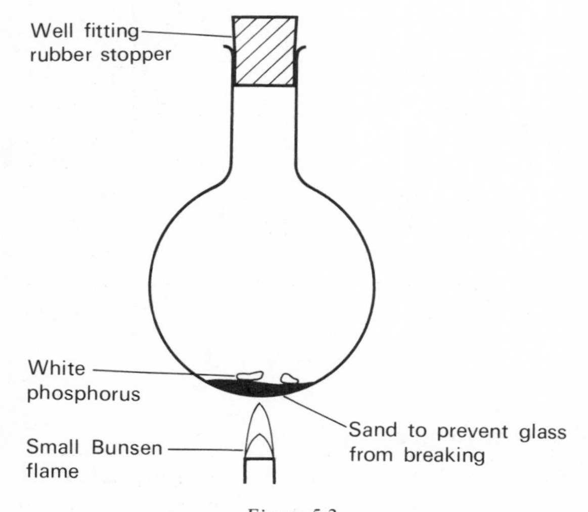

Figure 5.2
The combustion of phosphorus in a sealed vessel

5.3 The Proportion of the Air concerned in Combustion

The experiment at the end of the previous section suggested that part of the air is used up when a substance burns. In order to assess the decrease in volume accurately it is necessary to burn a substance in a sample of air in a closed space whose volume can change. The simplest way of doing this is to enclose the air in a jar over water: the water level will change as the volume of the air changes.

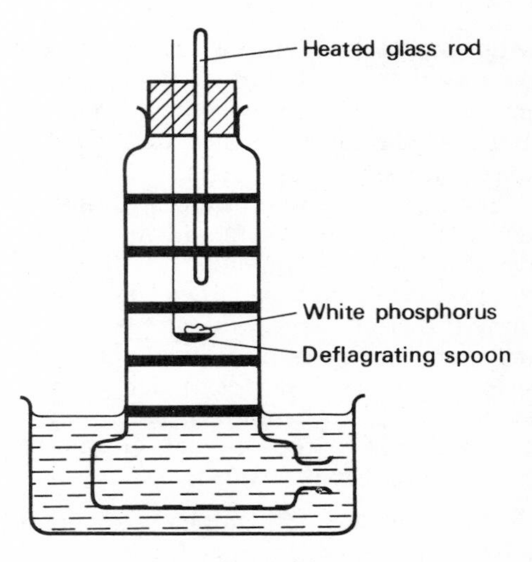

Figure 5.3
The combustion of phosphorus in air (*I*)

Experiment (*a*). A tall, narrow jar such as an absorption tower is graduated into five equal parts (see figure 5.3). The water levels inside and outside are equalized. The phosphorus is held on a deflagrating spoon and its combustion started by heating a glass rod and poking that through a second hole in the stopper to come into contact with the phosphorus. The combustion of the phosphorus gives clouds of white smoke and the heat of the reaction causes the air to expand and the water level to fall initially. However, as the air cools the level rises and eventually comes to rest one fifth of the way up the jar. If necessary, more water should be put into the bowl to make the levels inside and outside, and hence the pressures, equal.

Other substances can be burnt in the same manner but the proportion of air does not vary from one fifth. This suggests that there are at least two different gases present in air, one of which is used up when a substance burns. This gas, which supports combustion and is responsible for the increase in mass when a substance burns, is called oxygen. The gas which remains is mainly nitrogen. The substances formed when burning occurs are called oxides and the process of adding oxygen to a substance is called oxidation. Thus magnesium gives magnesium oxide, phosphorus yields phosphorus(V) oxide and the candle forms a mixture of hydrogen oxide (water) and carbon dioxide.

Experiment (*a*) above is not very accurate and the percentage of oxygen in the air can be found much more satisfactorily by letting the phosphorus burn slowly in air as in the following experiment:

Experiment (*b*). A piece of clean white phosphorus is placed carefully on the end of a long piece of thick copper wire and set up as in figure 5.4. The measured volume of air is trapped over water in a narrow graduated tube such as a burette. The water levels are made the same inside and outside when the initial and final volumes are recorded. The experiment takes a few days to complete. As the temperature and pressure are unlikely to be very similar upon the occasions when the volumes are read, a correction is made (see chapter 7). The phosphorus(V) oxide formed dissolves in the water. The experiment shows that 21% by volume of the air consists of oxygen.

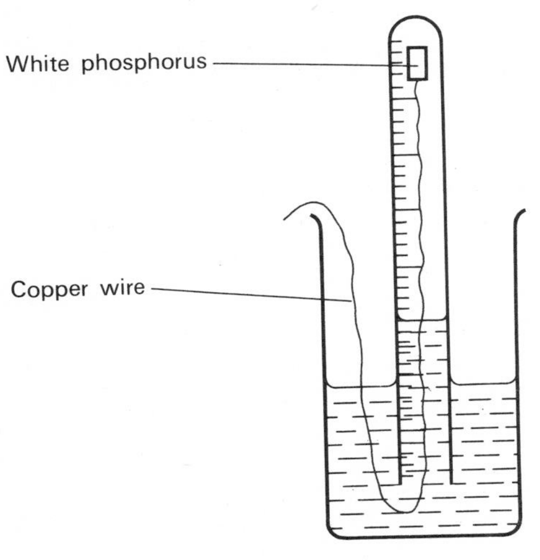

Figure 5.4
The combustion of phosphorus in air (II)

5.4 The Discovery of Oxygen

In 1774 Joseph Priestley (England, 1733–1804) discovered a red mercury compound which, when heated, gave off a gas that promoted the burning of a candle. The Swedish chemist C. W. Scheele discovered this gas in the same year but his results were not published until 1777.

Antione Lavoisier (France, 1743–94) has been called the Father of Modern Chemistry. His great achievements were the introduction of accurate methods of measurement and the importance he attached to quantitative experiments. He opposed the contemporary theory of combustion (known as the phlogiston theory) and in 1774, after a visit from Priestley, realized that oxygen was the active part of air for which he was looking.

To verify his suspicions Lavoisier heated about 100 g of mercury in a sealed retort for twelve days over a gentle fire. He observed that the volume of gas decreased by about one fifth and that a dark powder simultaneously appeared on the mercury. He took this powder and heated it by itself to a higher temperature. He regained the volume of gas lost above and showed that it was the vigorous supporter of combustion that Priestley had also found. He named the gas oxygen and stated the modern theory of burning.

$$\text{Mercury} + \text{Oxygen} \underset{\text{High temperature}}{\overset{\text{Moderate temperature}}{\rightleftharpoons}} \text{Mercury(II) oxide}$$

5.5 Burning: A Summary

When a substance burns in air it combines with the oxygen present in the air and thus gains in mass. If the products of burning happen to be gaseous there may be an apparent loss in mass.

The burning (or combustion) of a substance is a case of oxidation which is accompanied by the evolution of heat or light. Sometimes the reaction is very fast (hot white phosphorus) and sometimes it is very slow (the rusting of iron).

5.6 Rusting

Iron is the most widely used metal in the world but when it is exposed to the atmosphere it corrodes forming a red-brown powder which flakes off the article and so exposes more iron.

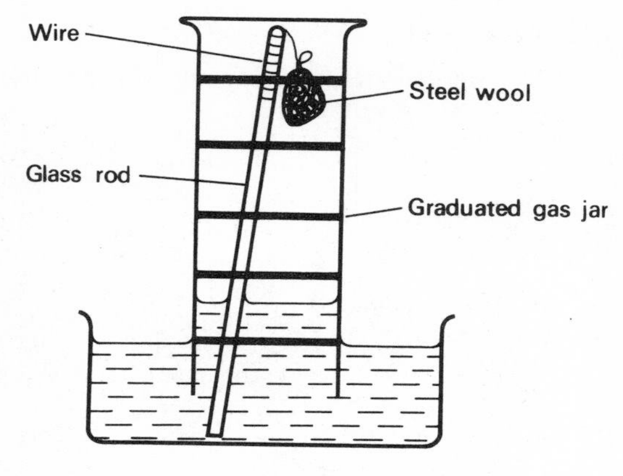

Figure 5.5
The rusting of steel (or iron)

Many methods are used to try to prevent this rusting of iron: painting, greasing, coating with other metals or plastic materials, etc. Rusting and burning take very different times but the same chemical process is involved: oxidation of the substance and hence there is a gain in mass.

Experiment (*a*). Some clean (non-greasy) iron filings are placed on a watch glass and weighed. They are then moistened with a small volume of water and left out in the laboratory for a week. After drying in an oven they are reweighed and found to have increased in mass.

It can also be shown that when rusting occurs some of the air is used up.

Experiment (*b*). A ball of steel wool (or bag of iron filings) is moistened and then placed in a graduated vessel as shown in figure 5.5. After a week the water levels inside and outside the vessel are equalized and it is found that one fifth of the air has gone. This immediately suggests that it is the oxygen of the air which has reacted with the iron.

Oxygen alone is not sufficient for rusting to occur: water is also necessary.

Experiment (*c*). Three test tubes each containing three clean bright iron nails are put in a rack, as implied in figure 5.6.
(i) To the first tube a few lumps of calcium chloride (granular, anhydrous) are added and the mouth of the tube is loosely plugged with cotton wool. The calcium chloride is a drying agent so here there is air but no water.
(ii) To the second tube water from which all the air has been boiled out is added and then this is covered with a layer of Vaseline (petroleum jelly) which prevents air from getting to the water. Here there is water but no air.
(iii) To the third tube ordinary tap water or distilled water, both of which contain dissolved air, is added. Here are both air and water.

After a week it will be found that only the nails in the third tube have rusted. Rusting is a complex reaction and may not always occur in the same manner. Rust is a hydrated oxide of iron. See also sections 15.6(e) and 26.7(c).

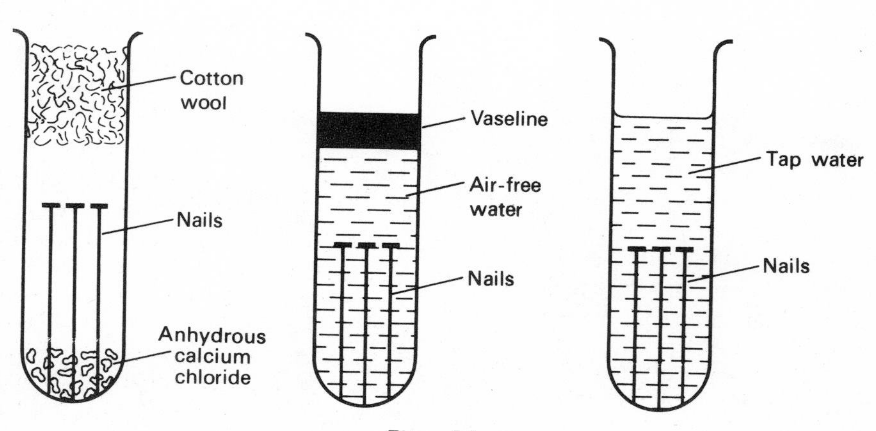

Figure 5.6
An examination of the conditions of rusting

II THE GASES OF THE AIR

5.7 Air is a Mixture

The presence of oxygen and nitrogen in the air has been discussed in the preceding sections. The presence of carbon dioxide is easily shown by drawing air through calcium hydroxide solution which then turns milky because of the formation of insoluble calcium carbonate. The composition of dried air (it commonly contains about 1% of water vapour) was shown in figure 1.3.

Air is believed to be a mixture of the gases concerned, rather than a compound, for the following reasons.

(*a*) When oxygen and nitrogen etc. are mixed in the right proportions no reaction is observed nor is there any heat or volume change. The mixture is indistinguishable from air: it has the same density and supports combustion to the same extent.

(*b*) The gases in air can be separated by physical means: air is cooled until it becomes a liquid and then allowed to boil yielding its components (see section 30.3).

(*c*) Although the composition of the atmosphere is very nearly constant slight variations are found from place to place, from time to time and from height to height: this is not so for compounds.

The nitrogen in the air dilutes the oxygen and so makes it suitable for breathing: pure oxygen can lead to pneumonia if breathed too long.

Nitrogen can be prepared in the laboratory from air as described in section 30.5. It will be contaminated with the noble gases but they are even less reactive than nitrogen itself.

5.8 The Uses of Air

Air is vital to life because it supplies oxygen in a suitably dilute form and the oxygen content of the upper atmosphere cuts off ultra violet light from the sun which would otherwise burn us (the oxygen is converted to ozone—see section 31.4). The atmosphere moderates the contrast in temperature between day and night and also burns up meteors arriving from outer space. People who work in deep mines or travel in submarines or high-flying aeroplanes must have a supply of air (or of oxygen).

The industrialist makes a lot of use of the atmosphere. From it he extracts oxygen, nitrogen and the noble gases. Everyone uses air to burn fuels; the industrialist also uses it to convert metallic sulphides to oxides (the first step often in the extraction of a metal), to convert sulphur dioxide to sulphur(VI) oxide (for sulphuric acid), to convert ammonia to nitrogen oxide (for nitric acid) and to burn the fuel in a blast furnace when making iron (4·5 tonnes of air for every tonne of iron). Compressed air is used for driving pneumatic drills.

5.9 The Solubility of Air in Water

Under average conditions 100 cm^3 of water will dissolve about 2 cm^3 of air—a small but nevertheless very important volume. The oxygen dissolved enables fish and plant life to exist. When water is heated, bubbles of air start to appear long before the water boils. The air can be collected as shown in figure 5.7 and then analyzed.

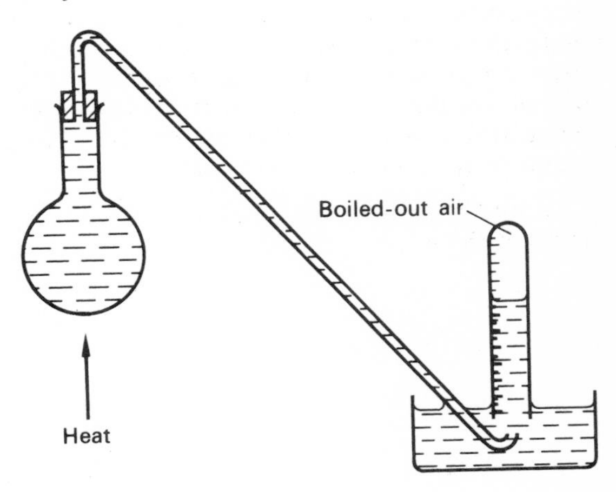

Figure 5.7
Boiling air out of water

Experiment. A round-bottomed flask is completely filled with water and fitted with a rubber stopper and delivery tube (which must not project through the stopper). The delivery tube and the graduated tube which acts as the receiver are both full of water. When the flask is heated until the water boils the air dissolved in the water is driven out into the receiver. The air can then be analyzed with phosphorus (see section 5.3); it is found to consist of one third oxygen because oxygen is more soluble than nitrogen.

This result is further evidence that air is a mixture and not a compound because a physical process (dissolution) has achieved a partial enrichment of oxygen.

5.10 Respiration

The similarity of combustion and rusting has been demonstrated, and now respiration can be shown to be of the same nature. All living things, whether animals or plants, need energy in order to exist and to grow. The energy they need is obtained when the food they have eaten or synthesized reacts with the oxygen they breathe in, i.e. the food is oxidized.

When sugar is burnt in the air, heat and light are produced, both of which are forms of energy. In the same way, when sugar is oxidized in the body, heat and other forms of energy are liberated, this time much more slowly.

In the lungs the air we breathe comes into intimate contact with the blood. The red blood cells (haemoglobin) combine loosely with the oxygen and carry it to all parts of the body. It oxidizes the products of digestion and so produces energy. The food we eat consists mainly of complicated compounds containing carbon, hydrogen and oxygen. When these are burnt, the carbon yields carbon dioxide and the hydrogen becomes hydrogen oxide (water)—see section 29.18.

Experiment (*a*). If a foodstuff, e.g. bread, is burnt on a deflagrating spoon in a jar of oxygen or air a film of water forms on the sides. If some calcium hydroxide solution is then added and the jar shaken the solution is observed to turn milky showing the presence of carbon dioxide.

In the same way when animals breathe the products of respiration are carbon dioxide and water. There is about 0·03% of carbon dioxide in the air inhaled and about 4% in the air exhaled; there is 21% of oxygen in the air inhaled and about 16% in that exhaled. There is sufficient oxygen in the exhaled air to be able to give mouth-to-mouth resuscitation to a patient suffering from asphyxia.

Experiment (*b*). Respiration can be studied using the apparatus shown in figure 5.8. Air is sucked in through one flask and then exhaled through the second. The calcium hydroxide solution in the second flask turns milky very soon.

5.11 Photosynthesis

Carbon dioxide is produced when oil, petrol, coal, town gas and wood are burnt and when living things respire. Oxygen is consumed in all these reactions. The proportions of oxygen and carbon dioxide in the atmosphere do not alter because there are ways in which the balance is restored. The vital agents are green plants. See also section 29.19, the carbon cycle.

Green plants take in oxygen through small pores in their leaves and give out carbon dioxide. This process is not essentially different from human respiration. However, plants take in carbon dioxide (by their leaves) and water (by their roots) and in sunlight convert them to sugars and starches. The catalyst is present in the green colouring matter of plants: it is called chlorophyll. As this catalytic reaction proceeds oxygen is released in quantities larger than those

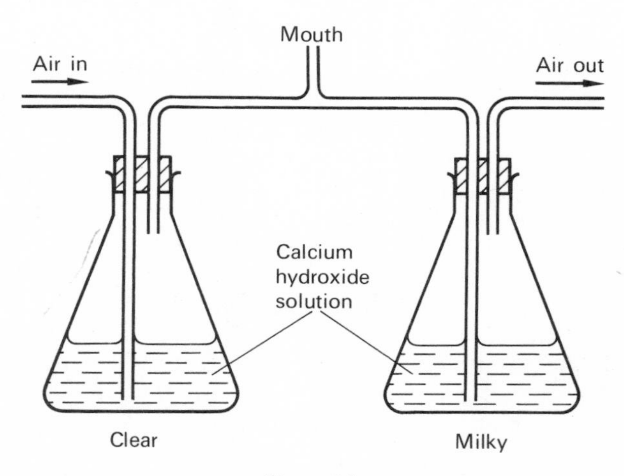

Figure 5.8
Breathing in and out

consumed by the plant breathing. The sugars and starches are then used to produce energy in the usual manner.

This building-up of a substance by light is called photosynthesis. It can be demonstrated using a suitable water plant, e.g. Canadian pond weed, so that the oxygen formed may be collected over water (see figure 5.9). The plant should be left in bright sunlight for a few days and the gas collected can then be tested with a glowing splint.

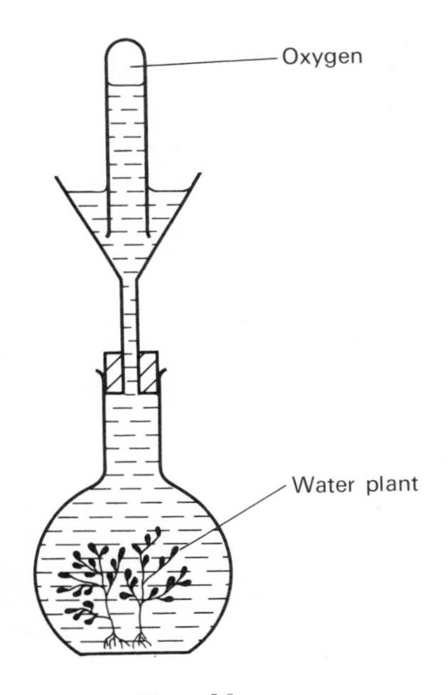

Figure 5.9
Photosynthesis

The plant breathes all the time but photosynthesis may only occur during the day. Some people like to move flowers out of a sick-room at night but the consumption of oxygen and production of carbon dioxide is very low. Much more sugar and oxygen is produced during the daytime by photosynthesis than is used by respiration in the course of both day and night.

5.12 The Noble Gases

Nitrogen can be prepared by removing oxygen, water vapour and carbon dioxide from air (see section 30.5). In 1892 Rayleigh found that the density of nitrogen so prepared was 1·2572 g/dm^3 whilst the density of nitrogen prepared from ammonium nitrite was only 1·2505 g/dm^3 (both at s.t.p.). The accuracy with which his experiments had been performed led him eventually, with the help of Ramsey, to the discovery two years later of argon which was identified by its spectrum and, in the following ten years, five similar gases were discovered (see section 13.6).

Helium is found in concentrations of up to 8% in natural gas in the U.S.A. and is one of the products of many radioactive changes. Neon, argon, krypton and xenon are all obtained by the very careful distillation of liquid air. Radon is a product of a few radioactive changes. The differences between these monatomic gases lie chiefly in their physical properties.

All the noble gases are used in discharge tubes: neon lights are a familiar medium for advertising and xenon is used in checking the light fastness of dyed cloth. They can also be used as the inert atmosphere in filament lamps: argon is often used but the extra expense of krypton for miners' cap lamps and xenon for lighthouses is worth while. Helium and argon are useful as inert atmospheres in the welding of reactive metals. Helium mixed with oxygen is safer than air for divers working at great depths underwater and is also used for patients suffering from asthma. Liquid helium (b.p. -269°C) is useful for obtaining very low temperatures.

5.13 Water Vapour in the Atmosphere

The amount of water vapour present in air is variable but very important: our physical comfort depends to a large extent on the relative humidity. The body controls its temperature by the evaporation of water from the skin. If the air is almost saturated with water vapour further evaporation is slow and we feel 'sticky' or say that the air is humid. On the other hand, when the air is hot and dry it will not feel at all oppressive.

The effect of water vapour on materials is dealt with in section 6.5.

III FLAMES

5.14 What is a Flame?

A flame is a reaction occurring at the boundary of two substances which gives out heat and light. The substance burnt (the fuel) may be a solid (coal or wood), a liquid (paraffin or petrol), or a gas (natural gas or ethyne). Usually the fuel burns in oxygen (as air) but oxygen is not essential: hydrogen will burn in chlorine. The type of reaction involved is an oxidation (see section 16.1) and the positions of fuel and oxidant can be reversed (see section 5.16).

5.15 The Bunsen Burner

The use of a Bunsen burner was mentioned in section 4.1.

(*a*) *The Non-luminous Flame* (*Airhole open*)

The flame, illustrated in figure 5.10, consists of three distinct zones.
(i) The inner colourless cone of unburnt gas.
(ii) A blue intermediate cone of incomplete combustion; this may be quite small if the air supply is good.
(iii) A pale blue outer cone of complete combustion.

Figure 5.10
The non-luminous Bunsen flame

Natural gas burns completely to give carbon dioxide and steam. The hottest part of the flame is just above the intermediate cone: here the temperature may reach 1200°C. The presence of relatively cold, unburnt gas in the inner cone may be shown in several ways.

Experiment (*i*). Balance a matchstick across the mouth of the Bunsen and light the gas. After a few moments, it will be found that the edges of the stick are charred, but that its centre is untouched.
Experiment (*ii*). Push a pin through a match just behind the head and place the match vertically in the centre of the Bunsen. Light the gas: the match does not ignite for several seconds.
Experiment (*iii*). Put one end of a thin glass tube in the inner cone and apply a light at the other. The gas burns at the end of the tube.

(*b*) *The Luminous Flame* (*Airhole closed*)

This flame, illustrated in figure 5.11, is much less definite in shape. Three zones may be distinguished in this quieter flame:
(i) A small dark inner cone of unburnt gas.
(ii) A large yellow region of incomplete combustion. The luminosity of this part of the flame is ascribed to small particles of hot carbon which will blacken a cold surface held here.
(iii) A thin, almost invisible, outer region of complete combustion utilizing more oxygen from the air.

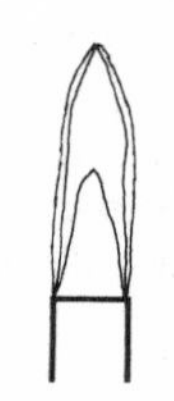

Figure 5.11
The luminous Bunsen flame

The luminous flame is only used when it is desired to produce carbon black (soot), a material added to rubber in large quantities to improve its resistance to wear. In gas fires and stoves there is a fixed airhole in the pipe bringing the gas supply to the burner so that they are similar to a Bunsen burner with the airhole open.

5.16 Striking Back and Reciprocal Combustion

If a Bunsen burner is lit with its airhole closed, then the gas supply turned full on and the airhole opened fully, and finally the gas supply turned down, the flame will jump from the top of the tube down to the jet. This is known as striking back. Gas and air have formed an explosive mixture so that the flame has travelled down the tube faster than the combustible mixture is supplied.

The striking back should be remedied rapidly before the tube becomes too hot: either by shutting the airhole (care should be taken—it may already be hot!) and hitting the rubber

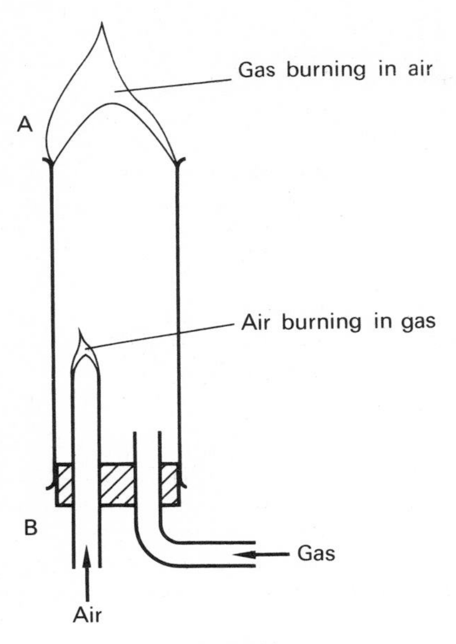

Figure 5.12
Reciprocal combustion

tube smartly, or by turning off both gas and air and relighting the Bunsen. Striking back is always liable to happen if the air supply is too great.

It is usual to speak of town gas burning in air but it is possible to reverse the usual positions. A wide glass tube, as in figure 5.12, is first set up with a cork at A. When gas is passed in until the tube is full a light is applied at the lower end of B: here town gas burns in air. When the cork is removed from A the flame runs up the tube B and burns there as shown: here air is burning in an atmosphere of gas. By lighting the gas issuing from the top of the wide tube at A it is possible to have both types of flame burning simultaneously.

This experiment shows that the terms combustible substance and supporter of combustion are only relative. The effect is called reciprocal combustion: the same chemical reaction occurs whenever air and gas are present, no matter in what manner they are mixed.

5.17 The Candle Flame

Candles are manufactured by allowing paraffin wax, a mixture of hydrocarbons of high boiling point, to solidify around a central wick made of string. When the wick is lit some of the wax melts and is absorbed by the wick. The heat of the flame converts some of this liquid wax to a vapour and this vapour burns.

A candle flame, as illustrated in figure 5.13, consists of three zones:

(i) A small dark inner zone of unburnt wax vapour.
(ii) A large luminous zone where combustion is incomplete.
(iii) A thin pale blue outer zone of complete combustion.

Complete combustion of the paraffin wax produces carbon dioxide and steam. It is easier to show that incomplete combustion produces steam and carbon than it is with a Bunsen burner.

Figure 5.13
The candle flame

QUESTIONS 5

Air and Combustion

1 Describe experiments to show the combustion of carbon, phosphorus and iron in oxygen. In each case describe what you would observe. [C]

2 (*a*) Give three reasons why air is considered to be a mixture of oxygen and nitrogen. (*b*) Of what importance is the oxygen dissolved in natural waters? (*c*) Of what importance is the carbon dioxide in the air? [C]

3 Write down (*a*) the approximate percentage of oxygen in the atmosphere; (*b*) the names of two neutral and almost insoluble gases present in the atmosphere; (*c*) two natural processes that remove carbon dioxide from the atmosphere. [C]

4 There is said to be a gain in mass whenever an element burns in air. How would you show that this is true for the burning of two of the following elements: magnesium, carbon, phosphorus? [C]

5 Describe two ways by which the rusting of iron can be prevented. [C]

6 Draw a diagram of a Bunsen burner, labelling its parts. Explain why a Bunsen flame will sometimes strike back, saying how you could bring this about. Why is it that the flame does not strike back beyond the burner? [C]

7 Explain carefully why magnesium gains in mass but a piece of coal appears to lose mass when each is burnt in air.

8 How would you show that: (*a*) iron filings, (*b*) phosphorus, (*c*) a candle gain in mass when each is burnt in air?

9 Describe three experiments you would carry out to determine the conditions under which iron rusts, and state clearly the conclusion to be reached from each experiment. Mention three different methods that are used to protect iron from rusting. [C]

10 How could you obtain a specimen of the air dissolved in tap water? What are the differences between this and ordinary air?

11 Describe an experiment to determine as accurately as possible the percentage of oxygen in a sample of air. Draw a diagram to illustrate the experiment. Explain why this percentage remains fairly constant although oxygen is being constantly used by all living organisms. [O]

12 For what reasons is air regarded as a mixture of its constituent gases? Describe an experiment to isolate a sample of atmospheric nitrogen by removing unwanted gases which are normally present. How does atmospheric nitrogen differ from pure nitrogen? Outline the method used industrially to obtain nitrogen from the air.

13 A stream of air is drawn slowly (*a*) through a U-tube immersed in a freezing mixture, then (*b*) through calcium hydroxide solution, and, lastly, (*c*) over heated copper turnings. Describe and explain what happens at each stage. [C]

14 What do you understand by the term *combustion*? Describe one experiment in each case to show that: (*a*) combustion may take place in the absence of oxygen and (*b*) the terms *combustible* and *supporter of combustion* are interchangeable. What experiments would show the differences between the luminous and the non-luminous flames of a Bunsen burner?

15 Describe an experiment to show that the terms *combustible substance* and *supporter of combustion* are interchangeable.

Explain briefly: (*a*) why a Bunsen burner strikes back under certain conditions; (*b*) why a non-luminous Bunsen flame is cleaner and hotter than the luminous flame.

16 The main constituent of 'Propagas' is the hydrocarbon propane (C_3H_8). Name the possible products of combustion of 'Propagas' if it is burned (*a*) in a limited supply of air, (*b*) in a plentiful supply of air. Represent the reaction in (*b*) by an equation.

If you are provided with a cylinder of 'Propagas' draw a diagram of the apparatus you would use, and describe the experimental procedure you would follow, in order to prove that the products of combustion in (*b*) were those which you have stated.

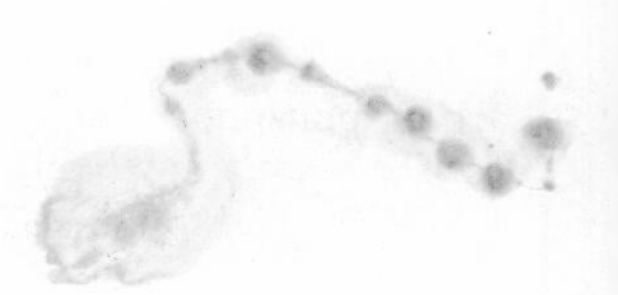

6 Water and Solubility

6.1 Introduction

Water is the commonest and most important of all compounds. It is essential for all forms of life and is the main constituent of all living organisms. Many living things, such as the jelly fish and the cucumber, contain over 90% water; everyday substances like soap and paper are useless if they are completely dried; food which has been completely dried cannot be eaten until it has been moistened. Industry requires water in large quantities for heating and cooling purposes, for generating steam to drive turbines, as a solvent and for its chemical reactions. In the United Kingdom the demand is for 300 dm^3 for every person every day: some of this is recycled but most of it joins the rivers flowing out to sea.

Water is the best solvent known and so when it is found in nature it is rarely pure but it is not always required in a very pure state. Supplies of water are obtained from rivers, wells and, if necessary, from the sea; London, for example, uses 140×10^8 dm^3 every day, two thirds being obtained from the River Thames, one sixth from the River Lea and one sixth from wells, whereas Manchester uses 6×10^8 dm^3 every day obtained from Lake District reservoirs. For further consideration of water supplies for drinking and industrial purposes see section 25.13.

In the laboratory very pure water may be required for many purposes, e.g. analysis, where the presence of other materials would be a serious disadvantage. The industrialist may also need very pure water for cooling electricity cables or for washing transistors where the presence of small controlled amounts of added substances are vital to their correct performance in electronic equipment. This pure water may be prepared by distillation (see section 4.11) or by using ion-exchange resins (see section 25.17).

6.2 The Physical Properties of Water

Water is such a common substance that many of its physical properties are taken as standards. The more important ones can be used to distinguish pure water from all other liquids.

(*a*) Pure water freezes at 0°C. Impurities lower the freezing point.
(*b*) Pure water boils at 100°C if the pressure is one atmosphere (101 325 Pa). At high pressures, e.g. in pressure cookers, water boils at a higher temperature; at low pressures, e.g. on top of Mount Everest, water boils at a lower temperature (there 72°C). Impurities usually raise the boiling point.
(*c*) Pure water has a density of 1000 kg/m^3 at 4°C.
(*d*) Pure water has a refractive index of 1·33.

6.3 The Reactions of Water with Metals

Water is a very stable substance in the sense that it shows little tendency to decompose into its elements, except at very high temperatures, but it is very reactive if brought into contact with some substances. Its behaviour with metals enables the metals to be arranged in an order of their reactivity (see section 15.6(c)).

(*a*) *With Potassium*

Potassium is a reactive metal which has to be kept under an oil (naphtha, similar to paraffin) to prevent it reacting with the atmosphere. Despite this, it is usually covered with a grey coating of oxide and carbonate and so shows its metallic lustre only when freshly cut. Holding the potassium by tongs on a porcelain tile enables a small piece to be cut (about the size of a pea) and this added to some water in a bowl. The potassium floats, melts and fizzes violently as it races round the bowl (stand clear). It reacts with the water to give a gas (hydrogen) which immediately ignites and burns here with a violet flame. The final solution turns red litmus blue showing that an alkali has been formed: potassium hydroxide.

$$\text{Potassium} + \text{Water} \rightarrow \begin{array}{c}\text{Potassium}\\ \text{hydroxide}\end{array} + \text{Hydrogen}$$

or in symbols:

$$2K + 2H_2O \rightarrow 2KOH + H_2\uparrow$$

(b) With Sodium

Sodium, like potassium, is kept under oil and can be added to water in a like manner. The reaction is not so violent (the hydrogen does not catch fire) but at the end there may be a small explosion, so stand clear again. The reaction follows a similar course.

$$\text{Sodium} + \text{Water} \rightarrow \begin{matrix}\text{Sodium}\\\text{hydroxide}\end{matrix} + \text{Hydrogen}$$

$$\text{or} \quad 2Na + 2H_2O \rightarrow 2NaOH + H_2\uparrow$$

(c) With Calcium

Calcium is a silvery white brittle metal which can be kept in a bottle but the metal soon becomes covered with a thin white film of its oxide. On adding a piece to water it sinks but after a few moments, during which the oxide film is dissolving, a stream of bubbles of hydrogen is evolved, the rate increasing then decreasing. The gas can be collected in an inverted test tube and ignited, when a mild explosion will be heard. The other product of the reaction, calcium hydroxide, is not very soluble in water and so a white precipitate may form after a while. The solution is alkaline and is used as a test for carbon dioxide.

$$\text{Calcium} + \text{Water} \rightarrow \begin{matrix}\text{Calcium}\\\text{hydroxide}\end{matrix} + \text{Hydrogen}$$

$$\text{or} \quad Ca + 2H_2O \rightarrow Ca(OH)_2 + H_2\uparrow$$

(d) With Magnesium

Magnesium, even if finely divided, hardly reacts with hot or cold water. However, magnesium will react with steam as can be demonstrated using the apparatus shown in figure 6.1. When the water is boiling in the flask the magnesium is also heated: the magnesium burns with a brilliant white light and the hydrogen burns at the hole in the tube. A white powder, magnesium oxide, is left in the tube, which may crack.

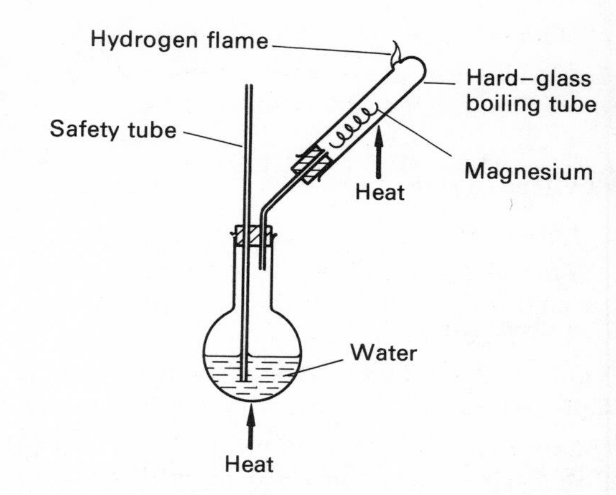

Figure 6.1
The combustion of magnesium in steam

$$\text{Magnesium} + \text{Steam} \rightarrow \begin{matrix}\text{Magnesium}\\\text{oxide}\end{matrix} + \text{Hydrogen}$$

or

$$Mg + H_2O \rightarrow MgO + H_2\uparrow$$

This reaction shows more clearly than the previous ones that water is an oxidizing agent.

(e) With Iron

Iron will not react with either hot or cold water in the absence of air (see section 5.6). When steam is passed over red-hot iron filings an unspectacular reaction occurs. The apparatus used is shown in figure 6.2. The gas collected has a smell: it is impure hydrogen and will burn with a mild explosion if ignited.

$$\text{Iron} + \text{Steam} \rightarrow \begin{matrix}\text{Triiron}\\\text{tetraoxide}\end{matrix} + \text{Hydrogen}$$

$$\text{or} \quad 3Fe + 4H_2O \rightarrow Fe_3O_4 + 4H_2\uparrow$$

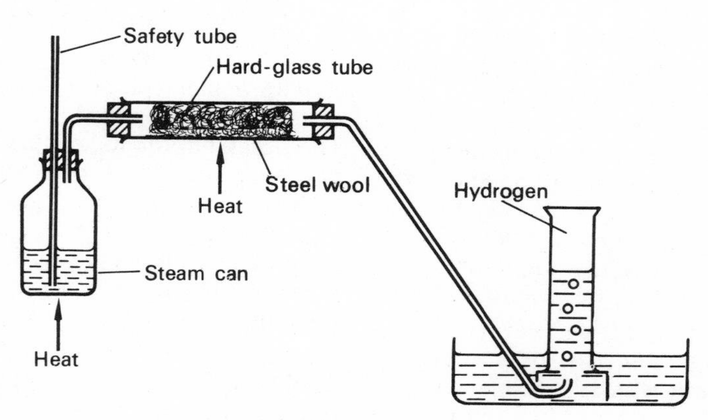

Figure 6.2
The reaction of steel (iron) with steam

(*f*) *With Other Metals*

Metals like lead and copper do not react with pure air-free water or steam under any conditions.

6.4 Other Reactions of Water

(*a*) *With Carbon*

Steam will react with white-hot coke to give a 1:1 mixture by volume of carbon monoxide and hydrogen, known as **water gas.**

$$C + H_2O \rightarrow CO + H_2\uparrow$$

(*b*) *With Oxides*

Water reacts with some metallic oxides to give alkalis (e.g. sodium and calcium oxides) and with many non-metallic oxides to give acids (e.g. sulphur(VI) oxide and carbon dioxide).

$$CaO + H_2O \rightarrow Ca(OH)_2$$
$$CO_2 + H_2O \rightarrow H_2CO_3$$

(*c*) *With Anhydrous Copper Sulphate*

Many crystalline substances which appear perfectly dry contain water, which is chemically combined but which can be driven off by heating. Water, loosely combined in this way, is called **water of crystallization** and the substances are referred to as hydrates. Copper sulphate when it crystallizes from solution forms blue crystals of the composition $CuSO_4,5H_2O$, copper sulphate-5-water or copper sulphate pentahydrate. These crystals when gently heated lose this water and crumble to a white powder called anhydrous (i.e. without water) copper sulphate. This white powder can be used as a test for the presence of water but it is not a test for the purity of water as any water-containing substance will restore the colour. However, it will readily distinguish water from ethanol, propanone or dimethylbenzene which are similar colourless liquids.

$$CuSO_4 + 5H_2O \rightleftharpoons CuSO_4,5H_2O$$

A comma or point is written to indicate that the water, although definite in quantity, is only loosely combined.

Cobalt chloride-6-water (pink) behaves similarly: it becomes blue on gentle heating. It is frequently used to colour the silica gel put in desiccators to keep substances under dry conditions: it enables the state of the drying agent to be assessed.

Many substances contain water of crystallization but not all of them can be successfully dehydrated by gentle heating. Some crystals, e.g. sodium chloride and sodium nitrate, are always anhydrous.

6.5 Water Vapour reacting with Solids and Liquids

If copper(II) oxide (black) or concentrated sulphuric acid is left out for a while in a room it acquires moisture from the atmosphere but there is no change of state: the copper oxide remains a solid, the acid a liquid. They are known as **hygroscopic** substances. Hygroscopic substances are used in desiccators for gently drying materials and for storing materials in a dry atmosphere.

If anhydrous calcium chloride is left out on a watch glass, then after a few days the white granules will be seen to have become a rather sticky concentrated solution. Calcium chloride has such an affinity for water that it attracts water out of the atmosphere and forms a saturated solution in it. The phenomenon is known as **deliquescence**. Calcium chloride-6-water can be crystallized out from aqueous solutions. Magnesium chloride and sodium hydroxide behave similarly. Ordinary table salt (sodium chloride) is deliquescent only because it contains magnesium chloride as an impurity. All deliquescent substances are thus hygroscopic and very soluble in water. Some of them, e.g. calcium chloride, are used as drying agents.

However, if sodium carbonate-10-water (decahydrate), otherwise known as washing soda, is left out in the atmosphere the large, clear, glossy crystals crumble to a fine white powder which on analysis is found to be sodium carbonate-1-water (the monohydrate). This phenomenon is known as **efflorescence**.

$$Na_2CO_3,10H_2O \rightarrow Na_2CO_3,H_2O + 9H_2O\uparrow$$

6.6 The Determination of the Percentage of Water of Crystallization in a Hydrated Salt

Barium chloride-2-water or magnesium sulphate-7-water are suitable.

Experiment. The steps in this simple experiment are:

(*a*) A clean dry crucible with its lid is weighed empty.
(*b*) Powdered crystals are put into the crucible, but not past the half way position, and the crucible is reweighed.
(*c*) The crucible is heated with the lid just ajar, gently at first and then more strongly with a Bunsen burner. The crucible should be supported in a pipe clay triangle. After a few minutes the lid is removed completely with tongs and put down where it will not get dirty.

(*d*) After heating strongly for 5 minutes the burner is turned off and the crucible allowed to cool with its lid on to prevent absorption of water from the air. Ideally it should be put in the dry atmosphere of a desiccator.
(*e*) The crucible, lid and residue are reweighed when they are cold.
(*f*) Finally they are reheated for 1 minute, cooled and reweighed to make sure that all the water has been driven off. This may have to be repeated until a constant mass is obtained.

Calculation

Mass of crucible and lid	= 15·20 g
Mass of crucible, lid and barium chloride crystals	= 18·32 g
Mass of crucible, lid and anhydrous substance	= 17·86 g
Mass after reheating	= 17·86 g
∴ Mass of crystals	= 3·12 g
∴ Mass of water of crystallization	= 0·46 g
∴ Percentage of water of crystallization	$= \frac{0{\cdot}46}{3{\cdot}12} \times 100$
	= 14·7

This calculation can be modified to determine the formulae of hydrated salts (see section 11.5).

6.7 The Composition of Water

Water is given the formula H_2O, i.e. two atoms of hydrogen have combined with one atom of oxygen.

(*a*) *Cavendish's Experiment*

Priestley, the discoverer of oxygen (1774), was the first to notice that a mixture of air and hydrogen could be exploded by an electric spark. He also observed that the walls of the vessel became covered with moisture. These experiments were repeated more accurately by Cavendish, the discoverer of hydrogen (1766). He used mixtures of hydrogen and oxygen and found that the explosion of a mixture in the ratio of two volumes of hydrogen to one volume of oxygen left nothing but a few drops of water. Oddly enough, it was left to Lavoisier to deduce that water is a compound of hydrogen and oxygen!

(*b*) *The Laboratory Synthesis of Water*

A stream of hydrogen from a Kipp's apparatus is carefully dried by anhydrous calcium chloride or concentrated sulphuric acid and when all the air has been driven out of the vessel it is ignited at a jet. The flame is allowed to play on a cold glass surface as in figure 6.3; the water vapour formed condenses and eventually drips on to the watch glass. The liquid can be proved to be water (*a*) qualitatively by using anhydrous copper sulphate and (*b*) quantitatively by measuring a physical property such as its melting point or boiling point.

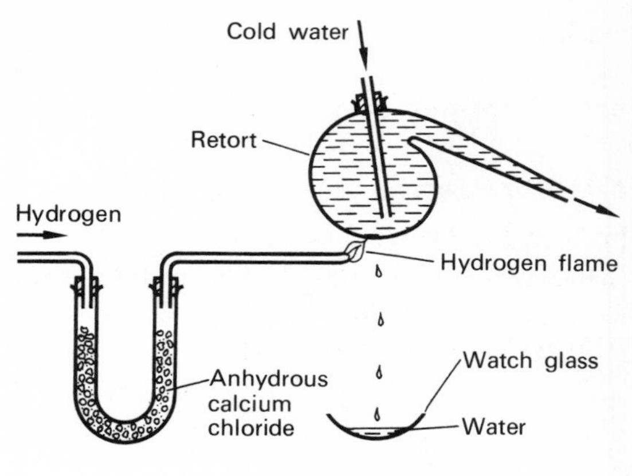

Figure 6.3
The laboratory synthesis of water

(*c*) *The Electrolysis of 'Water'*

If two pieces of platinum foil are connected to the opposite terminals of a cell and dipped into pure water nothing happens, i.e. water is an insulator. However, if a little dilute sulphuric acid or sodium hydroxide or sodium sulphate are added then an electric current flows quite readily and gases are evolved at the two electrodes (the points where the current enters and leaves the solution). These gases may be collected by using an apparatus such as the Water Voltameter (Hofmann) illustrated in figure 6.4.

The electricity is able to decompose the liquid and furthermore some heat is given out, (for further details see section 20.8(a)). The situation is rather different from an electric fire where a lot of heat is given out but no gases are seen to be evolved from the 'element'.

The current should be allowed to flow for a few minutes before collecting the gases to allow the solution time to become saturated. When the taps are closed and the gases collected the gas that comes off from the cathode is found to have twice the volume of the gas from the anode. The gas from the cathode burns with a mild explosion (it is hydrogen) and that from the anode reignites a glowing splint (it is oxygen). Thus water has been decomposed into its elements and the substance added (acid, alkali

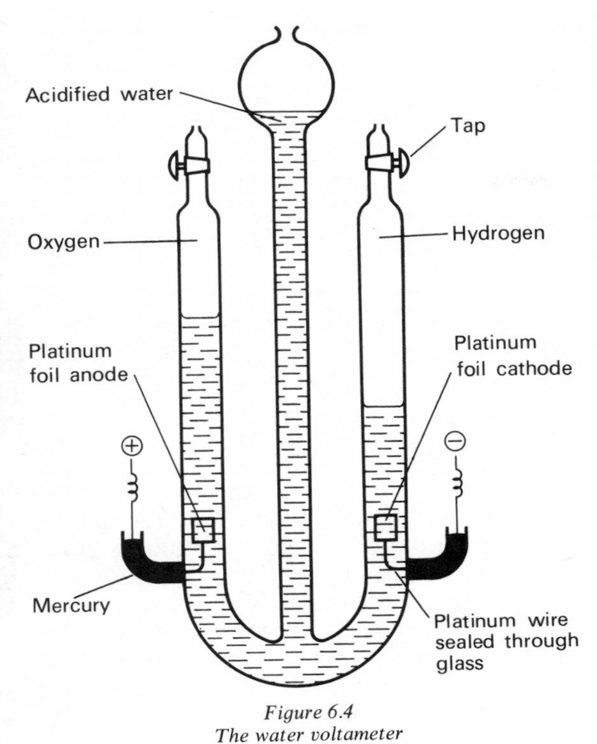

Figure 6.4
The water voltameter

This result confirms the result of the electrolysis:

2 volumes of hydrogen + 1 volume of oxygen → 2 volumes of steam

On stopping the stream of pentanol vapour the tube cools and the mercury (more must be poured in to keep the pressures equal) fills the eudiometer tube: the minute volume of water formed is inconspicuous.

Using Avogadro's Law (see section 7.11) the result of this experiment can be used to deduce the formula of water:

2 volumes hydrogen + 1 volume oxygen →2 volumes steam

can be rewritten

2 molecules hydrogen + 1 molecule oxygen →2 molecules steam

or 1 molecule hydrogen + $\frac{1}{2}$ molecule oxygen →1 molecule steam

The molecules of hydrogen and oxygen are both diatomic so

1 molecule of steam contains 2 atoms of hydrogen + 1 atom of oxygen

∴ the formula of steam (and hence water and ice) is H_2O and the equation is

$$2H_2 + O_2 \rightarrow 2H_2O$$

or salt) is unchanged, but has increased in concentration.

i.e. from water 2 volumes of hydrogen and 1 volume of oxygen are obtained.

(d) The Volume Composition of Steam

The apparatus used to determine the volume composition of a gas is known as a eudiometer; it is illustrated in figure 6.5. It consists of a stout tube fitted with electrical contacts so that a spark can be passed. Two volumes of hydrogen and one volume of oxygen are contained in the tube by mercury. The tube is surrounded by a jacket through which pentanol (b.p. 130°C) can be passed which will keep the steam in the gaseous state.

The mercury levels are equalized and the total volume recorded. The open limb should be lightly plugged to prevent mercury being spilt and then a spark passed using an induction coil. After a few minutes to allow the gas produced to cool to 130°C and mercury having been poured into the tube to equalize pressure inside and outside the apparatus, the volume of the gas can be measured: it is two thirds of the total volume of the starting materials.

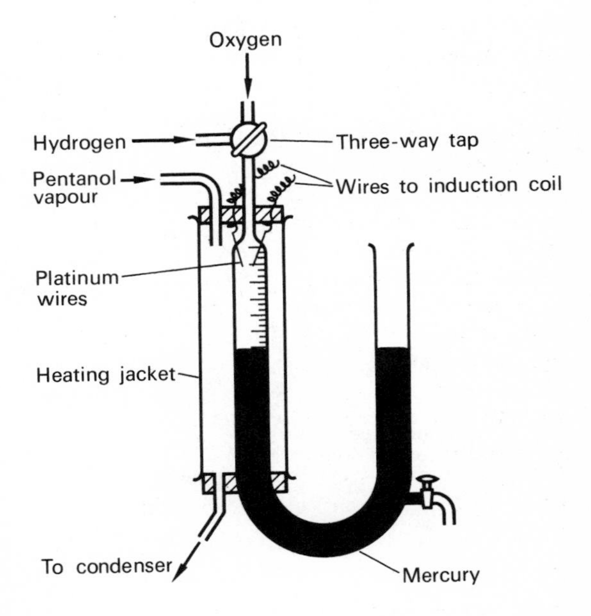

Figure 6.5
The volume composition of steam

(e) The Composition of Water by Mass

This experiment was first performed with any accuracy by the French chemist Dumas in 1842 (see figure 6.6). Balances were still not very accurate for weighing gases so he did not try to measure the masses of hydrogen and oxygen directly. Instead, he made hydrogen by the action of dilute sulphuric acid on zinc and after careful purification passed it over heated black copper oxide: this oxidized the hydrogen to water.

$$H_2 + CuO \rightarrow Cu + H_2O\uparrow$$

Most of the water formed condensed in a glass bulb and the remainder was absorbed in weighed U-tubes containing phosphorus(V) oxide (P_4O_{10}) or concentrated sulphuric acid which are both very powerful drying agents. He used seven tubes each 1 metre high. In the experiment, which lasted about 12 hours, over 1 kg of water was collected. The difference in mass of the copper oxide before and after gave him the mass of oxygen used and so, on subtraction of this from the mass of water formed, he found the mass of hydrogen. The guard tube at the end of the absorption train was to make sure that no water vapour from the atmosphere diffused back into the apparatus.

Dumas took elaborate precautions to dry and purify the hydrogen and did the experiment 19 times. The average of his results showed that the ratio by mass of hydrogen to oxygen is 1:7·98. More accurate experiments by Morley (U.S.A. 1895) have obtained the result 1:7·939. The constancy of the results in these experiments is good evidence for the Law of Constant Composition (see section 9.3).

6.8 Water is a Compound

That water is a compound and not a mixture of hydrogen and oxygen is based on the following evidence:

(*a*) the properties of water bear no close relation to those of hydrogen and oxygen,
(*b*) when water is formed from hydrogen and oxygen by sparking or burning there is a large energy change,
(*c*) water cannot be separated into hydrogen and oxygen by any simple physical means, and
(*d*) the composition of water by mass is always the same.

6.9 Solutions and Suspensions

When a little sodium chloride (common salt) is shaken up with water the solid particles disappear and a clear liquid is left. The salt has dissolved in the water and the resulting liquid is called a **solution** of salt in water.

However, when powdered calcium carbonate (chalk) is shaken with water, the liquid turns cloudy or turbid and, on standing, the little white particles separate out. The cloudy liquid is called a **suspension** of chalk in water. The chalk has not dissolved and the particles can be removed by filtering. In a true solution the dissolved material will not separate out on standing and cannot be removed by filtration. A solution will always be clear, though it may be coloured. Copper sulphate crystals and water give a blue solution, but this is quite different from the blue suspension made by shaking powdered blue chalk for the blackboard (actually dyed calcium sulphate) with water.

The difference between a solution and a

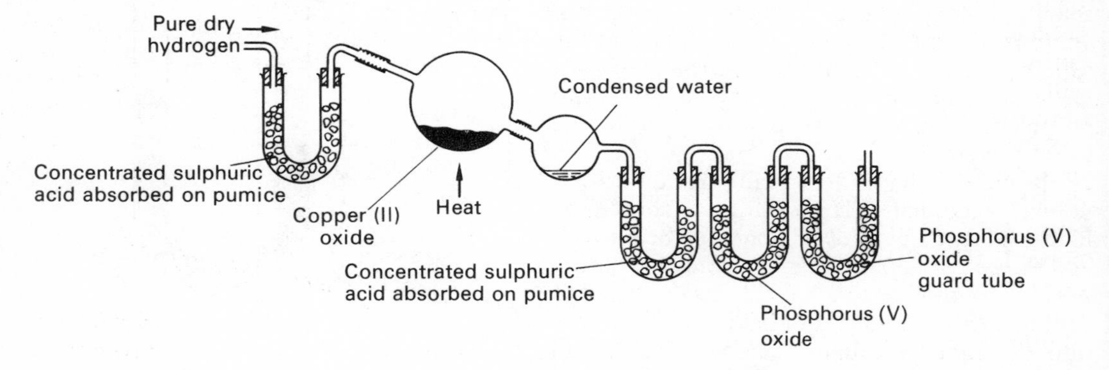

Figure 6.6
The composition of water by mass
(Dumas' experiment)

suspension is due to the size of the particles. In a solution the dissolved substance exists as very small particles and cannot be detected even with a very powerful microscope. In a suspension the solid substance, even if finely powdered, still consists of particles which are much larger: too large to mix freely with the liquid molecules, they thus settle out under the action of gravity.

When one substance dissolves in another substance it is said to be soluble. Salt is soluble in water but chalk is almost insoluble. Dissolution, the process of making a solution, is the act of spreading one substance out within another so that a sample taken from any part of the mixture has the same composition, i.e. a solution is homogeneous. The substance which dissolves is called the **solute** and the substance in which it dissolves is called the **solvent**. It is the solute which changes its state (e.g. solid salt becomes a liquid) and the solvent which keeps its state (e.g. water remains a liquid). Sometimes a distinction can be made on the grounds that the solute is the minor constituent and the solvent the major constituent but a distinction sometimes cannot be made on any grounds at all, e.g. an alloy such as solder is a solution, a mixture of lead and tin, and may vary from one extreme in composition to another.

Solute + Solvent → Solution

On the other hand,

Insoluble solid + Liquid → Suspension
(powder)

Water is the commonest and most important of all solvents and aqueous solutions are used most often by the chemist, as well as by industry and in the home. Some substances, e.g. chalk, glass, fat and copper, only dissolve to a very small extent in water and so are said to be **insoluble.** They may be soluble in other solvents: petrol dissolves most kinds of fat, and mercury will dissolve metals such as tin and silver (forming amalgams).

Solid solutes are not the only variety: a gas or a liquid will also dissolve in a liquid, e.g. air dissolves in water and wine is a solution of ethanol (and other materials) in water. The bottle labelled 'Dilute Hydrochloric Acid' contains a solution of a gas, hydrogen chloride, in water; while the bottle labelled 'Dilute Sulphuric Acid' contains a solution of a liquid, pure hydrogen sulphate, in water.

A solution is a homogeneous mixture of two or more substances.

A suspension is a heterogeneous mixture of a liquid and a finely divided but insoluble solid.

In chapter 2 it was considered that the process of making a solution is a physical change on the grounds that it is easily reversible and there is generally no great heat change. In the case of a solution of a solid in a liquid the solute may be recovered by careful evaporation of the solvent (see section 4.10) and the solvent may be obtained by distillation (see section 4.11). A heat change does occur but it is usually much smaller than when a chemical change occurs. When solids and liquids dissolve in liquids the solution usually becomes slightly colder. Energy is needed to separate the solid particles so that they can freely dissolve and this energy in the form of heat comes from the liquid, which consequently becomes cold. The effect is very noticeable for some substances, e.g. sodium thiosulphate and ammonium chloride. When gases dissolve in liquids the particles come closer together and energy in the form of heat is given out. Another example where heat is evolved is the addition of concentrated sulphuric acid to water (see section 18.1).

Dissolution is not always a physical change: all the common acids are electrovalent (see sections 17.3 and 14.4), but they are covalent when deprived of water (see section 14.3).

To the chemist the main importance of solutions is that they provide the most convenient way of bringing about a chemical reaction: most reactions are carried out in aqueous solution. Dissolved substances are finely divided and react readily when the solutions are mixed.

When a substance such as sodium chloride is added to water the melting point decreases (see section 24.15) and the boiling point of the solution increases: the effect of impurities on these physical properties, which are frequently used as criteria of purity (see sections 2.1 and 6.2), can lead to the determination of the relative molecular mass of a covalent compound or the extent of ionization of an electrovalent compound.

6.10 To Determine if a Given Solid is Soluble in a Given Liquid

The given liquid must be pure to start with, e.g. distilled water. The solid should be ground to a fine powder with a pestle in a mortar because small particles dissolve more quickly than large ones. The powder can then be added, with shaking, to a test tube half full of the liquid. If a clear solution results more of the solid can be added in small quantities at a time to find out roughly how soluble it is. If the solid is apparently insoluble the test tube should be warmed and gently shaken; most substances are more soluble as the temperature rises.

For a more accurate investigation, especially of a slightly soluble substance, excess of the finely-ground substance should be stirred and boiled for several minutes with the liquid. After filtering or centrifuging, the liquid should be evaporated to dryness. A solid residue will show that the solid was, to some extent, soluble. A simple experiment is to study distilled water, tap water and sea water by evaporating samples of the liquids on watch glasses placed over beakers of boiling water.

6.11 Saturated Solutions

If more and more sodium chloride is added to water, a point is eventually reached when no more will dissolve—the temperature should be kept constant. The solution is now said to be saturated. Unless there is an excess of a solid visibly present it is not obvious that a solution is saturated: the solution may be unsaturated or just saturated.

A solution is said to be saturated when, in the presence of excess of the solute, no more solute dissolves, the temperature remaining constant.

The situation is not static because particles of solute are constantly crystallizing out but they are being replaced at precisely the same rate by particles from the solid solute.

A given mass of a solvent will be saturated by different masses of different solutes. The mass dissolving depends not only on the solute but also on the temperature: the variations are great, as shown in figure 6.7.

The solubility of a solid in a solvent, at a given temperature, is the maximum number of grams of the solid that will dissolve in 1 kg (1000 g) of the solvent, in the presence of excess of the solid, at that temperature.

It is safer to quote the quantity of solvent than to omit it (some books consider 100 g of solvent).

6.12 The Determination of the Solubility of a Salt in Water

(a) The Preparation of a Saturated Solution

Dissolution is facilitated by finely powdering the substance in a mortar using a pestle and by stirring portions of the solute into distilled water as they are warmed. Even if a solid is soluble it may dissolve very slowly if the temperature is low so that it is advantageous to warm the solution above the required temperature and allow it to cool in the presence of excess solute. More crystals will probably separate out during the cooling. With some substances, e.g. copper sulphate, it is desirable not to boil the water. The solution must then be allowed to come to the required temperature, preferably in a thermostat (a constant temperature bath).

Grams (anhydrous) in 1 kg water at various temperatures (°C)

	0	10	20	30	40	50	60	70	80	90	100
Sodium chloride	357	358	360	363	366	370	374	378	381	386	392
Potassium chloride	280	310	345	375	400	430	455	485	510	540	565
Sodium nitrate	730	805	880	960	1 050	1 140	1 240	1 350	1 480	1 610	1 780
Potassium nitrate	135	210	315	455	625	845	1 080	1 370	1 680	2 030	2 450
Sodium chlorate (V)	800	—	990	—	1 170	1 355	1 430	—	—	—	2 300
Potassium chlorate (V)	30	50	75	105	140	190	240	300	380	460	540
Sodium hydroxide	420	1 020	1 090	1 190	1 290	1 450	1 740	—	3 160	—	3 470
Potassium hydroxide	970	1 030	1 120	1 260	1 370	1 400	1 450	—	1 590	—	1 780
Sodium carbonate	70	125	215	400	480	475	470	465	460	457	455
Potassium carbonate	1 050	1 080	1 100	1 140	1 170	1 210	1 270	1 330	1 400	1 480	1 560
Sodium hydrogen-carbonate	70	80	95	110	125	145	165	Decomposes			
Sodium sulphate	45	90	205	410	480	465	450	440	435	430	425
Sodium thiosulphate	525	610	700	850	1 040	1 700	2 070	2 420	2 480	2 540	2 660
Calcium hydroxide	1·85	1·75	1·65	1·55	1·40	1·30	1·15	1·05	0·95	0·85	0·80
Calcium sulphate	1·75	1·92	2·05	2·10	2·11	2·08	2·03	1·95	1·80	—	1·62
Sucrose	1 790	1 910	2 040	2 200	2 380	2 600	2 870	3 200	3 620	4 160	4 870
Copper(II) sulphate	145	175	205	250	285	335	400	455	550	675	755

Figure 6.7
The solubilities of substances in water

(b) A Rough Measurement of Solubility

A clean dry evaporating basin should be weighed empty and then containing a portion of the saturated solution. The solution must be obtained free from crystals of the excess solute either by careful decantation or by using a pipette fitted with a plug of glass wool in a piece of rubber tubing. The solution should then be evaporated to dryness very carefully to avoid losses of solute by spitting. The early stages can be carried out on a tripod and gauze over a Bunsen burner but the final stages are better carried out over a water bath or in a steam oven. The basin and solute are weighed when they are at room temperature again. They may be reheated and reweighed to check that the removal of water was complete.

Calculation

Mass of basin = 30·52 g
Mass of basin + saturated solution = 72·55 g
Mass of basin + solute = 38·27 g
Temperature of saturated solution = 17·5°C
∴ Mass of solute = (38·27 − 30·52) g = 7·75 g
∴ Mass of water = (72·55 − 38·27) g = 34·28 g

i.e. 34·28 g water dissolves 7·75 g solute

$$\therefore 1000 \text{ g water dissolves } \frac{7{\cdot}75 \times 1000}{34{\cdot}28} \text{ g solute} = 226 \text{ g solute}$$

The solubility of the solute at 17·5°C is 226 g in 1000 g of water.

(c) Another Method of Measuring Solubility

A clean dry boiling tube should be weighed empty and then containing 1–5 g of potassium chlorate—the experiment may be done on a class basis so that a graph of solubility as a function of temperature can be plotted from the results. To the crystals exactly 10 cm^3 (10 g) of distilled water should be added by means of a pipette and then the tube clamped in a beaker of water over a Bunsen burner. On warming the beaker of water, which serves to steady the temperature changes, the crystals dissolve: the temperature should be noted. The thermometer should be in the boiling tube: it can function as the stirrer or a copper wire stirrer may be used. Then the beaker is allowed to cool and the temperature at which crystals reappear is noted. The average temperature is that at which a known mass of solute is soluble in 10 g of water.

A modification is to start with about 4·5 g of crystals and, having found the temperature at which this mass of crystals is soluble in 10 cm^3 of water, to add 5 cm^3 of water and find the temperature when dissolution again occurs.

(d) The Accurate Measurement of Solubility

The only accurate method of assessing the solubility of a solid in a liquid is, for the moment, by volumetric analysis. This means that the method is restricted to acids and alkalis, oxidizing and reducing agents and to substances which by precipitation reactions will give silver halides.

Experiment: The Determination of the Solubility of Calcium Hydroxide in Water

About 100 cm^3 of a saturated solution of calcium hydroxide in water is prepared at a particular temperature—on a class basis a range of temperature can be covered. The saturated solution is filtered to remove the excess solid and then 25 cm^3 portions of the filtrate can be titrated with a standard solution (0·05 M) of hydrochloric acid using methyl orange or phenolphthalein as the indicator.

$$Ca(OH)_2 + 2HCl \rightarrow CaCl_2 + 2H_2O$$

The concentration of the calcium hydroxide can be calculated in mol/dm^3 and then in grams/dm^3.

6.13 Solubility Curves

When the solubility of a substance in water has been determined at a number of different temperatures it is best to record the results in the form of a graph because this (*a*) shows up irregularities most clearly, (*b*) makes some allowance for experimental errors, and (*c*) enables solubilities at other temperatures to be found. Some typical solubility curves are shown in figure 6.8 and others may be plotted using the values given in figure 6.7.

The solubilities of most salts increase rapidly as the temperature rises (see section 19.5). Sodium chloride is an exception because its solubility in water shows little variation with temperature. There are a few substances, e.g. calcium hydroxide, the solubility of which decreases as the temperature rises.

The solubility curve of sodium sulphate is composed of two portions with a sharp break between them. At low temperatures the hydrated salt, sodium sulphate-10-water, crystallizes out

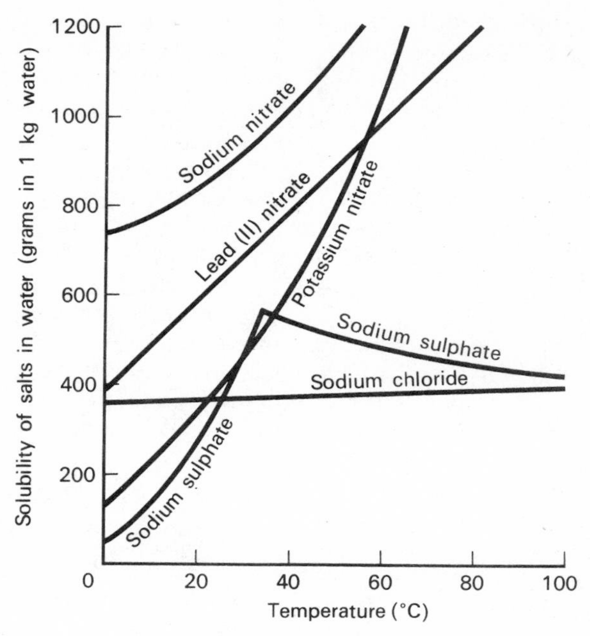

Figure 6.8
The solubilities of salts in water

from saturated solutions but at high temperatures the anhydrous salt does so. The transition temperatures between the two forms of solid crystals is 32·4°C. The determination of solubility gives a method of measuring transition temperatures, e.g. between the allotropes of sulphur.

The solubility curve of potassium chlorate(V) is given in figure 6.9 for further examination:

(*a*) The solubility at temperatures other than those at which experiments were done can be found, e.g. at 40°C the solubility is 140 g potassium chlorate in 1 kg water (point A).

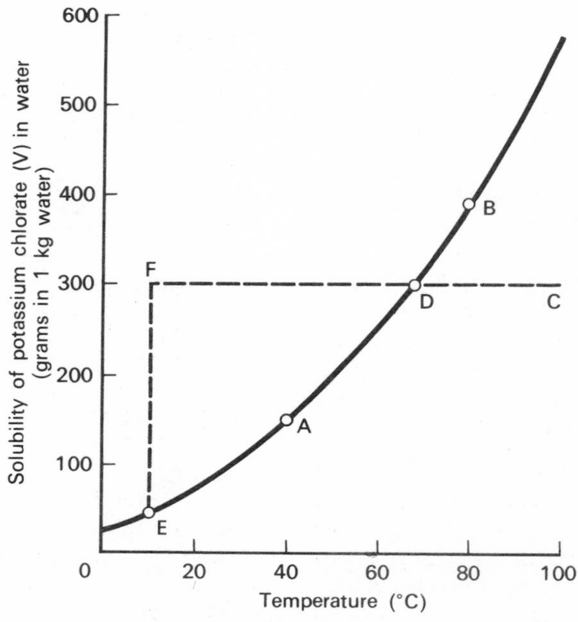

Figure 6.9
The solubility of potassium chlorate(V) in water

(*b*) The temperature at which given masses of solute and solvent will start to yield crystals can be determined, e.g. 380 g potassium chlorate in 1 kg of water will be saturated at 80°C (point B).
(*c*) The mass of crystals obtained on cooling a hot solution can be assessed, e.g. on cooling a solution containing 300 g potassium chlorate at 100°C (point C) to 15°C (point E) crystallization will start at 70°C (point D). The solubility at 15°C is 60 g crystals in 1 kg water: thus 240 g of crystals are obtained. (N.B. Sometimes the mass of solution rather than the mass of solvent is quoted: remember that mass is additive even if volume is not.)

6.14 Fractional Crystallization

When solutions of sodium chlorate and potassium chloride are mixed the solution contains the ions of sodium, potassium, chlorate and chloride. The solubility curves of four substances must be consulted to predict the behaviour of the mixture: sodium chlorate, sodium chloride, potassium chlorate and potassium chloride. Sodium chloride is the least soluble substance in hot water and so, on boiling the solution and allowing the steam to escape, sodium chloride will crystallize out first. After a while the solution is cooled and potassium chlorate, which is the least soluble substance in cold water, crystallizes out. The substances have been prepared by fractional crystallization.

In a similar manner potassium nitrate and sodium chloride can be prepared from potassium chloride and sodium nitrate.

When preparing a substance by crystallization all the solvent must not be boiled off because this would lead to all the substances (wanted and unwanted) in the solution crystallizing.

6.15 Supersaturation

The definition of solubility emphasises that excess solute must be present for a solution to be definitely saturated. If the solution of potassium chlorate in the section 6.13 was cooled to 15°C without any crystals appearing then it would be a supersaturated solution (point F). Supersaturated solutions are unstable and if a tiny crystal of the salt is added (called 'seeding' the solution) crystallization ensues and continues until the solution is truly saturated (point E). Sometimes shaking the supersaturated solution or a particle of dust causes the change.

A supersaturated solution is one in which the solvent contains more solute than is necessary to form a saturated solution at that temperature.

Supersaturation is shown most easily using substances with steep solubility curves, e.g. sodium thiosulphate. Sodium thiosulphate-5-water will dissolve on heating in its water of crystallization, or alternatively a little water can be added. On cooling the hot solution, however, no crystals reappear. If a crystal of the solute is added to the cold supersaturated solution rapid crystallization follows and the temperature rises considerably (dissolution absorbs heat, and crystallization evolves heat in this case).

6.16 The Solubility of Gases in Water

There are two outstanding differences between the solubilities of solids and of gases in liquids.
(*a*) Gases are less soluble in liquids as the temperature rises, e.g. air dissolved in water is driven out on boiling.
(*b*) The influence of pressure on the solubility of a gas in a liquid is great: as the pressure increases the mass of gas dissolving increases. A soda-water siphon contains a solution of carbon dioxide in water under pressure; ethyne is stored by dissolving it under pressure in propanone.

The solubilities of some gases in water at 15°C when their pressure is 760 mmHg are given below (vol./vol.); to allow for the effect of pressure and temperature on the volume of the gas the volumes given are standardized at 0°C and 760 mmHg:

Ammonia	770
Hydrogen chloride	458
Sulphur dioxide	47
Hydrogen sulphide	2·9
Chlorine	2·6
Carbon dioxide	1·0
Dinitrogen oxide	0·7
Nitrogen oxide	0·051
Oxygen	0·034
Hydrogen	0·019
Nitrogen	0·018
Helium	0·009

The fountain experiment can be used to illustrate the high solubility of ammonia and hydrogen chloride (see section 32.10). Most of the other gases except sulphur dioxide can be collected over water with very little loss.

6.17 Chromatography

The technique of paper chromatography was briefly mentioned in section 3.7; it depends upon relative solubility. An appreciable proportion of the mass of a piece of paper is of water which is loosely combined with the cellulose of the paper (see section 29.18). It can be shown that a substance may dissolve in two solvents simultaneously by putting iodine crystals into a separating funnel containing roughly equal volumes of 1,1,1-trichloroethane and water. After the layers have been separated each can be tested with starch, which will turn dark blue.

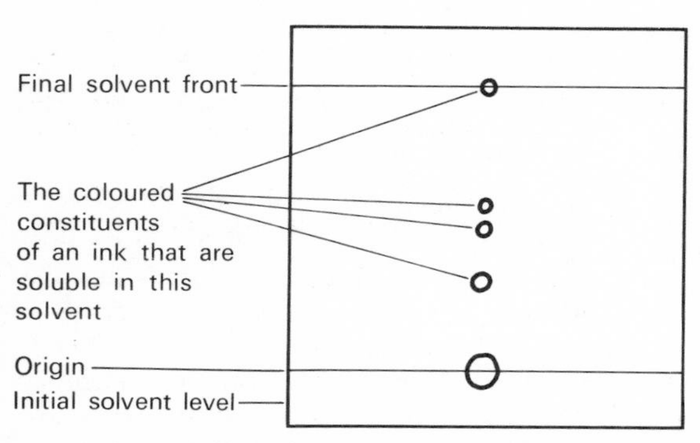

Figure 6.10
A paper chromatogram (ascending)

Similarly, as a solvent flows along a piece of filter paper the solutes present are soluble to different extents in the bound water (water trapped in the paper) and in the mobile solvent. Thus a separation is achieved. If an ink is examined it can easily be found whether it is a mixture or a single coloured substance. A black ink may give a result such as that illustrated in figure 6.10. It is advisable to put the drop of ink on to the paper at the origin with a capillary tube and to dry it if it shows signs of spreading. The solvent should be put in a tall beaker which can be covered to prevent too much evaporation; its depth should be less than the distance of the origin from the edge of the paper. The paper is stood in the solvent which will then rapidly rise up by capillary action. When the solvent approaches the top of the paper the paper is taken out and dried, the solvent front is marked and the spots of coloured substances ringed. A useful ratio to quote to describe a substance is its $\mathbf{R_f}$ value in that solvent, defined as

$$\mathbf{R_f} = \frac{\textbf{Distance flowed by solute}}{\textbf{Distance flowed by solvent}}$$

Ascending chromatography as described in the previous paragraph gives better results than radial chromatography if the concentration of a material is low. Descending chromatography gives results faster than ascending chromatography but is more difficult to carry out in the laboratory. At either side of the substance to be examined known substances can be put on the origin line to facilitate identification. The mobile solvents employed are frequently mixtures, e.g. ethanol, butanol and ammonia in water or propanone with hydrochloric acid. If one solvent fails to give an adequate separation of the substances in a mixture a two-way chromatogram is prepared: the mixture is put near one corner and one solvent run up the paper. The paper with substances spread along one edge is then put with that edge down into a second solvent which will then traverse the paper at right angles to the first. This method is very useful for the amino acids in a protein which are very hard to separate except by this means. The amino acids, like the majority of other substances, are colourless and so the chromatogram has to be 'developed'. Development in this case is carried out by spraying the paper with a reagent that forms coloured derivatives with the acids.

Ion exchange (see section 25.17(iii)) is a form of chromatography.

The techniques can be extended to gases: the fixed phase is put in a tube with a detector, e.g. a thermal conductivity device, at the far end. The sample is injected into the gas stream at the head of the column and when the constituents reach the meter a reading is given. In this way petrol, which is a very complex mixture, can be rapidly analyzed.

QUESTIONS 6

Water and Solubility

1 From the data in figure 6.7 plot the solubility curves of potassium chloride (*A*) and potassium nitrate (*B*). From your graphs answer the following questions:
(*a*) Above what temperature is *B* the more soluble salt?
(*b*) What happens when a mixture of 100 g of *B* with 100 g of water is heated to 80°C?
(*c*) What happens when the mixture in (*b*) is cooled from 80°C to 20°C?
(*d*) A mixture of 60 g of *A* and 40 g of *B* is added to 100 g of water and heated to 70°C until there is no further change, and the solution is then cooled to 10°C. Describe what will happen and calculate the percentage of *A* and of *B* in the crystals that are formed. [L]

2 (*a*) What is water of crystallization? Give the formulae for two substances containing water of crystallization. (*b*) Describe and explain what happens when an efflorescent substance is exposed to air. [O]

3 (*a*) Describe an experiment by which the composition of water by mass can be determined. Draw a simple diagram of the apparatus used and mention any precautions necessary to carry out the experiment safely and to obtain a reasonably accurate result.
(*b*) Explain carefully how rain water acquires temporary hardness in some districts.
(*c*) Calculate the percentage of water of crystallization in sodium thiosulphate crystals, $Na_2S_2O_3, 5H_2O$. [C]

4 Explain carefully what you understand by a *saturated solution.* How would you determine the solubility of potassium chloride at 20°C?

100 g of water at 15°C dissolve at saturation 37 g of sodium chloride and 25 g of potassium nitrate but at 70°C the corresponding masses are 38 g and 140 g per 100 g of water respectively. 100 g of a mixture of the above two salts, in equal proportions by mass, are shaken with 100 g of water at 70°C until equilibrium is reached. The solution is filtered hot. The hot filtrate is slowly cooled to 15°C and again filtered. The final filtrate is evaporated to dryness. What will be the masses and compositions of the three residues, assuming that no solution is left in either of the filter papers? [C]

5 Explain the formation of water (*a*) on the cool sides of a test tube in which sodium hydrogencarbonate is being heated, (*b*) at the exhaust of a motor car when the engine is first started from cold. Write equations for one reaction in each case in which either water or steam reacts with (i) a metal, (ii) a non-metal, (iii) a metallic oxide. State the conditions under which each reaction occurs. Diagrams are not required.

4·76 g of the hydrated chloride of a divalent metal *M*, of relative atomic mass 59, contain 2·60 g of the anhydrous salt. Find the number of molecules of water of crystallization in each molecule of the hydrated salt. [W]

6 The water used by industry, even when freed from suspended material by filtration, is seldom pure water. Choose one industry which uses large quantities of water and discuss some of the problems which arise from impure water. How might this industry deal with its water supply in order to overcome these problems?

7 (*a*) It is required to separate a mixture of three solid dyes; one red, one yellow and one blue. The following facts are known about the dyes. The blue and yellow dyes are soluble in cold water, while the red dye is insoluble. When an excess of aluminium oxide is added to a stirred, green aqueous solution of the mixed blue and yellow dyes and the aluminium oxide is filtered off and washed with water, it is found that the solid residue is yellow and the filtrate blue. When the yellow is stirred with ethanol and the mixture filtered, the solid residue is white and the filtrate yellow. Describe how you would obtain dry samples of the three dyes.
(*b*) A green dye is known to be a hydrate. Describe how you would determine the percentage of water of crystallization in the hydrate. [J]

8 Describe two different ways of producing a specimen of water from hydrogen. Draw diagrams to show the apparatus you would use and state four different tests you would perform on the product to prove that it was, in fact, pure water. [O]

9 Give a fully labelled diagram to show (*a*) how to prepare dry hydrogen and (*b*) pass it over dry copper oxide in order (*c*) to prove the gravimetric composition of water. Before setting up your apparatus how would you prove your copper oxide to be dry? What weighings would you make in order to find: (i) the mass of water formed in the experiment; (ii) the mass of oxygen used in the experiment; (iii) the mass of hydrogen that has reacted in the experiment? State the percentage composition of water as found by experiments of this kind. [J]

10 Dry hydrogen was passed over 3·18 g of black copper oxide heated in a hard-glass tube. When the reduction was complete, 2·54 g of copper was left and 0·72 g of water was collected. Calculate the composition of water by mass. Draw a labelled diagram of the apparatus necessary to perform the above experiment. (A description of the experiment is not required.)

11 Starting from potassium nitrate, explain with full practical details how you would: (*a*) prepare a saturated solution (in water) at room temperature, and hence (*b*) measure the solubility of potassium nitrate at room temperature. How does the solubility of potassium nitrate change with temperature? [J]

12 Describe an experiment by which you could determine the solubility of potassium chlorate at 30°C. Draw the solubility curve from the data given in figure 6.7. From the curve determine the solubility at 34°C, and hence whether 150 g of solution containing 18 g of solute would be saturated or not at this temperature. [J]

13 From the data given in figure 6.7, plot the solubility curve for potassium chlorate. What is the solubility at (*a*) 43°C; (*b*) 55°C? If 130 g of a solution saturated with potassium chlorate at 70°C were cooled to 10°C, what mass of crystals would be deposited? [J]

14 From the data in figure 6.7, construct solubility curves for potassium chloride and potassium nitrate on the same graph paper. Use your graphs to decide what you would expect to happen if a solution containing equal masses of these two substances is (*a*) left to evaporate at 15°C; (*b*) evaporated at 50°C until solid appears, then allowed to cool. [O]

15 Describe an experiment to show that carbon dioxide is more soluble in cold than in hot water, and another experiment to show that sodium nitrate is more soluble in hot than in cold water. State one important consequence of each of these effects. [O]

16 Construct the solubility curve of copper sulphate from the data given in figure 6.7. From your curve find: (*a*) the solubilities at 25°C and 70°C; (*b*) how much solid will crystallize out of solution if 100 g of solution, saturated at 100°C, are cooled from 100°C to 30°C. [O]

17 Explain how you would: (*a*) show that tap water contains dissolved gases; (*b*) find the boiling point of water in the laboratory; (*c*) find the solubility of potassium chloride in water at room temperature. [J]

18 Using the data given in figure 6.7 for the solubility of sodium sulphate plot a graph. Point out any abnormal characteristics of the graph and explain them.

19 Explain carefully what you understand by the term *solution*. Give as many examples as you can of the use of solvents and solutions in everyday life and in industrial processes.

20 What is meant by a *supersaturated solution*? How would you prepare a supersaturated solution of sodium thiosulphate? How could it be made to crystallize and what would you observe as it did so?

21 What do you understand by *water of crystallization*? Give the names and formulae of two crystalline compounds possessing water of crystallization and of two which have none. Describe an experiment which you could perform in the laboratory to determine the percentage of water of crystallization in a crystalline substance. [J]

22 Calculate the percentage of water of crystallization in Glauber's salt, $Na_2SO_4,10H_2O$. How could you check your result by experiment, and what precautions would you have to take to get an accurate result?

23 Explain what is meant by *water of crystallization*. Some pure zinc sulphate crystals were heated until all the water of crystallization had been driven off. Calculate from the following data the number of molecules of water of crystallization combined with one molecule of anhydrous zinc sulphate ($ZnSO_4$).

Mass of empty crucible = 18·74 g
Mass of crucible and crystals = 21·20 g
Mass of crucible and anhydrous salt = 20·12 g [J]

24 What is meant by *deliquescence* and *efflorescence*? Describe what happens when the following substances are left exposed to the atmosphere: (*a*) sodium; (*b*) iron(II) sulphate crystals. Give any necessary explanations. Samples of each of the following substances were weighed on watch glasses and left in the air for some time before being reweighed. State in each case whether there would be any gain or loss in mass, giving any necessary explanation: (i) calcium chloride; (ii) sodium carbonate-10-water; (iii) sodium nitrate.

25 Define *efflorescence*. When 2·50 g of a sample of sodium carbonate crystals which have partly effloresced were heated in a crucible to constant mass there was a loss in mass of 1·44 g. Find: (*a*) the percentage by mass of water in the crystals; (b) the mass of water combined with one mole of sodium carbonate, and hence the value of *x* in the formula Na_2CO_3,xH_2O. [J]

Atoms and Molecules

7 The Gas Laws and the Kinetic Theory

7.1 Introduction

The differences between solids, liquids and gases have been discussed in chapter 2.

The gas laws found by experiment are considered first, then they are explained in terms of the motion of particles and finally the consequences and importance of these laws are illustrated.

That a gas consists of particles which are in rapid random motion can easily be shown by simple experiments.

Experiments illustrating Gaseous Behaviour

(*a*) A test tube of carbon dioxide is held vertically for a few minutes, with its mouth to a test tube of air (uppermost). Then some calcium hydroxide solution is shaken in each tube and it turns milky.

(*b*) A test tube of hydrogen (uppermost) is held vertically for a few seconds with its mouth to a test tube of air. Upon ignition both tubes of gas give an explosion.

(*c*) Bromine and nitrogen dioxide are both red-brown poisonous gases but they can be studied in a fume cupboard. Gas jars containing these substances are placed below those of air: the progress of the coloured gases can easily be seen over a period of about an hour.

(*d*) Pieces of cotton wool soaked in concentrated ammonia solution and concentrated hydrochloric acid respectively can be corked in at opposite ends of a horizontal glass tube, with approximate dimensions of length 50 cm and diameter 3 cm. It will be observed that a white ring starts to form nearer the hydrochloric acid than the ammonia solution; this suggests that particles of ammonia travel faster than those of hydrogen chloride.

(*e*) Observe particles of dust undergoing rapid random motion where a shaft of sunlight into a room shows up some of the atmospheric pollution. The random convection currents of gas particles causing the zig-zag motion of the dust cannot be seen, only their effect is visible.

(*f*) **Brownian motion,** the random zig-zag path of small particles, can be observed under a microscope. 1 cm^3 of 0·1 M potassium carbonate solution is put into 200 cm^3 of water, and 1 cm^3 of 0.01 M lead ethanoate solution is put into another 200 cm^3 water. These two solutions are mixed at 40°C and, preferably after standing for a day, contain suitably small particles of lead carbonate for observation under a microscope. A powerful light with a green filter is shone through a sample of the solution under a microscope and the haphazard movement of the particles, due to their uneven bombardment by solvent particles, can be seen. If the bulk of the solution in a beaker is illuminated by a convergent beam of light a scattering effect is observed, known as the **Tyndall effect.**

7.2 Boyle's Law (1662)

The volume occupied by a fixed mass of gas is inversely proportional to the pressure, provided that the temperature remains constant.

Figure 7.1 shows the experimental situation: if the piston is moved slowly to the left the gas is compressed and the pressure increases as the volume decreases. Conversely, if the piston is moved slowly to the right, the gas expands and the pressure decreases as the volume increases. In both cases the mass of gas and the temperature remain constant.

In mathematical terms:

$$V \propto \frac{1}{p} \quad \text{(T constant)}$$

hence

$$pV = k_1$$

where k_1 is a constant dependent upon the mass of gas and the temperature.

For two given situations:

$$p_1V_1 = p_2V_2$$

Example: At a pressure of 4 atmospheres, a given mass of gas has a volume of 200 cm^3. What volume will it occupy at a pressure of 1 atmosphere, if the temperature remains unchanged?

$p_1 = 4$ atm $V_1 = 200$ cm^3
$p_2 = 1$ atm V_2 is unknown

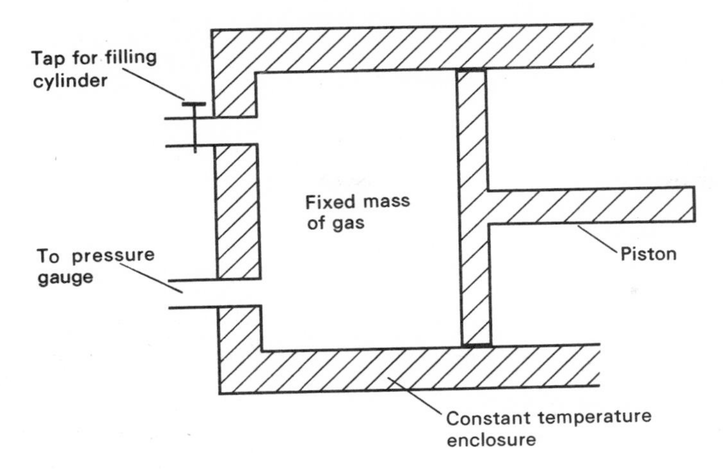

Figure 7.1
Apparatus for verifying Boyle's Law

By Boyle's Law:

$$4 \times 200 = 1 \times V_2$$

$$\therefore \quad V_2 = 800 \text{ cm}^3$$

The required volume is 800 cm^3. The units used must be consistent: all volumes in cubic centimetres, all pressures in atmospheres (or, in both cases, other convenient units). It is often useful to measure pressures in multiples of the atmospheric pressure which is usually about 760 mmHg (the standard pressure, 0.101 Pa).

7.3 Charles' Law (1787)

The volume of a fixed mass of gas is directly proportional to the thermodynamic temperature, provided that the pressure is constant.

Experiment shows that the volume of a gas increases by $\frac{1}{273}$ of its volume at 0°C for every degree rise in temperature (C for celsius; commonly, but erroneously, called centigrade), provided that the pressure is constant. This is an alternative form of Charles' Law. The opposite is true if the temperature is decreased: the volume of the gas decreases regularly and the graph (figure 7.2) can be extrapolated to zero volume at -273°C if no account is taken of the fact that in practice all gases on cooling liquefy and eventually solidify (except helium at atmospheric pressure).

The temperature -273°C is often referred to as absolute zero. The **thermodynamic or Kelvin scale** of temperature starts at this point and it is a simple matter to convert a temperature in °C to one in K (no° is written before the symbol) by adding 273. This scale of temperature is used in all calculations involving the variation of pressures and volumes of gases with temperature.

In mathematical terms Charles' Law is:

$$V \propto T \quad \text{(p constant)}$$

hence

$$\frac{V}{T} = k_2$$

where k_2 is a constant dependent upon the mass of gas and the pressure.

For two given situations:

$$\frac{V_1}{T_1} = \frac{V_2}{T_2}$$

Example: At atmospheric pressure a given mass of gas occupies a volume of 300 cm^3 at 27°C.

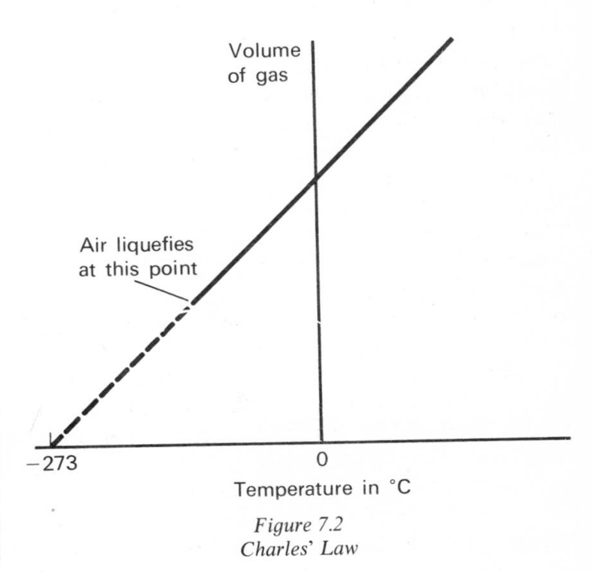

Figure 7.2
Charles' Law

What volume will it occupy at 127°C at the same pressure?

$V_1 = 300\ cm^3$ $\quad T_1 = (273+27)K$
V_2 is unknown $\quad T_2 = (273+127)K$

$$\frac{300}{300} = \frac{V_2}{400}$$

$$\therefore \quad V_2 = 400\ cm^3$$

The required volume is 400 cm^3.

The units used must be consistent: all volumes in cubic centimetres (usually), but all temperatures must be quoted on the thermodynamic scale.

7.4 The Law of Pressures

This law, sometimes referred to as the Constant Volume Law, is not ascribed to any particular scientist.

The pressure exerted by a fixed mass of gas is directly proportional to the thermodynamic temperature, provided that the volume is constant.

In mathematical terms:

$$p \propto T \quad (V \text{ constant})$$

hence
$$\frac{p}{T} = k_3$$

where k_3 is a constant dependent upon the mass of gas and the volume. For two given situations:

$$\frac{p_1}{T_1} = \frac{p_2}{T_2}$$

7.5 The General Gas Equation

The three laws discussed above may be combined into a single equation for a fixed mass of gas:

$$\frac{pV}{T} = k_4$$

where k_4 is a constant dependent upon the mass of gas.

For two given situations:

$$\frac{p_1V_1}{T_1} = \frac{p_2V_2}{T_2}$$

Example: A gas occupies 250 cm^3 at 27°C and 720 mmHg pressure. What volume will it occupy at 77°C and 750 mmHg pressure?

$p_1 = 720$ mmHg $\quad p_2 = 750$ mmHg
$V_1 = 250\ cm^3$ $\quad V_2$ is unknown
$T_1 = (273+27)K$ $\quad T_2 = (273+77)K$

By the general gas equation:

$$V_2 = 250 \times \frac{720}{750} \times \frac{350}{300}$$

$$\therefore \quad V_2 = 280\ cm^3$$

The required volume is 280 cm^3.

Because volumes of gases are so dependent upon temperature and pressure it is vital to have a standard set of conditions: those chosen are 0°C and 760 mmHg pressure, being designated as s.t.p. (standard temperature, the melting point of pure ice, and standard pressure, 101 325 Pa or N/m^2).

Example: A gas at 273°C under a pressure of 1600 mmHg occupies a volume of 190 cm^3. What is its volume at s.t.p.?

$p_1 = 1600$ mmHg $\quad p_2 = 760$ mmHg
$V_1 = 190\ cm^3$ $\quad V_2$ is unknown
$T_1 = (273+273)K$ $\quad T_2 = 273$ K

By the general gas equation:

$$V_2 = 190 \times \frac{1600}{760} \times \frac{273}{546}$$

$$\therefore V_2 = 200\ cm^3$$

The required volume is 200 cm^3.

7.6 Dalton's Law of Partial Pressures (1801)

The total pressure of a mixture of gases, which do not react with each other, is the sum of the pressures that each gas would exert if it alone occupied the volume of the mixture.

The pressures exerted by the individual gases are called partial pressures. Ordinary air contains 80% nitrogen and 20% oxygen by volume. If a vessel contains air at a total pressure of 1 atmosphere, the partial pressure of the nitrogen will be 0·8 atmosphere, and of the oxygen 0·2 atmosphere.

Example: A bulb containing 100 cm^3 of hydrogen at 2 atmospheres pressure and one containing 300 cm^3 of nitrogen at 1 atmosphere pressure are connected at room temperature. What will the total pressure be, if the temperature remains constant?

For hydrogen:

$p_1 = 2$ atm $\quad V_1 = 100\ cm^3$
p_2 is unknown $\quad V_2 = 400\ cm^3$

By Boyle's Law:

$$p_2 = \frac{2 \times 100}{400}$$

$$\therefore p_2 = 0{\cdot}5 \text{ atm}$$

For nitrogen:

$p_1 = 1$ atm $\quad V_1 = 300\ cm^3$
p_2 is unknown $\quad V_2 = 400\ cm^3$

By Boyle's Law:

$$p_2 = \frac{1 \times 300}{400}$$

$$\therefore p_2 = 0{\cdot}75 \text{ atm}$$

By Dalton's Law the total pressure

$= 0{\cdot}5 + 0{\cdot}75$

$= 1{\cdot}25$ atm

The required pressure is 1·25 atmospheres.

7.7 Collecting Gases over Water

When a gas is collected over water, and its volume is required accurately, as in the determination of relative molecular masses (see section 10.4), it is necessary first of all to be sure that the water levels inside and outside the collecting vessel are the same. When the levels have been equalized, the total pressure inside must be equal to the pressure outside, which is atmospheric pressure, and can be obtained from a barometer.

However, the gas collected over water will contain a certain amount of water vapour. This will exert its own pressure, dependent upon the temperature of collection, so that, by Dalton's Law:

Atmospheric Pressure = Total Pressure Inside
= Gas Pressure + Water Vapour Pressure

So, to get the true pressure of the gas, it is necessary to subtract the 'saturated vapour pressure' of the water. This will vary with temperature, the value being obtainable from tables.

Example: 76 cm³ of air were collected over water at a temperature of 15°C when the pressure was 733 mmHg. The saturation vapour pressure of water at 15°C is 13 mmHg. Calculate the volume of air at s.t.p.

For air alone:

$p_1 = (733-13)$ mmHg $\quad p_2 = 760$ mmHg

$V_1 = 76$ cm³ $\quad V_2$ is unknown

$T_1 = (273+15)$K $\quad T_2 = 273$ K

By the general gas equation:

$$V_2 = 76 \times \frac{273}{288} \times \frac{720}{760}$$

$$\therefore V_2 = 68 \text{ cm}^3$$

The required volume of gas is 68 cm³.

7.8 The Motion of Gases

More evidence that the particles of a gas are in continual motion comes from the fact that a gas tends to fill any space in which it finds itself and this 'spreading out' is, in most experiments, independent of gravity in a small space.

Experiments illustrating the motion of gases

(*a*) Hydrogen is the least dense of gases and so if a gas jar is filled with hydrogen and held mouth downwards it might be expected that none would escape. Yet after a while, if a burning splint is held to the jar an explosion is not forthcoming: the hydrogen has escaped.

(*b*) Bromine is a volatile red liquid giving a dense red-brown poisonous vapour. If, in a fume cupboard, a few drops of bromine are put at the bottom of a gas jar by means of a teat pipette they evaporate and the vapour fills the jar before escaping.

(*c*) Gases will also diffuse through porous material like unglazed earthenware or even rubber (plastic tubing may often be substituted for rubber if hydrogen is to be carried) and this experiment may be used to determine relative molecular masses.

(*d*) The gas jar in figure 7.3 is filled with hydrogen by putting the delivery tube from a Kipp's apparatus (containing zinc and dilute sulphuric acid) between the porous pot and gas jar. It is observed that a gas (mostly air) bubbles out through the tube into the beaker of water: air cannot diffuse out through the porous pot as rapidly as hydrogen is getting in and so the pressure inside the pot is momentarily above

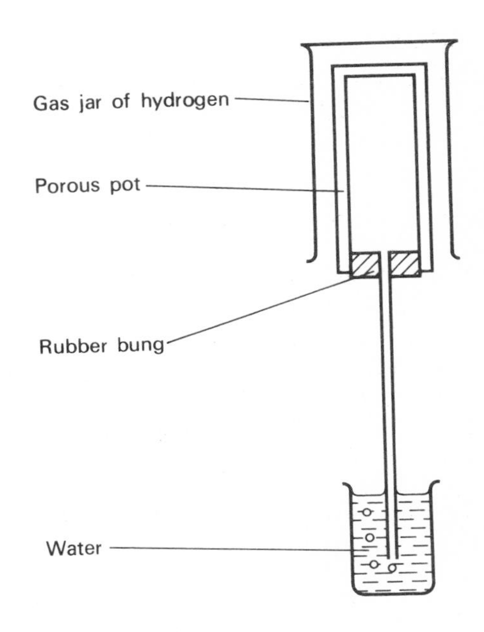

Figure 7.3
Diffusion

atmospheric and equilibrium is restored by the escape of some gas through the tube. If after a few minutes the gas jar is removed, the water level inside the tube is observed to rise rapidly and then slowly sink to its original level: hydrogen escapes from the pot faster than air can get in so the pressure is momentarily below atmospheric, but this is slowly rectified.

The experiment can be repeated using carbon dioxide instead of hydrogen: results in the reverse direction are obtained.

7.9 Graham's Law of Diffusion (1833)

The rate of diffusion of a gas at a constant pressure is inversely proportional to the square root of its density.

The rate (R) of diffusion is best measured by the volume (V) of gas diffusing through a given porous substance in a given time (t), e.g. in units such as cm^3/s. In mathematical terms:

$$R \propto \frac{1}{\sqrt{d}}$$

or

$$\frac{V}{t} \propto \frac{1}{\sqrt{d}}$$

Hence for two gases

$$\frac{R_1}{R_2} = \sqrt{\frac{d_2}{d_1}}$$

Oxygen has a density sixteen times that of hydrogen so that hydrogen diffuses four times as fast as oxygen.

Example: 60 cm^3 of oxygen diffuse through a porous pot in 10 seconds and 100 cm^3 of chlorine diffuse through the same pot in 25 seconds. If the density of oxygen is 16 times that of hydrogen, what is the density of chlorine compared to hydrogen?

Rate of diffusion of oxygen $= 6\ cm^3/s$
Rate of diffusion of chlorine $= 4\ cm^3/s$

By Graham's Law:

$$\frac{\text{Rate of diffusion of oxygen}}{\text{Rate of diffusion of chlorine}} = \sqrt{\frac{\text{Density of chlorine}}{\text{Density of oxygen}}}$$

$$\therefore \quad \frac{6}{4} = \sqrt{\frac{\text{Density of chlorine}}{16}}$$

$$\frac{6^2}{4^2} = \frac{\text{Density of chlorine}}{16}$$

Therefore the relative density of chlorine $= \frac{36}{16} \times 16 = 36$

7.10 Gay-Lussac's Law of Gaseous Combining Volumes (1808)

When gases react they do so in volumes which bear a simple ratio to one another, and to the volumes of the products formed if gaseous, all measurements being made at, or corrected to, the same temperature and pressure.

For example under the same conditions:

1 volume of hydrogen combines with 1 volume of chlorine to form 2 volumes of hydrogen chloride.

2 volumes of carbon monoxide combine with 1 volume of oxygen to form 2 volumes of carbon dioxide.

2 volumes of hydrogen combine with 1 volume of oxygen to form 2 volumes of steam.

In the last experiment, if the temperature (at 760 mmHg pressure) falls below 100°C, water, and not steam, is formed. No comparison can then be made with the volume of the product, as it is not gaseous.

7.11 Avogadro's Law (1811)

Equal volumes of all gases under the same conditions of temperature and pressure contain equal numbers of molecules.

The distinction between the molecule—the free particle—and the atom—the smallest possible particle of any element—was not realized until nearly fifty years after Avogadro put forward his idea. Until the work of Cannizzaro upon relative atomic masses convinced chemists that Dalton and Avogadro were both right, the greater success of Dalton's Atomic Theory led to a temporary rejection of Avogadro's hypothesis. Avogadro's Law is nowadays accepted as being as true of gases as any of the laws so far discussed.

The Avogadro constant is the number of particles in 0.012 kg of the carbon-12 isotope, i.e. $6{\cdot}02 \times 10^{23}$.

It is a number so large that the combined efforts of the population of the whole world could not attain it by counting at the rate of one per second for the whole of a lifetime even without sleeping! A mole of material contains this number of particles (see section 22.2).

7.12 The Kinetic Theory of Gases

In order that the experimental laws about gases discovered between 1662 and 1833 could be satisfactorily explained, Clerk Maxwell (1859,

Amadeo Avogadro, 1776-1856, Italy

He was first a professor of physics at Vercelli (1809) and then in 1820 became professor of mathematics at Turin. It was in 1811 that he put forward his hypothesis that 'equal volumes of all gases under the same conditions of temperature and pressure contain an equal number of particles'.

This hypothesis could not be reconciled with Dalton's Atomic Theory and the pre-eminence of Dalton led to disregard of the hypothesis. In 1858 the work of Cannizzaro upon relative atomic masses of substances forming gaseous compounds showed that the two ideas could both be true. The hypothesis should now be regarded as a law because it is as true a description of the properties of gases as Boyle's Law or Charles' Law. Scientists today appreciate the difference between an atom and a molecule of an element such as oxygen or nitrogen, but 150 years ago no one could suggest why two identical atoms should combine with one another.

England) and Clausius (1867, Germany) suggested a kinetic model based upon the following assumptions.

(*a*) A gas consists of particles in rapid random motion.

(*b*) The distances between particles are usually much greater than the dimensions of the particles.

(*c*) Particles of a gas exert no appreciable forces on one another except when they collide so that between collisions (either with one another or with the wall of the containing vessel) they move in straight lines with constant velocity.

(*d*) The heat energy of a monatomic gas consists completely of the kinetic energy of its molecules.

(*e*) The pressure exerted by a gas is due to the impacts of the particles on the walls of the containing vessel.

(*f*) The average speed of the particles of a gas at a constant temperature is unchanged by collisions.

They were able to prove that:

$$\mathbf{pV = \tfrac{1}{3}\,nmc^2}$$

where p is the pressure of the gas
V is the volume occupied by the gas
n is the number of particles
m is the mass of a particle
c^2 is the average of the squares of the velocities of the particles.

This algebraic expression is in agreement with the experimental laws described above, particularly at temperatures well above the boiling point of the substance. Extensions apply most accurately to a monatomic gas, i.e. one in which the terms atom and molecule are synonymous. The word particle can be replaced by the word molecule in the above discussion if a high degree of accuracy is not required.

When a gas cools, the forces of attraction, e.g. van der Waals' forces (see section 14.3), between the particles cause it to liquefy and at low temperatures the size of the particles of the gas is an appreciable part of the volume occupied by the gas.

7.13 The Hydrogen Molecule

In accordance with Gay-Lussac's Law it is found by experiment that at a given temperature and pressure:

1 volume of hydrogen + 1 volume of chlorine
→ 2 volumes of hydrogen chloride.

Avogadro's Law may be applied to this giving:

n molecules of hydrogen + n molecules of chlorine
→ $2n$ molecules of hydrogen chloride.

or

1 molecule of hydrogen + 1 molecule of chlorine
→ 2 molecules of hydrogen chloride.

The smallest amount of a compound that can be prepared is one molecule so

$\frac{1}{2}$ molecule of hydrogen + $\frac{1}{2}$ molecule of chlorine
→ 1 molecule of hydrogen chloride.

In other words it must be possible to have half a molecule of hydrogen and of chlorine: the number of atoms in the molecule is 2, 4 . . .

The second piece of experimental evidence is that when the hydrogen chloride is dissolved in water it gives a monobasic acid, i.e. one which has only one series of salts, unlike sulphuric, phosphoric and many other acids which form several series of salts. Thus it seems likely that hydrochloric acid contains only one atom of hydrogen in each molecule of the acid and in turn the hydrogen molecule must contain only two atoms—if the hydrogen molecule contained four atoms then each hydrogen chloride molecule would contain two atoms of hydrogen.

7.14 The Relative Density of a Gas

The least dense gas is hydrogen (0·000 09 g/cm^3) and there are many situations where a knowledge of the relative rather than the absolute densities of gases is sufficient or more useful.

$$\text{Relative density of gas} = \frac{\text{Mass of 1 volume of gas}}{\text{Mass of 1 volume of hydrogen}}$$

where all measurements are under the same conditions of temperature and pressure. The mass of 1 volume of a gas is its density.

$$\therefore \text{Relative density} = \frac{\text{Density of gas}}{\text{Density of hydrogen}}$$

The determination of the relative density of a gas is a way of determining the relative molecular mass of a gas (see chapter 10) but the practical details are not required at this stage.

Example: Calculate the relative density of air. Air consists of 80% nitrogen (R.D. = 14) and 20% oxygen (R.D. = 16); the proportions and relative densities of the minor constituents do not affect the result.

$$\text{R.D. of air} = \frac{(80 \times 14) + (20 \times 16)}{100}$$

$\therefore$ The relative density of air = 14·4.

7.15 The Determination of the Formulae of Gases

(a) Steam

See section 6.7(d)

(b) Hydrogen Chloride

It was shown in section 7.13 that one molecule of hydrogen chloride contains one atom of hydrogen. The relative density of hydrogen chloride is 18 and hence the relative molecular mass is 36. Thus hydrogen chloride contains one atom of chlorine also (1 + 35 = 36 is the only possible numerical deduction from the relative atomic masses).

(c) Nitrogen, Oxygen and Chlorine

Many gases have a relative atomic mass which is numerically equal to their relative density. As the relative density is half the relative molecular mass the relative atomic mass is also; hence these gases are diatomic.

(d) Carbon Dioxide

When carbon burns in excess oxygen the volume of gas, as observed under constant conditions, is unchanged. Thus

1 volume of oxygen + a very small volume of carbon
→ 1 volume of carbon dioxide

By Avogadro's Law, which applies to gases only:

1 molecule of oxygen → 1 molecule of carbon dioxide

Oxygen is diatomic so the formula of carbon dioxide is C_xO_2.

The relative density of carbon dioxide is 22; thus the relative molecular mass is 44.

Hence $12x + 32 = 44$

$x = 1$

i.e. the formula of carbon dioxide is CO_2.

The two experimental facts—the relative density of the gas and a volume relationship in a gaseous reaction—yield the formulae of most gases.

QUESTIONS 7

The Gas Laws and the Kinetic Theory

1 A boy reports finding that 1 dm^3 of hydrogen combines with exactly 16 dm^3 of oxygen, or with 35·5 dm^3 of chlorine. What chemical law would seem to be contradicted? What should the correct figures be in each case? [OC]

2 Semi-water gas may be considered as containing 20% hydrogen, 30% carbon monoxide and 50% nitrogen. If 40 cm^3 of this gas are exploded with 20 cm^3 of oxygen, what will be the composition of the final mixture of gases, all volumes being measured at room temperature and pressure? [C]

3 (*a*) The mass of 1 dm^3 of a gas at s.t.p. is 1·52 g. What is its relative molecular mass?
(*b*) By how many grams is 5·6 dm^3 of sulphur dioxide heavier than the same volume of carbon dioxide (both volumes being measured at s.t.p.)? [C]

4 10 dm^3 of dinitrogen oxide are passed over heated copper and the gas formed is collected. If the reaction goes to completion, and all volumes are measured at s.t.p., what is the volume of the gas collected and the mass of the copper(II) oxide formed? [C]

5 (*a*) 40 cm^3 of hydrogen sulphide are exploded with 100 cm^3 of oxygen. (i) If all volumes, before and after, are measured at s.t.p., what is the volume composition of the resulting mixture? (ii) If all volumes are measured at 200°C and normal pressure, what is the volume composition of the resulting mixture?
(*b*) On complete combustion in oxygen, 0·42 g of carbon yielded a quantity of carbon dioxide weighing 1·53 g and occupying 784 cm^3 measured at s.t.p. What is the equivalent (combining) mass of carbon corresponding to these results? Calculate also a possible value of the relative atomic mass of carbon from these results. [C]

6 State Gay-Lussac's Law of Volumes. The following results were obtained with a sample of the air expelled from tap water by boiling. When 100 cm^3 of the air were shaken with sodium hydroxide solution a decrease in volume of 3 cm^3 was observed. Excess hydrogen was then added to the residual gas and the mixture of gases was sparked. When the remaining gases had cooled they were found to occupy 90 cm^3 less than before sparking. All the volumes were measured at room temperature and pressure. Calculate the percentage composition by volume of the air which had been expelled from the tap water. How does the composition of this air differ from that of ordinary air? What explanations can you offer for these differences?

7 Describe the main differences between solids, liquids and gases. How are these explained by the kinetic theory of matter?

8 State Boyle's Law. A cylinder of hydrogen contains 10 m^3 at a pressure of 150 atmospheres. How large a balloon will it fill at atmospheric pressure, if the temperature does not change? How does the kinetic theory explain Boyle's Law?

9 In the following table—which refers to gases—calculate the missing quantities.

	Original Conditions			*Final Conditions*		
	Pressure (mmHg)	*Volume* (cm^3)	*Temperature* (°C)	*Pressure* (mmHg)	*Volume* (cm^3)	*Temperature* (°C)
(*a*)	760	1000	0	760	—	100
(*b*)	740	500	20	200	500	—
(*c*)	300	300	100	—	1000	100
(*d*)	800	250	15	760	—	0
(*e*)	650	120	−27	1000	60	—
(*f*)	760	1600	91	—	1500	0
(*g*)	770	750	112	870	—	17
(*h*)	750	5000	0	500	—	91

10 State Graham's Law of Diffusion and describe any experiments you have seen relating to the diffusion of gases. In 1 minute 250 cm^3 of oxygen diffuse through a porous plate. How long will it take 250 cm^3 of chlorine to diffuse through the same plate?

11 State Dalton's Law of Partial Pressures. Two glass bulbs each of volume 500 cm^3 contain respectively oxygen at a pressure of 1 atmosphere and nitrogen at a pressure of 0·5 atmosphere. If the two bulbs are joined what will be the partial pressures of the two gases in the mixture, and what will be the total pressure?

12 Explain why it is necessary to correct the volumes of gases collected over water. 50 cm^3 of oxygen are contained over water at 20°C. The barometric pressure is 750 mmHg and the saturated vapour pressure of water at 20°C is 17 mmHg. What is the volume of the oxygen at s.t.p.?

13 Illustrate by means of labelled diagrams two experiments to show that hydrogen diffuses more rapidly than air. Suggest a method, based on the rate of diffusion, by which you could find out if a sample of hydrogen contained a small percentage of air. If 150 cm^3 of carbon monoxide diffused through a porous plate in 1 minute, how long would it take 200 cm^3 of hydrogen to pass through the same plate under similar conditions of temperature and pressure?

14 A steel cylinder at 39°C is filled with hydrogen at 40 atmospheres pressure. The internal volume of the cylinder is 32 000 cm^3. Calculate (*a*) the volume of the hydrogen if measured at s.t.p., (*b*) the mass of hydrogen in the cylinder.

15 State the law relating to the volumes of gases which react together. Illustrate your answer by reference to the combination of (*a*) nitrogen and hydrogen to form ammonia; (*b*) hydrogen and chlorine to form hydrogen chloride. Use Avogadro's Law in conjunction with the latter example to show that the hydrogen molecule must contain at least two atoms. [J]

16 100 cm^3 carbon monoxide are sparked with 50 cm^3 oxygen. (*a*) Find the volume of the resulting gas. (All volumes are measured at the same temperature and pressure). (*b*) Find the relative density of the resulting gas. [J]

17 10 cm^3 of each of the following were burned in excess of oxygen: carbon monoxide, hydrogen sulphide. Calculate the volume of oxygen consumed and the volume of the product in each case. All volumes are measured at s.t.p. [OC]

18 State Gay-Lussac's Law of Combination of Gases by Volume. Describe briefly a laboratory experiment to illustrate the law. 10 cm^3 of a gas, containing carbon and hydrogen only, required 20 cm^3 of oxygen for complete combustion, yielding 10 cm^3 of carbon dioxide and some water vapour. Assuming the formula of the gas to be of type C_xH_y, write a balanced equation for the reaction and find the formula of the gas. (The gaseous volumes were measured under the same conditions of temperature and pressure.)

19 In a suitable apparatus at 130°C, 100 cm^3 of hydrogen were exploded with 100 cm^3 of oxygen. The volume was then found to be 150 cm^3. The apparatus was then cooled to room temperature when the water vapour condensed and it was found that the remaining gas was oxygen and that its volume was 50 cm^3. (Assume that all volume measurements have been corrected to s.t.p.) Assuming that both hydrogen and oxygen molecules are diatomic, deduce from this data the formula for steam. [J]

20 Name the gas remaining and state its volume (all volumes being measured at s.t.p.) when: (*a*) 50 cm^3 of hydrogen are exploded with 100 cm^3 of oxygen in a suitable apparatus; (*b*) 50 cm^3 of ozone are heated and then cooled to s.t.p.; (*c*) excess of sulphur is burned in 500 cm^3 of oxygen and then cooled to s.t.p. [J]

21 State Gay-Lussac's Law and explain its use in the following calculation. By boiling some water in a suitable apparatus, 30 cm^3 of dissolved air were expelled and collected in a eudiometer. 50 cm^3 of hydrogen were added, and after explosion, the volume of the remaining gas was found to be 50 cm^3. (Assume all volumes to be measured at room temperature and pressure). Calculate the percentage of oxygen in this sample of air. Explain how this result confirms that air is a mixture. [J]

22 What is meant by *diffusion* of a gas? What evidence does diffusion provide about the molecules present in the gas? Why does diffusion take place more rapidly in gases than in liquids? Describe experiments to show (*a*) that a gas or vapour does diffuse, (*b*) that hydrogen diffuses faster than air. [A]

23 The gaseous oxide of an element *A* contained 73% of oxygen. Under the same conditions, equal volumes of the oxide of *A* and oxygen diffused in 4·7 minutes and 4·0 minutes respectively. Calculate (*a*) the relative density and relative molecular mass of the oxide and (*b*) the relative atomic mass of *A*, given that the oxide contains only one atom of *A* in its molecule.

8 Crystals

8.1 The Properties of Crystals

All pure chemical compounds which are solids can be obtained as particles of varying size but perfectly definite shape. These particles are called crystals.

A crystal is a solid of regular shape enclosed naturally by flat faces.

The crystalline structure of a material may not be apparent in a fine powder even if a microscope is employed. Substances which appear to have no regularity of shape whatsoever are said to be amorphous. Some regularity may be found by using X-rays (see section 8.5) even if it is not apparent to the eye. Normally, glass does not have a crystal shape—it is a super-cooled liquid—but it may devitrify (crystallize) after a long while or upon a sudden mechanical shock, e.g. a car windscreen struck by a stone. Glass may be cut to resemble, for example, a diamond crystal but this is not a natural form.

Although the crystals of a given compound may be prepared in one particular manner, they may superficially appear different. It is very difficult to grow perfect large crystals of any substance and even small crystals may have irregular edges or faces which disguise their true form. Some ideal crystal shapes are shown in figure 8.1.

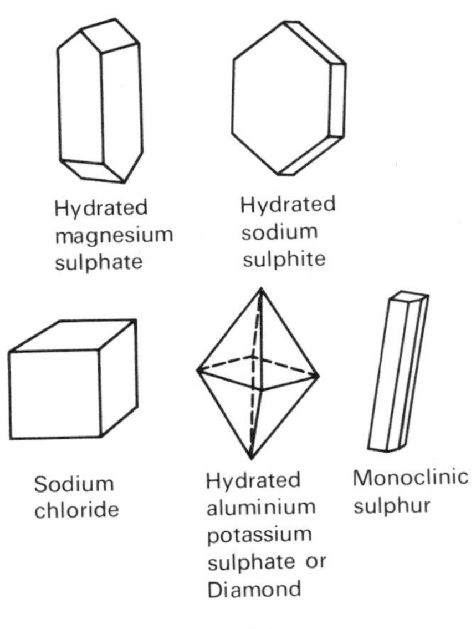

Figure 8.1
The shapes of crystals

When a crystal is broken it splits into smaller crystals with similar plane faces and sharp edges: usually it will split more easily in some directions than in others. Although diamond is the hardest natural substance it can be split along one of its cleavage planes.

The properties of a perfect crystal can be summarized thus:

(*a*) it has a definite geometric shape,
(*b*) it is bounded by plane faces and straight edges,
(*c*) the angles between its faces are constant,
and (*d*) it can be split into smaller crystals.

8.2 The Preparation of Crystals from Solutions

In section 6.13 it was stated that when a hot saturated solution is cooled the solid solute will separate out; it usually does so in the form of crystals. Rapid cooling produces a crop of very small crystals whereas slower cooling (helped by lagging the beaker) produces fewer but larger crystals. Good crystals may also be produced by allowing a cold saturated solution to evaporate slowly in a dish covered with a sheet of paper pierced with a few holes.

It is not easy to grow a large well-shaped crystal nor can it be done in a hurry. Two substances which give reasonable results are aluminium potassium sulphate (alum) and copper sulphate (**N.B.** poisonous and expensive). The procedure should be as follows:

(*a*) A cold saturated solution is prepared and filtered into a clean beaker.

(*b*) A small but well shaped crystal of the substance is suspended by means of a thread from a glass rod so that the crystal is immersed centrally in the solution (see figure 8.2).

(*c*) The beaker is covered with a sheet of paper pierced with a few holes and set aside in a place

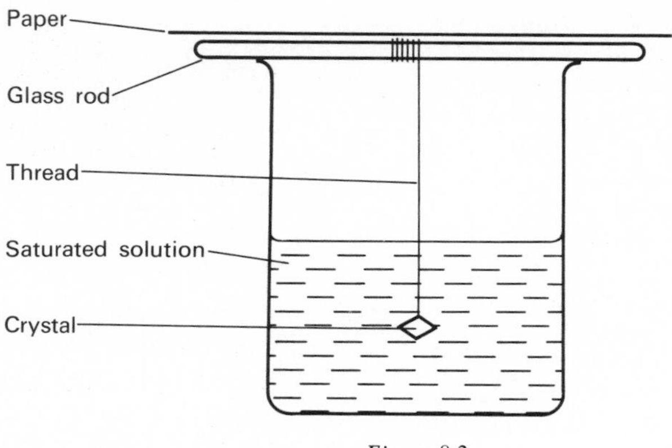

Figure 8.2
Growing a crystal

where the temperature is constant.

The solution will slowly evaporate and the crystal grows in size as the solid coming out of the solution is deposited preferentially on the crystal already present.

(*d*) If small crystals appear elsewhere they are removed.

(*e*) The saturated solution is renewed at intervals.

Impurities, which may be deliberately added, may alter the crystal structure of a substance (known as habit modifiers).

8.3 Other Ways of Making Crystals

(*a*) *From Liquids* (*Fused Solids*)

Crystals are formed when liquids solidify. Ice crystals grow on the top of puddles and ponds when the water starts to freeze because water has a maximum density at 4°C (see section 31.12). Crystals of monoclinic sulphur are obtained when the molten element is cooled (see section 31.5(b)). Most natural minerals crystallized directly from the liquid state or solution many millions of years ago.

(*b*) *From Gases*

Crystals are also produced when gases pass directly to the solid state without first becoming liquid. The crystals of hoar frost on a window-pane are formed in this way from water vapour. This method of producing crystals is used in the process of sublimation (see section 4.12), e.g. iodine and ammonium chloride.

8.4 Isomorphism, Allotropy and Polymorphism

Mitscherlich in 1819 observed that substances of similar chemical constitution had identical crystalline shape (or nearly so) and that they would crystallize out together from solution: he called them **isomorphous** substances. Well-known examples of isomorphous substances are the alums: aluminium potassium sulphate, ammonium iron(III) sulphate, chromium(III) potassium sulphate, etc. each of which crystallizes out from solution with twelve molecules of water of crystallization (see section 27.7(e)).

In a crystal the atoms or molecules are arranged in a definite pattern. Sometimes it is possible for the same substance to crystallize in several different ways, each corresponding to a different arrangement of the fundamental 'building bricks'. When this occurs with elements the different forms are called allotropes and the phenomenon is known as allotropy.

Allotropy is the property possessed by some elements of being able to exist in two or more different forms, all in the same physical state.

In certain cases the term is extended to cover the existence of different liquid or gaseous forms of the same element. Here the difference between the allotropes is usually due to different numbers of atoms in the molecule. Thus oxygen (O_2) and ozone (O_3) are allotropes.

Polymorphism is the existence of a compound in two or more different forms without changing its physical state. If the temperature is the governing factor in selecting the crystalline shape the substance is said to be **enantiotropic** e.g. iron (see section 26.6) and sulphur (see section 31.6) Ammonium nitrate exists in no less than five different forms. A change in crystal shape may be accompanied by a change in colour and an application of this is in heat sensitive paints.

If the temperature is not the governing factor, i.e. one form is always more stable than the other, then the substance is said to be **monotropic** e.g. carbon (see section 28.5), phosphorus (see section 30.6) and oxygen (see section 31.6). Calcium carbonate in the form of calcite is unstable with respect to the form aragonite.

A third type of allotropy, **dynamic allotropy** is shown by substances which exist in several forms the proportions of which change as the temperature changes, e.g. liquid sulphur (see section 31.5).

8.5 Ripple Tanks and X-rays

A crystal has a definite shape because the molecules or atoms or ions of which it is composed are arranged in a perfectly ordered pattern.

There is a limit to what can be seen by the naked eye even when it is aided by a microscope and this limit is reached before the dimensions of molecules are attained. The electron microscope enables further progress to be made. Electrons, particles much smaller than, and contained in, atoms, can be accelerated in an evacuated enclosure by means of electrical and magnetic fields. The fast-moving electrons hit the specimen under examination, are scattered and then focussed on to a photographic plate. A picture of single crystals of lead is shown in figure 8.3. What appears smooth to the eye and touch is thus 'seen' to be comparatively rough but the individual lead atoms cannot be distinguished.

If a single vibrator is set going in a ripple tank and a barrier, having several holes at regular intervals, is inserted in the tank a short distance away then the ripples beyond the barrier have a pattern which depends on the angle and the distance from the holes because

(*a*) a peak may arrive at a point from one hole at the same time as a trough arrives from another hole thus cancelling one another out,
(*b*) two peaks from separate holes may arrive at the same point at the same time thus giving a large wave,
(*c*) two troughs from separate holes may arrive at the same point at the same time thus giving a large trough.

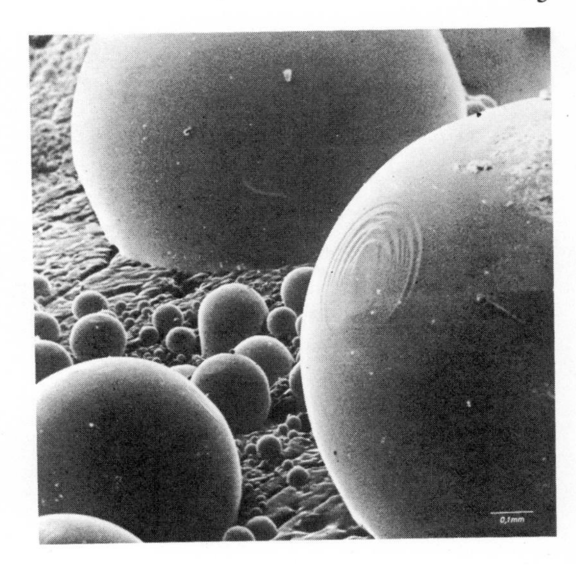

Figure 8.3
Lead spheres—single crystals taken by a scanning electron microscope

Viewed from above the pattern observed is two-dimensional.

X-rays are like light and radio waves in that they are a wave motion; the wavelength of X-rays is about 10^{-10} m. X-rays are generated whenever high speed electrons are suddenly stopped, for example, by a dense metal. When X-rays are shone through a crystal the effect is similar to the ripple tank experiment described above. The particles of the crystal correspond to the holes in the barrier. Many X-rays pass straight through the crystal but some are diffracted and give a pattern on a photographic plate placed around the crystal.

Figure 8.4 shows the result of passing X-rays

Figure 8.4
The X-ray diffraction pattern of sodium chloride

Figure 8.5
X-ray camera

of wavelength $1{\cdot}542 \times 10^{-10}$ m from a copper target through a crystal of sodium chloride for 15 minutes in a 3 cm camera. The following points may be noted.

(*a*) The main X-ray beam passed straight through giving a central dot.

(*b*) A rectangular shadow is cast by a metal object of known position to check that the film is in the correct position in the camera, seen in figure 8.5.

(*c*) The lines are conic sections because the crystal is a three-dimensional object in the path of the X-rays.

(*d*) The lines show some variation in intensity.

The connection between the wavelength of the X-rays (λ), the distance between the particles of the crystal (d) and the angle which the X-ray beam has been bent (2θ) was discovered by Lawrence Bragg. It is that reinforcement of the diffracted X-rays occurs giving a line whenever

$$\frac{2d \sin \theta}{\lambda} \text{ is an integer } (n).$$

This equation has been a most useful one in the investigation of crystal structures. It is not essential to have a large crystal—a powder will do. It is possible to identify compounds which simple analysis may not be able to do, e.g. a mixture of sodium chloride and potassium bromide can be distinguished from a mixture of sodium bromide and potassium chloride. For complex substances a computer is necessary to work out the results.

William and Lawrence Bragg, 1862–1942, 1890–1971, England

Father and son made their most important contribution to experimental knowledge in the field of X-ray crystallography.

William was a professor of mathematics and physics in Australia before returning to England in 1909: he went first to Leeds, then in 1915 to London.

Lawrence studied in Cambridge before starting his work in 1911 on the 'von Laue' phenomenon, that a crystal scattered X-rays in definite patterns. He discovered the relationship now known as the Bragg equation that $n\lambda = 2d \sin \theta$*. In 1919 he became professor of physics at Manchester, then after a spell at the National Physical Laboratory (1937) he moved to Cambridge. His later research was upon the structure of proteins.*

QUESTIONS 8

Crystals

1 Suppose that you are provided with two water-soluble ionic solids *A* and *B*, each in the form of a finely ground powder. Describe experiments you would carry out to investigate whether the arrangement of ions in solid *A* was similar to that of the ions in solid *B*.

2 Describe clearly three different ways of making crystals so that a non-scientist could complete the experiments successfully by following your instructions.

3 Describe the experiments you would make in order to show whether two substances had the same crystalline shape. What additional experiments would be required to show whether the two substances were isomorphic or not?

9 The Fundamental Laws and Dalton's Atomic Theory

9.1 Introduction

The early scientists gathered much **qualitative** information about the materials they studied. This information, although of value and interest, did not lead to any great advance in the study of chemistry. Much progress was made when experiments were done in a more systematic manner and were done **quantitatively** It may be useful to know that substance A when mixed with substance B gives a new substance C or that when substance X is heated it changes into substances Y and Z but it is much better to know this information and the quantities that are involved.

Black (1755) found the mass relationships between several materials (see section 1.5). Rey (1630) had made quantitative experiments on combustion but the main advance was made by Lavoisier (see section 5.4) when after his experiments he summed up his and other scientists' work in the statement that 'matter can neither be created nor destroyed, and in every chemical reaction there is just as much matter present before as after the reaction has taken place.' This was referred to in section 2.6 and is the first of the four basic laws of chemical combination discussed below. These four laws are sometimes referred to as the Laws of Stoichiometry, i.e. measurement of the numerical proportions in which substances react.

9.2 The Law of Conservation of Mass

In a chemical reaction the total mass of the reacting substances is equal to the total mass of the substances formed, i.e. matter can neither be created nor destroyed.

This may seem obvious—it had been presumed for many years previously—but it is important to verify the statement to establish quantitative chemistry on a firm basis.

(*a*) A sealed tube containing a small piece of white phosphorus should be weighed and then warmed gently where the bead of phosphorus rests so that combustion ensues. After the reaction is completed and the tube is cool again it is found to have the same mass as before. (Care should be taken in disposing of the tube.)
(*b*) Similar experiments to (*a*) can be carried out using metals such as copper or tin (which Lavoisier used in addition to mercury).
(*c*) In the laboratory the easiest experiments to carry out are ones based on the procedure adopted by Landolt (1893). A reaction is chosen in which there is a visible change, e.g. a colour change or the formation of a precipitate, and the reagents put in separate compartments of the apparatus. This may be done by having one reagent in a conical flask and the second in a test tube suspended vertically by a thread, the flask being sealed with a stopper. After weighing the flask and contents the thread is released, or the flask tipped, enabling the reagents to react, e.g.

Copper sulphate + Sodium carbonate
(blue solution) (colourless solution)

→ Copper carbonate + Sodium sulphate
(bluish-white precipitate) (colourless solution)

After a pause to allow the materials to cool back to room temperature the flask and contents are found to have the same mass as before.

9.3 The Law of Constant Composition

This is sometimes known as the Law of Definite (or Fixed) Proportions.

All pure samples of the same chemical compound contain the same elements united in the same proportions by mass.

Lavoisier seems to have assumed the truth of this statement and it was established as a law by Proust (1799).

There are two aspects to a verification of this law in the laboratory: the first is the preparation of samples of a substance, e.g. copper oxide, by several methods and the second is the analysis of those samples to see that they do have a constant composition however they have been made.

The black oxide of copper can be prepared in several ways starting from copper or a naturally occurring copper compound such as malachite or azurite (see section 26.2). If copper is used it should be put in a test tube in a fume cupboard and a little concentrated nitric acid added; the two ores mentioned can be dissolved in dilute nitric acid in the open laboratory. This gives a solution of copper nitrate which if it is not clear should be centrifuged or filtered. Then the solution is divided into three portions:

(*a*) The solution is boiled to drive off the water and then the crystals heated more strongly (in a fume cupboard).
(*b*) Sodium carbonate solution is added, the precipitate of copper carbonate is then centrifuged or filtered off and heated.
(*c*) Sodium hydroxide solution is added, the precipitate of copper hydroxide being centrifuged or filtered off and heated.

In each case the compound decomposes to give copper oxide: heating should be carried out until constant mass is attained. The samples should be stored in a desiccator because they are hygroscopic.

The samples are then reduced to copper by heating them in a stream of hydrogen or gas, the experiment being based on that done by Dumas (see section 6.7(e)). The combustion tube, which can be replaced by a test tube with a hole at one side near the bottom, should be clamped so that it slopes at 10° in order that any water condensing runs away from the hot glass. The gas should be allowed to expel the air, otherwise there may be an explosion. The calcium chloride tube (see figure 9.1) is not essential for the simplest possible experiment; if used it will collect all the water formed in the experiment. After the reduction of the oxide is complete the gas should be allowed to pass while the metal cools so that the metal oxide does not reform.

Copper oxide + Hydrogen → Copper + Water

or

$$\underset{\text{Black}}{CuO} + H_2 \rightarrow \underset{\text{Red}}{Cu} + H_2O$$

Results

		grams
Mass of boat	=	1.72
Mass of boat + copper oxide	=	2·92
∴ Mass of copper oxide	=	1·20
Mass of boat + copper	=	2·68
∴ Mass of copper	=	0·96
∴ Mass of oxygen	=	0·24
Mass of calcium chloride tube at start	=	11·55
Mass of calcium chloride tube at end	=	11·82
∴ Mass of water formed	=	0·27
∴ Mass of hydrogen that reacts	=	0·03

Calculation

$$\text{Percentage of copper in copper oxide} = \frac{0{\cdot}96}{1{\cdot}20} \times 100 = 80$$

$$\text{Percentage of hydrogen in water} = \frac{0{\cdot}03}{0{\cdot}27} \times 100 = 11$$

The percentage of copper in each sample of copper oxide can be calculated as shown and if the calcium chloride tube is incorporated in the apparatus the percentage of hydrogen in water can also be found: constant results within the limits of experimental error are obtained (see section 4.13).

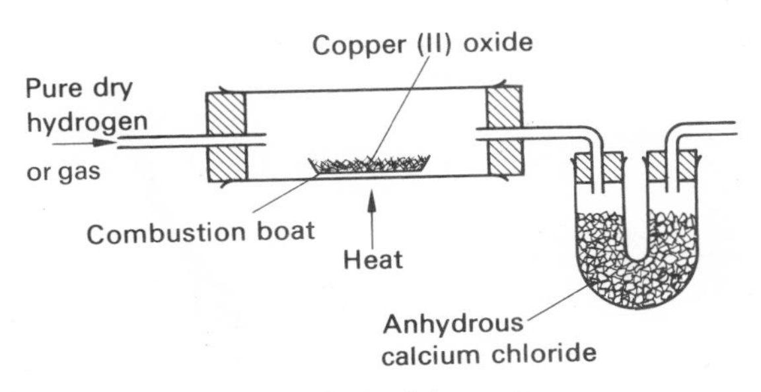

Figure 9.1
The reduction of a metal oxide

9.4 Combining or Equivalent Masses

Elements combine together in definite proportions by mass; compounds react similarly, e.g.

1 g of hydrogen combines with 8 g of oxygen,
8 g of oxygen combines with 12 g of magnesium,
12 g of magnesium reacts with 36·5 g of hydrogen chloride,
and 36·5 g of hydrogen chloride reacts with 40 g of sodium hydroxide.

For elementary purposes it may be taken that:

The combining or equivalent mass of an element is the number of parts by mass of it that will combine with or displace one part by mass of hydrogen or 8 parts by mass of oxygen or 35·5 parts by mass of chlorine.

The combining mass is a number; the sign ≡ is often used to express the equivalence.

9.5 Other Experiments Illustrating the Law of Constant Composition

(a) The Composition of Steam and of Water

In section 6.7(d) the result obtained was:

2 volumes of hydrogen ≡ 1 volume of oxygen
Density of hydrogen = 0·09 g/dm^3
Density of oxygen = 1·44 g/dm^3

Thus, taking 1 volume of gas to be 1 dm^3

2 × 0·09 g hydrogen ≡ 1·44 g oxygen
∴ 1 g hydrogen ≡ 8 g oxygen

The combining mass of oxygen is 8. This is very close to the result obtained by Dumas (see sections 6.7(e) and 9.3).

(b) The Combustion of Magnesium

See section 5.1(b): for more accurate results a drop of water should be added to decompose any magnesium nitride formed and the crucible and contents reheated to constant mass.

Results

		grams
Mass of crucible and lid	=	10·24
Mass of crucible, lid and magnesium	=	10·42
∴ Mass of magnesium	=	0·18
Mass of crucible, lid and magnesium oxide	=	10·54
∴ Mass of magnesium oxide	=	0·30
∴ Mass of oxygen	=	0·12

Calculation

(i) Percentage of magnesium in magnesium oxide $= \frac{0·18}{0·30} \times 100$ $= 60$

(ii) 0·12 g oxygen ≡ 0·18 g magnesium

∴ 8 g oxygen $\equiv \frac{0·18}{0·12} \times 8$ g magnesium

= 12 g magnesium

The combining mass of magnesium is 12.

(c) The Combustion of Carbon (Phosphorus, Sulphur, etc.)

The apparatus is shown in figure 9.2: it may be used for any substance which forms a volatile oxide and is particularly useful for showing the identity of the allotropes of an element.

Oxygen or air which has been dried and had the carbon dioxide removed by absorption in sodium hydroxide solution is passed over a known mass of the element in a boat in a combustion tube which is strongly heated. The carbon dioxide formed is absorbed in more sodium hydroxide (or potassium hydroxide) solution in previously weighed bulbs. The element does not need to burn completely away if the boat and contents are reweighed at the end.

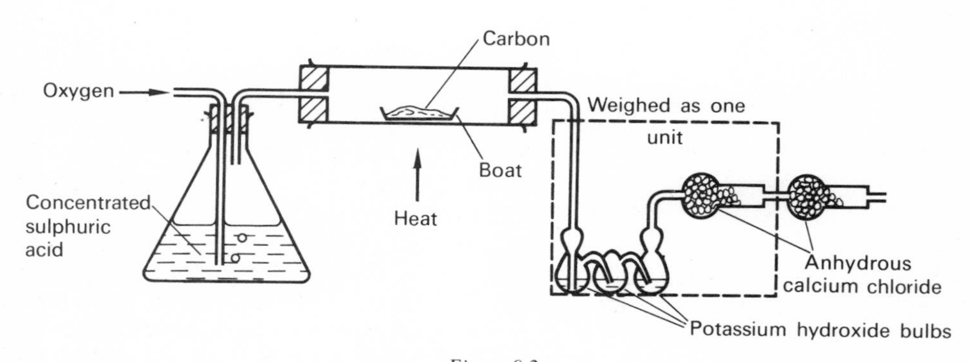

Figure 9.2
The synthesis of carbon dioxide

Results

	grams
Mass of boat	= 10·17
Mass of boat + carbon	= 11·25
∴ Mass of carbon	= 1·08
Mass of bulbs at start	= 33·72
Mass of bulbs at end	= 37·68
∴ Mass of carbon dioxide formed	= 3·96
∴ Mass of oxygen	= 2·88

Calculation

(i) Percentage of carbon in carbon dioxide $= \frac{1{\cdot}08}{3{\cdot}96} \times 100$

$= 27$

(ii) 2·88 g oxygen ≡ 1·08 g carbon
∴ 8 g oxygen ≡ 3 g carbon

The combining mass of carbon is 3.

(d) Indirect Combination with Oxygen

This method is suitable for metals such as copper, tin, zinc and lead. The metal is converted to its nitrate (or, in the case of tin, directly to an oxide) and then to its oxide as outlined in section 9.3. A suitable mass of copper to start the experiment is about 1 g: this is weighed into a boiling tube or an evaporating basin and then concentrated nitric acid added in sufficient quantities to dissolve the metal completely. The experiment should be performed in a fume cupboard because the oxides of nitrogen evolved are poisonous (see section 30.15). When the blue solution is heated water and surplus acid are evolved and, with care to avoid spitting, the solid copper nitrate left can be decomposed by further heating to the oxide. The black solid left is heated to constant mass; the calculation of the result is as in (b).

The experiments described above lead to combining or equivalent masses as well as illustrating, when done on a class basis, the Law of Constant Composition. Experiments done under the titles of reacting masses (see section 11.13), and of electrolysis (see section 20.9) also lead to the calculation of combining masses.

9.6 The Law of Multiple Proportions

If two elements combine to form more than one compound then the masses of one element which combine with a fixed mass of the other element are in simple ratio.

This law was deduced by Dalton (1803) before he published his Atomic Theory in full and it was first verified by Berzelius (1810). Berzelius did an experiment which is quite easy to repeat in the laboratory: he reduced the brown oxide of lead and the yellow oxide of lead to the metal by heating them separately in a current of hydrogen (gas will do). The procedure is the same as for the oxides of copper—discussed in section 9.3. The results are:

	Mass of lead	*Mass of oxygen*
Brown oxide	100 g	15·6 g
Yellow oxide	100 g	7·8 g

i.e. for a fixed mass of lead the masses of oxygen are in the ratio of 2:1.

The same result is obtained with the two oxides of carbon and of copper but the two chlorides of iron contain masses of chlorine in the ratio 2:3 and the two chlorides of phosphorus contain masses of chlorine in the ratio 3:5. Furthermore, if a compound forms two hydrates the masses of water of crystallization combining with a fixed mass of the compound are in a simple ratio.

9.7 The Law of Reciprocal Proportions

The proportion by mass in which two substances each combine with or displace a fixed mass of a third substance is a simple ratio of the proportion in which they combine with or displace one another.

This law, which is sometimes called the Law of Equivalent Proportions, applies to elements and to compounds and is based upon the work of Berzelius (with elements) and Richter (with acids and bases).

The law may also be stated thus:

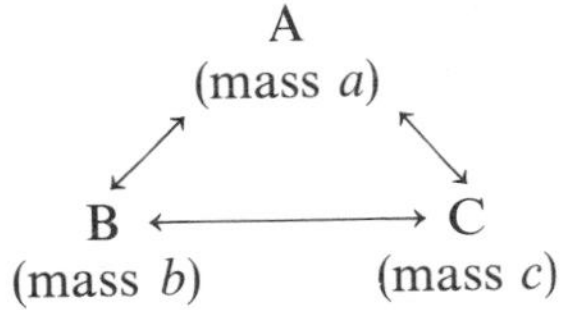

If *a* grams of A combine with or displace *b* grams of B
and if *a* grams of A combine with or displace *c* grams of C
then *b* grams of B combine with or displace *c* grams of C or a simple ratio of this relationship.

e.g. In methane: 1 g hydrogen + 3 g carbon.
In water: 1 g hydrogen + 8 g oxygen.
In carbon dioxide: 3 g carbon + 8 g oxygen; this is precisely the ratio predicted.
In carbon monoxide: 3 g carbon + 4 g oxygen; this is half the ratio predicted.

9.8 Dalton's Atomic Theory

Dalton did not invent the atomic theory nor revive it, but he put into words a theory which was, or could be, backed by chemical evidence. Democritus in 500 B.C. had talked about atoms and Newton had described matter as consisting of hard, massy, impenetrable movable particles. Dalton was led to the atomic theory by his interest in gases—he started by disproving the suggestion that air is a compound. Dalton emphasized that the mass of an atom is its most important property. The theory is summarized thus:

(*a*) **Matter is composed of a large number of atoms which are minute particles that cannot be split up, destroyed or created.**

(*b*) **All the atoms of the same element are identical in all respects; in particular they all have exactly the same mass. Different elements have atoms differing in mass.**

(*c*) **Chemical combination takes place between small whole numbers of atoms of the elements concerned, forming molecules** (he called them compound-atoms); **all the molecules of a compound are identical in all respects.**

N.B. An element is a substance which cannot be split up into two or more substances by chemical means (see section 3.1).

An atom is the smallest possible particle of an element that can exist (see section 3.2). Hence the atom is the smallest part of an element that can enter into or be expelled during chemical change.

9.9 The Explanation of the Laws of Chemical Combination

(*a*) *The Law of Conservation of Mass*

If atoms can neither be created nor destroyed during a chemical reaction but only rearranged then the total mass is unchanged.

(*b*) *The Law of Constant Composition*

Let element A have atoms each of mass a grams and element B have atoms each of mass b grams. Then if the molecule of the compound formed by A and B is AB then the total mass of one molecule is always $(a+b)$ grams, made up of a grams of A combining with b grams of B.

(*c*) *The Law of Multiple Proportions*

Let the elements have atoms of masses as in the previous paragraph but let there be two compounds of formulae AB and AB_2 respectively.

Then in AB a grams of A combine with b grams of B

and in AB_2 a grams of A combine with $2b$ grams of B

Thus the masses of B combining with a fixed mass of A (a g) are in the ratio 1:2, which is a simple ratio.

The *Law of Reciprocal Proportions* follows in a similar manner.

The laws can be explained whatever simple formulae are chosen for the compounds.

9.10 Problems arising from Dalton's Theory

(*a*) *Combining Masses*

When a compound has been studied the mass of one element combining with a given mass of the other element can be calculated but unless the formula for the compound is known this does not necessarily give the ratio of the masses of the atoms concerned. Thus the early tables of relative atomic masses were really tables of combining or equivalent masses. Dalton's theory offered no explanation for the regularity in the combining capacities of some elements, i.e. the different valencies, nor of the angular, rather than linear, shape of the water molecule and the pyramidal, rather than planar, shape of the ammonia molecule, etc.

(*b*) *Atoms and Molecules*

Despite his own and his contemporaries' work on gases, Dalton did not realize that the atoms of an element might exist freely in units of two or three and hence he did not accept Avogadro's Law.

(*c*) *Electrolysis*

Dalton's theory offered no explanation of the ease of electrolysis of some substances, e.g. the experiments done in the same decade by Davy yielding sodium and other metals, nor of the quantitative laws proposed by Faraday soon after.

To make these points is not to deny the value of Dalton's theory in the nineteenth century nor its application today to most of the problems met in the laboratory and in industry. Without his theory the development of chemistry would have been considerably impeded.

9.11 Modern Comments on Dalton's Theory

Atoms can be split: this is occurring all the time in substances which are naturally or artificially radioactive (see section 12.5).

The atoms of 75% of the known elements consist of more than one kind of atom, known as isotopes (see section 12.2). The nuclides (isotopes) of a given element are usually present in constant proportions and so when atoms are considered in bulk the average mass is constant.

There are many molecules, particularly in organic chemistry (see chapter 29), in which the ratio of the numbers of the atoms in the molecule of a compound is complex rather than simple. There are now many compounds known in which the ratio of the numbers of the atoms is variable: these **Berthollide or non-Daltonide compounds** may have useful electrical properties.

John Dalton, 1766–1844, England

Dalton was born in the Lake District and later he moved to Manchester where he became known as a teacher. Early in his life he commenced meteorological observations and kept a diary for 57 years. He described the nature of colour blindness, otherwise known as Daltonism, from which he suffered. In 1801 he published two important essays 'The Constitution of Mixed Gases' (proposing the Law of Partial Pressures) and 'The Expansion of Gases by Heat'.

Through his studies of the atmosphere and of other gases he developed the atomic hypothesis into a verifiable form which he published as the 'New System of Chemical Philosophy' in 1808. His earliest table of relative atomic masses was, however, put forward in 1803. The Law of Multiple Proportions was deduced from his 'Atomic Theory', then investigated experimentally and found to be true.

On the other hand he did not accept Gay-Lussac's Law of Gaseous Combining Volumes which was in conflict with his theory and science had to wait another 50 years before the differences between combining and relative atomic masses and between atoms and molecules were unravelled.

QUESTIONS 9

The Fundamental Laws and Dalton's Atomic Theory

1 State the law of multiple proportions. Explain how you would test the truth of the law for two of the oxides of lead, describing in detail how you would carry out the experiment. In such an experiment it was found that one oxide contained 90·7% and the other 92·8% of the metal. Show that the two values support the law of multiple proportions. [C]

2 Give the laws of constant composition and multiple proportions.

Two anhydrous sulphates of a metal contain respectively 37% and 28% of the metal. Calculate (*a*) its probable relative atomic mass and (*b*) the corresponding valencies. What other information would help to fix the relative atomic mass of the metal with more certainty? [C]

3 State the law of definite proportions (constant composition). What important conclusion about the nature of matter can you draw from it? Describe briefly a practical method of illustrating this law in the laboratory. Show that the following figures illustrate a different chemical law and state that law: oxide *A* contains 80% of copper and oxide *B* contains 89% copper. [A]

4 State (*a*) the Law of the Conservation of Mass (sometimes called the law of indestructibility of matter); (*b*) the Law of Multiple Proportions. Describe briefly an experiment you could perform to show the truth of the first of these laws. [J]

5 An element forms two oxides containing respectively 50% and 40% by mass of the element. Show how these oxides illustrate the Law of Multiple Proportions. [J]

6 State the Law of Definite Proportions. If 12 g of magnesium combine with 8 g of oxygen to form magnesium oxide, what mass of magnesium oxide could be formed from 1·8 g of magnesium?

7 On heating 12·25 g of potassium chlorate 4·80 g of oxygen were evolved. (*a*) What mass of potassium chloride was left? (*b*) Name and state the chemical law on which your calculation depends. [J]

8 An element forms three oxides containing respectively 63·6%, 30·4% and 46·7% by mass of the element. Show that these facts are in accordance with the Law of Multiple Proportions.

9 State the Law of Constant Composition.

Starting from metallic copper, outline the method by which you could obtain: (*a*) a solution of copper nitrate; (*b*) copper hydroxide. State how, from each of these products, it is possible to obtain copper oxide. How would you now proceed to use these two samples of copper oxide to verify the law you have stated? [J]

10 State the Law of Constant Composition. Show that the results given below demonstrate the Law.

Describe in detail the experiment by which the results could be obtained. Calculate the relative atomic mass of copper.

11 Weighed quantities of two oxides of the same metal were heated in a current of gas with the following results:

	grams
Mass of boat A empty	= 4·500
Mass of boat A with first oxide before heating	= 4·946
Mass of boat A with residue after final heating	= 4·914
Mass of boat B empty	= 5·000
Mass of boat B with second oxide before heating	= 5·462
Mass of boat B with residue after final heating	= 5·414

(*a*) Complete the calculation to show how the results are in accordance with the Law of Multiple Proportions. (*b*) Sketch the apparatus. [J]

12 State the Law of Multiple Proportions.

A metal forms two oxides, A and B. 1 g of A on reduction gave 0·881 g of the metal, while 1 g of B yielded 0·788 g of the metal. Show that these facts are in agreement with the Law of Multiple Proportions. What is the probable relative atomic mass of the metal? [OC]

13 List the main postulates of Dalton's Atomic Theory. How far can these be accepted in the light of modern work on atomic structure?

14 State the Law of Multiple Proportions and show that you understand its meaning by reference to the oxides of carbon. Two chlorides of a metal were found on analysis to contain respectively 37·6 and 54·2% of chlorine. Show that these figures are in accordance with the Law of Multiple Proportions.

15 Show how the following analytical data illustrate the Law of Equivalent Proportions:

	%C	%O	%H
Carbon dioxide	27·3	72·7	—
Methane	75·0	—	25·0
Water	—	88·9	11·1

16 By mass, water contains 11·11% hydrogen while hydrogen chloride contains 97·24% chlorine. In what ratio by mass would you expect chlorine and oxygen to combine?

Sample	*A*	*B*	*C*
Mass of boat empty, g	6·02	5·72	6·31
Mass of boat + copper(II) oxide before heating, g	8·07	7·97	8·16
Mass of boat + copper(II) oxide after heating, g	7·66	7·52	7·79

17 An experiment indicated that 4·5 g of an oxide of copper when completely reduced yielded 4·0 g of copper. Calculate the combining mass of copper from these results. Describe carefully how you would carry out the experiment referred to above, drawing special attention to those precautions that would increase the accuracy of your results. Draw a diagram of the apparatus you would use. [O]

18 Some copper was warmed in a crucible with a slight excess of concentrated nitric acid and the mixture was then strongly heated. The following weighings were obtained. Calculate the combining mass of copper.

	grams
Mass of clean dry crucible	= 17·65
Mass of this crucible and copper	= 18·61
Mass of crucible and residue after heating with nitric acid	= 18·98
Mass of crucible and residue after reheating	= 18·85
Mass of crucible and residue after again reheating	= 18·85

[J]

19 How would you find the percentage composition of copper(II) oxide? Two different oxides of copper were found to contain 80·0 and 88·9% of the metal respectively. Using these results, calculate the two combining masses of copper, and use them to illustrate a fundamental law of chemistry.

20 Two oxides of a certain element contain respectively 63·64 and 46·67% of the element. Calculate the two combining masses of the element and state a law of chemistry which they illustrate. What is the least possible relative atomic mass of the element, and assuming this to be the actual relative atomic mass, what are its valencies in the two oxides? [O]

10 Relative Atomic and Molecular Masses

10.1 Introduction

The importance of Dalton's Atomic Theory in the development of chemistry was discussed in the previous chapter. He emphasized that the mass of an atom is its most important property and this led to much research. That the various workers in this field obtained consistent answers for relative atomic masses is good evidence for the particular, i.e. atomic, theory of matter.

The absolute masses of atoms are very small and a comparative scale as suggested in section 1.5 is the most convenient. For all practical purposes the relative atomic mass scale based on hydrogen (H = 1) is sufficiently accurate.

The relative atomic mass of an element is the average mass of one atom of the element compared with the mass of one atom of the nuclide ^{12}C taken as 12.

The relative atomic mass is a number, and values for calculation purposes are quoted in figure 1.10.

The modern scale of relative atomic masses is based on a particular species of carbon atom, i.e. a nuclide (see section 12.2), and the values are quoted are relative to $^{12}_{6}C = 12$ in figure 10.1. The term relative atomic mass, A_r, has replaced the term atomic weight.

The relative molecular mass of an element or compound is the average mass of one molecule of the substance compared with the mass of one atom of the nuclide ^{12}C taken as 12.

For calculations, the relative molecular mass, M_r, is frequently obtained by summing the relative atomic masses in accordance with the formula of the substance.

The mass number of an isotope is the mass of one atom of that isotope to the nearest integer (see section 12.2).

10.2 Combining Mass and Valency.

Dalton knew that 1 g of hydrogen combined with 8 g of oxygen when forming water. In the absence of acceptable evidence to the contrary he therefore presumed the formula of water to be HO and that if an atom of hydrogen weighed one unit then an atom of oxygen weighed 8 units. The key to Dalton's dilemma was a quantity known as the valency (see section 11.3).

In water two atoms of hydrogen combine with one atom of oxygen, i.e. the valency of oxygen is two. 2 g of hydrogen will combine with 16 g of oxygen so if 2 g hydrogen represents 2 atoms of hydrogen then 16 g of oxygen represents one atom of oxygen, i.e. the relative atomic mass of oxygen is 16. In general terms:

Relative atomic mass

= Combining mass × Valency.

10.3 Dulong and Petit's Law (1819)

The product of the relative atomic mass of a solid element and its specific heat capacity is known as its atomic heat and is approximately 26 J $mol^{-1}\ °C^{-1}$.

The law is valid for most metals. It is applied as follows:

Tin has a specific heat capacity of $0{\cdot}23$ J $g^{-1}\ °C^{-1}$

The combining mass of tin is $59{\cdot}35$

∴ The relative atomic mass of tin is $\frac{26}{0{\cdot}23} = 113$

∴ The valency of tin $\frac{113}{59{\cdot}35} = 1{\cdot}9$

But the valency of an element must be a whole number and it is sensible to suggest that it is 2 in this case.

∴ The accurate relative atomic mass of tin

$= 2 \times 59{\cdot}35$
$= 118{\cdot}7$

10.4 Relative Molecular Mass and Relative Density

From section 10.1:

Relative molecular mass

$$= \frac{\text{Mass of 1 molecule of substance}}{\text{Mass of 1 atom of hydrogen}}$$

From section 7.14:

Based on the Assigned Relative Atomic Mass of $^{12}C = 12$

The following values apply to elements as they exist in materials of terrestrial origin and to certain artificial elements. When used with due regard to the footnotes, they are considered reliable to ± 1 in the last digit, or ± 3 if that digit is in small type.

Atomic Number	*Name*	*Symbol*	A_r	*Atomic Number*	*Name*	*Symbol*	A_r
89	Actinium	Ac	(227)	80	Mercury	Hg	200·5$_{9}$
13	Aluminium	Al	26·98154[a]	42	Molybdenum	Mo	95·9$_{4}$
95	Americium	Am	(243)	60	Neodymium	Nd	144·2$_{4}$
51	Antimony	Sb	121·7$_{5}$	10	Neon	Ne	20·17$_{9}$[c,e]
18	Argon	Ar	39·94$_{8}$[b,c,d,g]	93	Neptunium	Np	(237)
33	Arsenic	As	74·9216[a]	28	Nickel	Ni	58·70
85	Astatine	At	(210)	41	Niobium	Nb	92·9064[a]
56	Barium	Ba	137·3$_{4}$	7	Nitrogen	N	14·0067[b,c]
97	Berkelium	Bk	(247)	102	Nobelium	No	(259)
4	Beryllium	Be	9·01218[a]	76	Osmium	Os	190·2[g]
83	Bismuth	Bi	208·9804[a]	8	Oxygen	O	15·999$_{4}$[b,c,d]
5	Boron	B	10·81[c,d,e]	46	Palladium	Pd	106·4
35	Bromine	Br	79·904[c]	15	Phosphorus	P	30·97376[a]
48	Cadmium	Cd	112·40	78	Platinum	Pt	195·0$_{9}$
55	Caesium	Cs	132·9054[a]	94	Plutonium	Pu	(244)
20	Calcium	Ca	40·08[g]	84	Polonium	Po	(209)
98	Californium	Cf	(251)	19	Potassium	K	39·09$_{8}$
6	Carbon	C	12·011[b,d]	59	Praseodymium	Pr	140·9077[a]
58	Cerium	Ce	140·12	61	Promethium	Pm	(145)
17	Chlorine	Cl	35·453[c]	91	Protactinium	Pa	231·0359[f]
24	Chromium	Cr	51·996[c]	88	Radium	Ra	226·0254[f,g]
27	Cobalt	Co	58·9332[a]	86	Radon	Rn	(222)
29	Copper	Cu	63·54$_{6}$[c,d]	75	Rhenium	Re	186·207[c]
96	Curium	Cm	(247)	45	Rhodium	Rh	102·9055[a]
66	Dysprosium	Dy	162·5$_{0}$	37	Rubidium	Rb	85·467$_{8}$[c]
99	Einsteinium	Es	(254)	44	Ruthenium	Ru	101·0$_{7}$
68	Erbium	Er	167·2$_{6}$	62	Samarium	Sm	150·4
63	Europium	Eu	151·96	21	Scandium	Sc	44·9559[a]
100	Fermium	Fm	(257)	34	Selenium	Se	78·9$_{6}$
9	Fluorine	F	18·99840[a]	14	Silicon	Si	28·08$_{6}$[d]
87	Francium	Fr	(223)	47	Silver	Ag	107·868[c]
64	Gadolinium	Gd	157·2$_{5}$	11	Sodium	Na	22·98977[a]
31	Gallium	Ga	69·72	38	Strontium	Sr	87·62[g]
32	Germanium	Ge	72·5$_{9}$	16	Sulphur	S	32·06[d]
79	Gold	Au	196·9665[a]	73	Tantalum	Ta	180·947$_{9}$[b]
72	Hafnium	Hf	178·4$_{9}$	43	Technetium	Tc	(97)
2	Helium	He	4·00260[b,c]	52	Tellurium	Te	127·6$_{0}$
67	Holmium	Ho	164·9304[a]	65	Terbium	Tb	158·9254[a]
1	Hydrogen	H	1·0079[b,d]	81	Thallium	Tl	204·3$_{7}$
49	Indium	In	114·82	90	Thorium	Th	232·0381[f,g]
53	Iodine	I	126·9045[a]	69	Thulium	Tm	168·9342[a]
77	Iridium	Ir	192·2$_{2}$	50	Tin	Sn	118·6$_{9}$
26	Iron	Fe	55·84$_{7}$	22	Titanium	Ti	47·9$_{0}$
36	Krypton	Kr	83·80[e]	74	Tungsten	W	183·8$_{5}$
57	Lanthanum	La	138·905$_{5}$[b]	92	Uranium	U	238·029[b,c,e,g]
103	Lawrencium	Lr	(260)	23	Vanadium	V	50·9414[b,c]
82	Lead	Pb	207·2[d,g]	54	Xenon	Xe	131·30[e]
3	Lithium	Li	6·94$_{1}$[c,d,e,g]	70	Ytterbium	Yb	173·0$_{4}$
71	Lutetium	Lu	174·97	39	Yttrium	Y	88·9059[a]
12	Magnesium	Mg	24·305[c,g]	30	Zinc	Zn	65·38
25	Manganese	Mn	54·9380[a]	40	Zirconium	Zr	91·22
101	Mendelevium	Md	(258)				

Figure 10.1
Table of relative atomic masses, 1973

Notes

[a] Element with only one stable nuclide.

[b] Element with one predominant isotope (about 99–100% abundance).

[c] Element for which the atomic mass is based on calibrated measurements.

[d] Element for which variation in isotopic abundance in terrestrial samples limits the precision of the atomic mass given.

[e] Element for which users are cautioned against the possibility of large variations in atomic mass due to inadvertent or undisclosed artificial isotopic separation in commercially available materials.

[f] Most commonly available long-lived isotope.

[g] In some geological specimens this element has a highly anomalous isotopic composition corresponding to an atomic mass significantly different from that given.

Not confirmed: 104 Rutherfordium and 105 Hahnium.

Relative density

$$= \frac{\text{Mass of 1 volume of gas}}{\text{Mass of 1 volume of hydrogen}}$$

all measurements on the substance, which must be a gas, being made at the same temperature and pressure.

The restriction of the substance being a gas means that this approach to relative atomic masses is applied mostly to non-metals and it is therefore complementary to the method based on Dulong and Petit's Law.

Applying Avogadro's Law to the definition of relative density:

Relative density

$$= \frac{\text{Mass of } n \text{ molecules of gas}}{\text{Mass of } n \text{ molecules of hydrogen}}$$

$$= \frac{\text{Mass of 1 molecule of gas}}{\text{Mass of 1 molecule of hydrogen}}$$

$$= \frac{\text{Mass of 1 molecule of gas}}{\text{Mass of 2 atoms of hydrogen}}$$

The evidence for the diatomicity of hydrogen is given in section 7.13.

Thus

$$\frac{\text{Relative molecular mass}}{\text{Relative density}} = \frac{\text{Mass of 1 molecule of substance}}{\text{Mass of 1 atom of hydrogen}} \times \frac{\text{Mass of 1 molecule of substance}}{\text{Mass of 1 molecule of gas}}$$

$$\therefore \frac{\text{Relative molecular mass}}{\text{Relative density}} = 2$$

or **Relative molecular mass**

= 2 × Relative density.

10.5 Cannizzaro's Method of Finding Relative Atomic Masses (1858)

A molecule of a compound contains one or more atoms of each element within it. Thus the relative molecular mass of a compound contains one or more relative atomic masses of each element. If a large number of compounds of an element is studied, probably at least one compound will contain one atom of the element per molecule.

Thus, methane has a relative molecular mass of 16 (twice its relative density of 8), and on analysis is found to consist of 75% carbon. Hence 75% of 16, i.e. 12 units of mass in one relative molecular mass, are due to carbon. Extending the survey gives:

Compound	*Density in g/dm³ at s.t.p.*	*Relative density*	*Relative molecular mass*	*Percentage carbon in compound*	*Contribution of carbon to relative molecular mass*
Ethene	1·26	14	28	86	24
Ethoxyethane	3·33	37	74	65	48
Carbon dioxide	1·98	22	44	27	12
Trichloromethane	5·38	60	120	10	12
Propanone	2·61	29	58	62	36
Carbon monoxide	1·26	14	28	43	12

It thus seems reasonable to suggest that the relative atomic mass of carbon is 12, and that there is one atom of carbon in methane, carbon dioxide, trichloromethane and carbon monoxide, but two in ethene, three in propanone and four in ethoxyethane. Because the relative density is rarely determined with high accuracy, this method gives an approximate value for the relative atomic mass, and this enables the correct multiple of the accurate combining mass to be selected as before.

10.6 The Mass Spectrometer

(This section is best studied after chapter 12.)

Experiments on the passage of electricity through tubes containing gases at low pressures, e.g. as in a neon light, done by Goldstein (1886), Thomson (1910–20) and Aston (1919–27) led to the instrument known today as the mass spectrometer. In figure 10.2 the front and rear views of a machine are shown; a line diagram of the machine and a typical result are also given.

The substance under examination is ionized either by bombardment with high speed electrons or by electrical sparks between electrodes made of the material. The ions are positive because they consist of the nuclei of the atoms together with most but not all of their electrons. These ions are accelerated by attraction to a negatively charged plate in which there is a slit. The slit is adjustable down to 10^{-6} m (1 μm) and a narrow beam of positive particles travels very quickly through an electrostatic and then a magnetic field. These fields are adjusted so that

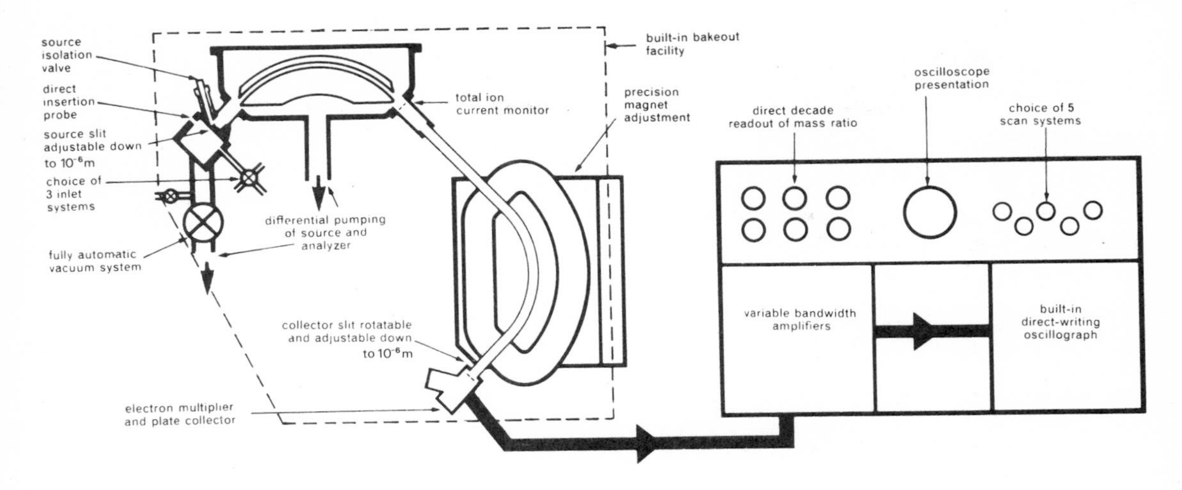

Structure of MS9 machine.

Front view of MS902 machine (electron bombardment type).

Rear view of MS902 machine.

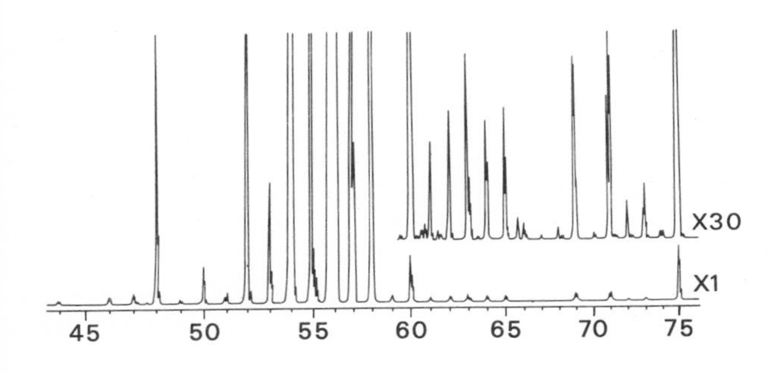

Electrically recorded scan of spectrum of steel standard SS54 obtained with radio-frequency spark discharge.
(MS702 machine; note 30 times enlargement of 60–75 part of spectrum.)

Figure 10.2
The mass spectrometer

Joseph John Thomson, 1856-1940, England

He was born in Manchester and in 1876 won a scholarship to Trinity College, Cambridge where he remained for the rest of his life. He was elected to a Fellowship of the Royal Society in 1883 and the following year became Cavendish Professor of Physics. He was the third to hold that office at Cambridge, his predecessors being Maxwell and Rayleigh, his successor Rutherford. An essay on vortex rings in 1883 showed his interest in atomic structure and in the next forty years he did a considerable amount of research on the discharge of electricity through gases. He showed that for cathode rays the ratio of mass to charge, regardless of the gas or the electrodes, was about one two-thousandth that obtained for hydrogen by electrolysis (published in 1897). Then he determined the charge on the particles and found it to be the same for gases and liquids: thus he suspected the particles were much smaller in mass than atoms.

He was awarded a Nobel Prize for Physics in 1906. During the years 1910–20 he did research on positive rays and his demonstration that they were deflected by magnetic and electrostatic fields led to the separation of isotopes by the mass spectrometer. In 1918 he was made Master of the college and in the following year he resigned his Professorship. In all he published 231 papers and 13 books. He had not set out to be a physicist: he won the scholarship at second attempt while passing the time waiting for a vacancy for an apprenticeship in making locomotives. That he finished his career as one of the legendary figures in physics, known universally as 'JJ', is perhaps an inspiration to those people who consider themselves early failures or who hesitate to change their jobs.

particles with the same ratio of mass to charge arrive at the same position and time at the collector. The electrical impulse received by the collector is shown on the oscilloscope and printed on graph paper.

The main peak given is that due to the molecule or the atom stripped of one electron; other peaks are given for more highly charged species. If an element is studied which consists of several isotopes then several peaks are given corresponding to the various isotopes each less one electron, e.g. for neon peaks are given for masses 20 and 22. The example in the illustration is for a steel so besides the main peak at 56 for the principal isotope of iron other peaks are seen for other isotopes of iron and for the other elements in the sample. When the resolving power is increased thirty-fold the minor constituents in the mass-range 60–75 are visible. The advent of the mass spectrometer has revolutionized the analysis of steel and hence the quality control of the largest metallurigcal industry in the world.

The mass spectrometer can be used to measure masses with an accuracy of 3 parts in a million and so is an invaluable tool for measuring relative atomic and molecular masses. Elements can be recognized by their characteristic isotope pattern and the structure of molecules found by studying the fragment peaks. The mass spectrometer uses very small samples of the material and can be used to isolate small samples of particular isotopes. Its use has led to the definition of atomic masses in terms of ${}^{12}_{6}C = 12$.

QUESTIONS 10

Relative Atomic and Molecular Masses

1 A metal M forms two sulphates MSO_4 and $M_2(SO_4)_3$. Its relative atomic mass is 96. (*a*) State its two valencies. (*b*) Find the two combining masses of the metal. (*c*) Write down the formulae of the two chlorides of M. (*d*) Find the approximate specific heat capacity of the metal M. [J]

2 A metallic chloride contains 34·45% of metal. Calculate the combining mass of the metal. If the valency of the metal is 3, what is the formula of the chloride and relative atomic mass of the metal? [OC]

3 The valency of a certain metal is 4, and 1·000 g of the metal forms 1·333 g of the oxide. What is the relative atomic mass of the metal? [J]

4 (*a*) The chloride of a metal X contains 71·0% of chlorine. What is the combining mass of the metal? (*b*) The specific heat capacity of this metal X is 0·67 $J\ mol^{-1}\ °C^{-1}$. What is its approximate relative atomic mass? (*c*) Hence write the formula of (i) the chloride; (ii) the oxide. [J]

5 When 0·5 g of a metal were placed in excess of acid, 100 cm^3 of hydrogen (measured at s.t.p.) were displaced. The specific heat capacity of the metal was found to be 0·24 $J\ mol^{-1}\ °C^{-1}$. What is the relative atomic mass of the metal? [J]

6 A gaseous compound containing only hydrogen and the element Y has a relative density of 40·5. The compound dissolves in water to give an acid. When a solution containing 2·025 g of the compound was allowed to react with excess of magnesium it was found that 280 cm^3 of hydrogen (measured at s.t.p.) was evolved. Find: (*a*) the relative molecular mass of the compound; (*b*) the number of atoms of hydrogen in one molecule of the compound; (*c*) the formula of the compound and hence the relative atomic mass of Y. [J]

7 A metal X can be both divalent and trivalent. The oxide in which it is divalent contains 78·7% of X. What is the relative atomic mass of X, and what is the mass of X in 100 g of the oxide in which it is trivalent? [OC]

8 (*a*) State the Law of Multiple Proportions.
(*b*) Two oxides of a metal M, in porcelain boats A and B, were reduced to the metal by heating in a current of hydrogen. The following weighings were recorded:
Mass of boat A and contents before heating, 3·728 g
Mass of boat A, 2·148 g. Find the mass of oxide.
Mass of boat A and contents after heating, 3·408 g
Mass of boat A, 2·148 g. Find the mass of metal.
Mass of boat B and contents before heating, 4·666 g
Mass of boat B, 2·536 g. Find the mass of oxide.
Mass of boat B and contents after heating, 4·426 g
Mass of boat B, 2·536 g. Find the mass of metal.
Show how the results confirm the law of multiple proportions.
(*c*) Given that the specific heat capacity of the metal is 0·42 $J\ g^{-1}\ °C^{-1}$, find the approximate relative atomic mass. [J]

9 Define combining mass of an element. A piece of a metal M weighing 2·25 g was converted to its oxide which weighed 3·45 g. Calculate the combining mass of the metal. In a second experiment, the same mass of metal M was heated in a stream of chlorine. What mass of metal chloride would be obtained? The relative atomic mass of M is 45. What is its valency? Write down the formulae of the oxide, nitrate and sulphate of the metal. [J]

10 Thomson found that his positive ray apparatus (a fore-runner of the mass spectrometer) gave two results when he studied neon: the line for mass 20 was roughly ten times as dark as that for mass 22. What is the relative atomic mass of neon?

11 Copper consists mainly of two isotopes (63 and 65) and has a relative atomic mass of 63·5. In what proportions are these isotopes present in the naturally occurring material?

12 When air is studied in a mass spectrometer the most intense lines obtained are for masses 28 and 32. Why? What other lines would you expect to find fairly easily?

11 Formulae and Equations

11.1 Symbols

Dalton, like earlier chemists, used a system of symbols to represent the atoms of different elements but to extend his system would be more of a hindrance than a help now that over 100 elements are known. The modern system was introduced by Berzelius, a Swedish scientist who lived at the same time as Dalton.

The symbol for an element represents one atom of the element and is shown by one or two letters, the first of which only is a capital.

Thus H stands for one atom of hydrogen, O for one atom of oxygen and C for one atom of carbon. There are only 26 letters in the alphabet so it is essential to use two letters for the majority of elements: thus Ca stands for one atom of calcium, Cd for one atom of cadmium, Ce for one atom of cerium, Cf for one atom of californium, Cl for one atom of chlorine, Cm for one atom of curium, Co for one atom of cobalt, Cr for one atom of chromium, Cs for one atom of caesium and Cu for one atom of copper. The Latin names for elements have often given rise to the symbols for the elements, e.g. Fe for iron, Ag for silver, Pb for lead and Sn for tin.

Symbols are often used in notes but they should not be used in sentences when the use of many atoms of those species is intended. Whereas H represents one atom of hydrogen, 2H represents two atoms of hydrogen and 3H three atoms of hydrogen.

11.2 Formulae

The distinction between the atom and the molecule of an element, which Dalton did not realize, was discussed in section 3.4. Hydrogen atoms in the gaseous state exist in pairs and the formula for a molecule is written H_2. This is not the same as writing 2H: hydrogen molecules are relatively inert compared with hydrogen atoms, but in both cases the numerical total is the same.

When a compound is formed the numbers of the atoms of the elements in one molecule may be the same or different.

The formula of a substance expresses the composition of a substance in terms of the relative numbers of atoms that have combined.

Each kind of atom is represented by its symbol and a small subscript gives the number present—unless there is only one, when it is omitted. One molecule of water contains two atoms of hydrogen and one of oxygen so the formula for water is written H_2O. Two molecules of water are written $2H_2O$, the two in front of the formula multiplying everything that follows it: $2H_2O$ means two molecules of water each of which contains two atoms of hydrogen and one atom of oxygen. It would be incorrect to write H_4O_2.

It may be convenient to write formulae in notes but it is not correct to write them in sentences. In writing formulae that contain several radicals (see section 11.3), the subscript multiplies everything within the bracket, e.g. $Pb(NO_3)_2$ means one molecule of lead nitrate containing one atom of lead and two nitrate radicals each containing one atom of nitrogen and three atoms of oxygen, making a total of one atom of lead, two of nitrogen and six of oxygen. Care must obviously be taken when assessing the numbers of atoms in several molecules of such a substance.

The order of writing symbols in a simple formula is to put the most metallic substance first and the most non-metallic substance last.

11.3 Valency

The valency of an element is the number of hydrogen atoms which will combine with or displace one atom of the element.

Hydrogen, the simplest and lightest element, has a valency of one.

In water there are two atoms of hydrogen for every atom of oxygen so in accordance with the definition, the valency of oxygen is two: the formula is written H_2O, the presence of a figure one after the oxygen being implied. Oxygen

combines with most of the elements known so that the valencies of most elements can be deduced if the formula of the oxide is known, e.g. zinc oxide, ZnO, so the valency of zinc is 2,
sodium oxide, Na_2O, so the valency of sodium is 1,
aluminium oxide, Al_2O_3, so the valency of aluminium is 3.

When a large number of compounds is studied it may be noticed that a set of atoms occurs as a unit in many of them, e.g. if various metals or their oxides are treated with dilute nitric acid, on evaporation of the solution compounds having formulae such as $NaNO_3$, CaN_2O_6, ZnN_2O_6, AlN_3O_9, BaN_2O_6 . . . may be isolated as crystals. If these formulae are rewritten as $NaNO_3$, $Ca(NO_3)_2$, $Zn(NO_3)_2$, $Al(NO_3)_3$, $Ba(NO_3)_2$ the pattern of behaviour is obvious: each of them contains the unit NO_3 which is known as the nitrate radical.

A radical is a group of atoms found as a unit in many compounds; it often does not have a separate existence.

A radical can be allotted a valency in the same manner as an element. The valency of a nitrate radical is one, the formula of nitric acid being HNO_3. The sulphate radical, SO_4, has a valency of two; the hydroxyl radical, OH, a valency of one. Ammonia (either the gas or a solution in water can be used) forms salts with many acids—see sections 14.4, 17.9, 30.11 and 30.12—and these salts contain the radical NH_4 which is known as the ammonium radical having a valency of one.

Some elements show several valencies e.g. lead 2 and 4, and sulphur 2, 4 and 6. For metals the valency is best shown in Roman numerals immediately after the word, e.g. lead(II) oxide is PbO and lead(IV) oxide is PbO_2, but in the past the lower valency has usually been implied by the ending -ous and the upper valency by the ending -ic e.g. plumbous oxide is PbO and plumbic oxide PbO_2. Such a system immediately runs into trouble when an element shows three valencies e.g. what will be the correct name for red lead Pb_3O_4? The answer is dilead(II) lead(IV) oxide or more obviously trilead tetraoxide. For non-metals the name is usually chosen to describe the formula e.g. phosphorus trichloride is PCl_3 and phosphorus pentachloride is PCl_5.

At this stage it is best to learn the valency table given in figure 11.1 so that formulae can be written. Later, in chapter 14, valency is interpreted in terms of electrons.

The prefixes mono-, di-, tri-, tetra-, penta- and hexa- are frequently used to convey the information that there are one . . . six atoms of a species.

11.4 The Deduction of Formulae from Valencies

Once the table of valencies of elements has been learnt continued practice enables the correct formula of a compound to be written. The fact that a formula can be written does not mean, however, that such a compound is easy to make or even exists, e.g. zinc and hydrogen do not react to give zinc hydride but the formula ZnH_2 might be deduced from the table.

If the periodic table (see chapter 13) of the elements is studied a connection between valency and the group number becomes obvious. Elements such as sodium, calcium, aluminium, which have valencies one, two and three respectively, are found in groups of the same number in the table. Elements such as chlorine, oxygen and nitrogen which often show valencies one, two and three respectively are found in groups numbered seven, six and five (N.B. $7+1$, $6+2$, $5+3$ all equal eight)

e.g. (*a*) In copper(I) iodide or cuprous iodide CuI, copper and iodine both have a valency of one.

(*b*) In zinc nitrate $Zn(NO_3)_2$, zinc has a valency of two and the nitrate radical of one (the two multiplies the numbers of all atoms in the bracket).

(*c*) In calcium phosphate $Ca_3(PO_4)_2$, calcium has a valency of two and the phosphate radical of three.

(*d*) In tin(II) sulphide or stannous sulphide SnS, tin and sulphur both have a valency of two so a compound is formed between one atom of each.

(*e*) In uranium(VI) fluoride or uranium hexafluoride UF_6, uranium shows a valency of six and fluorine of one.

(*f*) In hydrogen sulphide H_2S, sulphur shows a valency of two.

(*g*) In sulphur dioxide SO_2, sulphur shows a valency of four (sulphur(IV) oxide).

(*h*) In sulphur(VI) oxide SO_3, sulphur shows a valency of six (sulphur trioxide).

The valency of a radical is harder to perceive. In a sulphate, sulphur shows a valency of six, there are four oxygen atoms each having a

Valencies of metals and metallic radicals			
1	2	3	4
Sodium (Na) Potassium (K) Ammonium (NH_4) Silver (Ag) Copper (Cu) Mercury (Hg)	Calcium (Ca) Magnesium (Mg) Zinc (Zn) Copper (Cu) Iron (Fe) Lead (Pb) Mercury (Hg) Tin (Sn) Barium (Ba) Manganese (Mn)	Aluminium (Al) Iron (Fe) Chromium (Cr)	Tin (Sn) Lead (Pb) Manganese (Mn)

Valencies of non-metals and their radicals			
1	2	3	4
Chlorine (Cl) Bromine (Br) Iodine (I) Nitrate (NO_3) [nitrate(V)] Nitrite (NO_2) [nitrate(III)] Hydroxide (OH) Hydrogencarbonate (HCO_3) Chlorate (ClO_3) [chlorate(V)] Hydrogensulphate (HSO_4)	Oxygen (O) Sulphur (S) Sulphate (SO_4) [sulphate(VI)] Sulphite (SO_3) [sulphate(IV)] Carbonate (CO_3) [carbonate(IV)]	Nitrogen (N) Phosphorus (P) Phosphate (PO_4) [phosphate(V)]	Carbon (C) Silicon (Si) Sulphur (S)

N.B. Higher valencies are sometimes found, e.g. phosphorus displays a valency of 5 in such compounds as PCl_5 and P_4O_{10}, and sulphur a valency of 6 in SO_3.

Figure 11.1 Common valencies

valency of two and the overall valency of the radical is two. In a sulphite, sulphur shows a valency of four, there are three oxygen atoms each having a valency of two and the overall valency of the radical is again two. If the element shows more than two valencies in radicals then the terminations -ate and -ite are not sufficient. See page 362.

11.5 The Experimental Determination of Empirical Formulae

The analysis of black copper oxide (see section 9.3) has already been performed. The result is that the compound contains 80% of copper and 20% of oxygen. If a table of relative atomic masses is consulted (see figure 1.10, rough values will do) the relative masses of copper and oxygen are seen to be 64 and 16 respectively. On consideration of 100 g of the compound 80 g of copper represent $\frac{80}{64}$ (i.e. 1·25) of the mass of a mole of copper and 20 g of oxygen represents $\frac{20}{16}$ (i.e. 1·25) of the mass of a mole of oxygen. It is impossible, by Dalton's Atomic Theory, to have 1·25 atoms of an element. The percentages and the relative atomic masses are, however, relative quantities and so it is reasonable to deduce that in copper oxide there is one atom of copper to every atom of oxygen. Sometimes the ratio is not so obvious and then the smallest number obtained on dividing the percentage by the relative atomic mass is divided into the other numbers so obtained. The formula CuO is the simplest possible (or empirical) formula for the compound; in practice it may be Cu_2O_2 or Cu_3O_3 etc. and further evidence is really required e.g. from X-ray crystallography or from measurements of the relative density of the substance if it is a gas.

The empirical formula of a substance is the simplest formula of that substance: it expresses the relative numbers of the atoms of the elements.

The best way of presenting the calculation once the method is understood is in the form of a table:

Element	*% by mass*	*% ÷ A_r*	*Ratio of atoms*
Copper	80	$\frac{80}{64} = 1{\cdot}25$	1
Oxygen	20	$\frac{20}{16} = 1{\cdot}25$	1
Empirical formula = CuO			

It is not essential to have percentages in order to do the calculation of an empirical formula; it is sufficient to have the masses of the elements in a given mass of the compound, e.g. an acid of sulphur contains 6·1 parts by mass of hydrogen, 97·9 parts by mass of sulphur and 196 parts by mass of oxygen.

Element	*Parts by mass*	*Parts by mass divided by relative atomic mass*	*Ratio of atoms*
Hydrogen	6·1	$\frac{6{\cdot}1}{1} = 6{\cdot}1$	2
Sulphur	97·9	$\frac{97{\cdot}9}{32} = 3{\cdot}1$	1
Oxygen	196	$\frac{196}{16} = 12{\cdot}2$	4
Empirical formula = H_2SO_4			

Analyses can rarely be performed with very high accuracy and so on dividing the smallest number (3.1) obtained by the first division into the other (6.1 and 12.2) the nearest whole number obtained is taken.

The method of calculation can be extended to cases in which water of crystallization is present. In this case the percentage (or mass) of water is divided by the relative molecular mass e.g.

Unit	*% by mass*	*% ÷ relative atomic (molecular) mass*	*Ratio of Particles [÷ by smallest (0·41)]*
Barium	56·2	$\frac{56{\cdot}2}{137} = 0{\cdot}41$	1
Chlorine	29·0	$\frac{29{\cdot}0}{35} = 0{\cdot}82$	2
Water	14·8	$\frac{14{\cdot}8}{18} = 0{\cdot}82$	2
Empirical formula = $BaCl_2,2H_2O$			

11.6 Solution Methods of Determining Formulae

If the molar mass, M_r grams, of a substance is dissolved in water and made into 1000 cm^3 of solution it is referred to as a 1M solution, e.g. 1M HCl. By using two solutions of known concentration and finding precisely the point at which they react the formula of a product can be deduced. The two solutions may be for example an acid and an alkali (see chapter 17) or two which react to give a precipitate.

Experiment. To Determine the Formula of Lead Chloride

0·1 M solutions of lead nitrate and hydrochloric acid are used at room temperature. To 3 cm^3 of lead nitrate solution 1, 2, 3 . . . 9 cm^3 of hydrochloric acid are added. The height of the precipitate is measured after each addition, the test tube being centrifuged to compact the precipitate. On plotting a graph of the height of the precipitate against the volume of hydrochloric acid added it is seen that after a while the precipitate remains at a constant height: the reaction has been completed with the addition of 6 cm^3 of acid.

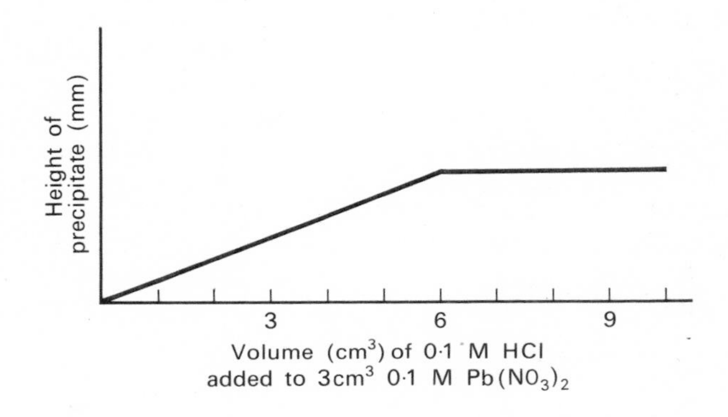

Figure 11.2

6 cm^3 of 0·1M hydrochloric acid contains twice as many chloride particles as 3 cm^3 of 0·1M lead nitrate contains lead particles, so the formula of lead chloride is $PbCl_2$.

11.7 The Experimental Determination of Molecular Formulae

The analysis of ethene shows that it contains 86% carbon and 14% hydrogen and so the empirical formula is CH_2. The molecular formula is thus $(CH_2)_n$ where n may be 1, 2, 3 etc. The empirical mass is 14 $(1 \times 12 + 2 \times 1)$. If the relative density of ethene is measured it is found to be 14 and this means that the relative molecular mass is 28 (see section 10.4). The relative molecular mass is thus twice the empirical mass and so the molecular formula is twice the empirical formula, i.e. C_2H_4.

See also isomerism, section 29.9.

11.8 Calculations involving Formulae

Once the formula of a substance has been derived using the valencies of the elements and radicals present other calculations can be performed. For this purpose a table of relative atomic masses is also required: in view of the accuracy one can attain in experiments, these masses only need to be approximate—as in figure 1.10. Relative molecular masses of elements and compounds are calculated by summing the appropriate numbers of the relative atomic masses.

(a) Calculation of Relative Molecular Masses

(i) Relative molecular mass of water, H_2O
$= (2 \times 1) + 16$
$= 18$

(ii) Relative molecular mass of nitrogen, N_2
$= 2 \times 14$
$= 28$

(iii) Relative molecular mass of copper(II) sulphate-5-water,
$CuSO_4,5H_2O = 64 + 32 + 64 + (5 \times 18)$
$= 250$

(b) Calculation of Percentage Composition

(i) Water has the formula H_2O. The relative molecular mass is 18. Of this total 2 parts are hydrogen and 16 parts are oxygen, so the proportion of hydrogen is $2 \div 18$ and of oxygen $16 \div 18$.

$$\% \text{ of hydrogen} = \frac{2}{18} \times 100 = 11{\cdot}1$$

$$\% \text{ of oxygen} = \frac{16}{18} \times 100 = 88{\cdot}9$$

It is better to calculate the separate percentages and then to check that they add up to 100 than it is to calculate one of them and then subtract that from 100.

(ii) Copper sulphate crystals have the formula $CuSO_4,5H_2O$ and the relative molecular mass is 250. The oxygen of the water could be included with the oxygen of the sulphate in calculating the percentage but in view of the behaviour of the crystals it is reasonable to calculate the percentage of water of crystallization of the substance:

$$\% \text{ of copper} = \frac{64}{250} \times 100 = 25{\cdot}6$$

$$\% \text{ of sulphate} = \frac{32}{250} \times 100 = 12{\cdot}8$$

$$\% \text{ of oxygen} = \frac{64}{250} \times 100 = 25{\cdot}6$$

$$\% \text{ of water} = \frac{90}{250} \times 100 = 36{\cdot}0$$

(iii) Calcium carbonate ($CaCO_3$) has a relative molecular mass of 100.

$$\% \text{ of calcium} = \frac{40}{100} \times 100 = 40$$

$$\% \text{ of carbon} = \frac{12}{100} \times 100 = 12$$

$$\% \text{ of oxygen} = \frac{48}{100} \times 100 = 48$$

(c) *Calculation of Mass of Substances*

(i) Calculate the mass of oxygen in 10 g of potassium nitrate (KNO_3). The relative molecular mass of potassium nitrate

$$= 39 + 14 + (3 \times 16)$$
$$= 101$$

In 101 g potassium nitrate (the **molar mass**) there are 48 g of oxygen.

$\therefore$ 10 g potassium nitrate contain $\frac{10 \times 48}{101}$ g of oxygen $= 4{\cdot}75$ g.

What is done in the laboratory using quantities in grams is often done in industry in tonnes (metric tons, 10^6 g).

(ii) Calculate the mass of water of crystallization in 10 g copper sulphate crystals ($CuSO_4$, $5H_2O$).

Relative molecular mass of crystals = 250

Contribution of water to M_r

$$= 90$$

In 250 g of the crystals there are 90 g of water

$\therefore$ in 10 g of the crystals there are $\frac{10}{250} \times 90$ g of water = $3{\cdot}6$ g

11.9 Equations

When a chemical change occurs new substances are formed and the change can often be briefly and conveniently represented by a chemical equation. The equation shows the reagents and the products.

(i) Magnesium and oxygen give magnesium oxide

or Magnesium + Oxygen → Magnesium oxide

Having stated the equation in words the second step is to write down the formulae of the reagents and products:

$$Mg + O_2 \rightarrow MgO$$

The reagent(s) are written on the left and the product(s) on the right hand side. An arrow showing the direction of the chemical change is preferable to an equal sign.

The third step, one which is necessary in the majority of situations, is to check that the equation balances. The oxygen molecules of the air each contain two atoms and each molecule must give rise to two molecules of magnesium oxide because each of the latter contain one atom of oxygen, so the equation is modified to give:

$$Mg + O_2 \rightarrow 2MgO$$

However, two molecules of magnesium oxide must be derived from two atoms of magnesium so the next modification gives:

$$2Mg + O_2 \rightarrow 2MgO$$

A recount now shows that there are two atoms of magnesium and two atoms of oxygen on each side of the equation: the equation has been balanced.

This equation is in accordance with the Law of Conservation of Matter and Dalton's Atomic Theory: during a chemical change the atoms of which matter is composed are neither destroyed nor created but merely rearranged.

N.B. (*a*) Never adjust the formula of a substance when attempting to balance the equation.

(*b*) A number in front of a formula multiplies every symbol that follows it.

(ii) Consider the reaction of zinc with dilute sulphuric acid, a reaction used to generate hydrogen, zinc sulphate solution being left.

$$\text{Zinc} + \text{Sulphuric acid} \rightarrow \text{Zinc sulphate} + \text{Hydrogen}$$

Substituting the formulae gives:

$$Zn + H_2SO_4 \rightarrow ZnSO_4 + H_2\uparrow$$

This time, counting the atoms of each element shows that the equation is already balanced.

(iii) Aluminium + Chlorine → Aluminium chloride

Formulae:

$$Al + Cl_2 \rightarrow AlCl_3$$

The aluminium atoms are already in balance but there are two chlorines on the left hand side and three on the right.

First adjustment:

$$Al + 3Cl_2 \rightarrow 2AlCl_3$$

The aluminium atoms are now twice as numerous on the right hand side as the left but the final adjustment is simple:

$$2Al + 3Cl_2 \rightarrow 2AlCl_3$$

(iv) Potassium chlorate → Potassium chloride + Oxygen

Formulae: $KClO_3 \rightarrow KCl + O_2\uparrow$

First step: $2KClO_3 \rightarrow KCl + 3O_2\uparrow$

Second step: $2KClO_3 \rightarrow 2KCl + 3O_2\uparrow$

(v) Calcium hydrogen carbonate → Calcium carbonate + Water + Carbon dioxide

Formulae:

$$Ca(HCO_3)_2 \rightarrow CaCO_3\downarrow + H_2O + CO_2\uparrow$$

The formulae yield a balanced equation.

11.10 Modifications to Equations

The equations quoted above are in the briefest possible form. It is possible to add s—for solid (sometimes c for crystal), l—for liquid, aq—for a solution in water and g—for gas. Also, an arrow downwards can be used to indicate a precipitate thrown out of solution and an arrow upwards a gas evolved. Hence the above equations can be written:

$$2Mg\,(s) + O_2\,(g) \rightarrow 2MgO\,(s)$$
$$Zn\,(s) + H_2SO_4\,(aq) \rightarrow ZnSO_4\,(aq) + H_2\,(g)$$
$$2Al\,(s) + 3Cl_2\,(g) \rightarrow 2AlCl_3\,(s)$$
$$2KClO_3\,(s) \rightarrow 2KCl\,(s) + 3O_2\,(g)$$
$$Ca(HCO_3)_2\,(aq) \rightarrow CaCO_3\,(s) + H_2O\,(l) + CO_2\,(g)$$

Often, however, sufficient emphasis can be conveyed by arrows:

$$2Mg + O_2 \rightarrow 2MgO$$
(smoke soon settles)
$$Zn + H_2SO_4 \rightarrow ZnSO_4 + H_2\uparrow$$
$$2Al + 3Cl_2 \rightarrow 2AlCl_3$$
(sublimes then solidifies)
$$2KClO_3 \rightarrow 2KCl + 3O_2\uparrow$$
$$Ca(HCO_3)_2 \rightarrow CaCO_3\downarrow + H_2O + CO_2\uparrow$$

Equations are usually written with metals first and non-metals second on the left hand side. On the right hand side the order of writing is usually: solids, liquids and substances in solution, finally gases.

The evolution or absorption of energy is not usually shown in the equation but may be written afterwards (see section 18.1).

A reversible reaction can be shown, by double arrows (see section 19.2):

$$3Fe + 4H_2O \rightleftharpoons Fe_3O_4 + 4H_2$$

11.11 Information not given in Equations

An equation can be written whether or not a reaction will occur in practice: the experimental knowledge must be gained. The equation does not state whether heating is needed to start a reaction or whether it will proceed spontaneously, i.e. no information is given about the rate of a reaction. 'Heat' may be written under the arrow.

Beyond indicating the phase (solid, liquid or gas) of a substance, the equation does not tell us what will be seen or smelt. Often under the equation words such as 'dilute' or 'concentrated' may be written to indicate the concentration of solutions; colour may also be recorded.

Often of interest are the structures of the substances involved and the course of the reaction, the mechanism by which the transformation occurs, but this information is not given in an equation.

11.12 Mass Calculations involving Equations

The first step is to write a **balanced** equation for the reaction: strictly speaking an equation cannot be unbalanced. The table of relative atomic masses is again useful.

(i) The equation for the combustion of magnesium is

$$2Mg + O_2 \rightarrow 2MgO$$

i.e. 2×24 g magnesium combine with 2×16 g oxygen to give $2 \times (24 + 16)$ g magnesium oxide. If the initial mass of magnesium is a fraction or a multiple of 48 g then a simple sum has to be performed.

Consider starting the reaction with 3 g magnesium, then $\frac{3}{48} \times 32$ g oxygen will be required for complete combustion and $\frac{3}{48} \times 80$ g magnesium oxide will be produced.

There is no need to calculate any quantities which are not mentioned in the question.

(ii) What mass of hydrogen will be produced

when 6·5 g zinc are dissolved in sufficient acid for a complete reaction?

$$Zn + H_2SO_4 \rightarrow ZnSO_4 + H_2\uparrow$$

Relative atomic masses:
65 1

Equation masses:
65 (2 × 1)
65 g of zinc → 2 g of hydrogen

Experimental masses:
6·5 g of zinc → 0·2 g of hydrogen

(iii) What is the percentage loss in mass when potassium chlorate is heated?

$$2KClO_3 \rightarrow 2KCl + 3O_2\uparrow$$

Relative atomic masses:
39,35·5,16 16

Equation masses:
(2 × 122·5) (3 × 32)

Experimental masses:
245 g 96 g

$$\% \text{ loss in mass} = \frac{96}{245} \times 100 = 39$$

(iv) What is the maximum yield of sodium hydroxide that could be obtained from 10·6 kg sodium carbonate by adding sufficient calcium hydroxide solution?

$$Na_2CO_3 + Ca(OH)_2 \rightarrow CaCO_3\downarrow + 2NaOH$$

Relative atomic masses:
23,12,16 23,16,1

Equation masses:
106 (2 × 40)
106 kg Na_2CO_3 → 80 kg NaOH

Experimental masses:
10·6 kg Na_2CO_3 → 8 kg NaOH

11.13 Experiments to Determine Reacting Masses

Many experiments have been considered already in Chapter 9 but two others that must be considered are:

(a) The Displacement of Hydrogen from an Acid

A reactive metal such as magnesium, aluminium, zinc or iron will react with dilute hydrochloric acid releasing hydrogen, e.g.

$$Mg + 2HCl \rightarrow MgCl_2 + H_2$$

or

$$Mg + 2H^+ \rightarrow Mg^{2+} + H_2$$

The mass of the metal is measured at the start of the experiment and the volume of hydrogen measured at the end: the apparatus may be as in figure 11.3(a) or (b). This volume must be measured at atmospheric pressure. The temperature of the room is also recorded so that the water vapour pressure can be found from tables. The tubes in the apparatus shown in figure 11.3(a) can be burette tubes if the flexible connection is as long as one burette. In the apparatus shown in figure 11.3(b) there must be some water in the measuring cylinder at the start in order that the siphon functions properly. If an aspirator is used instead of the bottle in the apparatus and a manometer put between the reaction vessel and the aspirator then it is an easy matter to run water out of the aspirator in order to keep the pressure inside the apparatus constant. See also figure 18.6.

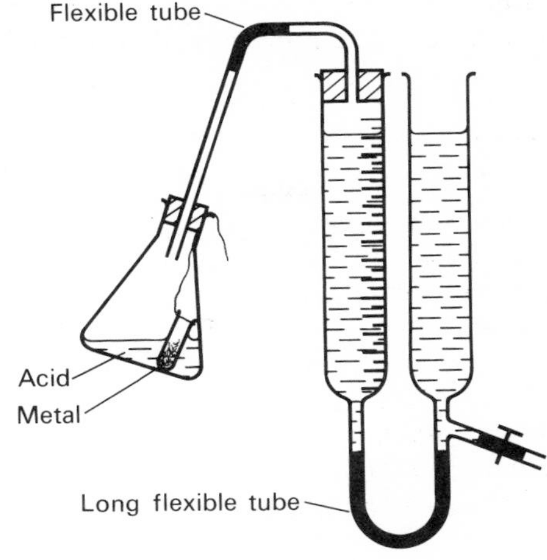

Figure 11.3(a)
Ostwald's gas burette

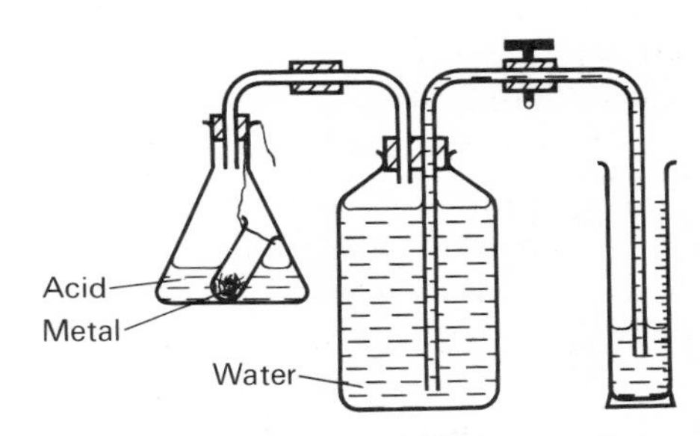

Figure 11.3(b)
An apparatus for measuring the volume of gas liberated in a reaction.

Results

Mass of magnesium	=	0·110 g
Volume of hydrogen	=	109 cm^3
Temperature	=	15°C
Barometric pressure	=	753 mmHg
Vapour pressure of water at 15°C	=	13 mmHg
Density of hydrogen at s.t.p.	=	0·09 g/dm^3

Calculation

Pressure of hydrogen $= 753-13$
$= 740$ mmHg

Volume of hydrogen at s.t.p. $= 109 \times \frac{740}{760} \times \frac{273}{288}$ cm^3
$= 101$ cm^3

$\therefore$ 101 cm^3 of hydrogen at s.t.p. displaced by 0·110 g of magnesium

$\therefore$ 0·00909 g of hydrogen displaced by 0·110 g of magnesium

$\therefore$ 1 g of hydrogen displaced by $\frac{1 \times 0{\cdot}110}{0{\cdot}00009}$ g of
$= 12.1$ g of magnesium

i.e. the combining mass of magnesium = 12·1 or, viewing the latter part of the calculation another way:

$$\frac{101}{22400} \text{ moles of hydrogen}$$
$$\equiv \frac{0{\cdot}110}{24} \text{ moles of magnesium}$$
$$\therefore \quad 0{\cdot}00451 \text{ moles of hydrogen}$$
$$\equiv 0{\cdot}00458 \text{ moles of magnesium.}$$

Thus within the limits of experimental error one mole of hydrogen is displaced by one mole of magnesium, a fact which enables the equation to be written.

$$Mg + 2HCl \rightarrow MgCl_2 + H_2\uparrow$$

(*b*) *The Displacement of a Metal from a Solution of one of its Salts by Another Metal*

This type of reaction is frequently studied qualitatively to compile the electrochemical series (see section 15.6), e.g. magnesium will displace silver from silver nitrate solution.

$$Mg + 2AgNO_3 \rightarrow Mg(NO_3)_2 + 2Ag\downarrow$$
or $$Mg + 2Ag^+ \rightarrow Mg^{2+} + 2Ag\downarrow$$

and zinc will displace copper from copper sulphate solution

$$Zn + CuSO_4 \rightarrow ZnSO_4 + Cu\downarrow$$
or $$Zn + Cu^{2+} \rightarrow Zn^{2+} + Cu\downarrow$$

A known mass of the metal is taken and added to a concentrated solution of the salt. When the reaction has ceased the precipitated metal is filtered off, washed and dried before weighing.

Results

Mass of magnesium = 0·21 g
Mass of silver = 1·89 g

Calculation

0·21 g magnesium $\equiv$ 1·89 g silver

$\therefore$ 24 g magnesium $\equiv \frac{1{\cdot}89 \times 24}{0{\cdot}21}$ g silver
$= 216$ g silver

The relative atomic masses of magnesium and silver are 24 and 108 respectively; thus 1 mole of magnesium displaces 2 moles of silver, as recorded in the equation above.

11.14 The Molar Volume of a Gas

The Molar Volume of a gas is 22 400 cm^3 (22.4 dm^3) at standard temperature and pressure (s.t.p.).

This is frequently called the Gram Molecular Volume because it is the volume that would be occupied by the relative molecular mass in grams of a gas at s.t.p., i.e. 1 mole. The volume can be deduced as follows:

Relative molecular mass
$= 2 \times$ Relative density

$$\frac{\text{Mass of 1 molecule of gas}}{\text{Mass of 1 atom of hydrogen}} = 2 \times \frac{\text{Mass of 1 volume of gas}}{\text{Mass of 1 volume of hydrogen}}$$

$$\frac{\text{Mass of } 6 \times 10^{23} \text{ molecules of gas}}{\text{Mass of } 6 \times 10^{23} \text{ atoms of hydrogen}} = 2 \times \frac{\text{Density of gas (in g/dm}^3)}{0{\cdot}09}$$

$$\frac{\text{Mass in grams of 1 mole of gas}}{\text{1 gram}} = 22{\cdot}2 \times \text{Density of gas in g/dm}^3$$

$\therefore$ Volume of gas = 22·2 dm^3

or 22·4 dm^3, using the accurate rather than the working definitions,

i.e. H = 1·008 not H = 1.

11.15 Volume Calculations involving Equations

Where gases are concerned it is difficult to make measurements of mass and comparatively easy to make volume measurements. The important quantity is 22 400 cm^3 which is the volume occupied by the molar mass of a gas at s.t.p. (standard temperature and pressure)—see section 7.5 and 7.11.

(i) 44 g CO_2, 2 g H_2, 32 g O_2 all, separately, occupy 22 400 cm^3 (22·4 l or dm^3) at s.t.p.

(ii) What volume of hydrogen is evolved when 6·5 g zinc are dissolved in an excess of acid?

$$Zn + 2HCl \rightarrow ZnCl_2 + H_2\uparrow$$

Equation quantities:

65 g Zn $\rightarrow$ 22 400 cm^3 H_2

Experimental quantities:

6·5 g Zn $\rightarrow$ 2 240 cm^3 H_2 (at s.t.p.)

(iii) What volume of carbon dioxide is produced on heating 1·68 g sodium hydrogencarbonate?

$$2NaHCO_3 \rightarrow Na_2CO_3 + H_2O + CO_2\uparrow$$

Equation quantities:
(2×84) g $NaHCO_3 \rightarrow 22\,400$ cm^3 CO_2
Experimental quantities:
1·68 g $NaHCO_3 \rightarrow 224$ cm^3 CO_2 (at s.t.p.)

If the conditions under which the carbon dioxide is collected are not standard conditions then the appropriate correction must be made.

QUESTIONS 11

Formulae and Equations

1 Calculate the percentage of water of crystallization in crystals of sodium sulphate ($Na_2SO_4,10H_2O$). [S]

2 What mass of magnesium hydroxide would be precipitated by adding excess of sodium hydroxide to a solution containing 12·3 g of Epsom salts ($MgSO_4,7H_2O$)? [S]

3 Calculate the percentage of water of crystallization in sodium sulphite crystals $Na_2SO_3,7H_2O$. [S]

4 Which of the following contains the greatest proportion of nitrogen by mass (*a*) ammonia NH_3, (*b*) ammonium sulphate $(NH_4)_2SO_4$, (*c*) hydrazine N_2H_4, (*d*) ammonium nitrate NH_4NO_3?

5 Calculate the simplest formula of a compound containing, by mass, 28·1% of iron, 35·7% of chlorine and 36·2% of water of crystallization. [C]

6 Calculate the simplest formula of a compound containing, by mass, 53·8% of iron and 46·2% of sulphur. [C]

7 (*a*) Calculate the percentage of water in crystals of washing soda, $Na_2CO_3,10H_2O$. (*b*) Calculate the volume of carbon dioxide evolved at s.t.p. when 71·5 g of crystals of washing soda are treated with excess of acid. [OC]

8 Calculate the percentage composition of the following compounds:
(*a*) CH_4 (*b*) CuO (*c*) C_2H_5OH (*d*) Na_2SO_3.

9 (*a*) Calculate the percentage by mass of phosphorus present in calcium phosphate, $Ca_3(PO_4)_2$. (*b*) Find the simplest formula of a compound containing 45·6% tin (Sn) and 54·4% chlorine. [C]

10 Crystalline barium chloride, $BaCl_2,2H_2O$, can be dehydrated by heating. Calculate the percentage by mass of water of crystallization in this compound. [W]

11 Calculate: (*a*) the percentage by mass of copper in copper sulphate pentahydrate ($CuSO_4,5H_2O$); (*b*) the volume of oxygen (at s.t.p.) required to burn completely 8 g of sulphur to sulphur dioxide; (*c*) the volume of oxygen required to burn completely 10 cm^3 of methane (CH_4) (measured at same temperature and pressure as the oxygen) ($CH_4 + 2O_2 \rightarrow CO_2 + 2H_2O$); (*d*) the volume of carbon dioxide (under the same conditions) formed in (c). [L]

12 Calculate: (*a*) the percentage of water of crystallization in magnesium sulphate-7-water, $MgSO_4,7H_2O$; (*b*) the volume of carbon monoxide (measured at s.t.p.) needed to reduce 111·5 g of lead(II) oxide (PbO) to lead; (*c*) the volume of the gaseous product of the reaction in (*b*) (also measured at s.t.p.). [L]

13 A pure hydrocarbon contains 80% of carbon. What is its empirical formula? If its relative molecular mass is 30 what is its molecular formula? [S]

14 Find the empirical formula of a substance which has the following percentage composition by mass: sodium 18·2, sulphur 12·7, oxygen 19·1, water of crystallization 50. [S]

15 Find the empirical formula of a substance having the following percentage composition by mass: Na 18·55, S 25·8, O 19·35; water of crystallization 36·3. [S]

16 On analysis a certain compound was found to contain iodine and oxygen in the ratio 127 g of iodine to 40 g of oxygen. Which of the following is the formula of the compound? (*a*) IO, (*b*) I_2O, (*c*) I_5O_2, (*d*) I_2O_5.

17 A crystalline substance was analyzed with the following results. 1·000 g of it was heated gently to constant mass; the residue weighed 0·703 g. The gas evolved was passed through a silica gel drying tube which gained in mass by 0·199 g, and then through potassium hydroxide solution which gained 0·098 g. Which of the following formulae best represents the composition of the crystals? Na_2CO_3,H_2O or $Na_2CO_3,NaHCO_3,2H_2O$ or $NaHCO_3$. [L]

18 A hydrate of copper sulphate contained 35·7% of copper. Find its formula. [OC]

19 Find the empirical formula of a compound which contains H 1·75, O 70·15, S 28·1%. [OC]

20 (*a*) A hydrocarbon contains 85·7% of carbon. Find its empirical formula. (*b*) 10 cm^3 of a hydrocarbon of empirical formula C_2H_5 gave 40 cm^3 of carbon dioxide when it was burned completely (all volumes measured at s.t.p.). Find its molecular formula. [OC]

21 Find the empirical formula of a compound *A* which contains C 26·7, H 2·22, O 71·08%. If *A* is a dibasic acid, what is its simplest possible molecular formula? [OC]

22 A compound contains C 54·5, H 9·1, O 36·4%. What is its empirical formula? [OC]

23 Calculate the empirical formula for a compound which contains 36·5% sodium, 25·4% sulphur and 38·1% oxygen. [A]

24 An oxide of titanium contains 40% oxygen. What is the simplest formula for the oxide? Write the formula for the corresponding chloride of titanium. [J]

25 Citric acid contains, by mass, 37·5% of carbon, 58·3% of oxygen, the remainder being hydrogen. Find its empirical formula. [W]

26 A certain element *E* of relative atomic mass 32 forms two oxides, oxide *A* containing 50% oxygen and oxide *B* containing 60% oxygen. Calculate the empirical formula of each oxide. [O]

27 (*a*) Define the terms (i) element, (ii) allotrope. (*b*) Calculate the empirical formula of a compound of oxygen with sulphur which was found by analysis to consist of 60% oxygen and 40% sulphur. [O]

28 An organic compound *X* contains, by mass, 12·8% of carbon, 2·13% of hydrogen and 85·1% of bromine: *X* has a relative molecular mass of 188. Find the molecular formula of the compound *X*. [C]

29 2·3 g of an anhydrous salt S that has a relative molecular mass of 161 was obtained by heating 4·1 g of the crystalline salt S,$x$$H_2O$. Calculate the value of x. [S]

30 (*a*) Complete and balance the following equations:
(i) $NaCl +$ $\rightarrow NaHSO_4 +$
(ii) $Ca(HCO_3)_2$ heat $\rightarrow$ $+$ $+$
(iii) $FeCl_3 + H_2S$ $\rightarrow$ $+$ $+$
(iv) $CaCO_3 + HCl$ $\rightarrow$ $+$ $+ CO_2$
(*b*) Calculate the empirical formula of a compound that has the composition 52·0% zinc, 9·6% carbon, 38·4% oxygen. [A]

31 A compound contains 22·0% carbon, 4·6% hydrogen and 73·4% bromine by mass. Its relative molecular mass is 109. Find (*a*) its empirical formula, (*b*) its molecular formula.

32 0·625 g of a mineral yield 175 cm^3 of hydrogen sulphide measured at s.t.p. after suitable treatment. Calculate the percentage of sulphur in the mineral. [S]

33 7·15 g of washing soda crystals ($Na_2CO_3,10H_2O$) are dissolved in excess of dilute hydrochloric acid. What volume of carbon dioxide, measured at s.t.p. is evolved, and what mass of sodium chloride would be obtained by evaporating the resulting solution? [S]

34 From the equation

$$4FeS_2 + 11O_2 \rightarrow 2Fe_2O_3 + 8SO_2$$

calculate (*a*) the volume of sulphur dioxide, measured at s.t.p., that would be obtained by the complete oxidation of 1 kg of iron sulphide, and (*b*) the volume of air, measured at s.t.p., used in the oxidation, assuming that air contains 20% of oxygen by volume. [C]

35 From the equation

$$Pb_3O_4 + 4HNO_3 \rightarrow PbO_2 + 2Pb(NO_3)_2 + 2H_2O$$

calculate (*a*) the mass of lead(IV) oxide formed from 100·0 g of trilead tetraoxide, (*b*) the minimum mass of nitric acid required to bring about the reaction in (*a*). [C]

36 From the equation

$$PbO_2 + 4HCl \rightarrow PbCl_2 + 2H_2O + Cl_2$$

calculate (*a*) the volume of chlorine, measured at s.t.p., liberated by the action of 9·56 g of lead(IV) oxide on hydrochloric acid, (*b*) the mass of lead chloride formed at the same time. [C]

37 From the equation

$$NaCl + H_2O + NH_3 + CO_2 \rightarrow NaHCO_3 + NH_4Cl$$

calculate (*a*) the total volume of ammonia and carbon dioxide (measured at s.t.p.) required to react with 117 g of sodium chloride, (*b*) the maximum mass of sodium hydrogencarbonate that would be formed if the reaction went to completion. [C]

38 Calculate the volume, at s.t.p., of gas evolved when 0·1 moles of pure calcium carbonate are heated to constant mass:

$$CaCO_3 \rightarrow CaO + CO_2$$

If this volume of gas were absorbed in excess of calcium hydroxide solution, calculate the mass of precipitate that could be formed. If this same volume of gas were decomposed into its elements (*a*) what mass of carbon would be released, (*b*) what volume, measured at s.t.p., of oxygen would be obtained? [OC]

39 The action of manganese(IV) oxide on concentrated hydrochloric acid can be represented by the equation

$$MnO_2 + 4HCl \rightarrow MnCl_2 + 2H_2O + Cl_2$$

Calculate (*a*) the mass of manganese chloride formed, and (*b*) the volume of chlorine produced at s.t.p. from the reaction of 8·7 g of manganese(IV) oxide with excess concentrated hydrochloric acid. [A]

40 Magnesium carbonate reacts with hydrochloric acid according to the following equation:

$$MgCO_3 + 2HCl \rightarrow MgCl_2 + CO_2 + H_2O$$

Suppose you were given an impure sample of magnesium carbonate in which the impurities did not evolve gases when treated with hydrochloric acid. Describe how you would find the percentage of magnesium carbonate in the impure sample by using the reaction given above, and measuring the volume of carbon dioxide formed. Give full experimental details and indicate how you would calculate your answer. [Nf]

41 Calculate the mass of zinc which must be dissolved in dilute sulphuric acid in order that 28·7 g of crystalline zinc sulphate $ZnSO_4,7H_2O$ may be obtained by evaporation of the resulting solution.

42 A compound contains 12·6% carbon, 3·2% hydrogen and 84·2% bromine by mass. Its relative density is 47·5. Find (*a*) its empirical formula, (*b*) its molecular formula.

43 Complete and balance the following skeleton equations:

(*a*) $Cu + ...H_2SO_4 \rightarrow CuSO_4 + ... + ...$
(*b*) $SO_2 + Cl_2 + ... \rightarrow H_2SO_4 + ...$
(*c*) $.....NaOH + ...NO_2 \rightarrow NaNO_2 + ... + ...$
(*d*) $..... + ...NH_4Cl \rightarrow .....NH_3 + CaCl_2 + ...$ [C]

44 The relative molecular mass of an anhydrous salt is 160 and the crystals contain 36% of water of crystallization. How many molecules of water are associated with one molecule of the anhydrous salt? [S]

45 The following experiment was used to determine the equation for the reaction between solutions of silver nitrate ($AgNO_3$) and potassium chromate (K_2CrO_4) in water (both are ionic compounds). 3 cm³ of 1 M potassium chromate solution were put into a tube and 1 cm³ of 1 M silver nitrate was added. The tube was shaken and then left to stand for a known time. The height of the precipitate of silver chromate was measured and is shown in the first column of the graph. The experiment was repeated using the same volume of the potassium chromate solution, increasing the volume of the silver nitrate solution by 1 cm³ each time. The heights of the precipitate, after standing for the same time, are shown in the graph.

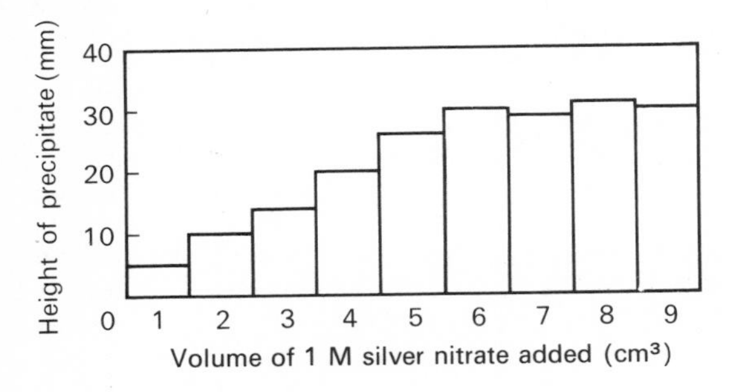

(*a*) What height of precipitate would you expect if 4·5 cm³ of 1 M silver nitrate had been added to 3 cm³ of 1 M potassium chromate? Assume the conditions to be the same. (*b*) What fraction of a mole of potassium chromate is present in 3 cm³ of a 1 M solution? (*c*) From the diagram deduce the volume of 1 M silver nitrate which will exactly react with 3 cm³ of 1 M potassium chromate. (*d*) What fraction of a mole of silver nitrate is present in your answer to (*c*)? (*e*) Deduce the ionic equation for the reaction between aqueous solutions of potassium chromate and silver nitrate. Show how you arrive at your answer. (*f*) Suppose that, instead of allowing the precipitates to stand, the tubes had been centrifuged for the same time. State and explain briefly how you would expect the resulting diagram of the heights of the precipitates (i) to differ from and (ii) to resemble the graph. [Nf]

46 Express fully in words as far as you can the information given by the following equations:

(*a*) $NaOH + HNO_3 \rightarrow NaNO_3 + H_2O$
(*b*) $ZnO + 2HCl \rightarrow ZnCl_2 + H_2O$
(*c*) $C + O_2 \rightarrow CO_2$
(*d*) $Al_2(SO_4)_3 + 3BaCl_2 \rightarrow 3BaSO_4 + 2AlCl_3$

47 The symbol X stands for an element. Below are given equations involving X and its compounds, but these equations are unbalanced. Insert numbers where necessary in order to balance each equation.

Example

Unbalanced equation $X_4O_{10} + H_2O \rightarrow H_3XO_4$
Balanced equation: $X_4O_{10} + 6H_2O \rightarrow 4H_3XO_4$

(*a*) $X + Fe_2O_3 \rightarrow X_2O_3 + Fe$
(*b*) $XCl_3 + H_2O \rightarrow X(OH)_3 + HCl$
(*c*) $X_2(SO_4)_3 + Pb(NO_3)_2 \rightarrow X(NO_3)_3 + PbSO_4$
(*d*) $X + KOH + H_2O \rightarrow K_3X(OH)_6 + H_2$ [J]

48 Taking the reaction given by the equation

$$Zn + H_2SO_4 + 7H_2O \rightarrow ZnSO_4,7H_2O + H_2$$

state all the information given by the equation. Describe fully how you would prove experimentally that the equation correctly represents the amount of hydrogen obtainable using a given mass of zinc.

49 Express fully in words as far as you can the information given by the following equations:

(*a*) $CuO + H_2SO_4 \rightarrow CuSO_4 + H_2O$
(*b*) $Ca(OH)_2 + 2HCl \rightarrow CaCl_2 + 2H_2O$
(*c*) $H_2 + S \rightarrow H_2S$
(*d*) $Fe_2(SO_4)_3 + 3SrCl_2 \rightarrow 3SrSO_4 + 2FeCl_3$

50 Rewrite the following equations using symbols and formulae:

(*a*) iron + sulphur → iron sulphide
(*b*) calcium oxide + water → calcium hydroxide
(*c*) potassium hydroxide + sulphuric acid → potassium sulphate + water
(*d*) copper sulphate + ammonium carbonate → copper carbonate + ammonium sulphate

51 X stands for an element. Below are given equations involving compounds of X, but these equations are unbalanced.

(*a*) Balance the equations:

$HXO_3 + HCl \rightarrow XOCl + H_2O + Cl_2$
$XO + H_2 \rightarrow XH_3 + H_2O$
$XH_3 + O_2 \rightarrow XO + H_2O$

(*b*) What are the valencies of the element X in its hydride and oxide? [J]

52 Express fully in words the meanings of the following symbols and formulae: $2Cl_2$; Hg; 3MgO; $4MnSO_4$; $Na_2CO_3,10H_2O$; $Pb(NO_3)_2$.

53 1 M solutions of a metal sulphate and of barium chloride were prepared. The following mixtures were made in small tubes:

	Vol. of sulphate solution (cm^3)	*Vol. of chloride solution* (cm^3)	*Vol. of water* (cm^3)
(i)	2	2	8
(ii)	2	4	6
(iii)	2	6	4
(iv)	2	8	2
(v)	2	10	nil

The tubes were centrifuged and the depths of the precipitates formed were measured and recorded as shown in the diagram.

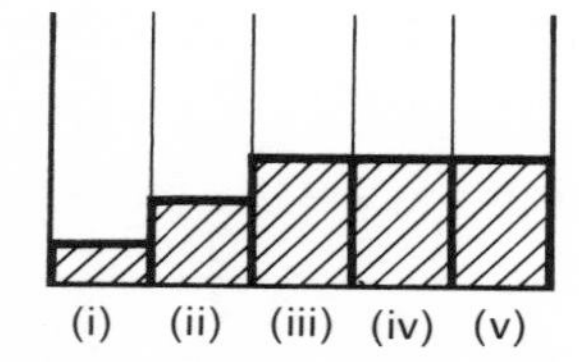

(*a*) Write the ionic equation for the reaction.
(*b*) What is the mass of barium chloride in 1 dm^3 of its solution?
(*c*) Why did the depth of the precipitate stay constant in (iii), (iv) and (v)?
(*d*) What volume of barium chloride solution reacts with 2 cm^3 of the sulphate solution?
(*e*) What volume of barium chloride solution reacts with 1 dm^3 of the sulphate solution?
(*f*) How many moles of barium chloride react with 1 mole of the sulphate?
(*g*) Using the symbol E for the unknown metal, write the formula for the sulphate and hence the equation for the reaction.
(*h*) 1 mole of the metal sulphate weighs 342 g. What is the relative atomic mass of E?
(*j*) How and why was the total volume of the mixture kept constant?
(*k*) What was the purpose of centrifuging the tubes? What other process might have been used? [J]

54 Write down the formulae of the following substances: three molecules of hydrogen, an atom of tin, one molecule each of silver nitrate, calcium chloride, sodium phosphate, zinc sulphide, iron(III) sulphate.

55 A metallic element, M, has valencies 2 and 3. Write down the formulae for its oxides, chlorides, carbonates, nitrates, phosphates, and hydroxides.

56 It is found that 4 g of oxygen, 7 g of the gaseous compound butene and 10·5 g of the gaseous element krypton all occupy the same volume at the same temperature and pressure. (*a*) What fraction of a mole is 4 g of oxygen? (*b*) Calculate the relative molecular masses of butene and krypton. (*c*) The empirical (simplest) formula of butene is CH_2. What is its molecular formula? (*d*) The relative atomic mass of krypton is 84. What is the atomicity of krypton? (*e*) Write a possible structural formula for butene. (*f*) Give the name and structural formula of another compound having the empirical formula CH_2. (*g*) Write the equation for the burning of butene in oxygen to give carbon dioxide and water. (*h*) If 10 cm^3 of butene are burnt according to your equation, calculate the volume of oxygen used and the volume of carbon dioxide formed (all volumes at the same temperature and pressure). [C]

57 What volume of carbon dioxide gas measured at s.t.p. would be evolved if 1·05 g of sodium hydrogen carbonate were (*a*) heated in a dry tube until no further loss of mass occurred; (*b*) reacted completely with dilute hydrochloric acid? [S]

58 0·145 g of a volatile liquid when vaporized occupied 61·1 cm^3 at 17°C and 755 mmHg. Determine the relative molecular mass of the liquid. The saturated vapour pressure of water at 17°C is 15 mmHg.

59 In an experiment the following results were obtained:

	grams
Mass of evacuated flask	= 48·500
Mass of flask filled with hydrogen at s.t.p.	= 48·540
Mass of flask filled with a gas at s.t.p.	= 49·060

Find: (*a*) the relative density of the gas; (*b*) the relative molecular mass of the gas; (*c*) what volume is occupied by 7·0 g of the gas at s.t.p. [J]

60 The density of a gas is 1·34 kg/m^3. Calculate its relative molecular mass.

61 Determine the relative molecular mass of a gas G, using the following experimental results:

	grams
Mass of evacuated bulb	= 56·40
Mass of bulb filled with hydrogen	= 56·88
Mass of bulb filled with gas G	= 64·56

Pressure and temperature were kept constant.

62 Potassium chlorate was heated in a test tube and the oxygen evolved collected:

	grams
Mass before heating	= 19·28
Mass after heating	= 18·88

300 cm^3 of oxygen were given at a temperature of 15°C and under a pressure of 740 mmHg. Determine the relative molecular mass of oxygen.

	Mass (kilogram)	*Charge (coulomb)*	*Approximate relative mass (H = 1)*	*Relative charge (e = 1)*
Proton (p)	$1{\cdot}6725 \times 10^{-27}$	$1{\cdot}602 \times 10^{-19}$	1	+1
Neutron (n)	$1{\cdot}6748 \times 10^{-27}$	0	1	0
Electron (e)	$9{\cdot}109 \times 10^{-31}$	$1{\cdot}602 \times 10^{-19}$	1/1837	−1
Hydrogen atom (H)	$1{\cdot}6734 \times 10^{-27}$	0	1	0

Nuclides are present in constant proportions in most samples of elements, hence the average atomic mass is constant. The mass spectrometer gives the numerical information, e.g. chlorine consists of 75 % of the 35 isotope and 25 % of the 37 isotope; hence the relative atomic mass is 35·5. The two isotopes can be designated as $^{35}_{17}Cl$ and $^{37}_{17}Cl$ the subscript being the atomic number and the superscript the mass number of that particular nuclide. The numbers of neutrons are eighteen and twenty respectively. About a quarter of the elements consist of one nuclide only: at the other extreme, tin has ten nuclides. Only in the case of hydrogen have the nuclides been given different names, because the masses are so different that differences can be detected in physical and chemical properties. The three forms of hydrogen are $^{1}_{1}H$ (protium, 'hydrogen'), $^{2}_{1}H$ (deuterium, heavy hydrogen, symbol D) and $^{3}_{1}H$ (tritium, symbol T). The hydrogen which is normally studied is a mixture of 6000 parts of protium to 1 part of deuterium, the proportion of tritium being negligible (it is radioactive, see below).

Nuclides of Hydrogen	*Nucleus*		*Electrons*
	Protons	*Neutrons*	
Protium	1	0	1
Deuterium	1	1	1
Tritium	1	2	1

The separation of isotopes is usually done by taking advantage of the slight differences in physical properties, e.g. the rate of diffusion of

Ernest Rutherford, 1871–1939, New Zealand (figure 12.1)

A scholarship brought Rutherford to Cambridge in 1895 as a student and then after spells in Canada and in Manchester he returned to Cambridge in 1919 to become professor of experimental physics.

He began investigations on the radioactive substances which Becquerel and the Curies had just discovered. With Soddy he developed the laws of nuclear disintegrations; with Geiger and Marsden he studied the nuclear structure of the atom. He was also associated with Moseley who obtained experimental evidence for atomic numbers, and Chadwick who discovered the neutron.

He published many scientific papers and made such a contribution to our knowledge of the structure of matter that he was buried beside Newton and Kelvin in London.

the gas or of discharge of ions in solution, or of the masses in a mass spectrometer, etc. The operation is quite often performed with a compound rather than the element itself. The differences in chemical properties between isotopes are too slight to be discussed here.

12.3 The Electron Shells

Surrounding the nucleus are sufficient electrons to make the atom electrically neutral. They are arranged in shells in which the electrons have different energies. Proceeding outwards from the nucleus the shells can be numbered 1, 2, 3 . . . or lettered K, L, M . . .

The radius of an atom, which is effectively the distance between the outermost electron and the nucleus, is about 10^{-10} m or 100 pm.

The capacities of the electron shells going outwards from the nucleus are 2, 8, 18 and 32, but a shell is not necessarily filled before the next one is started; the maximum number of electrons that is usually found in the outermost shell of an atom is 8. It will be seen that the family of elements known as the noble gases all have an octet of electrons in the outer shell (except helium where the first shell is full with two electrons). These elements which are unreactive are presumed to be so because of their electronic structures:

	K	L	M	N	O	P
He	2					
Ne	2	8				
Ar	2	8	8			
Kr	2	8	18	8		
Xe	2	8	18	18	8	
Rn	2	8	18	32	18	8

The shells of electrons are often drawn as circles, but this is a matter of diagrammatic convenience; the circles represent different energy levels—the smaller the radius of the circle the greater the energy binding the electron to the nucleus.

The evidence that the electrons in an atom are arranged in shells of different energy levels comes from the study of the light emitted when a hot substance (in a flame or when subjected to electric sparks) cools. The simplest illustration in the laboratory is the flame test (see section 21.10) in which sodium compounds are found to emit an orange-yellow colour. The energy emitted as light is due to an electron moving from a high energy orbit (far away from the nucleus) to a lower energy orbit (closer to the nucleus). By extending the observations beyond the range of the visible spectrum and by studying absorption as well as emission spectra it is possible to build up a complete picture of the energy levels for electrons in a sodium atom. If the energy supplied is great enough the atom ionizes. The first ionization energy of sodium is 515 kJ/mol.

The electronic energy levels of the hydrogen atom have been studied intensively (see figure 12.1) by means of gas discharge tubes; the ionization energy is 1310 kJ/mol.

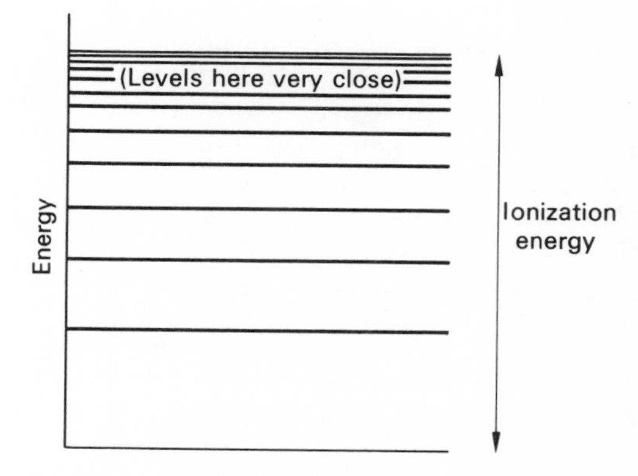

Figure 12.1
The electronic energy levels of the hydrogen atom.

12.4 The Electronic Structures of the Elements

Hydrogen has one electron in the K shell and helium two. These negatively charged particles are balanced by one and two positive particles

	K	L	M	N
H	1			
He	2			
Li	2	1		
Be	2	2		
B	2	3		
C	2	4		
N	2	5		
O	2	6		
F	2	7		
Ne	2	8		
Na	2	8	1	
Mg	2	8	2	
Al	2	8	3	
Si	2	8	4	
P	2	8	5	
S	2	8	6	
Cl	2	8	7	
Ar	2	8	8	
K	2	8	8	1
Ca	2	8	8	2

Figure 12.2
The electronic structures of the twenty lightest elements:

(protons) respectively in the nuclei. The numbers of neutrons associated with these numbers of electrons and protons may vary as has been seen when discussing isotopes. In a chemical reaction it is often only the outermost electrons of an atom that take part; in some cases the electrons in the penultimate (one before the outermost) shell may also be involved.

The K shell is full in the case of helium and so it is found that lithium, the next element, has three electrons arranged K2,L1. A similar situation occurs after neon where the L shell is full and thus sodium has eleven electrons arranged K2,L8,M1. (See figures 12.2 and 12.3.)

The half-life is an important characteristic of a radioelement.

Example: The half-life of tritium, $^{3}_{1}H$, is 12 years; that is to say that if one starts with 8 g of tritium, in 12 years one has 4 g, in 24 years 2 g and in 36 years 1 g. The half-lives of elements vary considerably:

^{238}U	$4{\cdot}47 \times 10^{9}$ years
^{226}Ra	$1{\cdot}60 \times 10^{3}$ years
^{24}Na	14·96 hours
^{110}Ag	24·4 seconds
^{212}Po	3×10^{-7} seconds

	Nature	*Speed*	*Penetrating Power*	*Ionization of Gases*
α	$^{4}_{2}He^{2+}$ (helium nucleus)	0·05 c	Low	High
β	$^{\ 0}_{-1}e$ (electron)	0·3 to 0·9 c	Medium	Low
γ	Electromagnetic waves of wavelength about 10^{-12}m	c	Very high	Fairly low

12.5 Radioactivity

The fact that lead atoms are created from radium and polonium, as mentioned in section 12.1, contradicts one of the postulates of Dalton's Atomic Theory (see section 9.8), but it must be understood that the basic change involved is physical and not chemical. Dalton's Atomic Theory and the Laws of Stoichiometry are the foundation of modern chemistry because in chemical processes no disintegration of the nuclei of the atoms takes place.

The principal radioactive elements are those that have an atomic number greater than eighty three. The phenomenon of radioactivity is a spontaneous disintegration of the nucleus; there is no known method of starting, stopping, accelerating or decelerating it. Although it cannot be said when a particular atom will break up, it can be observed that a definite fraction of the atoms disintegrate in a given time.

The half-life of an isotope is the time taken for half a given mass of the substance to disintegrate.

The types of radiation emitted by nuclei are listed in the table above; the c represents the speed of light.

In radioactive changes new elements are often produced; these changes are irreversible and are independent of the physical and chemical conditions. The release of energy is vast. Einstein predicted that this would be given by $E = mc^2$ and this was confirmed by studying the reaction

$$^{1}_{1}H + ^{7}_{3}Li \rightarrow ^{4}_{2}He + ^{4}_{2}He$$

in which the loss in mass corresponds to an energy release of $1{\cdot}67 \times 10^{12}$ joules. [$E = mc^2$, where c is the speed of light and m is the mass of matter destroyed; i.e. $E = 9 \times 10^{16}$ J/kg.]

Although the changes involved are physical in origin the chemist finds that radioactivity has many uses: the rates of reactions may be investigated; the analysis of substances carried out; the formation of unstable compounds followed (BiH_3); reactions which are not wanted stopped (sterilization of bandages); reactions

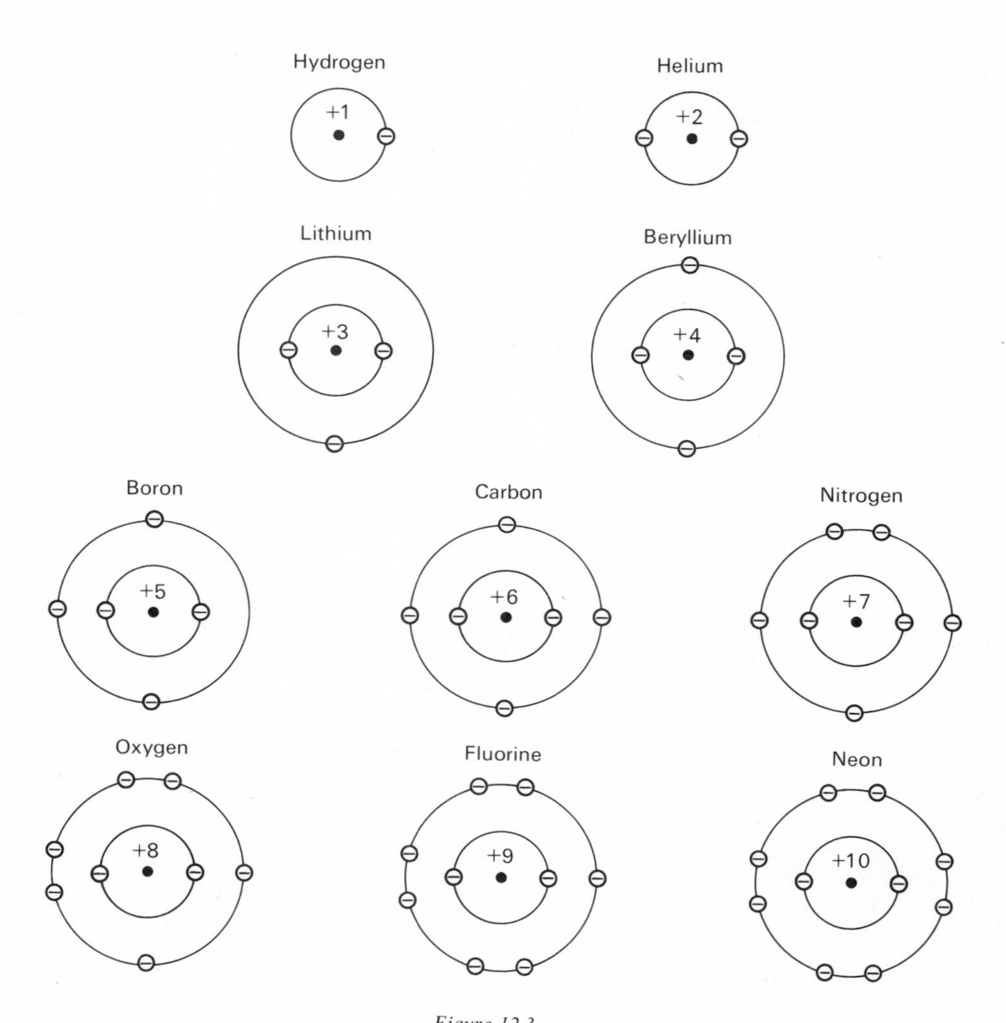

Figure 12.3
The electronic structures of the twenty lightest elements in diagrammatic form. The nucleus is shown as a central point together with its charge (its atomic number), the electrons by ⊖ orbiting the nucleus.

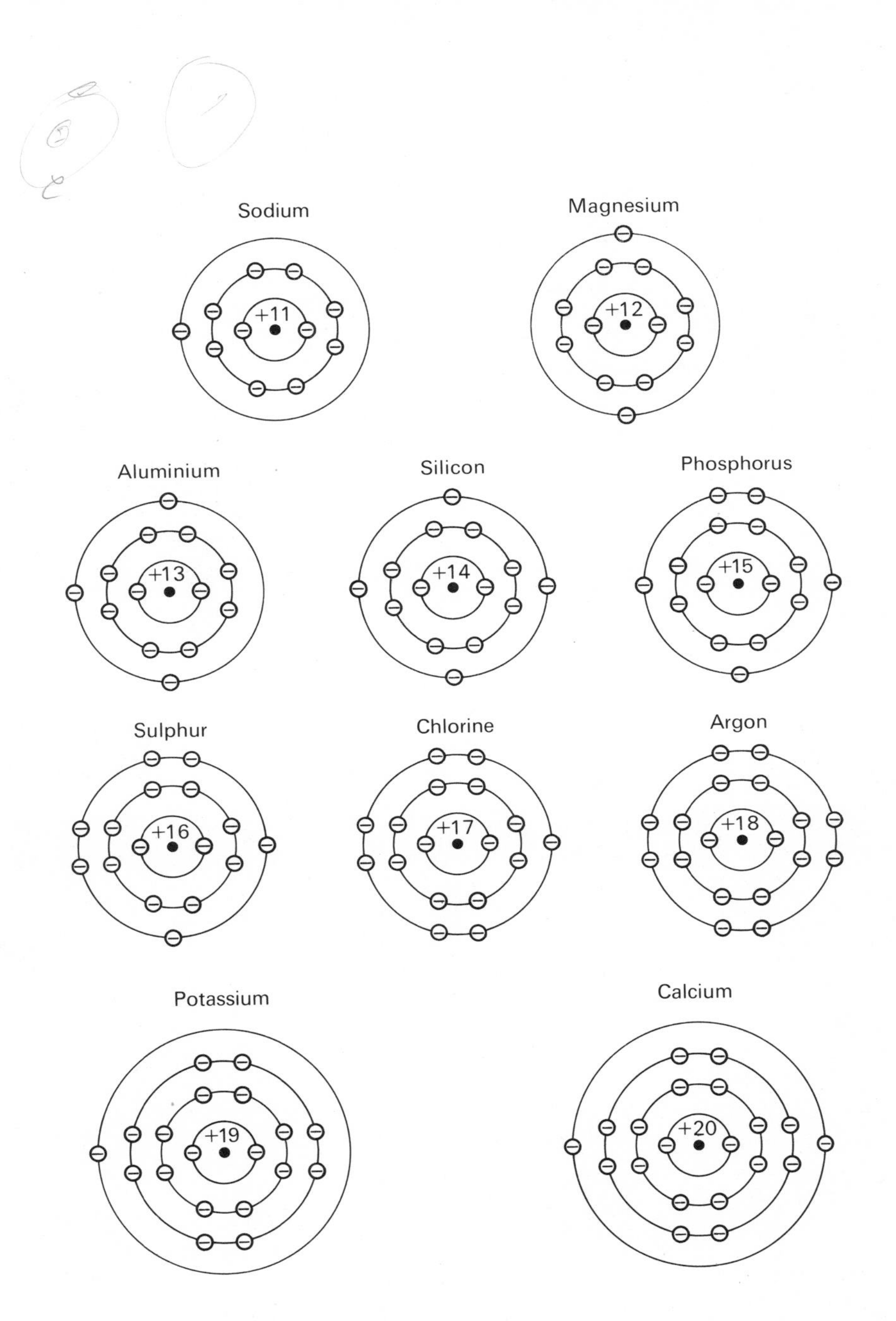

Figure 12.3 (cont.)

Pierre and Marie Curie, 1859–1906, 1867–1934, France

Pierre Curie was educated in Paris and became professor of physics and chemistry in 1895. Marie was educated in Poland and went to Paris for further studies; they married in 1895.

In 1896 Becquerel discovered radioactivity and fired by this discovery the Curies started upon some brilliant and painstaking research culminating in the discovery of two new elements, radium and polonium. The pitchblende they studied showed a radioactivity greater than could be accounted for by its uranium content, but the proportions of radium and its daughter element polonium were very small indeed.

Pierre was killed in 1906 and Marie succeeded him as professor of physics. She was the first woman to hold such a post and the first woman to win a Nobel prize—jointly in 1903 with her husband and Becquerel. She won a second Nobel prize, this time for chemistry, in 1911, an achievement which only Pauling has equalled. The work on radioactive substances eventually took its toll and she died of leukaemia.

which are required started (ethene plus hydrogen bromide).

The proportion of lead in a uranium mineral leads to the conclusion that the changes started about $4{\cdot}5 \times 10^9$ years ago, a figure that may represent the age of the earth as a planet. The proportion of ^{14}C in an organic object tells us how long ago the object, e.g. wood or paper, was part of a living system. The heat given out by the fission of uranium atoms is taken up by carbon dioxide in the U.K. 'magnox' reactors. The hot carbon dioxide converts the water to steam and hence the turbines of the generators are driven.

The absorption or scattering of radiation makes radioisotopes useful for thickness gauges in metal works and filling checkers for materials in containers. The emission of radiation makes radioisotopes useful in medicine, e.g. the progress of iodine-containing molecules to the thyroid gland in the neck; and in agriculture, e.g. the uptake and progress of fertilizers in plants. The penetrating power of γ-rays has led to their use in cancer treatment (^{60}Co) and in mobile units for checking the efficiency of welding (when the natural gas grid was laid in the U.K.).

QUESTIONS 12

Atomic Structure and Radioactivity

1 Complete the following table:

Particle	*Mass number*	*Atomic number*	*Protons*	*Neutrons*	*Electrons*
Oxygen atom	16	8			
Calcium ion		20		20	
Bromide ion	80		35		

2 How would you explain the phenomenon of radioactivity to a non-scientist? Give a brief description, suitable for such a person, of the main features of radioactivity (not the history of its discovery). In addition, describe an experiment by which you could demonstrate the usefulness of radioactivity for solving a chemical problem.

3 Give a brief account of the structure of the atom.
Show, by means of diagrams, the electronic structure of any three elements you select. [E]

4 Atom A has a mass number of 239 and atomic number 93. Atom B has a mass number of 239 and atomic number 94.
(*a*) How many protons has atom A?
(*b*) How many neutrons has atom B?
(*c*) Are atoms A and B isotopes of the same element? Explain.
(*d*) Explain whether a geologist would or would not be likely to locate compounds containing these atoms in the course of his work. [Sc]

5 In the early years of the last century it was believed:
(*a*) that all the atoms of an element were identical;
(*b*) that all relative atomic masses were integral (whole numbers);
(*c*) that the atom was indivisible.
To what extent have our modern ideas of the atom changed these beliefs? [NI]

6 What do you understand by the following: (*a*) an electron, (*b*) a proton, (*c*) a neutron, (*d*) the atomic number of an element?

7 Draw diagrams showing the electronic structure of the following elements: hydrogen, helium, carbon, nitrogen, oxygen, neon, sodium, magnesium, sulphur, argon and calcium.

8 What do you understand by the term *isotopes*? Illustrate your answer by diagrams showing the isotopes of: (*a*) hydrogen, (*b*) chlorine.

13 The Periodic Table

13.1 Introduction

An element is a substance which cannot be split up into two or more substances by chemical means. Today an element could be defined as a class of atoms all having the same atomic number.

At the time of Dalton (1808) only about a quarter of the elements now known had been discovered; at the end of the century about three quarters of the elements had been isolated. Today we possess the knowledge and the power to synthesize new elements, although their stability is such that they are not always obtained in large amounts.

Lavoisier (1790) and Berzelius (1820) attempted classifications which were essentially divisions into the permanent gases, metals, non-metals and substances suspected of being metal oxides. Elements such as arsenic and germanium which have metallic and non-metallic properties, present considerable difficulties in this respect: they are known as semi-metals (see section 15.2).

Whatever property is used in the classification must be independent of physical conditions, e.g. temperature or pressure. It would be useless to attempt to classify the elements, say by colour, as the colour varies with change of temperature. The relative atomic mass of an element is a constant, not dependent upon temperature, and was accepted originally as the basis of the classification of the elements.

13.2 Dalton and Döbereiner

Dalton's Atomic Theory expressed current opinion regarding the nature of atoms, and his insistence upon the importance of relative atomic masses gave a useful guide for classifications.

Döbereiner (1829) put forward his law of triads—'elements of similar character often possess relative atomic masses which are in arithmetical progression':

	Ca	Sr	Ba	40	88	137
and	Cl	Br	I	35	80	127

Later a fourth element was added to many triads.

13.3 Newlands' Law of Octaves

Newlands (1863) arranged the then known elements in order of increasing relative atomic mass. He noticed that the 'eighth one starting from a given one is a kind of repetition of the first'.

Element	Li	Be	B	C	N	O	F
Relative atomic mass	7	9	11	12	14	16	19

As one goes from lithium to fluorine the properties change:

lithium is a very reactive metal whose hydroxide is very basic,
beryllium is metallic but less so than lithium,
boron is a non-metal,
. . . and so on to
fluorine which is a typical non-metal.

Then the order of relative atomic masses continues:

Na	Mg	Al	Si	P	S	Cl
23	24	27	28	31	32	35

There is a drastic change in properties from the intensely non-metallic fluorine to the intensely metallic sodium.

But he knew that:
sodium resembles lithium,
magnesium resembles beryllium,
. . . and so on to
chlorine which resembles fluorine.

Newlands unfortunately left no gaps in the later part of his table for elements not yet discovered. Thus his scheme, though basically right (until the noble gases were discovered), was ridiculed.

13.4 Lothar Meyer

Lothar Meyer (1869) realized that a graphical approach to the problem of classification might overcome Newlands' difficulty, and he plotted the atomic volume (molar mass of the atoms divided by the density) against the relative atomic mass. The curve he obtained was periodic: similar elements occur in similar positions on the curve, a completed form of which is shown in figure 13.1. His approach to the classification was not very well known at that

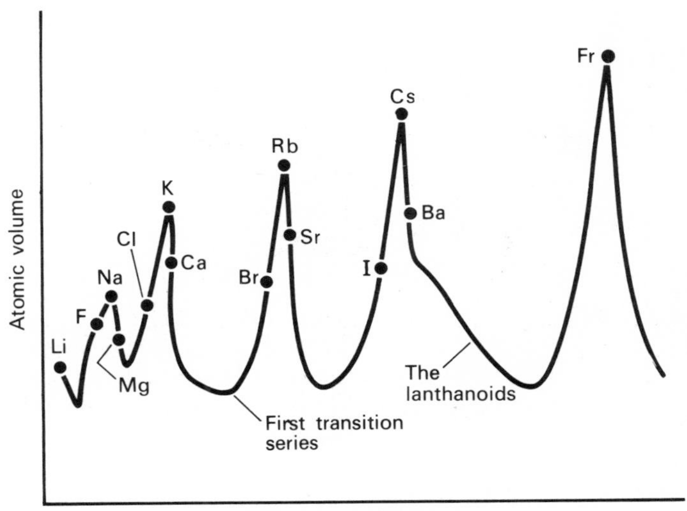

Figure 13.1
Lothar Meyer's Graph.

time and a more widely based approach was published by Mendeléeff.

13.5 Mendeléeff

Mendeléeff, in the same year as Lothar Meyer, but working independently, also realized the pitfalls of undiscovered elements and accordingly left gaps. He saw that similar elements were occurring at regular intervals, and he stated this in his law as follows:

"The properties of the elements (and of their corresponding compounds) are periodic functions of their relative atomic (and molecular) masses."

Not only did Mendeléeff leave gaps for elements that had not been discovered, but he even predicted their properties with astonishing accuracy. Across the table (e.g. sodium to chlorine) the properties change from metallic to non-metallic, whereas down the table (e.g. carbon to lead) the elements become more

Group 0	*Group 1*	*Group 2*	*Group 3*	*Group 4*	*Group 5*	*Group 6*	*Group 7*	*Group 8*
	E_2O EH	EO EH_2	E_2O_3 EH_3	EO_2 EH_4	E_2O_5 EH_3	EO_3 EH_2	E_2O_7 EH	EO_4 —
He	H Li	Be	B	C	N	O	F	
Ne	Na	Mg	Al	Si	P	S	Cl	
Ar	K Cu	Ca Zn	Sc Ga	Ti Ge	V As	Cr Se	Mn Br	Fe Co Ni
Kr	Rb Ag	Sr Cd	Y In	Zr Sn	Nb Sb	Mo Te	Tc I	Ru Rh Pd
Xe	Cs Au	Ba Hg	La* Tl	Hf Pb	Ta Bi	W Po	Re At	Os Ir Pt
Rn	Fr	Ra	Ac**					

*Lanthanoids or rare earths.
**Actinoids.

Figure 13.2
An early form of the periodic table

Dmitri Mendeléeff, 1834–1907, Russia

Early in his life he worked with Bunsen and met Cannizzaro. In 1866 he became professor of chemistry at Leningrad and three years later in his book 'The Principles of Chemistry' he proposed the Periodic Law. Others such as Newlands (England 1863) and Lothar Meyer (Germany 1869) made important contributions in this field, but as Lothar Meyer himself admitted 'the coping stone of the building which in the course of years was erected on the foundation of Döbereiner's Triads (1829) was, however, the work of Mendeléeff'. The most powerful application of the Periodic Law was in the prediction of the properties of several elements then unknown (gallium, scandium, germanium, rhenium) which when they were discovered proved to have the properties described up to twenty years previously.

metallic in properties. Thus, he predicted that 'Eka-Boron' (Sanskrit: one like boron) would be a metal, forming a sulphate $E_2(SO_4)_3$, and would have an oxide that would dissolve in acids but not in alkalis. He was correct (gallium was discovered in 1875).

The elements iron, cobalt, nickel; ruthenium, rhodium, palladium; osmium, iridium, platinum were placed in a special group (8), because they did not fit into the table (see figure 13.2). For example, after manganese (relative atomic mass 55) comes iron ($A_r = 56$), but in no way does it resemble the monovalent, highly reactive elements such as sodium or potassium in group I. The position was clarified when the structure of the atoms became known. He called these group VIII elements transition elements.

In the early groups it was noticed that the valency of the element was equal to the group number; in the later groups there are two principal valencies: the group number and eight minus the group number.

At the top of each group in his table Mendeléeff put the formula of the typical oxide and hydride formed by the elements concerned: the symbol E is used to stand for one atom of such an element.

13.6 The Noble Gases

Towards the end of the last century, whilst Rayleigh was redetermining the relative molecular masses of the gaseous elements by careful measurement of their relative densities, he discovered that the density of the nitrogen he obtained from the atmosphere was about 0·5% greater than that of the nitrogen he obtained from chemical compounds (see sections 5.12 and 30.5). This led Rayleigh and Ramsay to investigate atmospheric nitrogen and search for some unknown gas which must be present in air. Their work led to the discovery of not one gas but eventually to six gases: helium, neon, argon, krypton, xenon and radon. These were known as the inert or rare gases because of their lack of chemical activity and their comparative rarity in the atmosphere.

With the discovery of these inert gases, it was feared at first that they would not fit into the table, but it was found that they had relative atomic masses between hydrogen and lithium, fluorine and sodium, etc., and a group O was introduced without upsetting the remainder of the table.

In 1962 it was discovered that some of the inert gases would combine relatively easily with fluorine at 400°C in a nickel vessel, and so nowadays they are usually called the noble gases (by analogy with the noble metals).

13.7 The Modern Periodic Table

The atomic number is now known to be a more important property of an element than the relative atomic mass because the existence of isotopes affects the latter but not the former.

Hence four guide lines to the compilation of the table are to:

(*a*) put the elements in order of increasing atomic number,

(*b*) start a new line every time a new outer shell of electrons is started,

(*c*) put the elements into columns (groups or families) according to the number of electrons in the outermost shell,

and (*d*) for the transition elements (a term used more widely than Mendeléeff used it) put the elements in columns according to the number of electrons in the penultimate shell.

To overcome various objections to the eight column table, the periodic table usually used today is an eighteen column one (see figure 13.3).

Hydrogen is difficult to classify: it shows several similarities in chemical properties to sodium and several similarities in physical and chemical properties to chlorine.

A **group** is a vertical division of the periodic table containing elements of closely similar physical and chemical properties, e.g. the halogens. These properties are largely dependent upon the number of electrons in the outer shell of the atom. The forms in which these elements occur and the means of extracting the elements from their compounds are also similar. Proceeding down a group there is an increase in the tendency:

(*a*) to form electrovalent compounds containing positive ions,

(*b*) to show metallic character,

(*c*) to be reducing agents,

(*d*) to form basic oxides and hydroxides.

Therefore in the ensuing chapters the periodic table will frequently be drawn thus:

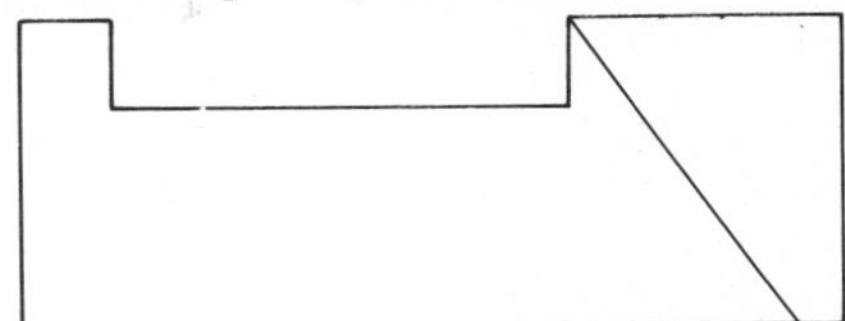

and discussion based upon it in this form (see sections 14.5, 15.2, 16.5, 17.17). For illustration and amplification of the properties of elements in a group see the introductions to chapters 24–28 and 30–32.

A **period or series** is a horizontal division of the periodic table. In a short period (1–3) the increase in electron content occurs solely in the outermost shell, while in a long period this increase also occurs in the penultimate shell. The change in physical and chemical properties is from a metal at the left hand side to a non-metal at the right (see section 15.2).

Group / Period	I	II											III	IV	V	VI	VII	0
1								1 H	2 He									
2	3 Li	4 Be											5 B	6 C	7 N	8 O	9 F	10 Ne
3	11 Na	12 Mg											13 Al	14 Si	15 P	16 S	17 Cl	18 Ar
4	19 K	20 Ca	21 Sc	22 Ti	23 V	24 Cr	25 Mn	26 Fe	27 Co	28 Ni	29 Cu	30 Zn	31 Ga	32 Ge	33 As	34 Se	35 Br	36 Kr
5	37 Rb	38 Sr	39 Y	40 Zr	41 Nb	42 Mo	43 Tc	44 Ru	45 Rh	46 Pd	47 Ag	48 Cd	49 In	50 Sn	51 Sb	52 Te	53 I	54 Xe
6	55 Cs	56 Ba	57 La*	72 Hf	73 Ta	74 W	75 Re	76 Os	77 Ir	78 Pt	79 Au	80 Hg	81 Tl	82 Pb	83 Bi	84 Po	85 At	86 Rn
7	87 Fr	88 Ra	89 Ac**	104 Rf?	105 Ha?	106 ?						112 ?						
			*	58 Ce	59 Pr	60 Nd	61 Pm	62 Sm	63 Eu	64 Gd	65 Tb	66 Dy	67 Ho	68 Er	69 Tm	70 Yb	71 Lu	
			**	90 Th	91 Pa	92 U	93 Np	94 Pu	95 Am	96 Cm	97 Bk	98 Cf	99 Es	100 Fm	101 Md	102 No	103 Lr	

Figure 13.3
A modern form of the Periodic Table.

The **transition elements (metals)** (see chapter 26) are members of long periods. The term covers the ten elements from scandium to zinc and those beneath them in the periodic table. The physical and chemical properties of the transition elements are more similar than those of successive members in a short period but not as similar as the members in a group. The increase in electron content between successive members occurs in the penultimate shell.

The elements immediately following lanthanum (57) and actinium (89) are often called the inner transition elements: the increase in electron content between successive members occurs in the antepenultimate shell (the two outermost shells are constant) and the members are therefore very similar to one another in physical and chemical properties.

The reactivity of metals, e.g. group I, the alkali metals, increases as the group is descended; the reactivity of non-metals, e.g. group VII, the halogens, decreases as the group is descended. As a period is traversed, e.g. sodium to chlorine, the reactivity falls and then rises again.

Lothar Meyer's curve (figure 13.1) could be replotted in modern terms (atomic volume against atomic number) but this would not be any more enlightening.

QUESTIONS 13

The Periodic Table

1 One row of the periodic table is given below. The numbers given along the lower line are the atomic numbers of the elements.

I	II	III	IV	V	VI	VII	O
11	$_{12}Mg$	13	14	$_{15}P$	$_{16}S$	$_{17}Cl$	18

Write the formula of one oxide of each of the following (*a*) magnesium, (*b*) phosphorus, (*c*) sulphur, (*d*) chlorine.

A gaseous element, *X*, is very unreactive and has a relative atomic mass of 39·9. Which position in the above row of the periodic table is appropriate for the element *X*? [J]

2 In this question you should use a copy of the periodic table (figure 13.4).

(*a*) Refer to the set of eight elements starting with titanium (22) and ending with copper (29). What general name can be given to the set of which these elements are members?

(*b*) Refer to the elements rubidium (37) and caesium (55). What general name can be given to the group of which these two elements are members?

(*c*) Give three examples of how the two sets of elements, given in answer to (*a*) and (*b*), differ from each other.

(*d*) State (i) the formula and (ii) the colour you would expect rubidium chloride to have.

(*e*) Describe briefly what you would expect to observe if a small piece of rubidium were dropped into water.

(*f*) Of the following elements some are metallic, some are non-metallic, and one could be classified as intermediate between metallic and non-metallic: caesium (55), phosphorus (15), fluorine (9), germanium (32), barium (56), cobalt (27). Classify these elements into these three categories. [Nf]

3 In answering this question you will need to refer to the information below:

The letters in the left hand column refer to elements, but they are not the symbols for the elements. In your answers you should use these letters.

(*a*) Two of the letters represent two allotropic forms of the same element. Which two letters are they?

(*b*) Which three of the elements could be metals?

(*c*) Which one of the elements could be a noble gas?

(*d*) Which one of the elements could be hydrogen?

(*e*) What is the volume of 1 mole of element E?

(*f*) 'Diamond and graphite both consist of a giant structure of carbon atoms.' Explain briefly (i) what is meant by a giant structure of atoms and (ii) in what respects the structures of diamond and graphite differ from each other. [Nf]

Element	*Melting point (°C)*	*Boiling point (°C)*	*Relative atomic mass*	*Density at 0°C (kg/m^3)*	*Electrical conductivity*	*Action with oxygen*
A	63	760	39	860	Good	Burns
B	98	880	23	970	Good	Burns
C	114	445	32	2070	Poor	Burns
D	119	445	32	1960	Poor	Burns
E	−39	357	201	14200	Good	Very slowly oxidized
F	−249	−246	20	0·9	Poor	None
G	−259	−253	1	0·09	Poor	Burns

4 Mendeléeff stated that 'some properties of the elements are a periodic function of their relative atomic masses'.

Describe some periodic functions of the elements, choosing as far as you are able both from 'chemical' properties and 'physical' ones. What is the connection between these properties and the structure of the atoms? [Nf]

5 Imagine that a new element named *trentium* has been recently discovered. It is a solid but is easily cut with a knife. When added to water it reacts vigorously and takes fire, forming a caustic, alkaline solution.

Answer the following questions concerning trentium.
(*a*) Name an element similar to it.
(*b*) How would you store it?
(*c*) Where would you place it in the electrochemical series?
(*d*) Is it a metal or a non-metal? Give your reason.
(*e*) What will be its valency?
(*f*) Assign to it a suitable chemical symbol. (Use this symbol where necessary in subsequent answers.)
(*g*) Write formulae for its chloride, sulphate and carbonate.
(*h*) Write an equation for its reaction with water.
(*i*) What do you consider will be the products of decomposition by heat of its nitrate? Give an equation.
(*j*) How many electrons will its atom have in its outer shell?
(*k*) When it combines with other elements what type of valency do you think it will have?
(*l*) From what compound and by what method do you consider it might have been isolated? [L]

6 Name two types of particles thought to be present in most atoms and indicate their relative masses and electrical charges (if any).

Describe briefly the structure of the atom of one named element other than hydrogen or helium.

Why do the properties of (*a*) chlorine and bromine, (*b*) sodium and potassium, closely resemble one another?

Show why the calcium ion has an electrovalency of two and the carbon atom a covalency of four.

7 These questions refer to the following periodic table of the first fifty-four elements. Some of the elements are shown by letters. (The letters are not the symbols of the elements.)
(*a*) Which of the lettered elements is a noble (inert) gas?
(*b*) State two of the lettered elements which are in the same group.
(*c*) Give the name of each halogen element together with the number of its position in the Table.
(*d*) Which one of the lettered elements would you expect to react most violently with chlorine?
(*e*) Which one of the lettered elements would you expect to form a compound with the composition 1 mole of the element: 3 moles of chlorine?
(*f*) What would you expect the formula of a compound of hydrogen with the element G to be? (Use G as the symbol of the element.)
(*g*) State briefly some of the properties and the structure which you would expect the oxide of the element F to have. [Nf]

						1	2										
3 A	4											5	6 E	7	8	9 H	10
11	12											13 D	14	15	16	17	18 I
19	20	21	22	23	24	25	26	27	28	29 C	30	31	32 F	33	34 G	35	36
37 B	38	39	40	41	42	43	44	45	46	47	48	49	50	51	52	53	54

14 The Electronic Theory of Valency

14.1 The Principle

The electronic theory of valency is based on the principle that when elements react they usually attain the electronic structure of a noble gas, i.e. their nuclei become surrounded by a stable octet of electrons. In the cases of hydrogen, lithium and beryllium, however, the nuclei may become surrounded by a pair. of electrons (structure of helium). These states may be achieved in two important ways:

(*a*) *Electrovalency*—a transfer of one or more electrons takes place. This was first described by Kossel (1916).

(*b*) *Covalency*—a sharing of electrons usually in pairs occurs. This was put forward by Lewis and Langmuir, also in 1916.

The number of electrons in the outermost shell of an atom often gives its maximum valency.

14.2 Electrovalency

Kossel realized that the noble gases had stable electronic configurations and that elements adjacent to them in atomic number would pass into this state either by loss or by gain of electrons. He explained the formation of sodium chloride as shown in figure 14.1.

The transfer of one electron results in charged particles or ions which are held together by electrostatic forces. The elements attain the electronic structures of their adjacent noble gases but the nuclei are not changed at all; the resemblances between the neon atom and the sodium ion are only in the number and position of the electrons.

A sodium chloride crystal when studied by X-rays is found to be cubic; the structure of the crystal of an electrovalent substance depends on the ratio of the radii of the ions and the size of their charges (see figure 14.2).

● represents one type of ion and o the other—the 'molecule' in this arrangement of ions continues to the boundaries of the crystal (the ● and the o could be interchanged on this diagram and it would still be a typical unit of sodium chloride). The crystal is face centred cubic with respect to each ion. No one sodium

Gilbert Lewis, 1875–1946, U.S.A.

Lewis studied in America and then in Germany before going to the Massachusetts Institute of Technology in 1905. In 1912 he became professor of physical chemistry at Berkeley.

In an article published in 1916 he suggested that electrons would have to be localized in certain directions in order to account for the geometry of molecules: he emphasized the importance in compound formation of the electron pair. In the same article he suggested that chemists should classify substances according to the type of valency exhibited rather than as inorganic or organic.

He proposed a general definition of acids and bases as electron pair acceptors and donors respectively—usually discussed in sixth form chemistry. He was the first to prepare and study pure heavy water, ${}^2_1H_2O$.

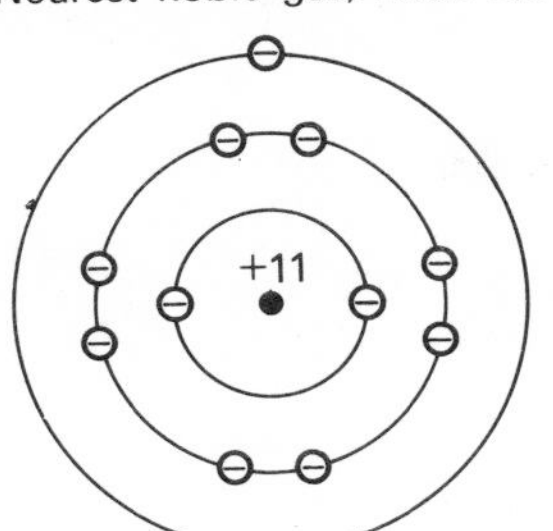

Both atoms electrically neutral

Chlorine atom (2,8,7)
Nearest noble gas, argon 2,8,8

Sodium atom loses one electron to the chlorine atom

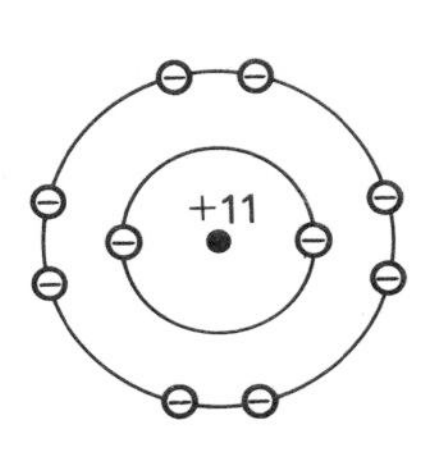

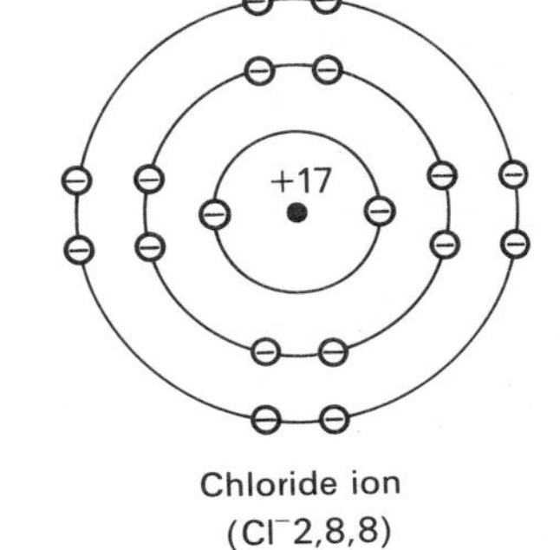

Loss of one electron (negative) gives a net positive charge of one. The sodium atom becomes a sodium *ion*.

Gain of one electron (negative) gives a net negative charge of one. The chlorine atom becomes a chloride *ion*.

This can be written as an equation

$$2Na + Cl_2 \rightarrow 2Na^+Cl^-$$

or in numerical terms

$$\begin{array}{llll} \text{Na } 2, 8, 1 & \xrightarrow{\text{transfer}} & Na^+ \ 2, 8 & \text{(like Ne)} \\ \text{Cl } 2, 8, 7 & 1\,e^- & Cl^- \ 2, 8, 8 & \text{(like Ar)} \end{array}$$

Figure 14.1
The synthesis of sodium chloride

is united to a particular chloride ion, but each sodium ion has six chloride ions near it and each chloride in turn has six sodium ions around it, thus the ionic ratio is 1:1. The formula quoted for an ionic substance is, strictly speaking, an empirical formula.

In the solid the ions vibrate. As the temperature rises this vibration increases until the electrostatic forces can no longer maintain the shape of the crystal and it is seen to melt. These electrostatic forces are so strong that the melting point of sodium chloride is high (800°C); electrovalent substances are usually solids at room temperature. In the liquid translational motion is also possible and there may be some rotational motion. A few 'ion-pairs' may escape and so the solid and the liquid have a vapour pressure. This pressure increases slowly as the temperature rises and at 1470°C the liquid boils. Dissolving sodium chloride in water also gives the ions translational freedom; the liquid and the solution conduct electricity (they are electrolytes, see section 20.7).

Other examples of electrovalency are the formation of the compounds shown in figure 14.3.

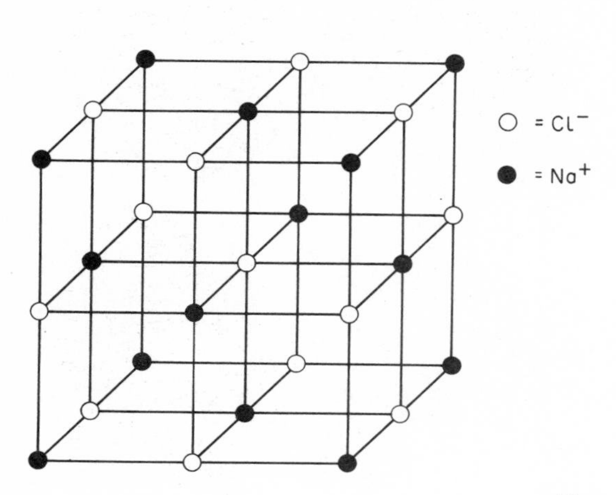

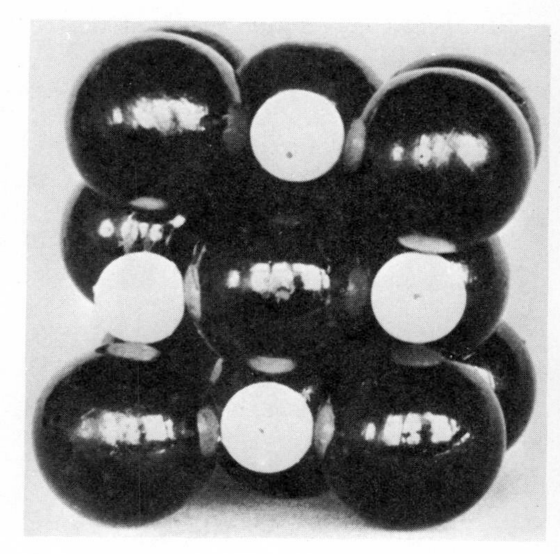

Figure 14.2
The crystal structure of sodium chloride

Example 1: Calcium Chloride

Ca	2, 8, 8, 2	transfer 2e⁻	Ca^{2+}	2, 8, 8	(like Ar)
2Cl	2, 8, 7		$2Cl^-$	2, 8, 8	(like Ar)

i.e. $Ca(s) + Cl_2(g) \longrightarrow Ca^{2+}(Cl^-)_2(s)$

Example 2: Calcium Oxide

Ca	2, 8, 8, 2	transfer 2e⁻	Ca^{2+}	2, 8, 8	(like Ar)
O	2, 6		O^{2-}	2, 8	(like Ne)

$Ca(s) + \frac{1}{2}O_2(g) \longrightarrow Ca^{2+}O^{2-}(s)$

Example 3: Sodium Oxide

2Na	2, 8, 1	transfer 2e⁻	$2Na^+$	2, 8	(like Ne)
O	2, 6		O^{2-}	2, 8	(like Ne)

$2Na(s) + \frac{1}{2}O_2(g) \longrightarrow (Na^+)_2O^{2-}(s)$

Example 4: Sodium Peroxide

2Na	2, 8, 1	transfer 2e⁻	$2Na^+$	2, 8	(like Ne)
2O	2, 6		O_2^{2-}	2, 8	(like Ne)

$2Na(s) + O_2(g) \longrightarrow (Na^+)_2O_2^{2-}(s)$

The two oxygen atoms of the peroxide ion (O_2^{2-}) are linked covalently (see section 14.3).

Example 5: Copper(II) Oxide

Cu	2, 8, 18, 1	transfer 2e⁻	Cu^{2+}	2, 8, 17	(not like any noble gas)
O	2, 6		O^{2-}	2, 8	(like Ne)

$Cu(s) + \frac{1}{2}O_2(g) \longrightarrow Cu^{2+}O^{2-}(s)$

In the case of transition elements, a noble gas electronic structure is not usually attained.

Example 6: Potassium Sulphide

2K	2, 8, 8, 1	transfer 2e⁻	$2K^+$	2, 8, 8	(like Ar)
S	2, 8, 6		S^{2-}	2, 8, 8	(like Ar)

$2K(s) + S(s) \longrightarrow (K^+)_2S^{2-}(s)$

Example 7: Aluminium Chloride

Al	2, 8, 3	transfer 3e⁻	Al^{3+}	2, 8	(like Ne)
3Cl	2, 8, 7		$3Cl^-$	2, 8, 8	(like Ar)

$Al(s) + 1\frac{1}{2}Cl_2(g) \longrightarrow Al^{3+}(Cl^-)_3(s)$

The ionic nature of this compound is only apparent at room temperature: on heating the substance becomes covalent. In general as one proceeds from left to right in the periodic table the compounds of the metals with a particular non-metal change from electrovalent to covalent: compare sodium, magnesium, aluminium and silicon chlorides.

Example 8: Caesium Chloride

Cs	2, 8, 18, 18, 8, 1	transfer 1e⁻	Cs^+	2, 8, 18, 18, 8	(like Xe)
Cl	2, 8, 7		Cl^-	2, 8, 8	(like Ar)

$Cs(s) + \frac{1}{2}Cl_2(g) \longrightarrow Cs^+Cl^-$

See figure 14.4 for a model of this structure.

Figure 14.3
Electrovalent compounds

No atom can lose or gain more than three electrons usually; not all compounds are formed by electrovalency. This theory, for example, is not able to explain the formation of carbon dioxide (CO_2), where both carbon (2,4) and oxygen (2,6) require electrons to make up their stable octets. Nor can it explain the formation of molecules such as chlorine (Cl_2) or oxygen (O_2).

14.3 Covalency

In covalency, molecules are formed by the combination of atoms which share electrons. There are no electrons gained or lost: there are no ions formed. Each atom, by sharing, usually attains the electronic arrangement of a noble gas. Each atom contributes one electron to the shared pair, thus setting up what is known as a covalent link. Chemical activity is once again due to the electrons in the outermost shell.

In these examples of covalent molecules the diagrams are restricted to showing only the outermost shell of electrons of the atoms concerned, i.e. those electrons most likely to be involved in the reaction. Although the electrons from one atom are shown by x and from the other by o the electrons cannot be distinguished from one another in practice. The symbol is used to represent the nucleus and the inner electrons of the atom.

Example 1. Chlorine molecule, Cl_2 (chlorine 2,8,7)

The nearest noble gas, argon, has the electronic arrangement 2,8,8. In the molecule one electron from the outer shell of each chlorine atom is shared by the outer shell of the other chlorine atom, giving each an electronic structure of 2,8,8.

$$\overset{\times\times}{\underset{\times\times}{{}_{\times}^{\times}\mathrm{Cl}^{\times}}} + \overset{\circ\circ}{\underset{\circ\circ}{{}_{\circ}\mathrm{Cl}_{\circ}^{\circ}}} \xrightarrow[2e^-]{\text{Share}} \overset{\times\times\ \circ\circ}{\underset{\times\times\ \circ\circ}{{}_{\times}^{\times}\mathrm{Cl}\,{}_{\circ}^{\times}\,\mathrm{Cl}_{\circ}^{\circ}}} \qquad \text{or Cl—Cl}$$

Atoms — Molecule

Each pair shared is represented by one covalent bond, shown as Cl–Cl. Hydrogen chloride is formed similarly: H–Cl.

Example 2. Oxygen molecule, O_2, (oxygen 2,6). Each oxygen atom is short of two electrons compared with the nearest noble gas, neon, therefore each atom in the molecule has to share two electrons, giving rise to two covalent bonds (a double bond) written as O=O (see section 31.6).

$$\overset{\circ\circ}{\underset{\circ\circ}{\mathrm{O}_{\circ}^{\circ}}} + \overset{\times\times}{\underset{\times\times}{{}_{\times}^{\times}\mathrm{O}}} \xrightarrow[4e^-]{\text{Share}} \overset{\circ\circ\circ\times\times}{\underset{\circ\circ\times\times\times}{\mathrm{O}\,{}_{\circ\,\circ}^{\times\,\times}\,\mathrm{O}}} \qquad \text{or O=O}$$

Figure 14.4
The structure of a caesium chloride crystal (simple cubic with respect to each ion)

Example 3. Carbon tetrachloride molecule, CCl_4, (carbon 2,4; chlorine 2,8,7)

$$4\ \overset{\times\times}{\underset{\times}{{}_{\times}^{\times}\mathrm{Cl}_{\times}^{\times}}} + \overset{\circ\ \circ}{\underset{\circ\ \circ}{\mathrm{C}}} \xrightarrow[\text{4 bonds}]{\text{Form}} \mathrm{CCl_4}\ (\text{C surrounded by four Cl, each sharing one }\times\circ\text{ pair})$$

Represented as:

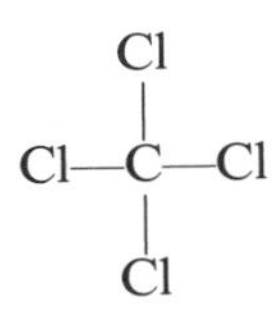

The carbon atom now has the electronic structure of neon and each of the chlorine atoms that of argon. Methane is similar (see figure 14.5). These molecules are tetrahedral.

Example 4. Carbon dioxide molecule, CO_2, (carbon 2,4; oxygen 2,6).

$$2\,\overset{\times\times}{\underset{\times\times}{\mathrm{O}_{\times}^{\times}}} + \overset{\circ}{\underset{\circ}{{}_{\circ}\mathrm{C}_{\circ}}} \xrightarrow[\text{double bonds}]{\text{Forms 2}} \overset{\times\times\ \times\ \ \times\ \times\ \times}{\underset{\times\ \ \ \times}{{}_{\times}^{\times}\mathrm{O}\ {}_{\circ}^{\circ}\mathrm{C}_{\circ}^{\circ}\ \mathrm{O}_{\times}^{\times}}}$$

It can be represented as O=C=O, and is a linear molecule. The carbon and the oxygen atoms now each have the electronic structure of neon.

Example 5. Hydrogen molecule, H_2, (hydrogen 1).

$$\mathrm{H}^{\times} + \mathrm{H}\circ \xrightarrow[2e^-]{\text{Share}} \mathrm{H}_{\circ}^{\times}\mathrm{H}$$

Represented as H—H. The hydrogen atoms now have the electronic structure of helium.

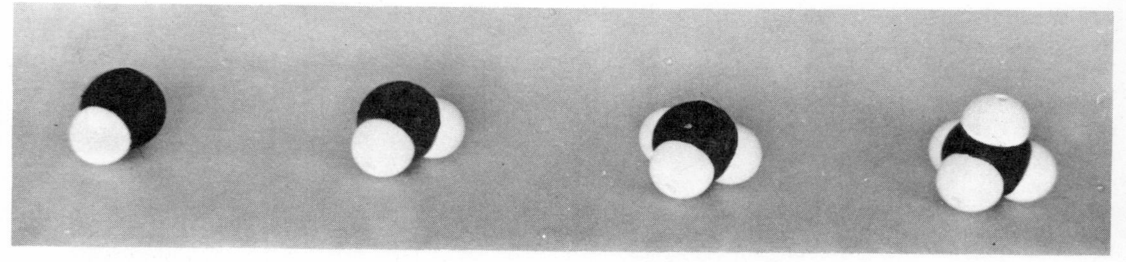

Figure 14.5 Models showing the structure of: Hydrogen chloride, Water, Ammonia, Methane

Example 6. Ammonia molecule, NH_3, (hydrogen 1; nitrogen 2,5).

$$3H^{\times} + \overset{\circ\circ}{N}{}^{\circ}_{\circ\,\circ} \xrightarrow[\text{3 bonds}]{\text{Forms}} H\overset{\times\circ}{\underset{\circ\times}{N}}H \quad (\text{with a third } H \text{ below } N)$$

Represented as:

$$\begin{array}{c} H \\ | \\ N\text{—}H \\ | \\ H \end{array}$$

The hydrogen atoms now have the electronic structure of helium and the nitrogen that of neon. The molecule is pyramidal. See figure 14.5.

Example 7. Nitrogen, (nitrogen 2,5).

$$\overset{\circ\circ}{N}{}^{\circ}_{\circ\,\circ} + \overset{\times\times}{N}{}^{\times}_{\times\,\times} \xrightarrow[\text{triple bond}]{\text{Forms a}} {}^{\circ}_{\circ}N \begin{smallmatrix}\times\\\circ\\\times\\\circ\\\times\\\circ\end{smallmatrix} N^{\times}_{\times}$$

The nitrogen atoms now have the electronic structure of neon.

Example 8. Water (hydrogen 1, oxygen 2,6).

$$2H^{\times} + \overset{\circ\circ}{\underset{\circ\circ}{O}}{}^{\circ}_{\circ} \xrightarrow[\text{2 bonds}]{\text{Forms}} H\overset{\times\circ}{\underset{\circ\times}{O}}{}^{\circ}_{\circ} \quad (\text{with a second } H \text{ below } O)$$

Each hydrogen atom now has the electronic structure of helium and the oxygen atom that of neon. The molecule is bent, not linear. Hydrogen sulphide is similar.

Example 9. Ethane (hydrogen 1, carbon 2,4).

$$6H^{\times} + 2\,{}^{\circ}\overset{\circ}{\underset{\circ}{C}}{}^{\circ} \xrightarrow[\text{7 bonds}]{\text{Forms}} \begin{array}{c} H\ \ H \\ H\overset{\times\circ}{\underset{\times\circ}{C}}\overset{\times\circ}{\underset{\circ\times}{C}}H \\ H\ \ H \end{array}$$

Each hydrogen atom now has the electronic structure of helium and each carbon atom that of neon. The molecule has the shape of two tetrahedra joined point-to-point (see figure 29.16).

Example 10. Ethene (hydrogen 1, carbon 2,4).

$$4H^{\times} + 2\,{}^{\circ}\overset{\circ}{\underset{\circ}{C}}{}^{\circ} \xrightarrow[\text{and 1 double bond}]{\text{Forms 4 single}} \begin{array}{c} H\ \ H \\ \times\circ\circ\times\circ \\ C\,{}^{\circ}_{\circ}\,C \\ \times\circ\circ\circ\times \\ H\ \ H \end{array}$$

Each hydrogen atom now has the electronic structure of helium; and the carbon atoms each that of neon by sharing four electrons between them, making a double bond. The angle between the bonds is 120°, and the molecule planar.

The crystal structure of a covalent compound depends on the ease with which the molecules can be packed in a three dimensional jig-saw. The forces which hold the molecules together in the crystal are weak van der Waals' forces, and so the melting and boiling points of covalent substances are much lower than those of electrovalent substances: they are usually volatile liquids and gases at room temperature.

Van der Waals' forces are the weak forces of attraction between particles when the atoms have combined to form molecules. It is permissible to talk of bonds and molecules in the case of covalent substances. Frequently the molecules are shown in rectangular or planar diagrams but the shape of many of the common substances is important (see figure 14.6).

H—Cl
Linear

$$\begin{array}{c} O \\ /\ \ \backslash \\ H\quad H \end{array} \qquad \begin{array}{c} N \\ /\,|\,\backslash \\ H\ |\ H \\ H \end{array} \qquad \begin{array}{c} H \\ | \\ C \\ /\,|\,\backslash \\ H\ |\ H \\ H \end{array}$$

Angles: 105° 107° 109°

Covalent substances do not conduct electricity either in the fused state or in solution. Allotropy arises when there are two possible ways of packing (*a*) the molecules of an element, e.g. sulphur, or (*b*) the atoms of an element constituting a molecule, e.g. carbon. Polymorphism arises when there are two ways of packing the molecules (or ions) of a compound (see section 8.4).

Unlike the above examples where covalent bonds join a few atoms together, many atoms may be joined giving rise to a giant or macro-molecule, e.g. diamond in which each carbon atom is linked to four others like the central one in this diagram:

```
      C
      |
      C
    / | \
   C  |  C
      C
```

Angles: 109°

and polythene:

```
  [ H     H     H ]
  [ |     |     | ]
  [ C  H  C  H  C ]
 /[ | \ | / | \ | / | ]\
H [ H  C  H  C  H ]  O—H
  [    |     |    ]
  [    H     H    ] n
```

Some of the properties of the macro-molecular substances are more akin to those of the electrovalent compounds than to the covalent ones discussed above.

Isomerism arises when there are two possible ways of linking together the atoms corresponding to a given molecular formula (see section 29.9).

The two electrons constituting a covalent bond are not equally shared if the nuclei are different, e.g. although in a chlorine molecule the bonding electrons are shared equally, in a hydrogen chloride molecule the bonding electrons are closer to the chlorine making the compound partially electrovalent.

14.4 Co-ordinate (Dative) Valency

In example 6 of the preceding section a pair of electrons not shared with any of the hydrogen atoms was apparent in the structure of ammonia. This **lone pair** is used to form a covalent link with a hydrogen ion (proton) giving an ammonium ion. This type of covalency in which one atom donates both electrons for the bond which the other atom accepts is known as co-ordinate valency: it may be shown by an arrow indicating the direction of donation.

```
   H                      [    H    ]+
   xo                     [    |    ]
 HxN x     +  H+   —→     [ H—N→H   ]
  o  x                    [    |    ]
   xo                     [    H    ]
   H
```

Ammonia molecule *Proton* *Ammonium ion*

The bond once formed is indistinguishable from the other three covalent bonds. The shape of the ion is tetrahedral like a methane molecule.

Under similar conditions, water, which has two lone pairs of electrons, forms a hydroxonium ion, H_3O^+, in which the oxygen is at the apex of a pyramid (compare ammonia).

```
  oo                  [  oo   ]+
 HxO o   +  H+  —→    [ HxO oH ]
  o  o                [  o  o  ]
  xo                  [  xo   ]
  H                   [  H    ]
```

Water molecule *Proton* *Hydroxonium ion*

This reaction shows the mechanism by which some covalent substances become electrovalent in aqueous solution, e.g. the anhydrous acids (see sections 17.2 and 20.7).

This type of bonding is very important amongst the transition elements. In copper(II) sulphate-5-water, some of the water molecules are linked to the central copper ion by co-ordinate valency. When ammonia solution is added there is first of all a pale blue precipitate (see section 26.11) but then a deep blue solution forms which contains ions in which ammonia molecules are linked by co-ordinate valencies to the copper. The water molecules of the hydrated ion have been replaced by ammonia molecules and the colour has deepened.

14.5 The general properties of electrovalent and covalent compounds

It must be remembered that the properties described here are those of typical members of a particular class, and that in many cases it is hard to decide whether a substance is electrovalent or covalent. A rough guide to whether two elements will combine by electrovalency or covalency can be given in terms of the periodic table:

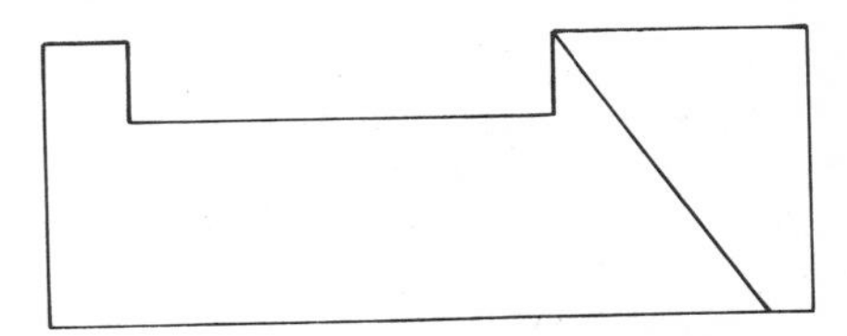

If the two elements are from the extremities of the table (particularly the bottom left and the top right) then electrovalency is exhibited. If the two elements are both from the top right hand corner then they combine by covalency. The third case, two elements from the left hand side, gives an alloy, the formation of which may involve a physical or chemical change beyond the scope of this book.

	Electrovalent	*Covalent (small molecules)*	*Covalent (macromolecules)*
Example	*Sodium Chloride* $NaCl$	*Ethane* C_2H_6	*Polythene* $H(CH_2)_nOH$
Melting point	High	Low	Intermediate to high
Boiling point	High	Low	Intermediate to high
Usual state seen	Solid	Volatile liquid or gas	Solid
Crystal structure	Depends on relative sizes and charges of ions	Depends on packing of molecules	Depends on packing of molecules
Crystal forces	Electrostatic between ions High	Van der Waals, between molecules Weak	Van der Waals, between molecules Weak
Hardness	High	Low	Low
Thermal stability	High	Low	Low
Solvents	Water	Organic solvents	Organic solvents (may only soften the material)
Electrolysis	When molten or in aqueous solution	Non-electrolytes	Non-electrolytes
Rate of reaction with compounds of same class	Fast	Intermediate	Slow

QUESTIONS 14

The Electronic Theory of Valency

1 The following table shows some properties of some elements and compounds.

Substance	*Melting point (°C)*	*Boiling point (°C)*	*Electrical Conductivity*		*Effect of air*
			Solid	*Liquid*	
P	−112	−108	Poor	Poor	No reaction
Q	800	1470	Poor	Good	No reaction
R	98	880	Good	Good	Burns when heated
S	44	282	Poor	Poor	Ignites, even at room temperature
T	Not known	3700	Poor	—	Burns when strongly heated
U	−111	46	Poor	Poor	Burns readily on slight warming

(The letters do not correspond with the symbols for the elements.)

From this list select
(*a*) a substance which could be a salt,
(*b*) a substance which could have a 'giant molecular' structure,
(*c*) a substance which is a liquid at room temperature,
(*d*) a substance which could have an ionic lattice structure.

What precautions would you take in storing the following? Give your reasons.
(*e*) Substance U.
(*f*) Substance S.
(*g*) The products of burning two of the substances in air would combine to form a salt. Which two?

[Nf]

2 The atoms of a certain element contain nuclei carrying a charge of nine positive units (atomic number is 9). Write down three facts about the chemistry of this element that can be deduced from this information. [OC]

3 (*a*) Name one covalent compound and give three properties deriving from its covalent structure.
(*b*) Explain why the elements sodium and potassium show very similar chemical properties. Quote two examples of this similarity. [OC]

4 From the following eight elements—sodium, nitrogen, zinc, potassium, lead, phosphorus, carbon and chlorine—write down (*a*) one pair that form a covalent compound, (*b*) one pair that readily combine to form an ionic compound, (*c*) an oxidizing agent, (*d*) one capable of showing more than one valency. [OC]

5 (*a*) Atoms of element Q carry a nuclear charge of 3, while the atoms of element R carry a nuclear charge of 8.

Write down (i) whether each element is a metal or non-metal, (ii) the probable formula of a compound of Q with R, (iii) the probable formula of a hydride of R.
(*b*) Explain how the law of multiple proportions provides evidence for the existence of atoms. [OC]

6 (*a*) An element Q, with non-metallic properties, has a relative atomic mass of 75 and is normally trivalent. (i) Write down the equation for the complete combustion of Q in oxygen. (ii) Calculate the mass of calcium that should combine with one gram of Q. (iii) State whether the solution in water of the oxide of Q is likely to be acidic, alkaline or neutral.
(*b*) Carbon tetrachloride, CCl_4, is a wholly covalent compound. Would you expect it to be: (i) volatile, (ii) miscible with water, (iii) a conductor of electricity? [OC]

7 Explain what is meant by (*a*) the mass number, and (*b*) the atomic number of an element. Describe briefly the structure of the atom of one named element other than hydrogen or helium. Explain the significance of the noble gases to modern ideas about valency. Explain what is meant by a covalent bond. Give the electronic formulae of two compounds, each containing one or more covalent bonds. [A]

8 Explain what is meant by (*a*) ionic bonding, (*b*) a covalent bond. Give an electronic formula for one compound containing ionic bonding and an electronic formula for one compound containing one or more covalent bonds. How do these formulae illustrate the valencies of the elements concerned?

What is an ionic lattice? Describe any one type, naming a compound which has that type of lattice. [A]

9 On the basis of modern views on the structure of the atom and on valency explain why (*a*) neon is chemically almost inert, (*b*) magnesium is divalent, (*c*) sodium and potassium have similar chemical properties, (*d*) an aqueous solution of sodium chloride conducts a current of electricity but an aqueous solution of ethanol (ethyl alcohol) does not.

What are the present views about the structure of a sodium chloride crystal? [A]

10 (*a*) The atomic number and relative atomic mass of nitrogen are 7 and 14, respectively. Give a labelled diagram showing the structure of this atom.
(*b*) Indicate the relative charges and masses of the particles making up the nitrogen atom.
(*c*) Ammonia is a covalent compound. Give a diagram showing how the atoms in the ammonia molecule are bonded. [A]

11 Explain simply the difference between electrovalency and covalency. Illustrate your answer by reference to sodium chloride and ammonia. Briefly describe two simple experiments which could help you to decide if the bonding in a given solid compound was electrovalent or covalent.

How do you account for the fact that liquid hydrogen chloride is neutral, whereas a solution of the gas in water is a strong acid? [S]

12 (*a*) What can you deduce about the structure of an atom if you are told that its atomic number is 20 and its relative atomic mass is 40?
(*b*) What type of valency do you expect to find in a crystalline solid which melts at 80°C, is insoluble in water, and is a non-electrolyte? Explain your answer.
(*c*) What are the constituent particles in a crystal of sodium chloride? How are they held together? Give a simple sketch to show the arrangement of the particles in the crystal. [S]

13 Give an account of the structure and shape of a molecule of ammonia. Draw the structures and shapes of methane (CH_4), water (H_2O), carbon dioxide (CO_2), ethene (C_2H_4) and phosphorus trifluoride (PF_3).

14 Name two compounds that are covalent when pure but produce ions when dissolved in water. For each compound, give the formulae of the ions formed in aqueous solution. [C]

15 Give simple diagrams to show: (*a*) the difference between a sodium atom and a sodium ion; (*b*) the difference between a chlorine atom and a chlorine ion; (*c*) the formation of sodium chloride. (Give a key to explain any symbols used.)

16 (*a*) Argon is a noble gas of atomic number 18 and its electrons are arranged in the pattern 2,8,8. It is the middle element in the sequence sulphur, chlorine, argon, potassium, calcium, the atomic numbers of which are respectively 16, 17, 18, 19 and 20.
(i) Write down the arrangement of the electrons in the atom of sulphur and in the atom of calcium. (ii) How many protons has the atom of chlorine? (iii) The relative atomic mass of potassium is 39; how many neutrons has the atom of potassium? (iv) The other four elements form ions with an electronic pattern similar to that of the argon atom. Write down the charge on each ion. (v) Write an ionic equation for the conversion of the sulphur atom to the sulphur ion. (vi) Write an ionic equation for the conversion of the potassium atom to the potassium ion. (vii) Hence write down the formula of potassium sulphide. What would be the formula of calcium sulphide? (viii) The potassium ion is considerably smaller than the chloride ion. Why do you think this statement is correct? (ix) The chlorine molecule has a covalent link. Give its structure. (x) The element carbon has a configuration 2,4 and forms covalent bonds. Explain very briefly why its chloride has the formula CCl_4.
(*b*) Hydrogen chloride is ionic in aqueous solution, but covalent in solvents such as benzene. Write down the two structures of hydrogen chloride. Explain briefly why a non-aqueous solution of hydrogen chloride does not affect litmus or give a precipitate with silver nitrate. [NI]

17 What are the main differences between the physical properties of sodium chloride and carbon tetrachloride? How far can these be explained from a knowledge of the electronic structures of these two compounds?

18 Show, in the form of a table, the chief properties of (*a*) electrovalent compounds, (*b*) covalent compounds. Classify each of the following under (*a*) or (*b*) and justify your answer in each case: HCl, $BaCl_2$, SO_2, CS_2, CH_4, CCl_4. [E]

19

I	II	III	IV	V	VI	VII	O
Li Na	Be Mg	B Al	C Si	N P	O S	F Cl	Ne Ar

The above two rows of the Periodic Table may help you to answer the questions below. One of the numbers, symbols or words given at the end of each of the following sentences completes the sentence correctly. Write the correct answer.
(*a*) The relative atomic mass of $^{14}_{7}N$ is (i) 7, (ii) 14, (iii) 98, (iv) 21, (v) 2.
(*b*) The number of neutrons in one atom of $^{23}_{11}Na$ is (i) 0, (ii) 11, (iii) 12, (iv) 23, (v) 253.
(*c*) A fluoride anion has the same number of electrons as (i) Be^{2+}, (ii) O, (iii) Cl^-, (iv) Ne.
(*d*) A sodium cation has a different number of electrons from (i) O^{2-}, (ii) F^-, (iii) Li^+, (iv) Al^{3+}.
(*e*) $^{35}_{17}Cl$ and $^{37}_{17}Cl$ are (i) allotropes, (ii) isomers, (iii) isotopes, (iv) molecules. [J]

20 An element M (relative atomic mass 40, atomic number 20) reacts with cold water to liberate hydrogen and forms a chloride whose relative molecular mass is 111. This chloride dissolves in water, and the solution conducts an electric current.
(*a*) Give simple diagrams to show and explain the difference between the M atom and the M ion. (*b*) State the type of valency which the chloride exemplifies. (*c*) Give the equation for the reaction of M with water. (*d*) Suggest a method by which you could prepare a small sample of M from the chloride. [S]

21 How does covalency account for the formation of the following molecules: methane (CH_4), ethene (C_2H_4), hydrogen (H_2) and propane (C_3H_8). Compare the properties of tetrachloromethane (carbon tetrachloride, CCl_4) and calcium chloride ($CaCl_2$). Explain these differences in terms of electrovalency and covalency.

22 A list of substances is given below.
Potassium iodide, sulphur dioxide, poly(ethene), calcium carbonate, lead(II) bromide, copper(II) chloride, sulphur, hydrogen, copper(II) oxide, hydrogen chloride.
From this list select:
(*a*) Two substances which are reasonably soluble in water.
(*b*) Two substances which would interact to give water as one of the products.
(*c*) Two substances which would conduct electricity when melted; state the products formed on electrolysis.
(*d*) Two substances whose solution in water would conduct electricity; state the products formed on electrolysis.
(*e*) Two coloured substances; state the colour of each. [L]

23 (*a*) Give four examples of the ways in which the chemical and physical properties of sodium and chlorine differ from those of sodium chloride.
(*b*) Mercury and ethanol are both liquids, yet only one of them conducts electricity well. Explain why this is so.
(*c*) Describe two important differences in the properties of the solutions obtained by dissolving (i) hydrogen chloride, and (ii) ammonia in water. [OC]

24 Suggest reasons for the low conductance and the low melting- and boiling-points of sucrose ($C_{12}H_{22}O_{11}$), iodoform (CHI_3), solid carbon dioxide (CO_2) and ice (H_2O).

15 Metals, Non-metals and the Electrochemical Series

15.1 Introduction

Metals to the layman are often distinguished by their physical properties: their lustre and strength, their capacity for being beaten into shape or pulled out into a wire. Early in man's search for tools and constructional materials there was a Bronze Age and an Iron Age. Metals such as iron, copper, gold, mercury, lead, tin, antimony and silver have been known for thousands of years but, except for iron, they are not present in very large quantities in the earth's crust. Where they occur they are conspicuous: this may be as the element or as a compound which is easily reduced to the element.

The modern chemist has eighty rather than eight metals at his disposal, and knows much more about mixing them to his advantage and making alloys than did his forebears. As the geochemists unravel the composition of the earth's crust and the industrialists improve the efficiency of the extraction processes, ores containing very minute proportions of the desired metal can be dealt with profitably.

However, the scientist has a different idea of what constitutes a metal from the layman. Thus under the heading 'metals' occur those elements which are good examples from the chemical point of view (sodium and calcium) and those which are good examples from the physical point of view (copper and iron).

An absolute distinction between metal and non-metal cannot be made; there are many borderline cases (aluminium, germanium). The decision must be made in view of the majority of the properties for a particular element, but it may be made differently for a particular property of a set of elements.

15.2 The General Properties of Metals and Non-metals

The properties of metals and non-metals tabulated overleaf are general properties. It is easy to quote examples illustrating these general properties; exceptions such as on the following list are also easy to find:

carbon (graphite) conducts electricity;
carbon (diamond) is hard;
mercury is a liquid metal (the only one);
carbon tetrachloride is stable to water;
sodium has a low melting point;
iodine is dense (more dense than titanium);
copper with many acids does not give hydrogen.

It will be seen that there are many parallels between this table of the general properties of metals and non-metals and the one describing the general properties of electrovalent and covalent compounds (see section 14.5).

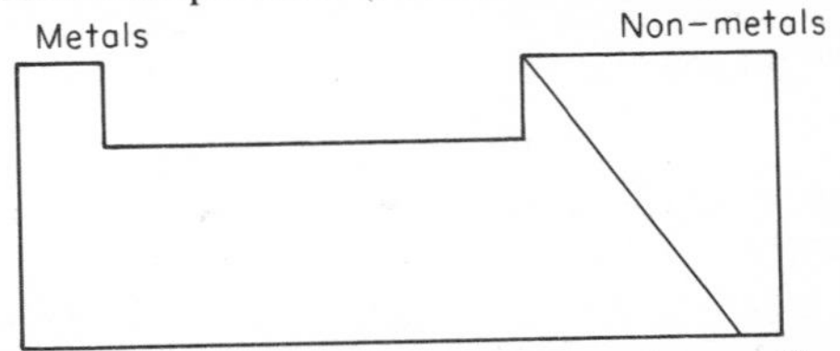

Metallic character increases on descending a group in the periodic table, and decreases on proceeding from left to right. The alkali metals (group I) are a fairly homogeneous group of metals, and the halogens (group VII) are a fairly homogeneous group of non-metals. But in groups IV and V the change from a non-metal at the top to a metal at the bottom of the group is particularly noticeable. As a definition of a non-metal one can take the statement:

A non-metal is an element which has more electrons in the outermost shell than its period number.

This means that the following elements are considered as non-metals:

					He
B	C	N	O	F	Ne
	Si	P	S	Cl	Ar
		As	Se	Br	Kr
			Te	I	Xe
				At	Rn

Physical Properties	*Metals*	*Non-metals*
Melting point	High	Low (unless a giant molecule)
Boiling point	High	Low (unless a giant molecule)
Hence usual state seen	Solid	Gas or volatile liquid but solid if giant molecule
Thermal conductivity	High	Low
Electrical conductivity	High, decreases as temperature increases	Low, increases as temperature increases
Solvents	Mercury, giving amalgams; other metals, giving alloys	Water and organic solvents
Density	High	Low
Electron emission	Readily on heating	Poor
Lustre	High, but may tarnish in moist air	Low if gaseous; yes if crystalline
Tensile strength	High	Low
Ductility	High	Low
Hardness	High	Low
Malleability	High	Low (brittle)
Sonority	High	Low
Chemical Properties		
Ions formed	Positive, cations; deposited at cathode in electrolysis	Negative, anions; may be deposited at anode in electrolysis
With acids	Give hydrogen and a salt: metal is oxidized	Oxidized by hot concentrated acids, no hydrogen released
With alkalis	Only a few react—they are amphoteric	Most react, some in a complex manner
Hydrides	Electrovalent or are complex	Covalent
Oxides	Electrovalent, basic; alkalis if they dissolve in water	Covalent, acidic; acids if they dissolve in water
Chlorides	Electrovalent, salts, stable to water	Covalent, not salts, react with water
Element-oxygen-hydrogen compounds	Bases (some alkalis)	Acids

This gives some precision to the position of the diagonal in the periodic table, but it must be viewed only as a guide. The term **semi-metals** can be used for the elements close to the boundary line: they are particularly hard to classify.

15.3 The Structures of Metals

Drude (1900) and Lorentz (1904) suggested that a metal consisted of an array (a network, a lattice, an arrangement) of cations (the nucleus and the inner shells of electrons of the atom) permeated by a cloud of electrons (the valency electrons). An analogy would be a collection of ball bearings in a lump of petroleum jelly: the ball bearings give the lump solidity and the jelly gives the lump cohesion.

The melting and boiling point and the density of a metal are high because the electrons hold the cations together tightly. The mobility of the electrons gives a metal its thermal and electrical conductivity: as the temperature increases the increased vibration of the cations hinders the movement of the electrons and so the electrical conductivity, which depends on the fairly free movement of these electrons, decreases. The electron 'cloud' is also responsible for the mechanical properties of a metal, e.g. ductility, tensile strength and malleability.

When metals are examined by X-rays it is found that the vast majority of them have one of these three structures at room temperature:

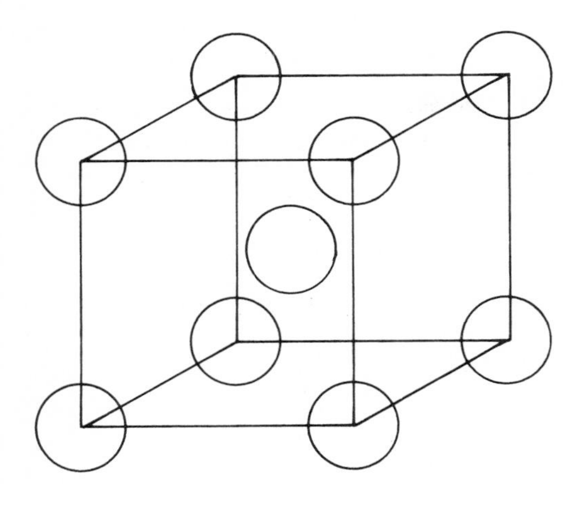

Figure 15.1
A body centred cubic crystal

(*a*) *Body centred cubic* (*b.c.c.*)

In a body centred cubic structure (see figure 15.1) the typical atom, for example, the one in the centre of the cube, has eight neighbours. Examples of this structure are iron and sodium.

(*b*) *Face centred cubic* (*f.c.c.*)

In a face centred cubic structure (see figure 15.2) the typical atom has twelve neighbours. Typical structures in this case are those of copper and calcium.

(*c*) *Close packed hexagonal* (*c.p.h.*)

In a close packed hexagonal structure (see figure 15.3) the typical atom again has twelve neighbours. Magnesium and zinc have structures like this.

The unity of metals as a set of elements comes from this close similarity in crystal structure.

On heating there is a tendency for the vibration of the electron cloud to overshoot the boundaries of the crystal, and in a vacuum the phenomenon of electron emission can be

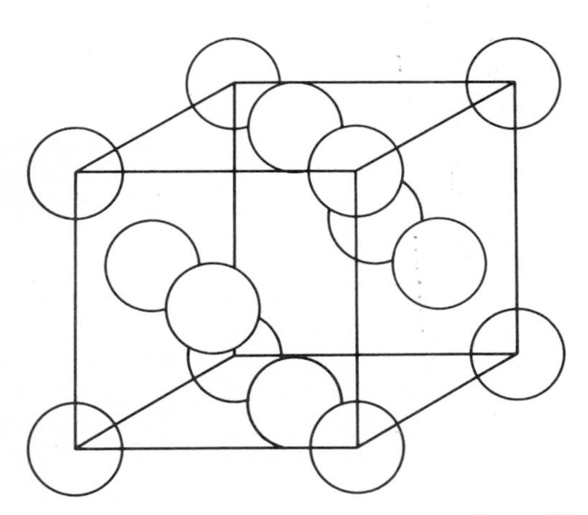

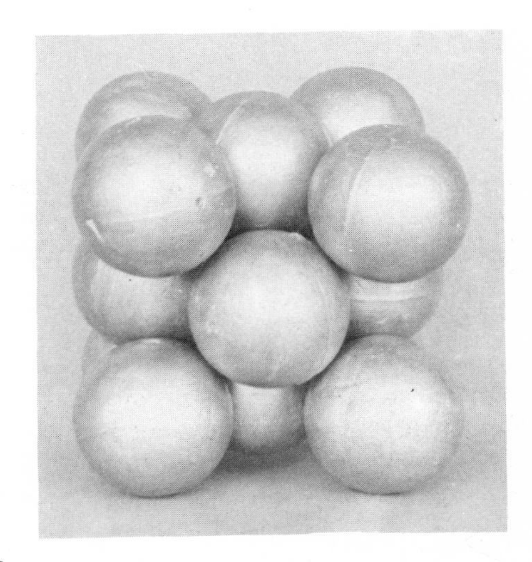

Figure 15.2
A face centred cubic crystal

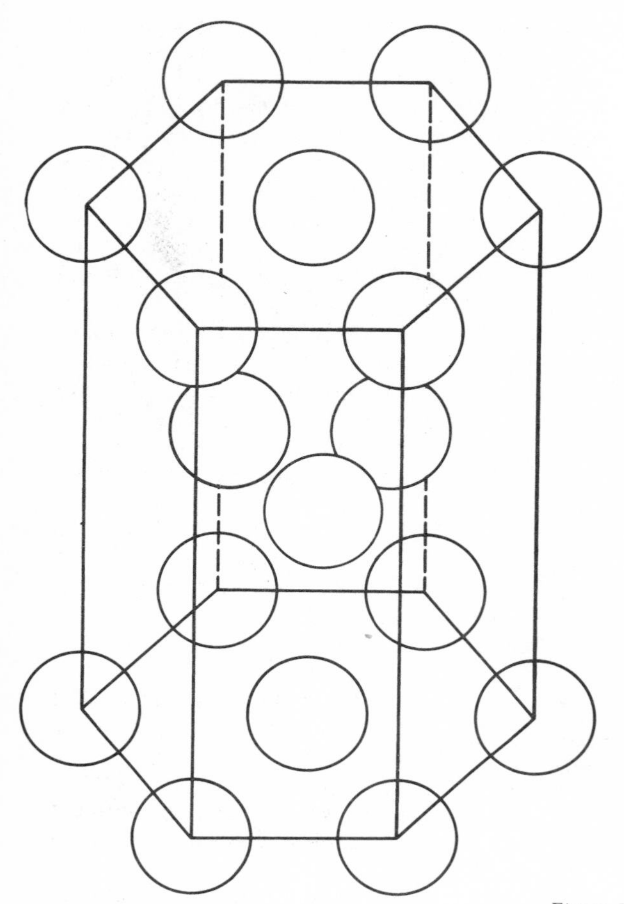

Figure 15.3
A close packed hexagonal crystal

detected (a radio valve, a cathode ray tube). A metal may lose electrons—

$$E \rightarrow E^{n+} + ne^-$$

—forming a cation: the chemical properties of metals are based upon this reaction, a case of oxidation.

15.4 The Structures of Non-metals

Some of the elements close to the borderline between metals and non-metals form macromolecules in which covalent bonds link all the atoms in the crystal; however, in the majority of cases, the structures of non-metals, as shown by X-rays, involve isolated molecules.

In group IV the first element carbon has two allotropes which are both macromolecules: diamond, in which the atoms are arranged tetrahedrally throughout the crystal, and graphite, in which the atoms in the separate layers are arranged hexagonally, the atoms in alternate sheets being vertically over one another (see figure 28.2). Silicon has the same structure as diamond.

In group V the typical elements are non-metals: nitrogen exists as diatomic molecules (N_2) and phosphorus, in its white allotrope, as tetraatomic molecules (P_4), but as a macromolecule in its red form (see section 30.6). The differences between the allotropes of phosphorus are greater than those between the allotropes of carbon because of this greater difference in the structures.

In group VI both oxygen and sulphur show allotropy. Oxygen (O_2) and ozone (O_3) are allotropes usually studied in the gaseous form (see section 31.6). The rhombic and monoclinic crystals of sulphur differ only in the packing of the S_8 molecules in the solid (see section 31.5).

In group VII the halogens all consist of diatomic molecules and in group O the noble gases are all monatomic.

The properties of most of these non-metals are similar because the molecules, whilst bound internally by covalency, are only held together externally by weak van der Waals' forces exerted over greater distances (see section 14.3). These forces, which are responsible for the cohesion of molecules in the solid and liquid forms of the elements, are easily overcome on heating, so that non-metals have low melting and boiling points. The densities of non-metals are lower than those of metals because of the greater spacing of the particles and, in general, the lower relative atomic masses.

The mechanical properties of non-metals are poor because of the weakness of the van der Waals' forces.

The electrons in molecules of non-metals are held more firmly than those in metal crystals. They are not able to move from molecule to molecule and so the non-metal is a poor conductor of heat and electricity. Many of the properties of non-metals are based upon their ability to attract electrons,

$$E + ne^- \rightarrow E^{n-}$$

an example of reduction. This may result in the formation of (*a*) an electrovalent compound containing a metal or (*b*) a compound which is partly electrovalent or (*c*) covalent molecules when two or more non-metal atoms form a molecule.

15.5 The Reactivity of Metals

Although the metals form a set of elements which has been considered so far as a whole, they differ greatly in their chemical reactivity.

The remarkable thing is that, whatever reaction is considered, the order of the metals is always the same (with a few minor exceptions). Thus, whether we consider oxidation, reaction with acids, ease of corrosion, or reaction with chlorine, sodium and potassium are always among the most reactive metals, while copper and gold are among the least reactive.

The reactivity can be connected with other properties of the elements. As might be expected, the unreactive metals, such as gold, silver, and copper are found uncombined in nature. The reactive metals, such as sodium and magnesium, form such stable compounds that it is very difficult to extract them from their ores by chemical means. So, although gold, silver, and copper have been known for centuries, the isolation of potassium, sodium, and magnesium has been carried out only since 1800.

'Reactivity' is, however, a vague term: a numerical result obtained by comparing metals would be better. For this purpose we can use the results of electrical experiments. Metals have a tendency to lose electrons and form positive ions: they are said to be electropositive. Also, in electrolysis (see chapter 20) some positive ions, e.g. sodium, are not discharged at the cathode whereas others, e.g. hydrogen, are. Similarly in a solution containing hydrogen and copper ions, the copper ions are more likely to be discharged. Sodium is said to be more electropositive than hydrogen whilst copper is less, i.e. sodium atoms lose electrons to form positive sodium ions in solution more readily than hydrogen atoms form hydrogen ions in solution and likewise hydrogen ionizes more readily than copper.

The tendency of metal atoms to form ions in solution cannot be measured in isolation and so it is assessed relative to hydrogen, i.e. the tendency of the reaction

$$H \rightarrow H^+ + e^-$$

to proceed is arbitrarily taken as zero. The hydrogen electrode is constructed by blowing hydrogen at 1 atm pressure on to a piece of platinum immersed in a 1M solution of hydrochloric acid; on the surface of the platinum the hydrogen molecules dissociate forming atoms.

The metal electrode is put into a 1M solution of one of its salts and the two solutions connected by a salt bridge (see figure 20.1). The electromotive force of this, and of similar cells constructed using other metals and the appropriate solutions, can be measured with a high resistance voltmeter. The e.m.f's obtained are often called electrode potentials, and the symbol used is $E^{\ominus}$.

There are more metals like sodium (more electropositive than hydrogen) than copper (less electropositive). The sign convention is that sodium and metals like it have negative values on the scale; when the metals are arranged in order a series, which agrees largely with that obtained by studying their reactivity, is obtained. It is called the electrochemical series and it is very useful as a means of classifying our knowledge about the chemistry of the metals.

It is possible but more difficult to obtain the e.m.f's of cells involving non-metals. The principles of the method can be extended to compounds which are oxidizing and reducing agents yielding redox potentials.

15.6 The Electrochemical Series

The series is obtained by measuring the electromotive forces of the following half cells against the standard hydrogen electrode at 25°C (in volts) i.e. the cell is

$$\text{Pt}, H_2 \mid H^+ \mid M^{n+} \mid M$$

where M^{n+} is the ion formed by a M atom.

K^+	K	−2·92	Pb^{2+}	Pb	−0·13
Ca^{2+}	Ca	−2·87	H^+	H	0·00
Na^+	Na	−2·71	Cu^{2+}	Cu	+0·34
Mg^{2+}	Mg	−2·37	Hg_2^{2+}	Hg	+0·79
Al^{3+}	Al	−1·66	Ag^+	Ag	+0·80
Zn^{2+}	Zn	−0·76	Pt^{2+}	Pt	+1·20
Fe^{2+}	Fe	−0·44	Cl	Cl^-	+1·36
Sn^{2+}	Sn	−0·14	Au^{3+}	Au	+1·50
Ni^{2+}	Ni	−0·25	F	F^-	+2·85

(a) The Extraction of Metals

The least electropositive metals, i.e. those which occur near the bottom of the electrochemical series, are found native, so that their extraction need not involve any chemical reaction, e.g. copper, silver and gold.

Most of the remaining heavy metals (e.g. iron, zinc, lead) can be obtained by reducing their heated oxides with carbon, carbon monoxide, or even hydrogen:

$$ZnO + C \rightarrow Zn + CO\uparrow$$
$$Fe_2O_3 + 3CO \rightarrow 2Fe + 3CO_2\uparrow$$
$$PbO + H_2 \rightarrow Pb + H_2O\uparrow$$

However, it can be a matter of considerable difficulty to convert the natural ore (which may not be the oxide but the sulphide or carbonate) into the oxide and to refine the impure metal.

The very reactive metals at the head of the series (e.g. sodium, magnesium, aluminium) are not usually obtained from their compounds by chemical reduction. More often the large amounts of energy required to liberate the free metals are supplied electrically; a molten (fused) compound of the metal in question is electrolyzed. Magnesium is produced by the electrolysis of magnesium chloride, aluminium is made by the electrolysis of its oxide.

(b) Reactions of Metals with Air

Gold and platinum do not tarnish or corrode at all when exposed to air. Silver is affected by traces of sulphur compounds, and gives black silver sulphide, but pure air has no action. Copper reacts with oxygen only on heating but slowly acquires a green 'patina' when exposed to ordinary air containing sulphur dioxide. Iron rusts rapidly, giving a brown hydrated oxide. Sodium and potassium are so reactive that they have to be kept under oil to prevent them coming into contact and reacting with the atmosphere.

From the series it would appear that metals such as zinc, magnesium, and aluminium should be rapidly attacked by the air. In fact, zinc is used for protecting iron, and both magnesium and aluminium can be safely used for constructional purposes. These metals react with the air to form a very thin protective film (usually of the oxide) which saves them from further attack. If the film is removed (as it can be from aluminium by rubbing with mercury) the metal reacts at once to give white, feathery growths of the oxide. These oxide films often lead to metals being considered to have been wrongly placed in the electrochemical series, but the physical method of obtaining the order of the elements is more reliable than the chemical method.

(c) Reactions of Metals with Water

Steam has no effect upon metals with positive electrode potentials, e.g. copper and silver, even when they are white-hot. Tin and lead react slightly at high temperatures, to give their oxides. Red-hot iron filings react easily with steam, but the reaction is reversible.

$$3Fe + 4H_2O \rightleftharpoons Fe_3O_4 + 4H_2\uparrow$$

Aluminium is again protected by its oxide film, but aluminium powder will burn in steam when heated as will magnesium:

$$Mg + H_2O \rightarrow MgO + H_2\uparrow$$

Calcium reacts slowly but steadily with cold water:

$$Ca + 2H_2O \rightarrow Ca(OH)_2 + H_2\uparrow$$

Sodium and potassium both react vigorously with cold water, in the latter case so much heat being evolved that the hydrogen ignites:

$$2Na + 2H_2O \rightarrow 2NaOH + H_2\uparrow$$
$$2K + 2H_2O \rightarrow 2KOH + H_2\uparrow$$

(d) Replacement (Displacement) Reactions

It is found that, in general, a metal high in the electrochemical series will displace a metal lower in the series from a solution of one of its salts. Thus, when an iron object is dipped into copper sulphate solution, some of the iron dissolves and is replaced by a thin layer of copper.

$$Fe + CuSO_4 \rightarrow Cu\downarrow + FeSO_4$$

It is better to write the reaction in terms of ions:

$$Fe + Cu^{2+} \rightarrow Cu\downarrow + Fe^{2+}$$

The iron atom is more electropositive than the

copper atom, i.e. it has a greater tendency to lose electrons and become an ion than the copper has to form a copper ion. The less electropositive copper ion is forced to accept the electrons and become a neutral atom.

In the same way, if a rod of zinc is immersed in lead nitrate or tin(II) chloride solution, beautiful lead and tin 'trees' are formed as the metals crystallize out of solution.

$$Zn + Pb^{2+} \rightarrow Pb\downarrow + Zn^{2+}$$
$$Zn + Sn^{2+} \rightarrow Sn\downarrow + Zn^{2+}$$

All the metals except lead above hydrogen in the series are able to replace it from dilute solutions of acids (hydrochloric acid, HCl, or sulphuric acid, H_2SO_4; but nitric acid, HNO_3, rarely gives hydrogen).

$$Mg + 2HCl \rightarrow MgCl_2 + H_2\uparrow$$

Or, writing it in terms of ions:

$$Mg + 2H^+ \rightarrow Mg^{2+} + H_2\uparrow$$

Metals below hydrogen cannot displace it from dilute solutions of acids.

The very reactive metals displace hydrogen, even from cold water, and they will do this rather than displace other metals from their salts, e.g. sodium does not displace iron from a solution of iron(II) sulphate but will react with the water releasing hydrogen.

(e) Voltaic Cells and Corrosion

If two metals are placed in contact in a solution of an electrolyte, the more electropositive metal (e.g. zinc) tends to force electrons into the less electropositive metal (e.g. copper) and zinc ions pass into solution:

$$Zn \rightarrow Zn^{2+} + 2e^-$$

So the more reactive (electropositive) metal (in this case zinc) is gradually dissolved. A simple voltaic cell can be devised using these two metals.

If the two metals are joined through a voltmeter the electromotive force (e.m.f.) of the cell can be measured. It is equal to the difference of the electrode potentials of the metals concerned. The electrode potential of copper is $+0{\cdot}34$ volt and the electrode potential of zinc is $-0{\cdot}76$ volt. Hence, the e.m.f. of this cell is $0{\cdot}34-(-0{\cdot}76) = 1{\cdot}1$ volts. (See section 20.3.)

Small voltaic cells are often set up naturally and are the source of much corrosion. If copper rivets were used to secure the steel plates of a ship's hull, the steel would rapidly corrode, as sea water is a very good electrolyte and steel (iron) is more electropositive than copper.

Similarly, if galvanized iron containers (zinc covered) are scratched, because of the moisture and acid gases in the atmosphere an electrolytic cell is set up—

Zinc | Acid | Iron

but since zinc is higher than iron in the electochemical series, the zinc passes into solution and not the iron,

$$Zn + 2H^+ \rightarrow Zn^{2+} + H_2\uparrow$$

Figure 15.4
The use of magnesium anodes to protect the steelwork at the stern of a tanker (before and after)

Element	Hydrogen on oxide	Carbon on oxide	Water on metal	Hydrochloric acid on metal
Potassium	No reduction of oxide	No reduction in the laboratory	Cold: lilac flame, fast	Violent reaction to give hydrogen (dangerous)
Calcium			Cold: slow	
Sodium			Cold: fast	
Magnesium			Cold: very slow even if metal in powder form Boiling water: fairly fast Steam: burns	Moderate reaction to give hydrogen (becoming slower from magnesium to tin and needing more concentrated acid)
Aluminium			Normally protected by surface film of oxide	
Zinc		Reduced	Steam: fairly fast	
Iron	Reversible reduction		Steam: reversible	
Tin	Reduction			
Lead			Metals unaffected by pure water or steam	Exceedingly slow reaction giving hydrogen
Copper	Rapid reduction			Hydrogen never given off. No attack in absence of oxidizing agents (attacked by oxidizing acids)
Silver	These oxides are decomposed by heat alone			
Gold				

N.B. The table can be extended to record the reactivity towards air, etc.

Figure 15.5
The electrochemical series and chemical properties

Thus initially, only the cover of the container corrodes away. In the case of a tin-plated iron container, once the container is scratched and the cell

Iron | Acid | Tin

set up, the iron passes into the solution, because iron is higher than tin in the electrochemical series. Thus in this case the container is eaten away and not the cover,

$$Fe + 2H^+ \rightarrow Fe^{2+} + H_2\uparrow$$

Sometimes the corrosion can be controlled, e.g. the rusting of buried oil pipelines can be prevented by attaching lumps of magnesium at intervals along their length. A cell is set up by reason of the impure water in the soil, and the more reactive magnesium dissolves in preference to the iron. A similar method is used to protect the steel framework of piers. See figure 15.4.

(*f*) *Stability of Compounds*

In general, the more electropositive metals will have the more stable compounds. Thus, calcium oxide (CaO) and magnesium oxide (MgO) cannot be reduced by carbon in the laboratory, while mercury(II) oxide (HgO) and silver oxide (Ag_2O), decompose on heating without any reducing agent.

Similarly sodium carbonate (Na_2CO_3) and potassium carbonate (K_2CO_3) do not decompose when heated; calcium carbonate ($CaCO_3$) decomposes with some difficulty, but most of the heavy metal carbonates lose carbon dioxide easily.

The stability of nitrates towards heat can also be considered. Whereas the nitrates of sodium and potassium only decompose to the nitrites on gentle heating:

$$2KNO_3 \rightarrow 2KNO_2 + O_2\uparrow$$

most other nitrates give the oxide, nitrogen dioxide, and oxygen:

$$2Pb(NO_3)_2 \rightarrow 2PbO + 4NO_2\uparrow + O_2\uparrow$$

The nitrates of silver, gold, and mercury decompose completely to metal:

$$2AgNO_3 \rightarrow 2Ag + 2NO_2\uparrow + O_2\uparrow$$

(*g*) *The Selective Discharge Theory*

If the electrochemical series for metals is

extended to non-metals and drawn thus

Ca^{2+}		
Na^{+}		
Mg^{2+}	SO_4^{2-}	
Al^{3+}	NO_3^{-}	
To ← Zn^{2+}	Cl^{-}	To
cathode Fe^{2+}	Br^{-}	→ anode
H^{+}	I^{-}	
Cu^{2+}	OH^{-}	

it can be stated that under comparable conditions using inert electrodes that the nearer an ion is to the bottom of this V-shaped table the more likely it is to be discharged in electrolysis.

QUESTIONS 15

Metals, Non-metals and the Electrochemical Series

1 Name a metal, different in each case, which (*a*) is monovalent, (*b*) forms a sulphate insoluble in water, (*c*) reacts vigorously with cold water, (*d*) when alloyed with copper forms brass, (*e*) forms an amphoteric oxide, (*f*) cannot be formed by the reduction of its oxide with hydrogen. [J]

2 'An important characteristic of metals is their ability to form positive ions.' Explain what you understand by this statement and, with reference to the electronic structures of sodium and magnesium, say what the formation of ions involves. Give three examples of chemical behaviour typical of metals, supporting your statements with equations for your chosen reactions. Name three elements which are classified as non-metals and give the formulae for their compounds with hydrogen. [O]

3 A new metal X is discovered and found to be between lead and copper in the electrochemical series and to have a valency of two. State the reactions (if any) you would expect when the following experiments are carried out with compounds of X. Give equations. (*a*) The carbonate is heated, (*b*) the nitrate is heated, (*c*) dilute hydrochloric acid is added to the metal, (*d*) hydrogen is passed over the heated oxide of the metal. Describe the methods you would use in the laboratory to prepare (i) the carbonate of the metal from its oxide, (ii) the sulphate of the metal from the powdered metal. [A]

4 (*a*) Explain what is meant by (i) an atom, (ii) an ion. (*b*) Give two chemical properties and two physical properties that distinguish a metal from a non-metal. [A]

5 Give the name and symbol of a metal (different in each instance) which fits the following information: (*a*) it must be kept out of contact with air or water; (*b*) it is prepared by electrolysis of its oxide; (*c*) limestone is one of its common compounds; (*d*) it has a green carbonate which, on heating, decomposes to give a black oxide. [S]

6 (*a*) Compare, in the form of a table, the chemical properties of metals with those of non-metals.
(*b*) Consider the metals calcium, iron, zinc and magnesium, with regard to their ability (i) to displace one another from solutions of their salts, (ii) to react with water, steam and hydrochloric acid. Place the four elements in a series to conform with their reactivity. What is this series called? [E]

7 How does galvanizing a dustbin help to prevent it rusting? An eccentric millionaire once had a dustbin gold-plated. Explain whether you think this was a good idea to prevent it from rusting.

When iron nails are placed in copper(II) sulphate solution, they become coated with copper. Write an ionic equation for this reaction. Selecting any other materials required, how would you make use of this reaction to produce an electric current? If there was excess copper(II) sulphate solution, explain whether the mass of copper produced would be greater than, equal to, or less than the mass of the original nails, assuming the latter to have reacted completely. [Sc]

8 Arrange the elements iron, sodium, lead, copper and zinc in order of the electrochemical series, starting with the most electropositive element first. Give two series of reactions which justify your classification, definitely stating what occurs in the case of each metal.

9 Name: (*a*) a metal which easily burns in air, (*b*) a metal which forms an insoluble sulphate, (*c*) a metal which would displace lead from a solution of lead nitrate, (*d*) a metal which does not displace hydrogen from cold water but does displace hydrogen from steam, (*e*) a metal which reacts with sodium hydroxide to produce hydrogen. [J]

10 The following is a list of metals: sodium, zinc, iron, lead, copper. From this list select: (*a*) two metals which will liberate hydrogen from water, (*b*) one metal which is used in the laboratory preparation of hydrogen, (*c*) one metal which will liberate hydrogen from an alkali, (*d*) one metal which is used in the laboratory preparation of nitrogen oxide, (*e*) one metal which will displace copper from copper(II) sulphate, (*f*) one metal which will displace lead from lead nitrate.

In each case describe briefly how you would bring about the reaction and give the equation, but do not give any details of apparatus. [J]

11 Here are five metallic elements arranged in the order of their chemical activity (the least electropositive metal is placed first): iron, sodium, calcium, potassium, rubidium (Rb). Rubidium has a valency of 1.
(*a*) What will be the action of heat on rubidium nitrate?
(*b*) Will rubidium carbonate be soluble? What will be the action of heat on the carbonate?
(*c*) Will the metal have any action on water?
(*d*) How would you attempt to prepare the metal from rubidium chloride? Give equations wherever you can.

12 Which hydroxides of the common metals react when heated alone and when heated with carbon? Which metals do not form hydroxides at all? Which hydroxides are soluble in water? Arrange the hydroxides in order of stability and compare your order with the electrochemical series given in figure 15.5.

13 State three characteristic chemical properties of metals which distinguish them from non-metals. What is an alloy? Name any two common alloys and the substances present in each of them. Give, in each case, two important uses of (*a*) lead (*b*) copper (*c*) tin, stating the property or properties on which the uses depend. [J]

14 'Diamond and graphite both consist of giant structures of carbon atoms whereas solid carbon dioxide is made up of carbon dioxide molecules.' Explain the meaning of the above statement and show how the properties of each substance can be related to its structure.

15 The U.K. Department of Trade and Industry estimated in 1971 that the material cost of corrosion was £1365 million each year. Of this, it was considered that £300 million could have been saved by better use of existing knowledge and techniques. If you were giving evidence to the Department, what would you recommend?

16 Give two physical and two chemical differences between a metal and a non-metal. The diagrams below represent the electron distributions in the atoms of six elements (e^- = electron).

In answering the following questions, you are not required to identify the atoms and you may use the letters A, B, C, etc. as symbols for the elements.
(*a*) Which of these elements would you expect to be (i) metallic, (ii) non-metallic, (iii) an inert gas?
(*b*) What would you expect to be the formulae of the compounds formed by (i) a with b, (ii) b with e, (iii) d with f? Indicate for each of these compounds whether the bonding will be electrovalent or covalent.
(*c*) Write down the atomic number of d.
(d) If there are 12 neutrons in the nucleus of d, what will be its relative atomic mass? [C]

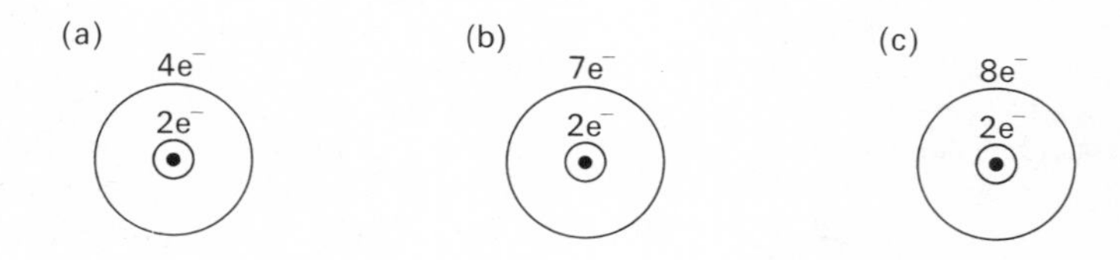

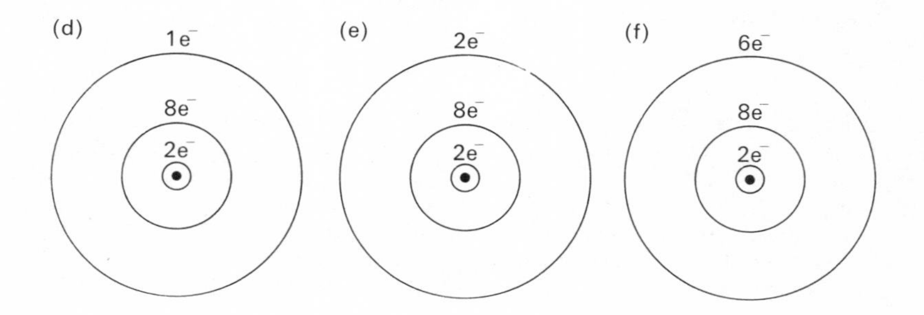

16 Oxidation and Reduction

16.1 Introduction

As the word suggests, the term oxidation was first applied to reactions in which oxygen is added on to an element or compound. The oxygen may come from the free gas, or it may be provided by an oxidizing agent. The burning of substances in air provides plenty of examples of the first type of reaction.

$$2Mg + O_2 \rightarrow 2MgO$$
$$CH_4 + 2O_2 \rightarrow CO_2 + 2H_2O$$

The action of steam on red-hot iron and the reaction between potassium nitrate and sulphur illustrate the second type.

$$3Fe + \underset{\text{Steam}}{4H_2O} \rightarrow Fe_3O_4 + 4H_2\uparrow$$
$$2KNO_3 + S \rightarrow 2KNO_2 + SO_2\uparrow$$

However, if oxidation is confined to these kinds of reactions, it becomes rather a restricted term of little general use, and it has, in fact, proved possible to include many other reactions under the same heading. A similar process of evolution has occurred in the use of terms like acid and base (see chapter 17).

(*a*) There is, for instance, a clear likeness between the burning of magnesium in oxygen and its burning in chlorine, or its explosive reaction with sulphur (dangerous).

$$2Mg + O_2 \rightarrow 2MgO$$
$$Mg + Cl_2 \rightarrow MgCl_2$$
$$Mg + S \rightarrow MgS$$

In all cases the magnesium has become attached to atoms of a non-metal.

(*b*) Iron(II) oxide changes at once to iron(III) oxide on exposure to air. This is clearly an oxidation as the end product has a greater percentage of oxygen in it than the original compound.

$$4FeO + O_2 \rightarrow 2Fe_2O_3$$

But this is analagous to the action of chlorine on iron(II) chloride to give iron(III) chloride.

$$2FeCl_2 + Cl_2 \rightarrow 2FeCl_3$$

In both cases the valency of the iron has increased from two to three, and the proportion of non-metal in the compound has increased.

In the same way iron(II) sulphate can be oxidized by nitric acid to iron(III) sulphate in the presence of dilute sulphuric acid.

$$6FeSO_4 + 3H_2SO_4 + 2HNO_3 \rightarrow 3Fe_2(SO_4)_3 + 4H_2O + 2NO$$

(*c*) The burning of hydrogen is an important reaction.

$$2H_2 + O_2 \rightarrow 2H_2O$$

Because hydrogen shows this strong affinity for oxygen, the removal of hydrogen from a compound has always been classed as an oxidation. For instance, when hydrogen sulphide burns in a restricted supply of air, or when turpentine burns in chlorine, oxidation of those substances takes place.

$$2H_2S + O_2 \rightarrow 2S\downarrow + 2H_2O$$
$$C_{10}H_{16} + 8Cl_2 \rightarrow 10C\downarrow + 16HCl\uparrow$$

Hydrogen is a non-metal, so it appears that the proportion of non-metal in the compound is being diminished, instead of increased, as above. However, hydrogen gives a positive ion and is discharged at the cathode in electrolysis, like a metal (see section 20.7); it is electropositive. Oxygen and chlorine, on the other hand, give negative ions and are discharged at the anode: they are electronegative.

Thus the feature common to all the reactions under consideration is that the substance oxidized has its electronegative part increased or its electropositive part diminished. When magnesium combines with a non-metal its electronegative part is increased from zero to one or two atoms, as appropriate.

(*d*) A fourth aspect of oxidation is obtained by considering the behaviour of the electrons: when magnesium burns in air each magnesium atom loses two electrons to an oxygen atom.

$$Mg\ (2,8,2) \rightarrow Mg^{2+}\ (2,8) + 2e^-$$
$$O\ (2,6) + 2e^- \rightarrow O^{2-}\ (2,8)$$

Similarly when iron is oxidized by hydrogen chloride there is a transfer of electrons.

$$Fe\ (2,8,14,2) \rightarrow Fe^{2+}\ (2,8,14) + 2e^-$$
$$2Cl\ (2,8,7) + 2e^- \rightarrow 2Cl^-\ (2,8,8)$$

This example shows the limitations of the

electronic approach where a covalent compound (hydrogen chloride) is concerned: the chlorine atom in hydrogen chloride is nominally credited with seventeen electrons the same as the isolated atom, but in fact, due to polarization within the molecule, it has a greater share of the two electrons bonding it to the hydrogen making its real electronic structure nearer to 2,8,8.

However, this approach does lead to the point, not mentioned previously, that reduction is, in all respects, the reverse of oxidation. The term was originally applied to reactions in which oxygen was removed from a compound. Reagents, such as hydrogen, carbon, or carbon monoxide, which could bring this about were called reducing agents. It was particularly important in connection with the extraction of metals from their ores.

$$CuO+H_2 \rightarrow Cu+H_2O\uparrow$$
$$Fe_2O_3+3CO \rightleftharpoons 2Fe+3CO_2\uparrow$$

The term was first extended to cover the addition of hydrogen to a substance. When hydrogen combines with oxygen or chlorine, it is still acting as a reducing agent, and oxygen and chlorine are being reduced.

$$2H_2+O_2 \rightarrow 2H_2O$$
$$H_2+Cl_2 \rightarrow 2HCl$$

It was further extended to cover the removal of any electronegative non-metal from a compound, as when iron(III) chloride is reduced to iron(II) chloride by the action of zinc in the presence of dilute hydrochloric acid or by sulphur dioxide.

$$Zn+2FeCl_3 \rightarrow 2FeCl_2+ZnCl_2$$
$$2FeCl_3+SO_2+2H_2O \rightarrow 2FeCl_2+H_2SO_4+2HCl$$

It will be noted that this corresponds to a reduction in the valency of the metal from three to two.

In all these reactions the substance reduced has had its electronegative part decreased or its electropositive part increased.

16.2 Simultaneous Oxidation and Reduction

From the aspects of oxidation and reduction given above, and from the examples which have been quoted it can be seen that they must always take place together. When the oxidizing agent is reduced the reducing agent must be oxidized.

For instance, consider the reaction between copper oxide and carbon monoxide:

Oxidized (CO → CO_2)

$$CuO+CO \rightarrow Cu+CO_2\uparrow$$

Reduced (CuO → Cu)

The copper oxide is reduced to copper and the carbon monoxide is oxidized to carbon dioxide.

Copper oxide is the oxidizing agent and carbon monoxide is the reducing agent.

Again, consider the action of concentrated sulphuric acid (oxidizing agent) on hydrogen bromide (reducing agent):

Oxidized (HBr → Br_2)

$$2HBr+H_2SO_4 \rightarrow 2H_2O+SO_2\uparrow+Br_2$$

Reduced (H_2SO_4 → SO_2)

The hydrogen bromide is oxidized to water and bromine, while the sulphuric acid is reduced to sulphur dioxide, water being a by-product.

It is a useful exercise to go through the examples given above, considering in each case which substance has been oxidized, which has been reduced, and into what each has been converted. Because of the simultaneous nature of reduction and oxidation the changes are frequently called redox reactions.

Thus **oxidation can be defined as:**

(*a*) **the addition of oxygen or any other non-metal,**
(*b*) **an increase in the proportion of the electronegative constituent, or decrease in the proportion of the electropositive constituent,**
(*c*) **the removal of hydrogen or a metal,**
(*d*) **the loss of electrons,**

and **reduction is the opposite.**

16.3 Oxidizing Agents and Tests for them

An oxidizing agent is a substance which can bring about oxidation.

A list of well-known oxidizing agents is given here. One equation illustrating their action is given in each case, and more can be found in chapters 23–32.

Oxygen
$$C+O_2 \rightarrow CO_2$$
Water (steam)
$$Mg+H_2O \rightarrow MgO+H_2\uparrow$$
The halogens
$$2Fe+3Cl_2 \rightarrow 2FeCl_3$$
Concentrated nitric acid
$$P+5HNO_3 \rightarrow H_3PO_4+H_2O+5NO_2\uparrow$$
Concentrated sulphuric acid
$$2HBr+H_2SO_4 \rightarrow 2H_2O+SO_2\uparrow+Br_2\uparrow$$
Hydrogen peroxide
$$PbS+4H_2O_2 \rightarrow PbSO_4+4H_2O$$

Metallic dioxides

$4HCl + MnO_2 \rightarrow MnCl_2 + 2H_2O + Cl_2\uparrow$

Chloric(I) [hypochlorous] acid and salts

$H_2SO_3 + HOCl \rightarrow H_2SO_4 + HCl$

Potassium manganate(VII) (or permanganate) with dilute sulphuric acid

$10FeSO_4 + 2KMnO_4 + 8H_2SO_4$
$\rightarrow 5Fe_2(SO_4)_3 + K_2SO_4 + 2MnSO_4 + 8H_2O$

Potassium dichromate (with dilute sulphuric acid)

$K_2Cr_2O_7 + 7H_2SO_4 + 6FeSO_4 \rightarrow K_2SO_4$
$+ Cr_2(SO_4)_3 + 3Fe_2(SO_4)_3 + 7H_2O$

The molecular equations quoted above can be simplified and the essential changes emphasized by writing ionic equations: all strong electrolytes are completely ionized in aqueous solution. Thus the ions occurring on each side of an equation can be eliminated, e.g. the reaction

$10FeSO_4 + 2KMnO_4 + 8H_2SO_4$
$\rightarrow 5Fe_2(SO_4)_3 + K_2SO_4 + 2MnSO_4 + 8H_2O$

involves mostly ionic substances, and so can be written

$10Fe^{2+} + 10SO_4^{2-} + 2K^+ + 2MnO_4^- + 16H^+$
$+ 8SO_4^{2-} \rightarrow 10Fe^{3+} + 15SO_4^{2-} + 2K^+$
$+ SO_4^{2-} + 2Mn^{2+} + 2SO_4^{2-} + 8H_2O$

and this can then be simplified to

$10Fe^{2+} + 16H^+ + 2MnO_4^- \rightarrow 10Fe^{3+}$
$+ 2Mn^{2+} + 8H_2O$

or

$5Fe^{2+} + 8H^+ + MnO_4^- \rightarrow 5Fe^{3+} + Mn^{2+}$
$+ 4H_2O$

Partial equations for the oxidizing agent and the reducing agent emphasize the transfer of electrons:

$Fe^{2+} \rightarrow Fe^{3+} + e^-$
To oxidizing agent

$MnO_4^- + 8H^+ + 5e^- \rightarrow Mn^{2+} + 4H_2O$
From reducing agent

There is no infallible test for an oxidizing agent, but it is possible to get a good idea of the nature of a given substance by trying the following:

(*a*) An oxidizing agent liberates iodine from an acidified solution of potassium iodide; the iodine gives a blue colour with starch solution: moist starch-iodide paper is a convenient reagent for this test.

$2I^- \rightarrow I_2 + 2e^-$
To oxidizing agent

(*b*) An oxidizing agent converts a sulphate(IV) [sulphite] into a sulphate(VI); a sulphate(VI), in the presence of hydrochloric acid, with barium chloride yields a white precipitate.

$SO_3^{2-} + H_2O \rightarrow SO_4^{2-} + 2H^+ + 2e^-$

(*c*) When an oxidizing agent is warmed with concentrated hydrochloric acid, chlorine is formed, which can be recognized by its colour and its bleaching effect on moist litmus paper.

$2Cl^- \rightarrow Cl_2 + 2e^-$
To oxidizing agent

16.4 Reducing Agents and Tests for them

A reducing agent is a substance which can bring about reduction.

Some common reducing agents are mentioned and illustrated in the following list; further information can be found in chapters 23–32.

Hydrogen

$PbO + H_2 \rightarrow Pb + H_2O$

Carbon

$CO_2 + C \rightarrow 2CO$

Carbon monoxide

$ZnO + CO \rightarrow Zn + CO_2\uparrow$

Potassium iodide (with dilute sulphuric acid)

$O_3 + H_2SO_4 + 2KI \rightarrow K_2SO_4 + H_2O + I_2$
$+ O_2\uparrow$

Hydrogen sulphide

$Cl_2 + H_2S \rightarrow 2HCl + S\downarrow$

Sulphur dioxide (and sulphites)

$O_2 + 2SO_2 \rightarrow 2SO_3$

Metals (especially very electropositive metals)

$Cr_2O_3 + 2Al \rightarrow Al_2O_3 + 2Cr$

Iron(II) salts

$Cl_2 + 2FeCl_2 \rightarrow 2FeCl_3$

Hydrogen iodide

$H_2SO_4 + 8HI \rightarrow 4H_2O + 4I_2 + H_2S\uparrow$

Ethene (ethylene)

$C_2H_4 + 3O_2 \rightarrow 2CO_2 + 2H_2O$

The following tests will indicate the presence of the majority of reducing agents, but not necessarily all.

(*a*) A dilute solution of potassium manganate(VII) (purple), acidified with dilute sulphuric acid, is decolorized by most reducing agents.

$MnO_4^- + 8H^+ + 5e^- \rightarrow Mn^{2+} + 4H_2O$
From reducing agent

(*b*) A dilute solution of potassium dichromate (orange), acidified with dilute sulphuric acid, is turned green by most reducing agents.

$Cr_2O_7^{2-} + 14H^+ + 6e^- \rightarrow 2Cr^{3+} + 7H_2O$
From reducing agent

(*c*) A reducing agent often converts an iron(III) salt to an iron(II) salt. There is a colour change from yellow to very pale green, but this is not usually definite enough to be conclusive: a confirmatory test may be necessary (see section 26.11).

$$Fe^{3+} + e^- \rightarrow Fe^{2+}$$

From reducing agent

16.5 Strengths of Oxidizing and Reducing Agents

Not all oxidizing or reducing agents can perform the same reactions. A powerful oxidizing agent, like acidified potassium manganate(VII), is able to oxidize practically all reducing agents, but a weaker oxidizing agent, like cold concentrated sulphuric acid, affects relatively few.

Sometimes a substance which normally behaves as an oxidizing agent may behave as a reducing agent, when it is in contact with a much more powerful oxidizing agent. Hydrogen peroxide is an oxidizing agent (see section 16.3), but it can be oxidized by silver oxide and then, of course, it is behaving as a reducing agent:

$$Ag_2O + H_2O_2 \rightarrow 2Ag + H_2O + O_2\uparrow$$

Also, potassium manganate(VII) (purple) in acidified solution is decolorized:

$$2KMnO_4 + 3H_2SO_4 + 5H_2O_2 \rightarrow K_2SO_4 + 2MnSO_4 + 8H_2O + 5O_2\uparrow$$

In a similar way, sulphur dioxide is usually a reducing agent (see section 16.4), but it can sometimes act as an oxidizing agent, e.g. the powerful reducing agent hydrogen sulphide is oxidized to sulphur:

Oxidized

$$SO_2 + 2H_2S \rightarrow 3S\downarrow + 2H_2O$$

Reduced

and burning magnesium is oxidized:

Oxidized

$$SO_2 + 2Mg \rightarrow S + 2MgO$$

Reduced

These examples show that the distinction between an oxidizing agent and a reducing agent is not hard and fast.

If only the elements are considered as redox reagents, the periodic table can again be used as a basis for classification:

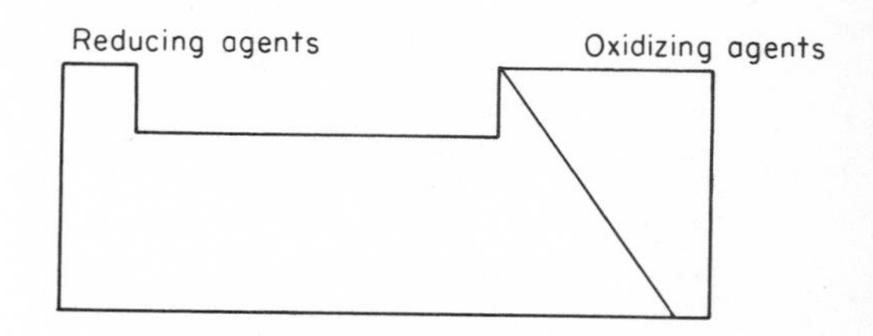

The relative strengths of reducing and oxidizing agents can be assessed quantitatively (redox potentials) in a similar manner to the determination of electrode potentials (see sections 15.5 and 20.2).

QUESTIONS 16

Oxidation and Reduction

1 Which of the following is the most appropriate ending for each sentence?

(*a*) Metal atoms form cations by
 (i) gaining protons.
 (ii) losing protons.
 (iii) gaining electrons.
 (iv) losing electrons.

(*b*) The discharge of cations during electrolysis is
 (i) oxidation.
 (ii) reduction.
 (iii) neither oxidation nor reduction. [J]

2 Define oxidation.

State what is oxidized and name the oxidation product in each of the following reactions:

(*a*) $SO_2 + 2HNO_3 \rightarrow H_2SO_4 + 2NO_2$

(*b*) $H_2O_2 + 2H^+ + 2I^- \rightarrow 2H_2O + I_2$ [J]

3 Titanium (Ti) forms three oxides in which it has valencies of 2, 3 and 4 respectively. Write formulae for these oxides, in this order.

Represent by equations

(*a*) the reduction of titanium(IV) chloride by sodium to form titanium, (*b*) the reduction by aluminium of titanium(IV) oxide to form titanium. [J]

4 (*a*) Name an oxidizing agent, and write an equation for a reaction in which it is used for oxidation.

(*b*) Name a reducing agent, and write an equation for a reaction in which it is used for reduction. [J]

5 Some metals form compounds in more than one 'oxidation state'—e.g. copper forms both copper(I) and copper(II) compounds. Select another metal known to you which forms compounds in two oxidation states, and discuss the methods by which:

(*a*) the metal may be converted into compounds of the lower oxidation state;

(*b*) compounds in the lower oxidation state may be converted into compounds in the higher oxidation state;

(*c*) the changes in (*a*) and (*b*) may be reversed.

[Nf]

6 Classify the following changes as oxidation or reduction, giving reasons:
(*a*) obtaining lead from lead(II) oxide;
(*b*) changing sodium sulphite into sodium sulphate.
(*c*) acting on magnesium with dilute sulphuric acid.
(*d*) acting on potassium iodide solution with hydrogen peroxide.

7 Briefly state one reaction in each case by means of which you could carry out the following conversions. Give the equations.
(*a*) Hydrogen molecules into hydrogen ions (H^+).
(*b*) Iodide ions (I^-) into free iodine.
(*c*) Sulphide ions (S^{2-}) into free sulphur.
(*d*) Chlorine molecules into chloride ions (Cl^-).
(*e*) Nitrate ions (NO_3^-) into nitrite ions (NO_2^-).
(*f*) Chlorate(V) ions (ClO_3^-) into chloride ions (Cl^-).
[L]

8 From the following list select one oxidizing agent and one reducing agent: sulphur dioxide, ethene, iodine, ammonia, potassium nitrate. Describe, in each case, one reaction illustrating the oxidizing or reducing property. [OC]

9(*a*) Write ionic equations for (i) the oxidation of iron(II) ions, (ii) the precipitation of barium sulphate, (iii) the neutralization of sulphuric acid with copper(II) oxide, (iv) the action of hot sodium hydroxide on ammonium chloride solution.
(*b*) Briefly describe how you would carry out reaction (i) and apply tests to show you had succeeded, and (*c*) explain for what special purpose reaction (ii) is used.
[O]

10 Describe and explain experiments (one for each section) by which you could convert (*a*) iron(III) ions to iron(II) ions, (*b*) hydrogen molecules to hydrogen ions, (*c*) chloride ions to chlorine molecules, (*d*) sulphite ions to sulphate ions. [A]

11 Study the equation $2Fe^{3+} + S^{2-} \rightarrow 2Fe^{2+} + S$
Now answer the following questions:
(*a*) How would you bring about this reaction in the laboratory?
(*b*) What would you observe as the reaction proceeded?
(*c*) What kind of reaction is this?
(*d*) What kind of change has been undergone by (i) the Fe^{3+} ions, (ii) the S^{2-} ion?
(*e*) Describe briefly one way in which a solution containing Fe^{2+} ions could be converted into one containing Fe^{3+} ions. [S]

12 What do you understand by the terms oxidation, oxidizing agent? Describe reactions in which the following substances are oxidized, and in each case state the oxidizing agent used and the conditions necessary. (*a*) ammonia, (*b*) iron(II) sulphate, (*c*) carbon monoxide, (*d*) hydrogen sulphide. [S]

13 Oxidation is often defined as increase in the positive valency of an element due to loss of electrons, and reduction as the reverse of oxidation. Explain how this statement applies in the case of the following reactions:
(*a*) the conversion of hydrogen ions to hydrogen molecules;
(*b*) the conversion of iron(II) ions into iron(III) ions;
(*c*) the liberation of sulphur from hydrogen sulphide;
(*d*) the conversion of sulphite ions into sulphate ions.
In each case describe an experiment in which these reactions occur, stating what reagents you use, and what you observe. [S]

14 Define the terms oxidation and reduction.

$$Cl_2 + 2H_2O + SO_2 \rightarrow 2HCl + H_2SO_4$$
$$Cl_2 + H_2S \rightarrow 2HCl + S\downarrow$$
$$SO_2 + 2H_2S \rightarrow 2H_2O + 3S\downarrow$$

From the above equations, deduce which of the three gases chlorine, sulphur dioxide, and hydrogen sulphide, is (*a*) the strongest oxidizing agent, (*b*) the strongest reducing agent. State clearly how you arrive at your answers.
Describe how you would carry out two reactions, the one using nitric acid and the other sulphuric acid, to demonstrate that each of these acids has oxidizing properties. In each case, explain why you classify the reaction as oxidation, and indicate how you would recognize the oxidized product of the reaction. [W]

15 Explain the meaning of the terms oxidation and reduction.
On the basis of your explanations state which substance is oxidized, and which substance is reduced, in each of the following equations:

$$C + H_2O \rightarrow CO + H_2$$
$$C_{10}H_{16} + 8Cl_2 \rightarrow 10C + 16HCl$$
$$2FeCl_3 + 2H_2O + SO_2 \rightarrow 2FeCl_2 + 2HCl + H_2SO_4$$

How would you demonstrate that both concentrated sulphuric acid and concentrated nitric acid possess oxidizing properties? One suitable reaction for each acid should be described and explained.
Describe two tests, which are commonly used, to find out whether a given substance is an oxidizing or a reducing agent. [W]

16 Define or explain the term reduction.
Describe and write equations for four different reactions (i.e. one reaction in each case) in which (*a*) an element is reduced, (*b*) an oxide is reduced, (*c*) a compound is reduced in solution, (*d*) reduction is used to manufacture an element on a large scale. (Manufacturing details are not required.) [C]

17 What do you understand by the term oxidizing agent?
You are provided with a solution of a chemical X in which is thought to be an oxidizing agent. Describe, with all the necessary practical details, two tests that you could apply to the solution and explain how the

results of the tests would justify your calling X an oxidizing agent. [C]

18 Define the terms acid and oxidizing agent.
Describe:
(*a*) two experiments (other than taste or the action of indicators) to show that nitric acid is an acid;
(*b*) two experiments to show that nitric acid is an oxidizing agent;
(*c*) one reaction in which hydrochloric acid acts as a reducing agent. [C]

19 Explain carefully what you understand by the term oxidation. The following have been termed redox reactions. In each case state
(*a*) the products of the reaction,
(*b*) the substance which has been oxidized,
(*c*) why you consider the substance has been oxidized:
(i) the action of dilute hydrochloric acid on zinc;
(ii) the action of steam on magnesium metal;
(iii) the action of zinc on aqueous copper sulphate solution;
(iv) the production of water gas from carbon and steam;
(v) the action of concentrated nitric acid on copper;
(vi) the action of concentrated hydrochloric acid on lead(IV) oxide.
Write equations for reactions (v) and (vi). [NI]

20 What do you understand by the terms oxidizing agent and reducing agent?
Classify the following substances as oxidizing agents or reducing agents and give in outline in each case a reaction supporting your statement: nitric acid, sulphur dioxide, chlorine, carbon monoxide, hydrogen sulphide. Illustrate your answer by equations. [J]

21 What do you understand by the term reduction?
Give three examples of reactions in which reduction occurs, using only materials selected from the following list: sulphur dioxide, carbon monoxide, carbon, hydrogen sulphide, an iron oxide, manganese(IV) oxide, a lead oxide, hydrochloric acid, iron(III) chloride solution.
State clearly in each example which substance is oxidized and which is reduced.
Describe briefly: (*a*) one experiment to show the reducing action of nascent hydrogen; (*b*) one experiment to show the oxidizing action of hydrogen peroxide solution. [J]

22 Name two substances which can act both as reducing agents and oxidizing agents, and give equations to show how they react.

23 In the following equations state, giving your reasons, which substances have been oxidized and which have been reduced:
(*a*) $2FeCl_3 + H_2S \rightarrow S\downarrow + 2FeCl_2 + 2HCl$
(*b*) $Fe_2O_3 + 3CO \rightarrow 2Fe + 3CO_2$
(*c*) $SnCl_2 + HgCl_2 \rightarrow Hg\downarrow + SnCl_4$
(*d*) $SO_2 + 2H_2S \rightarrow 3S\downarrow + 2H_2O$

24 When acidified potassium manganate(VII) solution is mixed with aqueous potassium iodide, a reaction takes place which may be represented as:

$$MnO_4^-(aq) + 8H^+(aq) + 5e^- \rightarrow Mn^{2+}(aq) + 4H_2O(l)$$
$$2I^-(aq) \rightarrow I_2(aq) + 2e^-$$

This reaction may be carried out without allowing the solutions to come into direct contact. Describe how you would do this and explain what happens. Give a sketch of your apparatus. [Nf]

25 Define oxidation in terms of electron transfer. Discuss the application of this definition to the following: (*a*) dissolving a metal in acid, (*b*) burning sodium in chlorine, (*c*) the change in oxidation state of a metal ion, (*d*) the rapid corrosion of a galvanized iron water tank when a copper water pipe is connected to it.

26 Describe simple test tube experiments, a different one in each case, by which you could convert (*a*) ammonium ions into ammonia, (*b*) sulphite ions into sulphate ions, (*c*) copper(II) ions into copper, (*d*) zinc into zinc ions. Name the reagents used in each case, and describe a test which you would apply to show that the required conversion had been carried out. [J]

17 Acids, Bases and Salts

17.1 Introduction

The word acid usually suggests to the non-chemist a corrosive substance with mysterious powers of eating its way through metals, clothes, and flesh. If something can stand up to the 'acid test' it must be very resistant indeed. To the chemist, however, a vague suggestion of corrosive power is not enough to define a group of compounds. There are many substances which can react in this way, and when they are carefully examined they are found to belong to several different categories.

The first property which was associated with acids was the possession of a sharp taste, and the word acid does in fact mean sour. For thousands of years it has been known that wine is liable to go sour and form vinegar, and that sour milk and unripe lemons affect the palate in a similar unpleasant way. Moreover, vinegar was known to be capable of attacking iron, and so possessed the corrosive power which later came to be regarded as characteristic of acids also. The effervescence produced when an acid reacts with a carbonate was also recognized as one of the tests for an acid.

One other property of vinegar is its ability to produce colour changes in certain vegetable dyes. For example, cabbage pickled in vinegar is bright red, while the natural vegetable is purple.

As more and more substances became known, the early chemists gradually began to give the name acid to those which showed properties similar to those of vinegar, sour milk, and lemon juice. The alchemists had discovered it was possible to make more powerful acids from mineral sources. For example, sulphuric acid was made by heating crystals of green vitriol (iron(II) sulphate) and condensing the vapour evolved. These mineral acids showed the reactions of vinegar to an enhanced degree, and made it easier to decide which were the important ones.

All the same, 'acid' remained a vague term until improved analytical methods and the introduction of the atomic theory made it possible to determine formulae. Lavoisier noticed that many acids could be made by dissolving the oxides of non-metals (e.g. sulphur, carbon, phosphorus) in water, and assumed that all acids contained oxygen. The discovery of hydrochloric acid and other similar compounds destroyed this theory. It was then realized that that all acids contained hydrogen and that, in their reactions with metals or carbonates, this hydrogen was replaced by a metal. Thus it became possible to give a definition of an acid which, although incomplete, could serve as a starting-point for further research.

17.2 Acids

An acid was thus originally defined as a compound containing hydrogen, some or all of which could be replaced by a metal, forming a compound known as a salt.

(*a*) Hydrochloric acid (HCl) contains one replaceable hydrogen atom per molecule of acid, and can be converted into sodium chloride ($NaCl$).

(*b*) Sulphuric acid (H_2SO_4) contains two replaceable hydrogen atoms per molecule of acid, and can be converted into either sodium sulphate (Na_2SO_4) or sodium hydrogensulphate ($NaHSO_4$).

(*c*) Ethanoic (acetic) acid (CH_3COOH), contains four hydrogen atoms per molecule of acid, but only one of these is *acidic*, and so ethanoic acid can be converted only to sodium ethanoate (acetate) CH_3COONa.

N.B. Sodium has a vigorous and **dangerous** reaction if added directly to an acid, and the above conversions are carried out indirectly (see section 17.11).

Acids possess many properties in common: by the ionic theory this results from the fact that they all yield hydrogen ions when dissolved in water.

$$HCl \rightarrow H^+ + Cl^-$$
$$H_2SO_4 \rightarrow 2H^+ + SO_4^{2-}$$
$$CH_3COOH \rightarrow H^+ + CH_3COO^-$$

A strong acid, like nitric, hydrochloric, or

sulphuric, ionizes almost completely. A weak acid, like ethanoic or boric, is only slightly ionized.

These hydrogen ions are hydrated and are frequently described as hydroxonium ions (H_3O^+) but this is a simplification of a situation which may involve $H_9O_4^+$. It should be known that most ions are hydrated even though this is not usually acknowledged in equations: it is therefore misleading to suggest that only hydrogen ions are hydrated.

A modern definition of an acid is as follows:

An acid is a solution of a compound which contains hydrogen (hydroxonium) ions as the only positive ions.

N.B. On this definition, sodium hydrogensulphate ($NaHSO_4$) is not an acid, because there are positive sodium ions as well as hydrogen ions in the crystal and in solution.

$$NaHSO_4(s) \text{ is } Na^+ + H^+ + SO_4^{2-}$$

Nevertheless, it possesses many of the characteristics of an acid, and is called an **acid salt** (see section 17.14).

17.3 Properties of Acids

The role of the solvent is vital to the exhibition of acid properties. When pure, solids like citric acid, liquids like sulphuric acid and gases like hydrogen chloride are not, strictly speaking, acids: they are in a covalent state but become ionic substances when dissolved in water and show very similar properties. Hydrogen chloride dissolved in dimethylbenzene is not an acid: there are no hydrogen ions formed.

The extent of ionization varies with the dilution; the term strong is used for acids (and bases) which are largely ionic and the term weak for those which are only partially ionized in solution. The 98% sulphuric acid usually supplied to a laboratory is concentrated but weak; the 0·4% hydrochloric acid in a human stomach is dilute but strong.

(*a*) All acids have a sour taste in dilute solution. This can be observed in lemon juice (citric acid), vinegar (acetic acid), and sour milk (lactic acid).

(*b*) Solutions of acids will usually turn blue litmus red. Some very weak acids, like carbonic acid, may only just affect the litmus. Substances like litmus, which change colour in acid or alkaline solution, are called **indicators**. Other useful indicators are phenolphthalein, which stays colourless in acid solution, and methyl orange, which turns pink in acid solution.

(*c*) Many acids will give hydrogen with the more reactive metals like magnesium, zinc, and iron. Sodium and potassium react with dangerous violence. The hydrogen of the acid is replaced by the metals, thereby forming a salt.

$$Mg + H_2SO_4 \rightarrow MgSO_4 + H_2\uparrow$$
$$Zn + 2HCl \rightarrow ZnCl_2 + H_2\uparrow$$

or

$$\underset{\text{(or Zn)}}{Mg} + 2H^+ \rightarrow \underset{\text{(or }Zn^{2+})}{Mg^{2+}} + H_2\uparrow$$

Nitric acid reacts with most metals, but hardly ever gives hydrogen (see figure 17.1).

(*d*) Almost all acids will react with a carbonate (e.g. washing soda—sodium carbonate) or a hydrogencarbonate to give a fizz (effervescence) of carbon dioxide. This may be used as a test for an acid.

$$\underset{\text{White powder}}{Na_2CO_3} + 2HCl \rightarrow \underset{\text{Colourless solution}}{2NaCl} + H_2O + CO_2\uparrow$$

$$\underset{\text{Green powder}}{CuCO_3} + H_2SO_4 \rightarrow \underset{\text{Blue solution}}{CuSO_4} + H_2O + CO_2\uparrow$$

$$\underset{\text{White powder}}{NaHCO_3} + HCl \rightarrow \underset{\text{Colourless solution}}{NaCl} + H_2O + CO_2\uparrow$$

or

$$2H^+ + CO_3^{2-} \rightarrow H_2O + CO_2\uparrow$$

(*e*) Acids are neutralized by bases to give salts and water only (see section 17.8)

17.4 Preparation of Acids

There are two important general methods for preparing acids:

(*a*) If an oxide of a non-metal is soluble in water it will dissolve to give an acid solution.

Thus, carbon dioxide gives carbonic acid.

$$CO_2 + H_2O \rightleftharpoons H_2CO_3 \rightleftharpoons H^+ + HCO_3^- \rightleftharpoons 2H^+ + CO_3^{2-}$$

Sulphur dioxide gives sulphurous acid, while sulphur trioxide gives sulphuric acid.

$$SO_2 + H_2O \rightleftharpoons H_2SO_3 \rightleftharpoons H^+ + HSO_3^- \rightleftharpoons 2H^+ + SO_3^{2-}$$

$$SO_3 + H_2O \rightleftharpoons H_2SO_4 \rightleftharpoons H^+ + HSO_4^- \rightleftharpoons 2H^+ + SO_4^{2-}$$

Phosphorus(V) oxide gives phosphoric acid.

$$P_4O_{10} + 6H_2O \rightarrow 4H_3PO_4$$

Non-metallic oxides are often called acid anhydrides because they react with water to give acids.

(*b*) If a salt of a volatile acid is heated with a non-volatile acid the volatile acid is driven off and can be condensed or dissolved in water. A volatile substance is one that has a low boiling point and therefore readily vaporizes. Hydrogen chloride is very volatile and is expelled when sodium chloride is warmed with the non-volatile sulphuric acid (concentrated).

$$NaCl + H_2SO_4 \rightarrow NaHSO_4 + HCl\uparrow$$

b.p. 338°C b.p. −85°C

17.5 Bases

A base is a compound which will react with an acid to give a salt and water only.

Bases are either the oxides or the hydroxides of metals. When a base reacts with an acid the reaction is called a neutralization. The following reactions are all neutralizations:

$$CuO + H_2SO_4 \rightarrow CuSO_4 + H_2O$$

Black solid → Blue solution

$$NaOH + HCl \rightarrow NaCl + H_2O$$

Colourless solutions → Colourless solution

$$ZnO + 2HNO_3 \rightarrow Zn(NO_3)_2 + H_2O$$

White solid → Colourless solution

All neutralizations can be summarized as:

Base + Acid → Salt + Water

i.e. $O^{2-} + 2H^+ \rightarrow H_2O$

and $OH^- + H^+ \rightarrow H_2O$

or $OH^- + H_3O^+ \rightarrow 2H_2O$

17.6 Alkalis

Most bases are insoluble in water. However, there are five common bases which are fairly soluble: these are sodium hydroxide (NaOH), potassium hydroxide (KOH), barium hydroxide ($Ba(OH)_2$), calcium hydroxide ($Ca(OH)_2$), and aqueous ammonia (NH_3 in H_2O). These soluble bases are called alkalis. They are all strong bases i.e. highly ionic, except ammonia (frequently written NH_4OH) which is weak.

An alkali is a base which is soluble in water.

Alkalis have a number of properties in common because they all give the hydroxide ion (OH^-) in solution.

NaOH is $Na^+ + OH^-$

$Ca(OH)_2$ is $Ca^{2+} + 2OH^-$

NH_4OH is $NH_4^+ + OH^- \rightleftharpoons NH_3 + H_2O$

17.7 Properties of Alkalis

(*a*) The strong alkalis (NaOH, KOH) are good solvents for oil and grease, and hence are dangerous to handle; their solutions have a slippery, soapy feel because they react with the skin.

(*b*) Solutions of alkalis will affect indicators. They turn red litmus blue, phenolphthalein pink, and methyl orange remains yellow. If a solution of an alkali is added to a solution of an acid coloured pink with litmus, the colour will change to blue as soon as the neutralization is complete (see section 22.5).

(*c*) Alkalis usually have little action on metals, but the strong alkalis react readily with aluminium to give hydrogen. This shows that to give hydrogen with a metal is not an infallible test for an acid.

$$2Al + 6NaOH + 6H_2O \rightarrow 2Na_3Al(OH)_6 + 3H_2\uparrow$$

Sodium aluminate

or

$$2Al + 6OH^- + 6H_2O \rightarrow 2Al(OH)_6^{3-} + 3H_2\uparrow$$

(*d*) Like all bases, alkalis will neutralize acids, giving a salt and water only.

(*e*) Solutions of alkalis will precipitate the insoluble hydroxides of other metals from solutions of their salts. Thus, when sodium hydroxide is added to copper sulphate solution, a blue precipitate of copper hydroxide is formed.

$$CuSO_4 + 2NaOH \rightarrow Cu(OH)_2\downarrow + Na_2SO_4$$

or $Cu^{2+} + 2OH^- \rightarrow Cu(OH)_2\downarrow$

17.8 Salts

A salt is a compound in which the replaceable hydrogen of an acid has been wholly or partly replaced by a metal.

In sodium chloride (NaCl) the hydrogen atom of hydrochloric acid (HCl) has been replaced by an atom of sodium. In copper sulphate ($CuSO_4$) both the hydrogen atoms of sulphuric acid (H_2SO_4) have been replaced by one atom of copper but in sodium hydrogensulphate ($NaHSO_4$) only one of the hydrogen atoms has been replaced.

Salts are named after the metallic and acidic radicals they contain. All salts derived from hydrochloric acid are called chlorides, all salts from nitric acid are nitrates, all salts from sulphuric acid are sulphates, all salts from sulphurous acid are sulphites [sulphate(IV)].

Hydrochloric acid *HCl*	*Nitric acid* *HNO_3*	*Sulphuric acid* *H_2SO_4*
Sodium chloride NaCl	Sodium nitrate $NaNO_3$	Sodium sulphate Na_2SO_4
Magnesium chloride $MgCl_2$	Magnesium nitrate $Mg(NO_3)_2$	Magnesium sulphate $MgSO_4$
Aluminium chloride $AlCl_3$	Aluminium nitrate $Al(NO_3)_3$	Aluminium sulphate $Al_2(SO_4)_3$

Acid	Dilute	Concentrated
Hydrochloric	Both give hydrogen with: magnesium, aluminium, zinc, iron, tin No action: lead, copper, silver, mercury	Both give hydrogen with: magnesium, aluminium, zinc, iron, tin Very slow action: lead, copper, silver, mercury
Sulphuric	Hydrogen evolved with: magnesium, zinc, iron, tin (slowly), aluminium (powder) No action: lead, copper, silver, mercury	Sulphur dioxide evolved if hot with: magnesium, aluminium, zinc, iron, tin, lead, copper, silver, mercury
Nitric	Less than 1 M solution, gives hydrogen with: magnesium A 1 to 6 M solution, gives nitrogen oxide with: magnesium, zinc, iron, tin, lead, copper, silver, mercury	Renders aluminium and iron passive More than a 6 M solution, gives nitrogen dioxide with: magnesium, zinc, tin, lead, copper, silver, mercury
N.B. The action of acids on potassium, calcium and sodium is very vigorous and often dangerously explosive, and should not be carried out		

Figure 17.1
The action of the mineral acids on the common metals

In these examples all the hydrogen of the acid has been replaced by a metal: they are normal salts (see sections 17.14 and 17.15).

In a few cases it is possible to replace the hydrogen of an acid directly with the metal concerned. Dilute hydrochloric acid and dilute sulphuric acid react safely with metals like magnesium, zinc and iron.

$$Mg + 2HCl \rightarrow MgCl_2 + H_2\uparrow$$

or

$$Mg + 2H^+ \rightarrow Mg^{2+} + H_2\uparrow$$

$$Fe + H_2SO_4 \rightarrow FeSO_4 + H_2\uparrow$$

or

$$Fe + 2H^+ \rightarrow Fe^{2+} + H_2\uparrow$$

Dilute nitric acid reacts with most metals (even such unreactive ones as copper and silver) to give nitrates, but it does not give hydrogen. Further reactions occur which usually result in brown fumes of nitrogen dioxide (NO_2) being produced (see figure 17.1).

17.9 Solubilities

Before discussing the various methods available for the preparation of salts it is necessary to summarize information about solubilities because this property usually governs the method.

The following are **soluble** in cold water:

(*a*) All nitrates.
All hydrogencarbonates.
All sodium, potassium and ammonium salts.

(*b*) All chlorides except
silver chloride ($AgCl$),
mercury(I) chloride (Hg_2Cl_2),
(lead chloride ($PbCl_2$), is soluble in hot water, but is usually prepared by precipitation).

(*c*) All sulphates except
barium sulphate ($BaSO_4$),
lead sulphate ($PbSO_4$),
(calcium sulphate ($CaSO_4$), slightly soluble, usually prepared by precipitation).

The following are **insoluble** in water:

(*a*) All carbonates except
sodium carbonate (Na_2CO_3),
potassium carbonate (K_2CO_3),
ammonium carbonate [$(NH_4)_2CO_3$].

(*b*) All hydroxides except
sodium hydroxide ($NaOH$),
potassium hydroxide (KOH),
ammonia (NH_3)—(see section 30.12),
barium hydroxide [$Ba(OH)_2$]—slightly soluble,
calcium hydroxide [$Ca(OH)_2$]—slightly soluble.

17.10 A Summary of the Methods of Preparing Salts

(*a*) *Soluble Salts*

(i) Add the metal to the acid until a little remains unchanged. This method is limited to the reactive metals, but is a general method for nitrates. To prepare the salts of unreactive

metals (e.g. copper) from the metal the method may proceed via the nitrate.
(ii) Add the base to the acid—this is a general method.
(iii) Add the carbonate to the acid—this also is a general method.

Method (i) is a case of oxidation of the metal. Methods (ii) and (iii) can also be considered in terms of an acid reacting with either an alkali (these are done by titration) or an insoluble base (of which excess is used) and are both known as neutralization reactions.

(*b*) *Insoluble Salts*

These are prepared by ionic association or precipitation, a reaction sometimes referred to as double decomposition. If an insoluble salt has to be prepared from an insoluble base or carbonate the method must proceed via a soluble salt.

(*c*) *Special Methods*

Some chlorides, sulphides etc. are prepared by direct combination (synthesis).

17.11 Preparation of Soluble Salts

Experiment (*a*). *The Preparation of Zinc Sulphate Crystals from Metallic Zinc*

Some dilute sulphuric acid is put in a beaker and a few pieces of granulated zinc added. If the reaction is slow a few drops of copper sulphate may be added and the beaker warmed. When the evolution of hydrogen slackens, the hot solution is filtered from the unchanged zinc and impurities into an evaporating basin. The solution is evaporated carefully to about half bulk on a tripod and gauze and set aside for some hours to cool. White, needle-like crystals of zinc sulphate ($ZnSO_4$, $7H_2O$) are slowly formed. These are filtered from the *mother liquor*, washed with a little distilled water, and dried between pieces of filter paper.

$$Zn + H_2SO_4 \rightarrow ZnSO_4 + H_2\uparrow$$
or
$$Zn + 2H^+ \rightarrow Zn^{2+} + H_2\uparrow$$

N.B. The action of the copper sulphate may be explained as follows: because copper is below zinc in the electrochemical series, a little copper will be displaced by the zinc.

$$Zn + Cu^{2+} \rightarrow Zn^{2+} + Cu$$

The copper and zinc in acid solution form a series of small voltaic cells from which hydrogen is much more easily evolved than from pure zinc alone.

Similar methods may be used to make green crystals of iron(II) sulphate ($FeSO_4,7H_2O$) and white crystals of magnesium sulphate ($MgSO_4,7H_2O$). In these cases it is not necessary to add copper sulphate. The iron(II) sulphate solution must not be heated, because it may oxidize to iron(III) sulphate ($Fe_2(SO_4)_3$).

However, in the majority of cases it is not possible to replace the hydrogen of an acid directly, and other methods have to be found.

Experiment (*b*). *The Preparation of Copper Sulphate Crystals from Copper(II) Oxide* (*which is an Insoluble Base*)

Some dilute sulphuric acid is warmed in a beaker on a gauze and small quantities of copper(II) oxide added with constant stirring. The black oxide is added until it is definitely in excess and then the hot solution filtered from the excess copper oxide into an evaporating basin. The blue solution is evaporated slightly, then left to cool until blue crystals of $CuSO_4$, $5H_2O$ have formed. The mother liquor is decanted off, the crystals washed with a small quantity of distilled water, and then dried between filter paper.

$$CuO + H_2SO_4 \rightarrow CuSO_4 + H_2O$$
or
$$CuO + 2H^+ \rightarrow Cu^{2+} + H_2O$$

A great number of salts can be made by this method using either the oxide or hydroxide of the metal concerned. Thus colourless lead nitrate may be made from yellow lead(II) oxide.

$$PbO + 2HNO_3 \rightarrow Pb(NO_3)_2 + H_2O$$
or
$$PbO + 2H^+ \rightarrow Pb^{2+} + H_2O$$

Magnesium sulphate may be made from magnesium hydroxide.

$$Mg(OH)_2 + H_2SO_4 \rightarrow MgSO_4 + 2H_2O$$
or
$$Mg(OH)_2 + 2H^+ \rightarrow Mg^{2+} + 2H_2O$$

Experiment (*c*). *The Preparation of Sodium Chloride Crystals from Sodium Hydroxide* (*which is a Soluble Hydroxide—an Alkali*)

$$NaOH + HCl \rightarrow NaCl + H_2O$$

Because the base is soluble, it is impossible to add it to excess and then to filter off that excess as in (b). It is necessary to find some means of deciding when the neutralization is complete, and for this purpose litmus or some other suitable indicator can be used.

25 cm^3 of dilute sodium hydroxide are placed in a conical flask and a few drops of litmus added. Dilute hydrochloric acid is run in from a burette until the litmus just changes from blue to purple (the neutral shade). The volume needed

is noted and this same volume added to another 25 cm^3 of sodium hydroxide, this time without the litmus. (Alternatively the litmus-coloured neutral solution is boiled with animal charcoal—see section 28.7). The neutral solution is evaporated until crystals start to appear. On cooling the crystals must be extracted, washed and dried as described above.

This method, known as titration, may be used for all sodium, potassium and ammonium salts. For potassium nitrate, potassium hydroxide and nitric acid are used.

$$KOH + HNO_3 \rightarrow KNO_3 + H_2O$$

or $$OH^- + H^+ \rightarrow H_2O$$

For ammonium sulphate, ammonia solution and sulphuric acid are used.

$$2(NH_3) + H_2SO_4 \rightarrow (NH_4)_2SO_4$$

or $$NH_3 + H^+ \rightarrow NH_4^+$$

In preparing ammonium salts by this method methyl orange must be used as indicator instead of litmus.

Experiment (*d*). *The Preparation of Salts from Carbonates*

Another general method of making soluble salts is to use the carbonate of the metal concerned and a suitable acid. Carbonates are salts, not bases, because they give carbon dioxide, water and a new salt with acids. A few carbonates, such as those of sodium, potassium and ammonium, are soluble and hence other salts are prepared by method (c)—neutralization. However, the majority of carbonates are insoluble and method (b) above is applicable. A large beaker should be used as the carbon dioxide produced often causes excessive frothing.

Carbonate + Acid → Salt + Water + Carbon dioxide

$$CaCO_3 + 2HCl \rightarrow CaCl_2 + H_2O + CO_2\uparrow$$

The equations for the preparation of lead nitrate and copper sulphate from their respective carbonates are:

$$PbCO_3 + 2HNO_3 \rightarrow Pb(NO_3)_2 + H_2O + CO_2\uparrow$$

$$CuCO_3 + H_2SO_4 \rightarrow CuSO_4 + H_2O + CO_2\uparrow$$

or $$CO_3^{2-} + 2H^+ \rightarrow H_2O + CO_2\uparrow$$

17.12 Preparation of Insoluble Salts

Lead(II) sulphate is insoluble, and it is not possible to make it by the action of lead(II) oxide (also insoluble in water) on dilute sulphuric acid (in a manner similar to copper sulphate above): the reaction is incomplete, as a layer of the insoluble lead sulphate coats each particle of the oxide. Similarly, there is little effervescence and little reaction when dilute sulphuric acid is added to lead(II) carbonate.

Insoluble salts are precipitated by the method of **ionic association** otherwise known as double decomposition.

Experiment (*a*). Two soluble compounds are used, one of which contains the metallic and the other the acidic half of the salt required. To make lead sulphate, lead nitrate and sodium sulphate are used. When solutions of these are mixed it appears that there is an interchange of radicals and the insoluble salt is precipitated (hence the name double decomposition).

$$Pb(NO_3)_2 + Na_2SO_4 \rightarrow PbSO_4\downarrow + 2NaNO_3$$

Any soluble lead salt and any soluble sulphate (including dilute sulphuric acid) would serve equally well. The precipitate is filtered off, washed with distilled water, and dried. Filtration is quicker if the mixture is first heated.

In terms of the ionic theory it can be seen that the essential reaction is simply:

$$Pb^{2+} + SO_4^{2-} \rightarrow PbSO_4\downarrow$$

This view of the reaction leads to the term ionic association. The sodium and nitrate ions are left in solution and do not associate (making it double decomposition) unless the filtrate is evaporated.

Ionic association is the only satisfactory method for preparing many insoluble salts. Other examples include:

Silver chloride

$$AgNO_3 + HCl \rightarrow AgCl\downarrow + HNO_3$$

or $$Ag^+ + Cl^- \rightarrow AgCl\downarrow$$

Lead chloride

$$Pb(NO_3)_2 + 2HCl \rightarrow PbCl_2\downarrow + 2HNO_3$$

or $$Pb^{2+} + 2Cl^- \rightarrow PbCl_2\downarrow$$

Barium sulphate

$$BaCl_2 + K_2SO_4 \rightarrow BaSO_4\downarrow + 2KCl$$

or $$Ba^{2+} + SO_4^{2-} \rightarrow BaSO_4\downarrow$$

Lead chloride is quite soluble in hot water, so the solution must not be warmed. The first of these reactions is the general test for a chloride and the third reaction the general test for a sulphate.

Experiment (*b*). If it is necessary to prepare an insoluble salt from an insoluble base or carbonate (e.g. calcium sulphate from calcium carbonate) the insoluble compound must first be converted to a soluble salt and then the method of ionic association used.

For instance, the calcium carbonate could be dissolved in dilute hydrochloric acid, and dilute sulphuric acid then added to the solution of calcium chloride formed.

$$CaCO_3 + 2HCl \rightarrow CaCl_2 + H_2O + CO_2\uparrow$$
or $$CO_3^{2-} + 2H^+ \rightarrow H_2O + CO_2\uparrow$$
and $$CaCl_2 + H_2SO_4 \rightarrow CaSO_4\downarrow + 2HCl$$
or $$Ca^{2+} + SO_4^{2-} \rightarrow CaSO_4\downarrow$$

Similarly, lead chloride would be prepared from lead oxide via the nitrate.

$$PbO + 2HNO_3 \rightarrow Pb(NO_3)_2 + H_2O$$
or $$PbO + 2H^+ \rightarrow Pb^{2+} + H_2O$$
and $$Pb(NO_3)_2 + 2HCl \rightarrow PbCl_2\downarrow + 2HNO_3$$
or $$Pb^{2+} + 2Cl^- \rightarrow PbCl_2\downarrow$$

17.13 Preparation of Salts by Synthesis

Certain classes of salts which contain only two elements (e.g. chlorides, bromides, iodides, sulphides) can sometimes be made directly from the elements concerned.

If a piece of sodium is heated on a deflagrating spoon and plunged into a jar of chlorine gas it will burn with a yellow flame, and white clouds of sodium chloride are formed, which deposit crystals on the walls of the jar.

$$2Na + Cl_2 \rightarrow 2NaCl$$

If dry chlorine is passed over heated iron filings or aluminium foil in a hard-glass tube volatile brown iron(III) chloride or white aluminium chloride is formed, which can be condensed in a cooled vessel.

$$2Fe + 3Cl_2 \rightarrow 2FeCl_3$$
$$2Al + 3Cl_2 \rightarrow 2AlCl_3$$

For further experimental details see section 26.13.

It has already been shown (see section 3.6) that when iron filings and sulphur are heated in a test tube a vigorous reaction occurs, with the formation of black iron(II) sulphide.

$$Fe + S \rightarrow FeS$$

Direct combination is mainly of importance in preparing chlorides in the anhydrous state. It is easy to make crystals of magnesium chloride ($MgCl_2,6H_2O$) by dissolving magnesium ribbon in dilute hydrochloric acid and evaporating the solution to the point of crystallization. These crystals, however, decompose on heating to give magnesium oxide and hydrochloric acid, and it is impossible to get anhydrous magnesium chloride ($MgCl_2$) in this way. But the action of dry chlorine on heated magnesium will give the anhydrous salt.

$$Mg + Cl_2 \rightarrow MgCl_2$$

If a metal (e.g. iron or tin) forms more than one chloride the action of chlorine usually causes the metal to exhibit the higher valency (i.e. to give iron(III) chloride or tin(IV) chloride), whereas hydrogen chloride causes it to exhibit the lower valency (i.e. to give iron(II) chloride or tin(II) chloride).

17.14 Acid Salts

If an acid contains only one atom of replaceable hydrogen in its molecule it is said to be *monobasic* or *monoprotic* (e.g. HCl, HNO_3). If the molecule contains two atoms of replaceable hydrogen (e.g. H_2SO_4) it is *dibasic*; if three atoms of replaceable hydrogen (e.g. H_3PO_4) it is *tribasic*, and so on.

The basicity of an acid is the number of replaceable hydrogen atoms in one molecule of the acid.

It should be emphasized that the basicity of ethanoic (acetic) acid (CH_3COOH) is one, although there are four hydrogen atoms in the molecule, because only one of the atoms is replaceable by a metal.

When all the hydrogen of an acid has been replaced by a metal a normal salt is formed. Thus Na_2SO_4 is normal sodium sulphate, and Na_3PO_4 is normal sodium phosphate. However, it is possible for only part of the hydrogen to be replaced, and the resulting compounds are called acid salts.

An acid salt is one in which only part of the replaceable hydrogen of an acid has been replaced by a metal and on dissolving in water yields both the hydrogen and the metal as cations.

Thus sodium hydrogensulphate ($NaHSO_4$), sodium dihydrogenphosphate (NaH_2PO_4) and disodium hydrogenphosphate (Na_2HPO_4) are acid salts.

Acid salts resemble acids in so far as they release hydrogen ions in solution.

$$NaHSO_4 \text{ is } Na^+ + H^+ + SO_4^{2-}$$

They will not necessarily react acid to indicators. For example, although sodium hydrogensulphate reacts acid to litmus, sodium hydrogencarbonate ($NaHCO_3$) reacts alkaline to litmus.

Experiment. The Preparation of Sodium Sulphate and Sodium Hydrogensulphate

$$NaOH + H_2SO_4 \rightarrow NaHSO_4 + H_2O$$
$$2NaOH + H_2SO_4 \rightarrow Na_2SO_4 + 2H_2O$$

It can be seen that the preparation of the normal salt requires twice as much alkali as the preparation of the acid salt. Alternatively, to produce the acid salt rather than the normal salt, requires twice as much acid for the same volume of alkali taken. A solution of the normal salt, sodium sulphate, is prepared by neutralization, using the method described in section 17.11. Then if a volume of sulphuric acid equal to that already used is now added, and the solution evaporated, crystals of sodium hydrogensulphate ($NaHSO_4,H_2O$) are obtained, which are different in shape from those of sodium sulphate ($Na_2SO_4,10H_2O$).

The fact that two different salts can be prepared in this way shows that sulphuric acid is dibasic or diprotic.

17.15 Basic Salts

When sodium carbonate solution is added to a solution of copper sulphate the precipitate formed does not have the expected composition ($CuCO_3$). It appears to contain some $Cu(OH)_2$ and is called a basic carbonate.

A basic salt is one formed by combination of the normal salt with the oxide or hydroxide of the metal.

Basic salts usually have rather indefinite compositions, but are often important. Thus basic lead carbonate [$Pb(OH)_2,2PbCO_3$] is the pigment, white lead, which has been used as the starting material for many paints.

Basic carbonates are usually precipitated by sodium carbonate. The normal carbonates are more likely to be obtained if sodium hydrogencarbonate is used as the precipitating agent.

The acidity of a base is the number of hydrogen atoms (of an acid) which will react with one molecule of the base.

17.16 The pH Scale

The indicators mentioned in section 17.3 are sometimes found to give variable results, e.g. sodium hydrogencarbonate solution is neutral to phenolphthalein but alkaline to methyl orange, whereas boric acid (H_3BO_3) is neutral to methyl orange but acidic to phenolphthalein. A 'universal indicator' can be made of a mixture of simple indicators which shows a range of colours as the acidity or alkalinity of a solution is varied: distilled water is taken as neutral.

Water is very slightly ionized

$$H_2O \rightleftharpoons H^+ + OH^-$$

but it is inconvenient to measure the hydrogen ion concentration in gram/dm^3 i.e. mol H^+/dm^3 (because H = 1) and better to consider it on a logarithmic scale which reduces a large range of values to a comprehensible set of numbers:

less than 7	acidic
7	neutral
more than 7	alkaline

The pH of a solution is the negative logarithm of the hydrogen ion concentration in mol/dm^3.

$$pH = -\log_{10}[H^+]$$

Thus pH = 1 indicates a very acidic solution while pH = 4 shows a solution to be fairly acidic. On the other hand pH = 10 corresponds to a fairly alkaline solution and pH = 14 to a very alkaline solution. Precision can thus be given to the terms strong and weak as applied above to acids and alkalis, the concentration or dilution of the substance being known. It is instructive to use pH paper to study the nature of foodstuffs stored at home.

QUESTIONS 17

Acids, Bases and Salts

1 In this question you will be asked to refer to the following:

A Dry hydrogen chloride.
B A solution of hydrogen chloride in dry methylbenzene.
C A solution of hydrogen chloride in water.
D Pure ethanoic (acetic) acid.
E A solution of ethanoic (acetic) acid in water.
F A solution of potassium hydroxide in water.

The solutions are all of the same molecular concentration.

(*a*) State three of them which would not change the colour of dry blue litmus paper (indicate your answer by writing the appropriate letters).

(*b*) Which two of them would you expect to be the best conductors of electricity?

(*c*) An excess of B was shaken with some of F. Two layers were formed and they were separated by means of a separating funnel. The aqueous layer, which was found to be acidic, was evaporated and yielded a white solid (G).

(i) Name G.

(ii) Explain concisely the changes which take place when excess of B is shaken with F. Give appropriate equations.

(*d*) Brønsted defined an acid as a proton (hydrogen ion) donor, and a base as a proton acceptor. Explain these definitions concisely by referring to the changes which take place when B is shaken with water. [Nf]

2 What do you understand by a salt? Give three different methods by which salts can be prepared.

Describe in detail how you would prepare samples of potassium nitrate and lead chloride, using a different method in each case. Describe a test for each salt to show that you have prepared a nitrate and a chloride. [C]

3 State, with essential experimental details, how you would prepare specimens of (*a*) copper(II) oxide, from a solution of copper sulphate; (*b*) calcium sulphate, from calcium carbonate; (*c*) crystals of iron(II) sulphate, from metallic iron. [C]

4 Study the following table:

Solubility in Water

Compound	*Carbonate*	*Oxide*	*Hydroxide*	*Chloride*	*Nitrate*	*Sulphate*
Caesium	Soluble	Soluble	Soluble	Soluble	Soluble	Soluble
Cadmium	Insoluble	Insoluble	Insoluble	Soluble	Soluble	Soluble
Strontium	Insoluble	Soluble	Soluble	Soluble	Soluble	Insoluble

You are given the data above and dilute solutions of hydrochloric acid, nitric acid, sulphuric acid, water, the usual indicators, and any apparatus you require.

Describe carefully how you would prepare (*a*) crystals of caesium chloride from caesium hydroxide; (*b*) cadmium hydroxide from cadmium oxide; (*c*) strontium sulphate from strontium carbonate.

5 (*a*) In preparing crystals of sodium sulphate, it is a bad technique to evaporate the solution to dryness, or nearly to dryness, but to obtain a substantial sample of sodium chloride, it is better to evaporate to dryness. Explain why these two different techniques are necessary.

How would you obtain good crystals of sodium chloride from a concentrated solution?

(*b*) How would you prepare crystals of sodium nitrate, starting from sodium carbonate and dilute nitric acid?

Describe the action of heat on sodium nitrate crystals and name the products. [C]

6 How would you prepare samples of:
(*a*) barium sulphate from barium chloride;
(*b*) copper(II) sulphate from copper(II) carbonate;
(*c*) potassium sulphate from potassium hydroxide? [Sc]

7 Give four properties by which you could recognize a given solution as an acid. If you were told that the acid was dibasic, what would this mean and how could you verify the statement by actual experiment? [J]

8 Give the names of three acids and three bases, and the names of the products when the first acid combines with the first base, the second acid with the second base, and the third acid with the third base. Give full experimental details showing how you would isolate the pure crystalline product in one of your examples. [J]

9 Give the name and formula, and a brief description of one alkali and one acid, and describe in detail how you would prepare a crystalline salt from them.

Describe how you would prepare (*a*) an insoluble hydroxide; (*b*) an insoluble salt.

10 State the main characteristics of bodies known respectively as acid, base, salt and alkali.

Give the name, formula and brief description (e.g. white solid, colourless liquid, etc.) of one member of each class.

Describe fully one reaction in which an acid acts on an alkali, showing how the resulting main product is obtained.

11 What do you understand by the terms acid, base and salt? Give the names and formulae of two substances belonging to each class.

How would you prepare (*a*) iron(II) sulphate crystals from iron filings; (*b*) sodium sulphate crystals from sodium hydroxide solution, given a supply of dilute sulphuric acid?

12 State four different methods which may be used to prepare salts.

Describe briefly how you would prepare pure dry specimens of (*a*) copper sulphide, starting from copper sulphate; (*b*) calcium sulphate, starting from calcium carbonate; (*c*) zinc sulphate crystals, starting from zinc carbonate. [J]

13 Describe how you would prepare, in a reasonably pure, dry state, two of the following salts: zinc sulphate, starting from zinc; lead sulphate, starting from lead(II) oxide; magnesium carbonate, starting from magnesium sulphate; potassium nitrate, starting from solid potassium hydroxide. [O]

14 What do you understand by the terms: (*a*) acid (*b*) acid salt (*c*) normal salt? Give the name and formula of one example of each.

Given a supply of sodium hydroxide solution, copper(II) oxide, and dilute sulphuric acid, describe with full experimental detail, how you would prepare crystalline samples of (*a*) an acid salt of sodium; (*b*) a normal salt of copper. What is the action of heat on the product obtained in (*b*)?

15 Describe in detail how you would obtain from the appropriate carbonate: (*a*) calcium chloride suitable for drying gases; (*b*) crystalline sodium chloride; (*c*) crystalline lead chloride. [C]

16 Starting from sodium hydroxide, describe how you would prepare solid specimens of (*a*) sodium carbonate; (*b*) sodium hydrogencarbonate; (*c*) sodium sulphate: (*d*) sodium hydrogensulphate.

17 Describe experiments by which you would prepare (*a*) an acidic oxide from oxygen; (*b*) a basic oxide from oxygen; (*c*) a normal salt from a basic oxide; (*d*) an acid from one of its salts.

18 What is meant by a basic salt? Describe the preparation of any one basic salt that is of commerical importance.

19 Tartaric acid is a crystalline solid which is soluble in water. Describe and explain three tests you would attempt with this substance to prove that it is an acid, stating clearly what you would expect to observe in each test. Write equations, where possible, using the simplified formula H_2X for the acid, which is dibasic. How would you prepare from the acid a pure dry specimen of lead tartrate (an insoluble salt)? Tartaric acid contains, by mass, 32% of carbon, 64% of oxygen, the remainder being hydrogen. Find its empirical formula. The relative molecular mass of the acid is 150. Deduce (*a*) the molecular formula of the acid, (*b*) the relative molecular mass of lead tartrate. [W]

20 The element strontium (Sr) is in Group II of the periodic table (Be, Mg, Ca, Sr, Ba, Ra). Describe with essential practical details how you would prepare from strontium carbonate (*a*) strontium nitrate crystals ($Sr(NO_3)_2,4H_2O$), (*b*) strontium sulphate ($SrSO_4$).

21 All acids are soluble, but this is not true of either bases or of salts. Suggest substances which illustrate the methods outlined in the table below, and write equations for all reactions.

Acids	*Base (or substance behaving as such)*	*Salt*	*Method*
	Soluble	Soluble	
A1	B1 (an alkali)	S1	Titrate, evaporate, filter, rinse and dry.
A2	B2 (a carbonate)	S2	
	Soluble	Insoluble	
A3	B3 (an alkali)	S3	Precipitate, filter, rinse and dry.
A4	B4 (a nitrate)	S4	
	Insoluble	Soluble	
A5	B5 (a metal)	S5	Filter off excess base, evaporate, filter, rinse and dry.
A6	B6 (an oxide)	S6	
A7	B7 (a carbonate)	S7	
	Insoluble	Insoluble	
A8	B8 (an oxide or hydroxide)	S8	Proceed via soluble salt, precipitate, filter, rinse and dry.
A9	B9 (a carbonate)	S9	

Miscellaneous and Analysis

18 Energy Changes and Rates of Reactions

18.1 Energy Changes

When concentrated sulphuric acid is carefully poured into cold water there is a large rise in the temperature but the opposite is observed when ammonium nitrate is dissolved in water: the temperature decreases. To make sodium thiosulphate crystals dissolve in water at an appreciable rate the water should be heated but when a supersaturated solution is seeded heat is released as crystallization occurs.

Reactions that are described as chemical are also associated with energy changes. These are usually much greater than those associated with common physical changes, but much less than those accompanying nuclear reactions (see sections 2.3 and 12.5).

An exothermic reaction is one in which energy is given out.

An endothermic reaction is one in which energy is taken in.

The combustion of a fuel is a most important exothermic reaction. Most common reactions are exothermic. The energy evolved is usually measured by electrical means and so it is better to state the change in joules than in calories. The conversion factor is

$$1 \text{ calorie} = 4{\cdot}184 \text{ joules}.$$

The energy change for a reaction may be quoted thus:

(*a*) $C + O_2 \rightarrow CO_2 \quad \Delta H = -406 \text{ kJ/mol}$

This means that when 12 g of carbon (as charcoal) react completely with 32 g of oxygen to give 44 g of carbon dioxide the energy evolved is 406 kilojoules. The substances are presumed to be in their normal states or, to avoid misunderstanding, s, l and g can be inserted (see section 11.10). See figure 18.1.

(*b*) $N_2 + O_2 \rightarrow 2NO \quad \Delta H = +184 \text{ kJ}$

This means that when 28 g of nitrogen react with 32 g of oxygen to give 60 g of nitrogen oxide the energy absorbed is 184 kilojoules. Energy changes measured at constant pressure are called enthalpy changes.

A compound which is formed from its elements with the evolution of energy is called an exothermic compound, and one which is formed from its elements with the absorption of heat is an endothermic compound.

Exothermic compounds, e.g. water, sodium chloride, hydrogen chloride, sulphuric acid, methane, are usually very stable and may be difficult to decompose into their elements. The amount of heat needed to decompose them must be the same as the heat given out when they

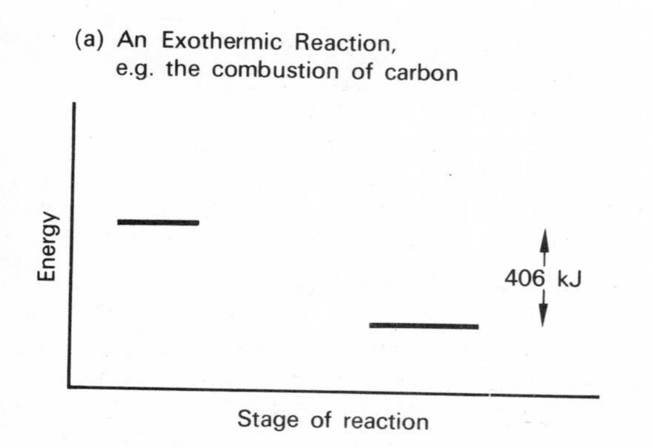

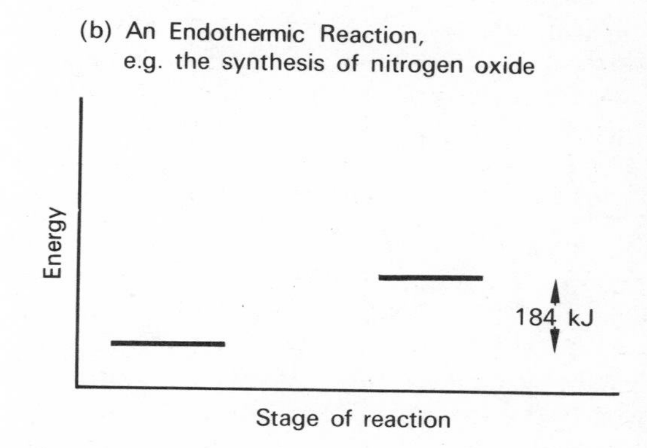

Figure 18.1
Energy changes in reactions

were formed. If this is large, it will be possible to provide enough energy only by using very high temperatures or by electrolysis.

Endothermic compounds, e.g. ethene, ethyne, hydrogen iodide, dinitrogen oxide and trioxygen, tend to be unstable and to decompose easily at quite low temperatures, and heat is given out as they decompose. Thus a glowing splint reignites in dinitrogen oxide because it is hot enough to decompose the gas which yields nitrogen (67%) and oxygen (33%), a mixture richer in oxygen than air. All these endothermic substances would decompose almost completely into their elements at room temperature if left long enough, but the speed at which they would do so is so slow that they can be kept almost indefinitely (with the exception of trioxygen).

18.2 The Measurement of Energy Changes

(a) The Bunsen Burner

The energy output of a Bunsen burner can be assessed roughly by putting it, after lighting and adjusting to a moderate non-luminous flame, under a 250 cm^3 beaker containing about 100 cm^3 of water: using a thermometer and a clock the rate at which the temperature rises can be found.

$$\text{Energy output} = \text{Mass of water} \times \text{Specific heat capacity of water} \times \text{Rate of rise of temperature}$$

The specific heat capacity of water is 4·184 kilojoules $kilogram^{-1}$ $°C^{-1}$.

(b) Liquid into Gas

The heating of the water in section (a) can be continued until approximately half the water has been boiled away. The experiment is then stopped, the time the water has been boiling is noted and the volume of water remaining is found using a measuring cylinder. The energy needed to vaporize 1 mol of water (18 g) can then be found: this is called the enthalpy of vaporization (41 kJ/mol).

(c) Liquid into Solid

The rate of cooling of a hot liquid, e.g. naphthalene, m.p. 80°C, and then the time it takes to solidify can be found: the enthalpy of fusion can be calculated. Alternatively, ice cubes can be dropped into warm water and the experiment stopped when the ice cubes on melting have caused the temperature to fall below that of the room: the enthalpy of fusion of ice is 6 kJ/mol.

(d) Acid and Alkali

When an acid is added to an alkali neutralization occurs and it is accompanied by the evolution of heat. The experiment can be performed using 50 cm^3 of a 1M solution of sodium hydroxide and adding 50 cm^3 of a 1M solution of hydrochloric acid. The enthalpy for the neutralization of a strang acid by a strong alkali (see section 17.3) is 57 kJ/mol because in all examples fulfilling these conditions the essential reaction is

$$H^+ + OH^- \rightarrow H_2O$$

(e) A Thermometric Titration

The previous experiment can be adapted to perform a titration without an indicator. A known amount of an alkali is put into a beaker and stirred with the thermometer. Small known portions of the acid are added and the temperature recorded after each addition: the portions should be added in fairly quick succession. The temperature rises whilst neutralization is occurring but thereafter only random small variations occur. From a graph the volumes of acid and alkali for neutralization can be found.

(f) A Displacement Reaction

A weighed portion of zinc dust can be added to a concentrated solution of copper sulphate. The solution must still be blue at the end of the reaction: all the zinc reacts and there is an excess of copper sulphate. The temperature change can be measured and the energy change in the reaction calculated.

18.3 Intermolecular Forces

Van der Waals' forces are the weak forces which hold molecules together in the liquid and solid states (see section 14.3). The enthalpies of fusion and evaporation of water are 6 and 41 kJ/mol respectively: much more energy is required to convert the liquid to a gas than the solid to liquid. This difference is also apparent when the volume occupied by 1 mol of water (18 g, 6×10^{23} molecules) in the three states is considered:

$$20\ cm^3\ \text{ice} \rightleftharpoons 18\ cm^3\ \text{liquid} \rightleftharpoons 30\,600\ cm^3\ \text{steam (at 100°C)}$$

A large amount of energy is needed to separate the molecules as they proceed from the liquid to the gaseous phase. If the molecules are constrained by a high pressure to be close together and then the pressure released the molecules separate and the temperature falls.

The refrigerator depends upon this change: the liquid under pressure expands through a valve and passes along a pipe through the interior of the machine cooling the contents of the cabinet. The gas is then compressed, energy being released in accordance with the work done, on the outside of the machine: the warm liquid is cooled by the room before returning to the valve. Ammonia and sulphur dioxide have been replaced in refrigerators by organic fluids which are much less toxic and less obnoxious if they escape.

Similar considerations apply when air is to be liquefied preparatory to being fractionally distilled to separate it into its components (see section 30.3).

18.4 The Speed of a Reaction

Many chemical changes seem to occur as soon as the reagents are mixed, e.g. a white precipitate of barium sulphate appears the moment dilute sulphuric acid is added to barium chloride solution. The speed of most precipitation reactions in inorganic chemistry is too fast to be measured easily. Some reactions, e.g. the rusting of iron, occur so slowly that experiments would be long drawn out affairs.

Other reactions take place at moderate speeds and are easily investigated in the laboratory, e.g. zinc dissolving in dilute sulphuric acid, marble chips (calcium carbonate) dissolving in dilute hydrochloric acid and the catalytic decomposition of hydrogen peroxide using manganese(IV) oxide.

Industrial chemists are frequently concerned with reactions which occur only slowly; many of these involve gases, e.g. the manufacture of sulphuric acid (see section 31.21), the manufacture of ammonia (see section 30.9).

There are three important factors which influence the rate of a reaction.

(*a*) The supply of energy, e.g. heat and light.
(*b*) The state of the reagents, e.g. the size of solid particles or the concentration of a solution or the pressure of a gas.
(*c*) The presence of a catalyst.

18.5 The Supply of Energy

The particles of a substance at most temperatures above 0 K (−273°C) are vibrating and rotating and, especially if the temperature is high enough for the substance to be a liquid or a gas, the particles will be moving about from one position to another (translational motion).

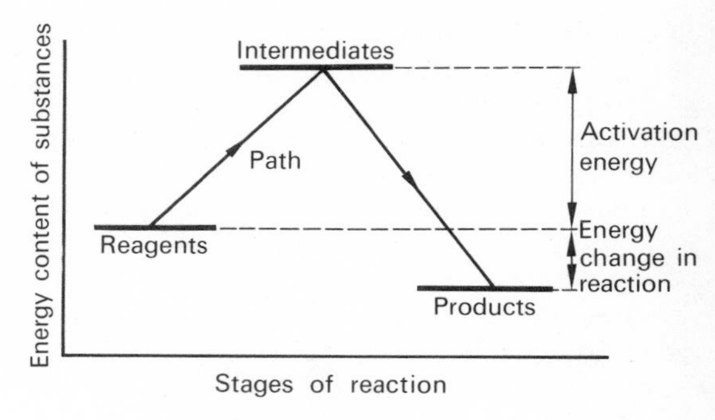

Figure 18.2
Activation energy

As the temperature increases all these movements become more violent: at high temperatures, covalent bonds may be broken and substances atomized.

There are some reactions that proceed spontaneously, e.g. most precipitations and the neutralization of an acid by a base (see section 17.11). Some substances, especially finely divided ones, are pyrophoric, i.e. they take fire spontaneously in air, e.g. white phosphorus. However, many of the reactions involving covalent molecules and metals do not take place unless these substances are heated: gas does not burn until ignited, a safety match does not catch fire until the heat coming from friction between it and the box raises the temperature, petrol does not burn until an electric spark raises the temperature of the vapour and magnesium does not burn until it has been heated above its melting point.

Although all these reactions evolve energy they do not proceed spontaneously: energy has to be supplied before they will occur. This energy is known as the **activation energy** of the reaction: it is the energy needed to break old bonds before new ones can form and a schematic diagram can be constructed (see figure 18.2). It may not be possible to isolate the intermediate compounds (activated complexes). More energy is regained when the new bonds form than was absorbed when the old bonds broke and so the reaction overall is exothermic. The equation for a reaction does not convey any information whatsoever about its mechanism.

The particles of a substance have various energies, the average energy being a measure of the temperature, and, as the temperature increases, more and more particles acquire the activation energy which is necessary before a reaction will take place. A rise of 10°C is usually sufficient to double the rate of a reaction because it doubles the number of particles possessing the activation energy.

Light, like heat, is a form of energy: many reactions will go faster if exposed to sunlight or a bright artificial light. A mixture of equal volumes of hydrogen and chlorine will not react at all in the dark; they react slowly in diffuse daylight but in sunlight or the light from burning magnesium the reaction is so rapid that an explosion occurs.

Other reactions in which the rate is affected by light include photosynthesis (see section 5.11) and photography. A photographic film contains silver bromide and where light falls on it silver is formed and bromine released into the film. Development intensifies the image and then fixation stabilizes the image: the 'negative' is now ready for printing a 'positive' of the picture.

In the manufacture of the insecticide 'Gammexane' the reagents, benzene and chlorine, are illuminated by ultra violet lamps.

18.6 The State of the Reagents

When the solubility of a substance is investigated (see section 6.10), it is found that to dissolve the substance rapidly it should be finely divided. A substance in the form of a powder will usually react more readily than the same substance in lumps, e.g. a magazine may be difficult to burn until the pages are ripped apart and crumpled. In general the finer the state of subdivision the larger the area of the reagent exposed and hence the greater the chance for the molecules to react. Materials in the form of very fine powders or dusts often create serious fire and health hazards. Coal dust in mines may be dangerously explosive and great care has to be taken to avoid sparks; asbestos dust in factories can injure the lungs of the workers there.

Reactions between two solids are often very slow and so to give the particles mobility they are made into solution, usually in water. In solution the particles are separated and move freely, and more easily come into contact with many other particles than when they were in the solid state.

If the concentration of a solution is increased, i.e. the number of particles in a given volume is increased, the chances of their meeting and reacting is increased. For example moderately concentrated hydrochloric acid will dissolve aluminium powder more rapidly than will the dilute acid. The precise relationship between the concentration of a substance and the rate of its reaction must be found by experiment: the Law of Mass Action, which has misled many people, has been relegated to section 33.6.

In gaseous reactions an increase in the pressure will cause a decrease in volume thus compressing the molecules into a smaller space. Hence an increase in pressure will result in an increase in concentration and in rate of reaction. The danger of explosions is greater when dealing with gases: mixtures of town gas and air, or hydrogen and oxygen explode violently when ignited. An **explosion** is an extremely rapid reaction in which a large amount of energy is liberated very quickly. The gases formed cause the generation of pressure waves which do most of the damage and produce the noise.

18.7 The Effect of a Catalyst

When potassium chlorate(V) is heated strongly in a test tube it melts and gives off oxygen.

$$2KClO_3 \rightarrow 2KCl + 3O_2\uparrow$$

This reaction is, however, slow and needs a high temperature. If the crystals are heated very gently until they have just melted, no oxygen appears but the addition of a little manganese(IV) oxide, a black powder, produces rapid evolution of the gas. Manganese(IV) oxide alone will only give off oxygen at a much higher temperature and in this experiment is found to be unchanged in mass at the end: in some way it has made it easier for the potassium chlorate to lose oxygen. It is said to have acted as a catalyst.

A catalyst is a substance which increases the rate of a chemical reaction but remains itself unchanged in mass and chemical composition at the end of the reaction.

Experiment. To Show that Manganese(IV) Oxide Catalyzes the Decomposition of Potassium Chlorate(V)

(*a*) Potassium chlorate(V) alone and potassium chlorate mixed with manganese(IV) oxide are put in separate hard glass test tubes embedded in sand on a sand bath. The sand bath is heated and oxygen is found to come from the mixture before the pure substance. Oxygen does come from both samples: it is only the rate of the reaction which has altered, not the final result.

(*b*) Some potassium chlorate to which a weighed amount of manganese(IV) oxide has been added is heated in a hard-glass test tube. After the reaction has ceased the tube is allowed to cool, and then water is added and the tube reheated: the potassium chloride (and any remaining potassium chlorate) dissolves. The suspension of manganese(IV) oxide is filtered or centrifuged off, washed and dried: its mass is found to be the same as before.

(c) The manganese(IV) oxide is unchanged by the reaction: samples when warmed with concentrated hydrochloric acid yield chlorine both before and after the oxygen-yielding reaction.

A similar experiment can be performed using hydrogen peroxide and manganese(IV) oxide.

$$2H_2O_2 \rightarrow 2H_2O + O_2\uparrow$$

The increasing use of catalysts is one of the most important factors in the development of modern chemistry and especially of chemical industry. Many of the reactions governing the production of vital substances can only be made to proceed at reasonable speeds if a catalyst is used. Enzymes, which are catalysts, associated with living organisms, are discussed in sections 29.13 and 29.18. The action of a catalyst may be compared very roughly to the role played by the lubrication oil in machinery. The machine may not work at all satisfactorily without the oil but if the oil is present at the start the machine will work and the oil remains at the end.

Usually only very little of the catalyst is required. This is particularly true when a solid is catalyzing the reaction between two gases, e.g. the synthesis of ammonia (see section 30.9). Although the catalyst is unchanged chemically at the end of the reaction there may be physical changes, e.g. if the manganese(IV) oxide in the above experiment is added in lumps it may end up as a fine powder.

There are a few cases known where changing the catalyst does seem to change the course of the reaction, e.g. ethanol vapour when passed over copper at 300°C loses hydrogen to give ethanal (acetaldehyde)

$$CH_3CH_2OH \rightarrow CH_3CHO + H_2$$

but when passed over aluminium oxide loses water to give ethene

$$CH_3CH_2OH \rightarrow CH_2{\equiv}CH_2 + H_2O$$

It is only that the proportions of the four substances alter: the equations are approximations.

Catalysts may have their efficiency improved by promoters, e.g. potassium hydroxide and aluminium oxide are added to iron for the Haber process. Catalysts may also be poisoned: many of the substances poisonous to human beings such as arsenic and sulphur compounds, carbon monoxide and cyanides, will inhibit the action of a catalyst.

There are some substances which slow down chemical reactions: they are better referred to as inhibitors rather than as negative catalysts because often their effectiveness depends on the quantity. Tetraethyllead is added to petrol as an anti-knock agent (see section 24.6); it hinders the pre-ignition of the air and petrol vapour mixture by the heat of compression before the mixture is sparked.

18.8 The Types of Catalyst

(a) Surface Catalysts

Faraday (1834) was the first to realize the importance of the surface when a solid was catalyzing a reaction between two gases. 1 g of a finely divided powder such as titanium(IV) oxide for paint, or charcoal for an adsorbent (see section 28.5) may have a surface area of 1000 m^2. Many metals, especially the transition elements (see chapter 26) are catalysts and good adsorbers of gases, e.g. hot palladium on cooling will adsorb nearly 1000 times its own volume of hydrogen. In **adsorption** the substance added remains on the surface of the first substance: it is sometimes hard to distinguish this from **absorption** where the added material penetrates into the body of the first material.

Name of Process	*Reagents*	*Product(s)*	*Catalyst*	*Section*
Steam reforming of natural gas	CH_4, H_2O	H_2, CO	Ni	23.3
Water–gas shift reaction	CO, H_2O	H_2, CO_2	Fe_2O_3	23.3
Catalytic cracking of petroleum	Large molecules	Smaller molecules	Al_2O_3	29.2
Sabatier–Senderens	H_2, oil	Fat	Ni	29.15
Haber	N_2, H_2	NH_3	Fe	30.9
Ostwald	NH_3, O_2	NO, H_2O	Pt	30.17
Contact	SO_2, O_2	SO_3	V_2O_5	31.20

Figure 18.3
Cases of surface catalysis

Name of Process	Reagents	Product(s)	Catalyst	Section(s)
—	H_2, Cl_2	HCl	H_2O	23.6, 32.6
—	CO, O_2	CO_2.	H_2O	28.9
Fermentation	Starch, H_2O	C_2H_5OH, CO_2	Enzymes	29.13, 29.18
Hydrolysis	An ester, H_2O	Alcohol, acid	Acid	29.14
—	NH_3, HCl	NH_4Cl	H_2O	30.11, 32.12
Lead Chamber	SO_2, H_2O, O_2	H_2SO_4	NO	31.21
—	H_2S, SO_2	S, H_2O	H_2O	31.14, 31.18
PF plastics	Phenol, methanal	A plastic	Acid	2.1, 29.20

Figure 18.4
Cases of chemical catalysis

The catalyst is in a different phase from the reactants. On the surface of the catalyst the reacting molecules are present in a greater concentration than they are when in the gaseous phase and for this reason, and other more complex factors, the reaction proceeds faster. Some industrial reactions of importance are shown in figure 18.3.

(b) Chemical Catalysts

In this case the catalyst is in the same phase (state) as the reactants. The catalyst is often needed in a larger proportion than for surface catalysis to give rapid results and it may form an identifiable compound with the reagents which in turn reacts to yield the products, e.g. X and Y reacting to give Z, a reaction catalyzed by C, might proceed thus

$$X + C \rightarrow XC$$
$$XC + Y \rightarrow C + Z$$

The thermal decomposition of potassium chlorate discussed in the previous section is an example of this type of catalysis. Water is a catalyst in many reactions: it is almost universal and so it goes unnoticed. See figure 18.4.

18.9 The Investigation of the Rate of a Reaction

(a) By Measuring the Decrease in Mass when a Gas is Evolved

$$CaCO_3 + 2HCl \rightarrow CaCl_2 + H_2O + CO_2\uparrow$$

A suitable pair of reagents for an elementary investigation is dilute hydrochloric acid and calcium carbonate in the form of marble chips; (an alternative is granulated zinc with dilute sulphuric acid in the presence of a little copper (II) sulphate solution). At a convenient time some calcium carbonate is added to the acid in a conical flask and a cotton wool plug put loosely in the neck of the flask. If placed on a top-pan balance the mass can be recorded at intervals and a graph plotted of mass against time to illustrate the rate of the reaction (see figure 18.5). The reaction is rapid to start with but then slows down as the marble chips become smaller, i.e. lower surface area, and as the acid becomes more dilute. The temperature should be recorded.

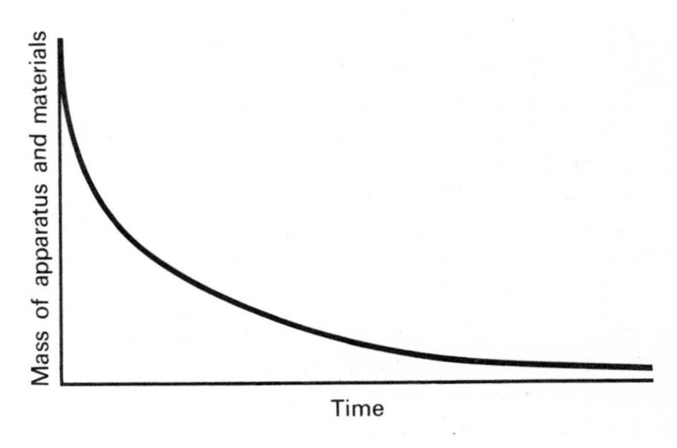

Figure 18.5
The graph for experiment 18.9(a)

(b) By Measuring the Volume of Gas Evolved

A suitable apparatus is as illustrated in figure 11.3 or, if syringes are available, as in figure 18.6. At a convenient time the reaction is started by mixing the reagents or by mixing the catalyst and the reagent, e.g. hydrogen peroxide is added to manganese(IV) oxide, the latter being dispersed in water. The volume of gas evolved is recorded at various times and a graph plotted: once again the reaction starts fast, slows down and then stops (figure 18.7). The temperature should be recorded.

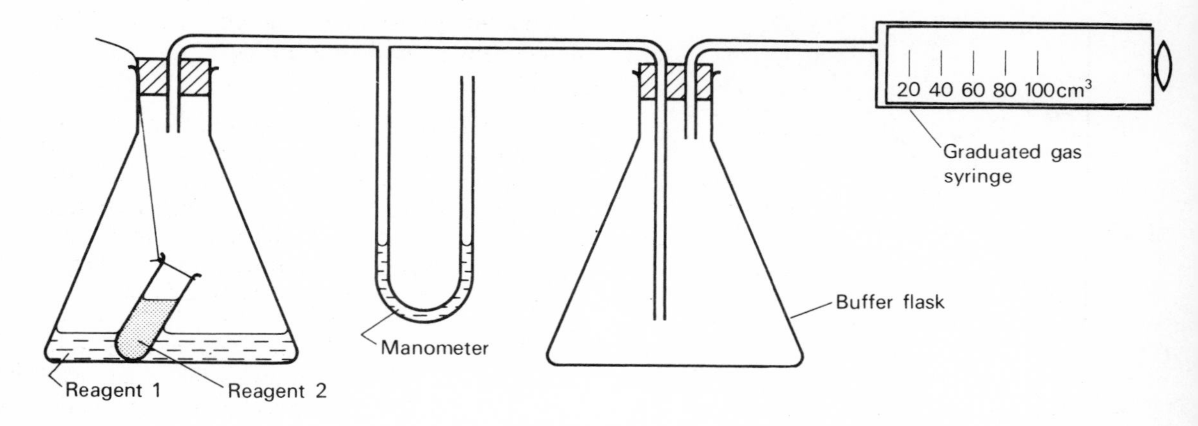

Figure 18.6
The measurement of the volume of a gas evolved

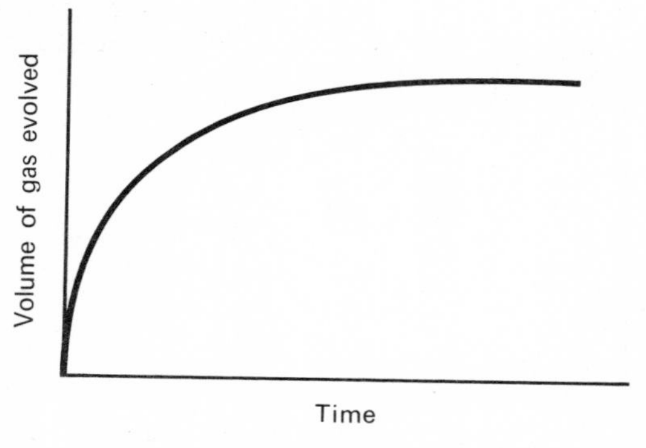

Figure 18.7
The graph for experiment 18.9(b)

(c) *By Measuring the Time for Disappearance of a Colour*

A suitable reaction for the investigation of the effect of temperature upon the rate of a reaction is the redox reaction of potassium manganate(VII) with ethanedioic acid containing dilute sulphuric acid. Two boiling tubes containing the reagents are warmed in a beaker of water and then the contents mixed, the time for the disappearance of the purple colour being measured. By studying the time taken at various temperatures (40–90°C) the statement that the rate of a reaction is doubled when the temperature is raised by 10°C can be verified.

A very similar experiment is to dissolve small weighed pieces of magnesium in dilute hydrochloric acid and to time their disappearance.

(d) *By Measuring the Time for Appearance of a Colour*

Oxidizing agents liberate iodine from acidified potassium iodide solution and iodine gives a blue coloration with starch solution. A little sodium thiosulphate solution is added so that the iodine liberated initially is never apparent, then, when all the sodium thiosulphate has reacted, the solution becomes coloured and the time taken to obscure writing on a piece of paper beneath the beaker, in which the reaction is being performed, can be found. By varying the quantities of the potassium iodide and the oxidizing agent, e.g. hydrogen peroxide, the dependence of the rate of the reaction upon the molar concentrations of the reagents can be found. The results of the experiment should be plotted on a graph as in figure 18.8; t^{-1} is a measure of the rate of the reaction because the quantity of iodine formed in each case is constant. The temperature should be recorded.

(e) *By Measuring the Time for Precipitation to Occur*

When sodium thiosulphate solution is acidified the solution becomes opaque because of the precipitation of sulphur.

$$Na_2S_2O_3 + 2HNO_3 \rightarrow S\downarrow + 2NaNO_3 + H_2O + SO_2$$

Once again a convenient way of assessing the rate of reaction is to measure the time taken to obscure writing on a piece of paper beneath the beaker in which the reaction is being performed. Various quantities of a moderately

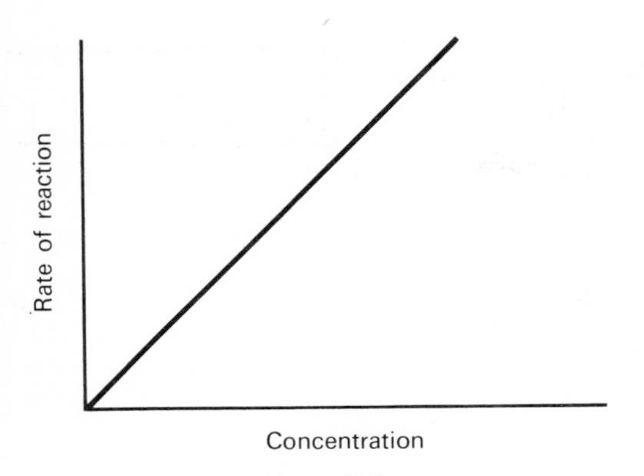

Figure 18.8
The relationship between the rate of the reaction and the concentration of one reagent

concentrated solution of sodium thiosulphate are put in beakers with distilled water to make the total volume the same in each case. At a convenient time the same quantity of dilute nitric acid is added to each beaker and time for precipitation to occur is found. The graph is the same as in figure 18.8 when the concentration of sodium thiosulphate is considered. The temperature should be recorded.

(f) By Titration

The hydrolysis of an ester yielding an alcohol and an acid can be investigated by titrating the acid liberated with sodium hydroxide solution using phenolphthalein as the indicator. If the catalyst used is dilute hydrochloric acid part of the titration is to neutralize this. A large volume of dilute hydrochloric acid is put in a thermostat, the temperature of which should be recorded, and a small sample removed and titrated. At a convenient time the ester, e.g. methyl methanoate, is added and at intervals more samples containing the catalyst, reactants and products removed for titration. The graph of the titres against the time is as in figure 18.7.

$$HCOOCH_3 + H_2O \rightarrow HCOOH + CH_3OH$$

Methyl methanoate (formate) — Methanoic (formic) acid — Methanol

If a conductance bridge is available, it may be used instead of titration to follow the course of the reaction.

QUESTIONS 18

Energies and Rates of Reactions

1 Choose three of the following and for each describe fully one reaction in which the speed of reaction is increased by (*a*) increase of temperature, (*b*) increase of pressure, (*c*) the absorption of light, (*d*) the presence of a named catalyst. A different reaction must be selected for each section. [A]

2 Explain what is meant by an exothermic reaction. Give examples of exothermic reactions occurring between (*a*) a gas and a solid, (*b*) a liquid and a solid, (*c*) two gases. State the conditions under which these reactions take place and the equations for the reactions. [A]

3 'The interaction of sodium hydroxide and hydrochloric acid is an *exothermic* reaction. What do you infer from this statement, and how, in the course of the preparation of sodium chloride crystals, could you demonstrate its truth? Give one example in each instance of exothermic reactions occurring between (*a*) a solid and a gas, (*b*) two gases. [S]

4 What do you understand by the term *photosynthesis*? Explain briefly how this reaction takes place naturally. Mention one other chemical reaction which can be brought about by means of light energy. [S]

5 (*a*) Explain briefly what you understand by photosynthesis.
(*b*) What is a catalyst? Write equations for two catalyzed reactions of industrial importance and for each reaction name the catalyst used. Describe experiments you would carry out to compare the effectiveness of two given metallic oxides when used as catalysts in the decomposition of hydrogen peroxide solution. [C]

6 (*a*) If you found, on carrying out the preparation of a gas by the action of an acid on a solid compound, that the reaction was too slow, suggest two ways, other than the use of a catalyst, by which you could make it faster.
(*b*) Dilute sulphuric acid was added to an excess of granulated zinc. It was noticed that hydrogen was evolved very slowly at first. The rate of evolution of gas increased rapidly for about two minutes and then gradually decreased until the reaction stopped. Attempt to explain these observations. [C]

7 A crystal of pure calcium carbonate weighing 7·50 g was placed in a flask with 50 cm^3 of dilute hydrochloric acid. The flask was kept at constant temperature and the carbon dioxide evolved was collected in a graduated vessel. The volume of carbon dioxide was recorded at 20-minute intervals and corrected to s.t.p. Some of the calcium carbonate remained undissolved at the end of the experiment. The results of the experiment are given in the following table:

Time from start of reaction (min)	*Volume of carbon dioxide corrected to s.t.p. (cm^3)*
20	655
40	910
60	1 065
80	1 100
100	1 120
120	1 120

(*a*) Write the equation for the reaction between calcium carbonate and hydrochloric acid.
(*b*) 655 cm^3 of gas were evolved during the first 20-minute interval. Write down the volumes of gas evolved (i) during the second 20-minute interval (20 to 40 minutes), (ii) during the third 20-minute interval (40 to 60 minutes), (iii) during the fourth 20-minute interval (60 to 80 minutes). Explain the variation in the volumes of gas formed during these intervals.
(*c*) Why is there no increase in volume of gas after 100 minutes?
(*d*) Plot the graph of the volume of carbon dioxide evolved against the time. How much gas was evolved after (i) 30 minutes, (ii) 50 minutes? [C]

8 (*a*) Describe and write equations for (i) one reaction of industrial importance in which heat is absorbed, (ii) one reaction of industrial importance in which heat is produced, (iii) two reactions in which light energy is absorbed.
(*b*) What is meant by the calorific value (heat of combustion) of a fuel? A fuel such as coal and a food such as bread are used in very different ways. Explain why the calorific value is an important characteristic of both the fuel and the food. [C]

9 Comment upon the reaction:

$$H_2(g)+Cl_2(g) \rightarrow 2HCl(g) \quad \Delta H = -184\ kJ$$

Give an explanation of the energy change in terms of the chemical bonds involved. Is the energy change consistent with the Law of Conservation of Energy?

10 The reaction between hydrogen peroxide and potassium iodide, in the presence of sulphuric acid, can be represented by the ionic equation:

$$H_2O_2(aq)+2I^-(aq)+2H^+(aq) \rightarrow 2H_2O(l)+I_2(aq)$$

This slow reaction can be followed by the darkening of the solution due to the formation of iodine which remains in solution in the presence of iodide ions. Copper(II) ions are said to act catalytically on the reaction. Describe in detail how you would test the truth of this statement experimentally.

11 Strips of magnesium ribbon of uniform width are reacted with excess dilute hydrochloric acid: the rate of evolution of hydrogen is found to be:

Length of ribbon (mm)	10	20	30	40	50	60	70
Rate of evolution of hydrogen (cm^3/min)	1·1	1·8	2·7	3·6	4·6	5·4	6·4

Draw a graph and state what conclusion you reach. Estimate the rate of evolution of hydrogen from pieces of magnesium ribbon (*a*) 35 mm, (*b*) 75 mm long under the same conditions.

12 Describe briefly how you would determine the heat of the reaction:

$$H^+(aq)+OH^-(aq) \rightarrow H_2O(l)$$

How would you modify your experiment to determine the volume change in the production of 1 mol of water.

13 Ammonia reacts with hydrogen chloride to form ammonium chloride and the reaction can be represented by the following energy level diagram:

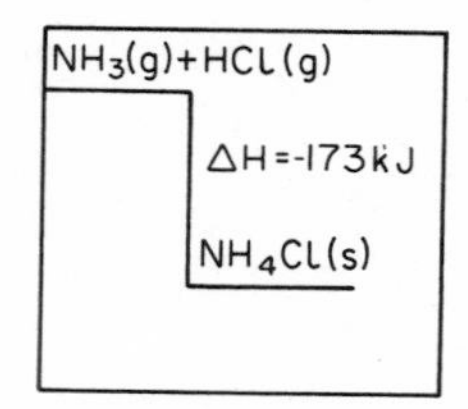

(*a*) Suppose 1 mol of ammonia was mixed with 2 mol of hydrogen chloride. Calculate the energy change which would take place. State whether the change would be exothermic or endothermic.
(*b*) When 1 mol of ammonia is dissolved in water to form a dilute solution, 35·2 kJ are given out. Draw an energy level diagram to represent the change.
(*c*) When 1 mol of hydrogen chloride is dissolved in water to form a dilute solution, 72·5 kJ are given out. When the solutions of ammonia and hydrogen chloride are mixed to form a solution of ammonium chloride, 52·5 kJ are given out. Calculate ΔH for the reaction:

$$NH_3(g)+HCl(g)+aq \rightarrow NH_4Cl(aq)$$

(*d*) Briefly describe an example of the following series of changes which is in use at the present time: a chemical change which gives out heat energy, the heat energy being changed to electrical energy and the latter used to do mechanical work. [Nf]

14 Describe an experiment, which you have performed in the laboratory, to show how temperature affects the rate of a chemical reaction. You should describe what you did, the observations and measurements which you made, and should explain how your observations and measurements lead to your conclusions.

15 This question is based on the following investigation into the heat evolved if 50 cm^3 quantities each containing 4 g of sodium hydroxide were mixed with various volumes of a solution of an acid. 50 cm^3 of the sodium hydroxide solution were put into a plastic bottle and the temperature taken. 25 cm^3 of the solution of acid (at the same temperature) were added and the maximum rise in temperature of the mixture was noted. From these observations the amount of heat evolved was calculated. The experiment was repeated using 50 cm^3 of sodium hydroxide solution each time with increasing volumes of the acid. The results are shown in the diagram:

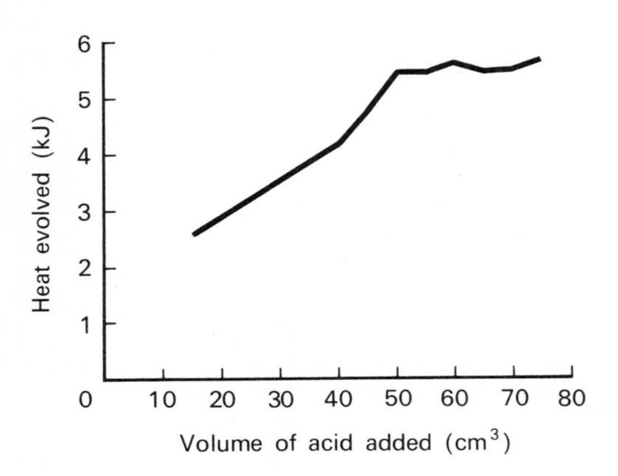

(*a*) Estimate the amount of heat given out if 100 cm^3 of a sodium hydroxide solution of the same concentration had been mixed with 35 cm^3 of the acid solution.
(*b*) What volume of the acid solution will exactly react with 50 cm^3 of the sodium hydroxide solution?
(*c*) How much heat is given out when the volumes of acid and sodium hydroxide solution shown in your answer (*b*) are mixed?
(*d*) How many moles of sodium hydroxide are contained in 50 cm^3 of solution?
(*e*) Calculate the heat which would be given out if a solution containing 1 mol of sodium hydroxide in 500 cm^3 reacted completely with the acid solution.
(*f*) Estimate the heat given out when a solution containing 1 mol of barium hydroxide, $Ba(OH)_2$, reacts completely with the acid solution. Give brief reasons for your answer.
(g) When the experiments were repeated using 4 g of sodium hydroxide dissolved in 10 cm^3 of solution, it was found that the heat given out was always a little higher than that shown in the diagram. Suggest a reason for this. [Nf]

16 The table of observations set out below is taken from an investigation, undertaken in a school laboratory, into the rate of reaction between magnesium and hydrochloric acid. The acid was put into a flask, its temperature taken, and a weighed amount of magnesium added. The contents of the flask were swirled during the reaction and the time taken for the magnesium to disappear was noted. The final temperature of the acid was noted.
(*a*) The results of most of the experiments could be used to plot a graph which would indicate how the rate of reaction varies with the concentration of the acid, but three of the experiments would be unsuitable for this purpose. State the letters of the unsuitable experiments and say briefly why they are unsuitable.
(*b*) From the remaining experiments draw a graph showing the reciprocal of the time on the vertical axis and the concentration of the acid on the horizontal axis.
(*c*) From your graph estimate the time it would take for 0·10 g of magnesium ribbon to react with 40 cm^3 of 0·6M hydrochloric acid starting at 21°C.
(*d*) Why is the temperature rise in most of the experiments approximately constant?
(*e*) State concisely one defect this method of investigating the relationship between rate of reaction and concentration may have.

Experiment	*Concentration of acid mol/dm³*	*Volume of acid cm³*	*Temperature °C*		*Mass of magnesium g*	*Form of magnesium*	*Time in seconds*	$\frac{1}{Time}$ s^{-1}
			Before	*After*				
A	0·5	40	21	31	0·10	ribbon	500	0·002
B	0·6	40	21	32	0·10	powder	50	0·020
C	0·7	40	21	32	0·10	ribbon	250	0·004
D	0·8	40	21	33	0·10	ribbon	160	0·006
E	0·9	40	21	39	0·20	ribbon	230	0·004
F	1·0	40	21	33	0·10	ribbon	100	0·010
G	1·0	40	35	44	0·10	ribbon	50	0·020
H	1·1	40	21	33	0·10	ribbon	75	0·013
I	1·2	40	21	32	0·10	ribbon	45	0·022
J	1·3	40	21	32	0·10	ribbon	40	0·025
K	1·5	40	21	31	0·10	ribbon	30	0·033

17 The graph below shows the evolution of carbon dioxide in an experiment in which a considerable excess of small marble chips was added to 10 cm^3 of hydrochloric acid. The gas volume was measured with a gas burette at room temperature (16°C) and normal pressure. The equation for the reaction is

$$\underset{\text{Marble}}{CaCO_3(s)} + 2HCl(aq) \rightarrow CaCl_2(aq) + H_2O(l) + CO_2(g)$$

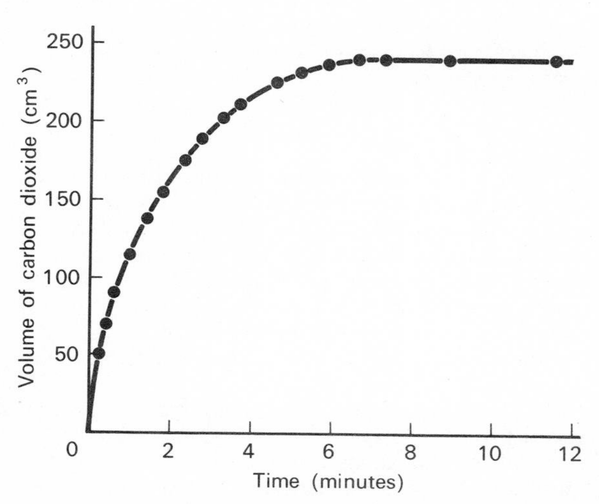

(*a*) Why is the curve steep at first, becoming less steep?
(*b*) Calculate the total number of moles of carbon dioxide given off in this reaction. (1 mol of any gas occupies about 24 dm^3 at 16°C and normal pressure.)
(*c*) How many moles of hydrogen chloride must have been used up?
(*d*) What was the concentration of the acid solution?
(*e*) Carefully sketch in curves on a graph and label them clearly (i), (ii) and (iii) to indicate what you think the results would have been: (i) if the experiment had been carried out with the same materials and quantities, but with the acid initially at 50°C; (ii) if the concentration of the acid had been only half as great, but other conditions were as in the original experiment; (iii) if the marble had consisted of a single cube instead of many small chips (other conditions as in the original experiment). [Nf]

18 Some experiments were done on the dissolving of iron wire in 1·0 M hydrochloric acid. 1·40 g of iron was used for each experiment and the volume of hydrogen given off was measured at room temperature and pressure. The results were tabulated as follows:

	A	B	C	D	E	F	G
Volume of acid (cm^3)	10	25	35	45	60	70	80
Volume of hydrogen (cm^3)	121	298	422	540	600	602	598

(*a*) Which two of the following properties would be most useful in confirming that the gas given off was hydrogen? Colour, smell, density, pH measurement, flammability.
(*b*) Plot a graph of volume of gas evolved against volume of 1·0 M hydrochloric acid used.
(*c*) Use your graph to find the volume of 1·0 M hydrochloric acid which will just dissolve 1·40 g of iron.
(*d*) What volume of 1·0 M hydrochloric acid will just dissolve 1 mol of iron?
(*e*) What volume of gas would be evolved if 30 cm^3 of 0·2 M hydrochloric acid were added to 1·40 g of iron?

In another experiment, gaseous hydrogen chloride was passed over heated iron wire. The hydrogen formed was burnt and pale yellow crystals of an iron chloride were formed. It was found that 0·280 g of iron formed 0·635 g of the iron chloride. When this chloride was dissolved in water, the solution had the same properties (colour, tests with reagents, etc.) as the solution obtained when iron was dissolved in 1·0 M hydrochloric acid.
(*f*) Calculate a formula for the iron chloride.
(*g*) (i) Write an equation for the reaction between iron and hydrochloric acid. (ii) Write down briefly the evidence you have used in order to write down this equation. [Nf]

19 Hydrogen peroxide undergoes catalytic decomposition yielding oxygen when manganese(IV) oxide is added. The following results were obtained at 16°C in the laboratory:

Time (minutes)	5	10	15	20	30	40	50	60	75	90	115	120
Volume (cm^3)	18	27	36	43	53	61	67	73	79	83	86	86

Plot a graph of the volume of oxygen against the time.
(*a*) Why were readings taken more often in the early part of the experiment?
(*b*) Why was the experiment stopped after 120 minutes?
(*c*) What does the graph tell you about the rate of evolution of oxygen?
(*d*) What is the ratio of the rates of the reaction after 9 minutes and after 42·5 minutes?
(*e*) What is the ratio of the rates of the reaction when 10 cm^3 and when 55 cm^3 of oxygen have been evolved?

20 A small flask was connected to a gas syringe by means of a stopper and delivery tube. 30 cm^3 of water and 0·5 g of manganese(IV) oxide were placed in the flask. 5 cm^3 of hydrogen peroxide were added, the flask was quickly stoppered and readings of the volume of gas in the syringe were recorded every 10 seconds.

Time	0	10	20	30	40	50	60	70	80
Volume (cm^3)	0	18	30	40	48	53	57	58	58

(*a*) On graph paper plot a graph of the volume shown on the syringe against the time in seconds. Label this curve 'A'. When the reaction was complete, the stopper was removed and the syringe was emptied of gas. Without emptying the flask, another 10 cm^3 of water and 5 cm^3 of hydrogen peroxide were added, and the experiment repeated exactly as before. On the same sheet of graph paper and using the same axes as were used for the previous graph, sketch a second graph to

show how the volume of gas collected would vary with time in this second experiment. Label this second curve 'B'. Explain the reasons for choosing the line you have drawn.
(*b*) In this reaction, the manganese(IV) oxide acts as a catalyst. Explain fully how you would attempt to prove this.
(*c*) Write an equation and state the conditions for another reaction of manganese(IV) oxide in which it does not act as a catalyst. What part is played by the oxide in this reaction? [J]

21 What do you understand by *synthesis*? Illustrate your answer by referring to the formation of (*a*) iron(II) sulphide, (*b*) ammonia, (*c*) carbon dioxide, stating briefly the conditions under which each synthesis is carried out. What energy change is common to these reactions? [O]

19 Reversible Reactions

19.1 Reactions that go to Completion

Many reactions do not give a 100% yield of the products expected from the equation: side reactions occur in which the reagents are converted into unexpected and unwanted materials or else there are losses from the apparatus. The reactions which usually do go effectively to completion without any special precautions include the following:

(a) Reactions in which a Solid is Precipitated

When barium chloride solution is added to a solution of a sulphate there is a precipitate of barium sulphate.

$$BaCl_2 + H_2SO_4 \rightarrow BaSO_4\downarrow + 2HCl$$

or $$Ba^{2+} + SO_4^{2-} \rightarrow BaSO_4\downarrow$$

Likewise $$AgNO_3 + HCl \rightarrow AgCl\downarrow + HNO_3$$

or $$Ag^+ + Cl^- \rightarrow AgCl\downarrow$$

(b) Reactions in which a Gas is Liberated

When zinc is dissolved in excess dilute acid, hydrogen is evolved and the reaction continues until all the zinc has dissolved.

$$Zn + 2HCl \rightarrow ZnCl_2 + H_2\uparrow$$

or $$Zn + 2H^+ \rightarrow Zn^{2+} + H_2\uparrow$$

Likewise

$$Na_2CO_3 + 2HCl \rightarrow 2NaCl + H_2O + CO_2\uparrow$$

or $$CO_3^{2-} + 2H^+ \rightarrow H_2O + CO_2\uparrow$$

In these reactions the gas evolved must be able to escape freely from the reagents. However, there are cases in which the reaction appears to stop half way and will not go to completion.

19.2 Reversible Reactions

The reactions just described cannot be made to go in the reverse direction: no amount of hydrochloric acid will dissolve barium sulphate, and hydrogen does not convert zinc chloride solution back to zinc.

Lavoisier (1774) found that when mercury was heated in a closed volume of air it was partly converted to red specks of mercury(II) oxide.

$$2Hg + O_2 \rightarrow 2HgO$$

Then on stronger heating the mercury(II) oxide decomposed again to mercury and oxygen.

$$2HgO \rightarrow 2Hg + O_2\uparrow$$

Another example is the reaction of steam with red-hot iron which yields triiron tetraoxide [iron(II) iron(III) oxide] and hydrogen (see section 6.3).

$$3Fe + 4H_2O \rightarrow Fe_3O_4 + 4H_2$$

If hydrogen is passed over triiron tetraoxide then steam is produced and iron is left (see figure 19.1).

$$Fe_3O_4 + 4H_2 \rightarrow 3Fe + 4H_2O$$

A third variation of this experiment is to heat steam and iron in a closed vessel—this is not easily demonstrated in the laboratory—and an examination of the contents of the vessel, whenever it is done, shows that four substances are present and that they are present in the same proportions.

The situation may be written

$$3Fe + 4H_2O \rightleftharpoons Fe_3O_4 + 4H_2$$

A state of balance has been achieved and

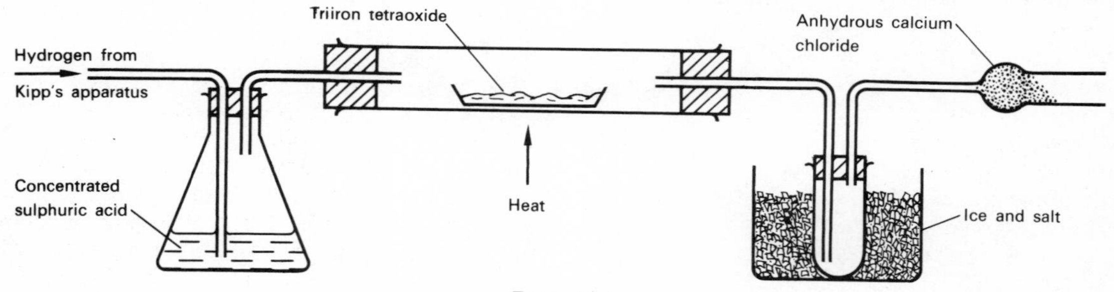

Figure 19.1
The reaction of hydrogen and triiron tetraoxide

neither reaction under these circumstances will go to completion.

A reversible reaction is one which can be made to go in either direction by changing the conditions under which it is carried out.

At equilibrium the rate of the forward reaction (left to right) is equal to the rate of the backward reaction (right to left, in the equation as written above). There is now no apparent change and a state of dynamic equilibrium is said to exist as in liquid sulphur at one particular temperature (see section 8.4 and 31.5).

Equilibrium can only be reached if all the substances concerned are prevented from escaping. In the combustion tube experiment fresh steam sweeps the hydrogen formed away from triiron tetraoxide, or the fresh hydrogen sweeps the steam formed away from the iron. Under those circumstances the reaction proceeds to completion.

Similarly, if calcium carbonate is heated in a limekiln (see section 25.8), the carbon dioxide escapes, being pushed out by fresh air, but if calcium carbonate is heated in a closed container the reaction is found to be reversible.

$$CaCO_3 \rightleftharpoons CaO + CO_2$$

The reaction of an organic acid with an alcohol is another important example of a reversible reaction. The conditions may be chosen either to prepare the ester (see section 29.14) or to hydrolyze it (see section 29.15).

19.3 Thermal Decomposition and Thermal Dissociation

When a compound is split up by the action of heat the reaction may be reversible or non-reversible. When the reaction is non-reversible it is called **thermal decomposition**. Examples of this type include the action of heat upon potassium chlorate and lead nitrate:

$$2KClO_3 \rightarrow 2KCl + 3O_2\uparrow$$

$$2Pb(NO_3)_2 \rightarrow 2PbO + 4NO_2\uparrow + O_2\uparrow$$

No amount of cooling or prolonged contact will bring about the reverse reaction.

When the reaction is reversible it is called **thermal dissociation** The action of heat upon ammonium chloride, iodine (and most non-metals if heated strongly), dinitrogen tetraoxide, calcium carbonate and phosphorus(V) chloride brings about dissociation.

$$NH_4Cl \rightleftharpoons NH_3 + HCl$$

$$I_2 \rightleftharpoons 2I$$

$$N_2O_4 \rightleftharpoons 2NO_2$$

$$CaCO_3 \rightleftharpoons CaO + CO_2$$

$$PCl_5 \rightleftharpoons PCl_3 + Cl_2$$

At high temperatures ammonium chloride is almost completely dissociated but on cooling ammonium chloride is reformed. The density measured at high temperatures is half the value obtained at low temperatures. There are twice as many particles at high temperatures and so, by Avogadro's Law, the volume occupied is twice that at low temperatures (even when the volumes are corrected to s.t.p.). The mass of material has not altered so the density has been halved.

The dissociation of dinitrogen tetraoxide or phosphorus(V) chloride (or the bromide) can be demonstrated by sealing some in a bulb and then warming the bulb. At room temperature the materials are colourless but they form coloured substances on heating; on cooling they become colourless again.

19.4 Le Chatelier's Principle (1888)

In any reversible reaction the position of equilibrium will alter as the various factors controlling the reaction are changed. There are several important industrial processes which involve reversible reactions, e.g. the manufacture of hydrogen, ammonia and sulphur trioxide, where the conditions have to be very carefully chosen to get the maximum possible yield with the minimum expenditure of energy. These factors are

(*a*) the temperature
(*b*) the concentrations of the substances, or, in gaseous reactions, the pressure.

The result of altering either of these can be predicted by Le Chatelier's Principle.

If any of the factors affecting a reversible reaction is changed, the system will react to diminish the change.

Thus the reaction follows the path of least resistance: if the pressure is increased the system adjusts so that the volume occupied is less, and if the system is heated the position of equilibrium shifts so that heat is absorbed.

The presence or absence of a catalyst does not alter the position of equilibrium. The rate of attainment of equilibrium is speeded up if a catalyst is available but the yield is not altered at all.

19.5 The Effect of Changing the Temperature

All reactions are accelerated as the temperature increases but in a reversible reaction the effect of increasing the temperature differs proportionately according to whether the backward or the forward reaction is considered.

Consider the synthesis of ammonia in the Haber process (see section 30.9).

$$N_2 + 3H_2 \rightleftharpoons 2NH_3 \qquad \Delta H = -92 kJ$$

It is advantageous from the point of view of obtaining a high yield to keep the temperature as low as possible because the reaction in proceeding from left to right evolves heat, but from the point of view of accomplishing the reaction speedily the temperature should be as high as possible. Thus a compromise is used (350°C) and a catalyst added to achieve a rapid attainment of equilibrium at that temperature. A similar choice has to be made in the manufacture of sulphur(VI) oxide (see section 31.20) and of hydrogen (see section 23.3).

The opposite case is of an endothermic reaction, e.g.

$$N_2 + O_2 \rightleftharpoons 2NO \qquad \Delta H = +184 kJ$$

(see section 30.7). The reaction in proceeding from left to right absorbs heat and increasing the temperature increases both the rate of obtaining and the yield of nitrogen oxide; a catalyst is not used. The gases that come from the electric arc are cooled suddenly to prevent the reaction from right to left occurring which would evolve heat. The preparation of ozone is similar (see section 31.4).

19.6 The Effect of Changing the Concentrations

When steam is passed over hot iron in a combustion tube the hydrogen formed is driven out of the tube so the reverse reaction cannot occur (see section 19.2).

In the hydrolysis of an ester (see section 18.9) water is added in excess of the quantity required by the equation in order that the yield of acid and alcohol may be high, and so the reaction becomes effectively one way, not reversible.

When an ester is being made (see section 29.14) the alcohol, being cheaper than the acid, is usually employed in a quantity in excess of that required by the equation so that a high yield is obtained.

In the second stage of the manufacture of hydrogen (see section 23.3) the steam is added in excess (2.5 volumes not 1) so that the carbon monoxide in the reaction mixture is almost eliminated.

19.7 The Effect of Changing the Pressure

In a gaseous reaction increasing the pressure increases the rate of reaction because the molecules are brought closer together (see section 18.6). If the reaction is reversible and involves a change in the number of gaseous molecules (which, by Avogadro's Law, means a change in the volume occupied), the position of equilibrium will also be affected.

Thus, in the reaction to make nitrogen oxide, altering the pressure does not alter the yield because there are two molecules of gas on each side of the equation.

$$N_2 + O_2 \rightleftharpoons 2NO$$

In the synthesis of ammonia four molecules of gas yield two molecules of gas.

$$N_2 + 3H_2 \rightleftharpoons 2NH_3$$

An increase of pressure therefore, by Le Chatelier's Principle, increases the yield of ammonia (see section 30.9). The limitations are that energy has to be expended to compress the reagents and that the engineering difficulties increase.

Higher pressures than atmospheric are not usually employed in the manufacture of sulphur(VI) oxide because although the equation

$$2SO_2 + O_2 \rightleftharpoons 2SO_3$$

shows there is a diminution in volume, air rather than oxygen is usually employed and the true decrease is from 7 to 6 volumes not 3 to 2, thus making compression uneconomic (see section 31.20).

The steam reforming of natural gas (see section 23.3) is performed at 10 atmospheres pressure despite the fact that the desired reaction involves an increase in the volume because the effect of increased concentration on the rate is more important.

QUESTIONS 19

Reverisble Reactions

1 Describe one reversible and one irreversible chemical reaction. In the case of the reversible reaction you choose, show how the conditions of the experiment may be adjusted so as to make it proceed almost to completion in one direction. [OC]

2 The reaction between steam and iron is said to be *reversible*. What is meant by this? Describe how you would carry out this reaction experimentally, and how you would show it to be reversible. State two other reversible reactions and give the equations for them. [A]

3 When solutions of calcium chloride and ethanedioic (oxalic) acid ($H_2C_2O_4$) are mixed a reversible reaction occurs and a precipitate of calcium ethanedioate is seen:

$$\underset{\text{Solution}}{CaCl_2} + \underset{\text{Solution}}{H_2C_2O_4} \rightleftharpoons \underset{\text{Solid}}{CaC_2O_4} + \underset{\text{Solution}}{2HCl}$$

Which of the following additions dissolve the precipitate of calcium ethanedioate:
(*a*) calcium chloride solution, (*b*) ethanedioic acid, (*c*) concentrated nitric acid, (*d*) concentrated hydrochloric acid?

4 Two equilibria may be summarized by the following equations:

$$CO_2(g) + 2H_2O(l) \rightleftharpoons HCO_3^-(aq) + H_3O^+(aq)$$
$$NH_3(g) + H_2O(l) \rightleftharpoons NH_4^+(aq) + OH^-(aq)$$

Use this information to deduce what will happen in the two situations described below. Explain briefly how you arrive at your deductions (appropriate equations may help you to do this).
(*a*) An aqueous solution of hydrobromic acid (a strong acid) is added to sodium hydrogencarbonate ($NaHCO_3$, a soluble ionic compound).
(*b*) An aqueous solution of barium hydroxide (a strong base) is added to ammonium iodide (NH_4I, a soluble ionic compound). [Nf]

5 What is meant by (*a*) electrolytic dissociation, (*b*) thermal decomposition, (*c*) thermal dissociation, (*d*) an exothermic reaction, (*e*) a reversible reaction? Into which one or more of the above classes may the following reactions be placed?

(i) $N_2 + 3H_2 \rightleftharpoons 2NH_3$ $\quad \Delta H = -92$ kJ
(ii) $2HgO \rightarrow 2Hg + O_2$ $\quad \Delta H = +90$ kJ
(iii) $N_2O_4 \rightleftharpoons 2NO_2$ $\quad \Delta H = +59$ kJ

6 Explain the difference between thermal dissociation and thermal decomposition, giving one example of each to illustrate your explanation. [J]

7 Explain what is meant by *an exothermic reaction* and *an endothermic compound*. The reaction for the formation of water gas is

$$H_2O + C \rightleftharpoons CO + H_2 \qquad \Delta H = +130 \text{ kJ}$$

What will be the effect of (*a*) increased pressure and (*b*) increased temperature on the yield of water gas?

20 Electrochemistry

20.1 The Conduction of Electricity

An electric current, which is a flow of electrons, will pass through a metal (a conductor) but not usually through a non-metal (a non-conductor or an insulator, e.g. sulphur). A direct current, as obtained from a cell, (a battery consists of several cells), or a d.c. generator consists of a continuous flow of electrons in one direction, rather like the flow of molecules of water in a river. The cell is an electron pump. The usual supply of electricity to a house is alternating current: the direction of this current changes 100 times every second (it is said to have a frequency of 50 Hertz).

Both forms of electricity have heating and magnetic effects but only direct current is useful for producing permanent chemical changes when it passes through some pure liquids and solutions.

20.2 The Simple Voltaic Cell

The first cell was used by Volta in 1796–99.

If two different wires, e.g. one of copper and one of zinc, are connected to an ammeter and dipped into dilute sulphuric acid then the ammeter shows that a current is flowing. Bubbles of hydrogen form on the copper and interfere with the movement of particles in the acid so that the current decreases rapidly (this is known as **polarization**). A high resistance voltmeter, measuring the relative reactivity of the two metals, can be substituted for the ammeter but, whilst a reading is obtained for a longer time, bubbles of gas accumulate on the zinc because it will react with the acid whether or not the circuit is complete (this is known as **local action**).

The voltmeter measures the force with which the electrons are propelled around a circuit; an ammeter measures the numbers of electrons passing that given point in a circuit every second.

An improvement on the simple cell discussed above can be obtained by substituting sodium sulphate solution for the acid but for reproducible measurements involving a large number of metals a cell such as that illustrated in figure 20.1 must be set up. Each metal dips into a solution of one of its salts of 1 M concentration. These solutions are prevented from mixing by the 'salt bridge' of sodium sulphate solution (in more accurate work potassium nitrate or chloride which are more expensive are used). Local action is eliminated and polarization is unlikely if a high resistance voltmeter is used to assess the potential difference in the cell, known as the electromotive force (e.m.f.). See also section 15.5.

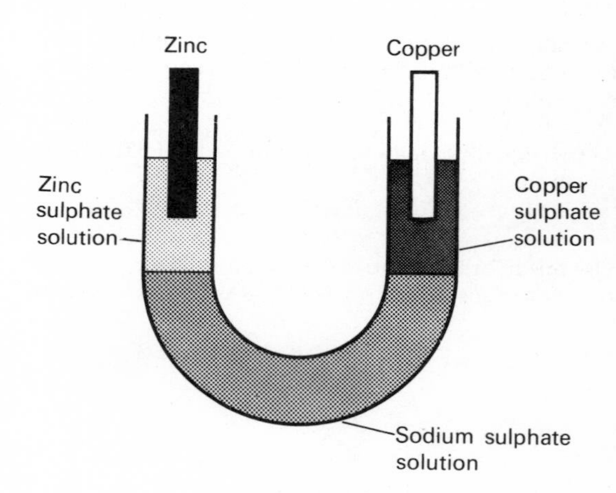

Figure 20.1
An improved version of the simple cell

The reactions in the simple cell are that the zinc ionizes and dissolves

$$Zn \rightarrow Zn^{2+} + 2e^- \text{ (anode)}$$

and that hydrogen is liberated on the copper:

$$2H^+ + 2e^- \rightarrow H_2\uparrow \text{ (cathode)}$$

The wires, or carbon rods, which carry the current in and out of the solution are known as **electrodes.** The **anode** is the one at which conventional current enters the cell, and the **cathode** that at which that current leaves the cell. The **electrolyte** is the solution which conducts the electric current; reactions occur at the electrodes as the current flows.

The electronic current leaves the anode

(usually coloured blue or black) of the cell and passes round the outside circuit to the cathode (usually coloured brown or red) of the cell. The anode plate or rod is called the negative terminal, and the cathode the positive terminal, because the conventional current is regarded as a flow of positive electricity from the copper to the zinc, rather than as a stream of negative electrons in the opposite direction.

20.3 The Daniell and Leclanché Cells

To overcome the defects of the simple cell various modifications have been made notably by Daniell (1836) and Leclanché (1866).

The Daniell cell has a zinc rod dipping into dilute sulphuric acid (which soon becomes contaminated with zinc sulphate solution) and a porous pot separates this solution from copper sulphate solution into which a piece of copper dips. The porous pot, unglazed porcelain, takes the place of the salt bridge. At the zinc anode the reaction is the same as in the simple cell.

$$Zn \rightarrow Zn^{2+} + 2e^-$$

However, at the copper cathode the reaction is

$$Cu^{2+} + 2e^- \rightarrow Cu\downarrow$$

and so few hydrogen ions from the acid reach here that there is no polarization (see section 20.2). There is some local action so that it is advisable to dismantle the cell after use and fill it with water, to prevent air bubbles being trapped in the pot. The e.m.f. is 1·15 volts and the cell can be represented by

$$Zn \mid Zn^{2+} \mid Cu^{2+} \mid Cu.$$

In the Leclanché cell (see figure 20.2), a mixture of manganese(IV) oxide (the depolarizer), carbon black (to make it conducting), zinc chloride, and ammonium chloride solutions (the electrolytes), are collectively known as the 'black paste'. The black paste is the cathode and touches the carbon rod. The anode is the zinc case and it is lined with paper impregnated with ammonium chloride solution (the 'white paste'). The carbon rod (a current collector) is given a brass cap, to improve contact between parts of the circuit, and the zinc case a tinplate jacket to guard against leaks.

The zinc anode (negative) ionizes.

$$Zn \rightarrow Zn^{2+} + 2e^-$$

In the cathode, the manganese(IV) oxide is reduced to the trivalent state as MnOOH or Mn_2O_3, H_2O. Further complex reactions occur involving the ammonium chloride. The reactions are slow so that the cell is best used for intermittent work.

In the wet version, some ammonia and hydrogen escape. The corrosive nature of ammonium chloride on the zinc case may be responsible for the cell leaking.

The e.m.f. of the Leclanché cell in its wet or dry forms is nearly 1·6 volts.

20.4 The Lead Accumulator

The simple cells described in the two previous sections have a limited life. In 1859 Planté invented the lead accumulator which after functioning for a time can be recharged for further use.

The negative pole (the anode) is lead and the positive pole (the cathode) is lead(IV) oxide on lead, the electrolyte being moderately concentrated sulphuric acid (5 M). When the cell is connected to an external circuit the following reactions occur.

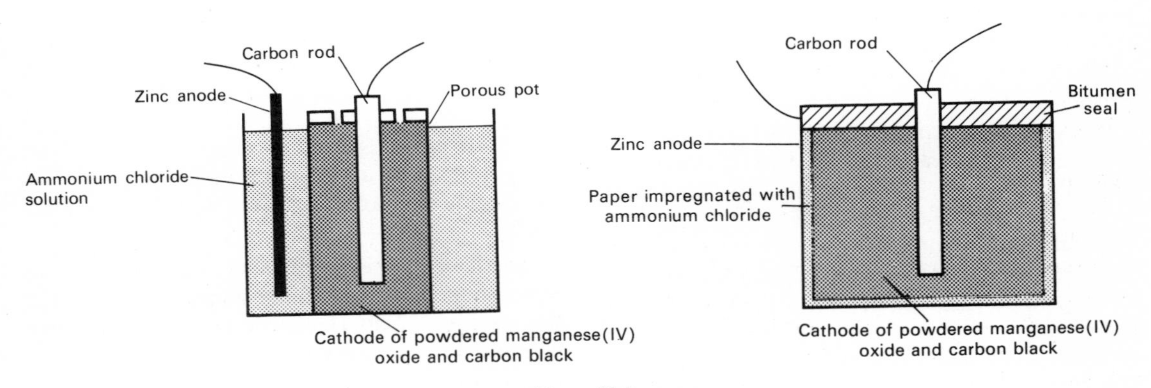

Figure 20.2
The Leclanché wet and dry cells

At the anode:

$$Pb \rightarrow Pb^{2+} + 2e^-$$

At the cathode:

$$PbO_2 + 4H^+ + 2e^- \rightarrow Pb^{2+} + 2H_2O$$

The e.m.f. is 2 volts.

The electrolyte therefore becomes dilute sulphuric acid and its density falls 1·28 to 1·12 g/cm^3: a simple check on the cell can be made with a hydrometer. If the cell is over-used there is a danger of excessive formation of lead sulphate. On recharging the cell the above reactions are reversed.

20.5 The Fuel Cell

The fuel cell was discovered by Grove in 1841 and for many years research has been done to try to find a practical design for such a cell to give continuous operation.

If two test tubes containing hydrogen and oxygen respectively are inverted over two graphite or metal electrodes partially submerged in dilute sulphuric acid and then a high resistance voltmeter put across these electrodes a potential is recorded of 1·23 volts. The cell suffers from polarization.

At the anode:

$$H_2 \rightarrow 2H^+ + 2e^-$$

At the cathode:

$$O_2 + 2H_2O + 4e^- \rightarrow 4OH^-$$

These reactions are followed by

$$H^+ + OH^- \rightarrow H_2O$$

so that the total concentration of these ions in the electrolyte is unaltered.

20.6 The Conduction of Electricity through Liquids and Solutions

Most solid conductors are metals (graphite is an exception) and liquid metals, e.g. mercury, are also good conductors. The current is again composed of a stream of electrons.

The electrolysis of water has been referred to in section 6.7(c). Pure liquids such as ethanol, dimethylbenzene and propanone, and solutions such as sugar in water and hydrogen chloride in dimethylbenzene, are found to be non-conductors. The circuit used is as in figure 20.3. The substances and solutions are referred to as non-electrolytes. However, molten sodium chloride and molten lead bromide, and solutions of many substances, e.g. sodium chloride (a salt), sodium hydroxide (an alkali) and hydrogen chloride (an acid), in water are found to be conductors: they are electrolytes. The solutions

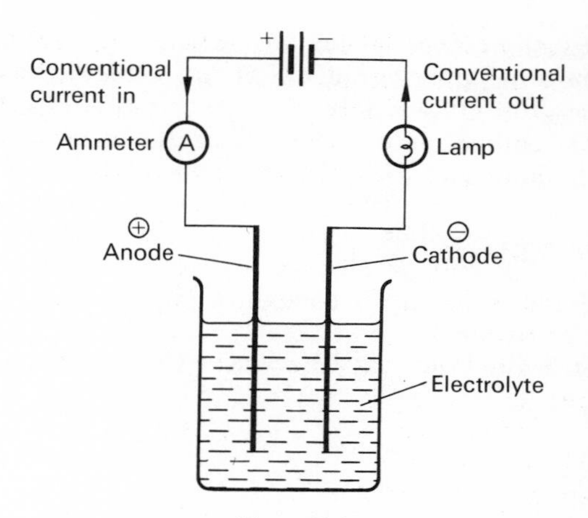

Figure 20.3
A circuit for an electrolysis experiment

can be studied in a U-tube, rather than a beaker, to prevent the electrodes touching.

Several cells (a battery) may be required to enable a current to flow through the electrolyte, the minimum voltage depending on the ions and the nature and distance apart of the electrodes involved (see section 15.5). The ions carry the current through the electrolyte. Ions are charged particles: they may be positive (cations) formed by a metal atom or radical losing electrons or negative (anions) formed by a non-metal atom or radical gaining electrons.

The electrode connected to the negative pole (the anode) of the cell is the cathode of the voltameter. The electrode connected to the positive pole (the cathode) of the cell is the anode of the voltameter. The conventional current flows from the positive pole to the negative pole of the cell, but the electronic current flows in the opposite direction. The ammeter, showing quantitatively the rate of passage of electrons, and the lamp, showing the rate qualitatively, may be placed anywhere in the circuit. Electrolysis occurs at the electrodes in the voltameter, the vessel in which the electrolyte is placed.

Electrolysis is the passage of an electric current through a fused substance or a solution accompanied by chemical reactions at the electrodes.

An electrolyte is a substance which in the fused state or in solution, usually aqueous, will permit the passage of an electric current accompanied by chemical reactions at the electrodes.

The reactions only occur at the electrodes and only whilst the current is flowing. The reactions involved are always two of the following: the dissolution of the metal anode, the deposition of a metal from the solution on to the cathode, the liberation of a non-metal at the anode or of hydrogen at the cathode.

20.7 The Ionic Theory

Some substances consist of ions—they are electrovalent (see section 14.2). When they are in a molten (fused or liquid) state the ions are free to move so that they will conduct an electric current. Reactions occur at the anode which yield electrons—oxidation of some species, atom, ion or molecule. At the cathode reduction occurs—reactions occur in which electrons are taken in (see section 16.1). Thus, although the electrons themselves do not directly pass through the electrolyte, as they do through a metallic conductor, the current is carried by the ions.

Other substances are covalent but they ionize when they dissolve in water and thus become electrolytes, e.g. hydrogen chloride and hydrogen ethanoate (see section 14.4). If the degree of ionization is high the substance is a strong electrolyte; if the degree is small the substance is a weak electrolyte, e.g. hydrochloric and ethanoic acids respectively.

The third group of substances is covalent and remains covalent even when dissolved in water; it consists of non-electrolytes (see section 14.3).

The ions wander around in the electrolyte: negative ions are attracted towards the positively charged electrode, the anode; positive ions are attracted towards the negatively charged electrode, the cathode. These electrodes have become charged because of the 'pumping' action of the cell upon the electronic population.

Hydrogen and metal atoms become positively charged, they become cations, because they form ions by loss of electrons, these ions usually having the electronic structure of the nearest noble gas. Non-metals and acid radicals become negatively charged, they become anions, because they form ions by gaining electrons. The number of electrons lost or gained by a species is its valency.

Ionization is the process of forming ions: it occurs whenever atoms or molecules combine by the transfer of electrons.

Electrolytic dissociation occurs when the ions of an ionic compound separate; the reverse is electrolytic association which occurs when an ionic compound is converted to crystals. When hydrogen chloride dissolves in water it reacts to give ions and those ions separate.

$$HCl \rightleftharpoons H^+ + Cl^-$$

or

$$HCl + H_2O \rightleftharpoons H_3O^+ + Cl^-$$

Hydroxonium ion

The reaction at an electrode depends on several factors:

(*a*) The ions that are present. Water is ionized to a slight extent and its ions may be more important than those of the electrolyte. See the electrochemical series, section 15.6.

(*b*) The concentrations of the ions that are present. Different results can be obtained by changing from dilute to concentrated solutions or to molten substances.

(*c*) The electrodes. Changing the electrodes may change the products of an electrolysis.

20.8 Some Examples of Electrolysis

(*a*) *Dilute Sulphuric Acid* (*Electrolysis of 'water'*)

Cathode: platinum, copper or carbon
Anode: platinum or carbon
In the electrolyte:

$$H_2O \rightleftharpoons H^+ + OH^-$$
$$H_2SO_4 \rightleftharpoons H^+ + HSO_4^- \rightleftharpoons 2H^+ + SO_4^{2-}$$

(The hydrogen ions are hydrated, e.g. H_3O^+)
At the cathode: electrons come from the cell and positive ions (H^+) are attracted.

$$H^+ + e^- \rightarrow H$$
$$2H \rightarrow H_2\uparrow$$

At the anode: electrons are taken away by the cell and negative ions (HSO_4^-, OH^-, SO_4^{2-}) are attracted: the hydroxide ion is discharged [see section 15.6(*g*)].

$$OH^- \rightarrow OH + e^-$$
$$2OH \rightarrow H_2O + O$$
$$2O \rightarrow O_2\uparrow$$

If 4 electrons go round the circuit:

$$4H^+ + 4e^- \rightarrow 2H_2\uparrow$$
$$4OH^- \rightarrow 2H_2O + O_2\uparrow + 4e^-$$

The result is the decomposition of water into its elements, two volumes of hydrogen being evolved for every volume of oxygen.

$$2H_2O \rightarrow 2H_2\uparrow + O_2\uparrow$$

Oxygen is more soluble in water than hydrogen so the apparatus should be left running for a while to saturate the electrolyte if measurements are to be made. More water ionizes to replace the ions lost. As the electrolysis proceeds the electrolyte becomes more concentrated.

(b) Dilute Hydrochloric Acid

As in (a) except that the reaction

$$HCl \rightleftharpoons H^+ + Cl^-$$

provides most of the ions. It is again the electrolysis of water.

(c) Concentrated Hydrochloric Acid

Cathode: carbon
Anode: carbon

In the electrolyte: as in (b)
At the cathode: as in (a)

$$2H^+ + 2e^- \rightarrow H_2\uparrow$$

At the anode: in the concentrated solution the chloride ions are more readily discharged than the hydroxide ions

$$Cl^- \rightarrow Cl + e^-$$
$$2Cl \rightarrow Cl_2\uparrow$$

Thus the overall result is the decomposition of hydrochloric acid:

$$2HCl \rightarrow H_2\uparrow + Cl_2\uparrow$$

(d) Copper Sulphate Solution

Cathode: copper (platinum or carbon could be used)
Anode: copper

In the electrolyte: $H_2O \rightleftharpoons H^+ + OH^-$
$CuSO_4$ is $Cu^{2+} + SO_4^{2-}$
At the cathode: $Cu^{2+} + 2e^- \rightarrow Cu\downarrow$
At the anode: although sulphate and hydroxide ions are attracted the reaction that occurs is

$$Cu \rightarrow Cu^{2+} + 2e^-$$

Thus for every atom of copper deposited on the cathode one dissolves at the anode. The electrolyte, which should be slightly acidified with dilute sulphuric acid for the best results, remains unchanged.

This process is the basis of the purification of copper (see section 26.5). If an inert anode such as platinum or carbon is substituted for the copper one then the situation is as in (a) and oxygen is released: in this case the blue colour of the solution disappears as the electrolyte becomes sulphuric acid.

(e) Dilute Sodium Chloride Solution

Cathode: platinum or carbon
Anode: platinum or carbon

In the electrolyte: $H_2O \rightleftharpoons H^+ + OH^-$
NaCl is $Na^+ + Cl^-$
At the cathode: hydrogen ions are discharged rather than sodium ions so the situation is the same as in (a);

$$2H^+ + 2e^- \rightarrow H_2\uparrow$$

At the anode: hydroxide ions are discharged rather than chloride ions so the situation is again as in (a):

$$4OH^- \rightarrow 4e^- + 2H_2O + O_2\uparrow$$

It is another case of the electrolysis of water.

(f) Concentrated Sodium Chloride Solution

Cathode: carbon or platinum
Anode: carbon

In the electrolyte: as in (e)
At the cathode: as in (e)

$$2H^+ + 2e^- \rightarrow H_2$$

At the anode: as in (c)

$$2Cl^- \rightarrow Cl_2 + 2e^-$$

The electrolyte thus becomes sodium hydroxide solution (see section 24.8) but reactions with the chlorine may ensue.

(g) Molten Sodium Chloride

Cathode: iron
Anode: carbon

Electrolyte: NaCl is $Na^+ + Cl^-$
Often other chlorides are added to lower the melting point (see the Downs process, section 24.3).
At the cathode: $Na^+ + e^- \rightarrow Na$
At the anode: $2Cl^- \rightarrow Cl_2 + 2e^-$

(h) Sodium Hydroxide Solution

Cathode: platinum or carbon
Anode: platinum or carbon

In the electrolyte: $H_2O \rightleftharpoons H^+ + OH^-$
NaOH is $Na^+ + OH^-$
At the cathode: as in (e):

$$2H^+ + 2e^- \rightarrow H_2\uparrow$$

At the anode: negative ions are attracted:

$$4OH^- \rightarrow 4e^- + 2H_2O + O_2\uparrow$$

This is the third case which is effectively the electrolysis of water: the electrolysis of a dilute solution of an acid, alkali or salt yields hydrogen and oxygen.

20.9 Faraday's Laws of Electrolysis

About 30 years after Volta discovered the cell, and following the time in which Davy first performed electrolytic experiments yielding sodium and potassium, metals hitherto unknown, Faraday performed quantitative experiments. He passed a known current for a known time and weighed the substances liberated at the electrodes. He discovered two relationships.

(a) **The mass of a substance liberated in electrolysis is proportional to the quantity of electricity passed.**

Quantities of electricity are measured in coulombs. If a current of 1 ampere flows for 1 second the quantity of electricity is 1 coulomb,

i.e. Coulombs = Amperes × Seconds

If M is the mass of the substance
t is the time in seconds
i is the current in amperes
and Z, the proportionality constant, known as the electrochemical equivalent of the substance concerned,

then $$M = Z \times i \times t$$

(*b*) **The masses of different substances liberated by the same quantity of electricity are proportional to their relative atomic masses divided by the valencies of their ions.**

(or since relative atomic mass = combining mass × valency, mass liberated is proportional to combining or equivalent mass).

If a substance forms singly charged ions the quantity of electricity carried by a given number of particles is half the quantity carried by the same number of doubly charged ions.

An alternative way of quantitatively describing the behaviour of electrolytes is to express the above laws in one statement:

The Faraday constant is the quantity of electricity required to liberate in electrolysis a mole of monovalent ions.

The value of the Faraday constant is 96 500 C/mol, which is 26·80 ampere-hours.

Example 1. In an electrolysis experiment the current flowing round the circuit caused the deposition of 0·050 g copper and 0·027 g chromium respectively in two voltameters.

Copper ions are divalent and chromium ions trivalent. The relative atomic masses of copper and chromium are 64 and 52 respectively so their combining masses are $\frac{64}{2}$ and $\frac{52}{3}$ respectively.

Then $$\frac{\text{Mass of chromium}}{\text{Mass of copper}} = \frac{0{\cdot}027}{0{\cdot}050} = 0{\cdot}54$$

$$\frac{\text{Combining mass of chromium}}{\text{Combining mass of copper}} = \frac{52}{3} \times \frac{2}{64} = 0\ 54$$

The results of the experiment support Faraday's second law.

Example 2. An electric current was passed for some time through (*a*) a voltameter containing dilute sulphuric acid and (*b*) a solution of copper sulphate. In (*a*) 82·5 cm^3 of hydrogen at 750 mmHg and 15°C were liberated; in (*b*) 0·220 g of copper were deposited. If hydrogen ions carry unit charge what do copper ions carry?

Volume of hydrogen at s.t.p.

$$= 82{\cdot}5 \times \frac{750}{760} \times \frac{273}{288}$$
$$= 77{\cdot}0 \text{ cm}^3$$

Mass of hydrogen

$$= 0{\cdot}00693 \text{ g}$$

i.e. there are 0·00693 moles of hydrogen atoms formed from its ions.

Number of moles of copper atoms

$$= \frac{0{\cdot}22}{64}$$
$$= 0{\cdot}00344$$

Thus $$\frac{\text{valency of copper}}{\text{valency of hydrogen}} = \frac{\text{number of moles of hydrogen}}{\text{number of moles of copper}}$$
$$\frac{0{\cdot}00693}{0{\cdot}00344}$$
$$= 2$$

i.e. Copper ions are Cu^{2+}

20.10 The Explanation of Faraday's Laws

When electrolysis is in progress the number of electrons being liberated by negative ions at the anode is equal to the number being taken up by positive ions at the cathode.

(*a*) *The First Law*

If the current is doubled the number of electrons flowing along the wire in a given time is also doubled. Hence the number of ions reacting at the electrodes is also doubled and so the masses of substances are doubled.

Also, if the current is allowed to flow for twice as long, the total number of ions discharged and hence the masses of substances liberated are again doubled. Thus the masses liberated are proportional to current and time, i.e. to the quantity of electricity.

(*b*) *The Second Law*

Consider the electrolysis of solutions of copper sulphate, silver nitrate and sulphuric acid in a series of voltameters. Electrons must be flowing at the same rate throughout the circuit. So for every one copper ion, Cu^{2+}, that is discharged, two silver ions, Ag^{+}, and two hydrogen ions, H^{+}, are discharged. The relative atomic masses of copper, silver and hydrogen are 64, 108 and 1 respectively so the masses of the substances will be in the ratio of 64:216:2 or 32:108:1, i.e. in the ratio of their combining masses.

Michael Faraday, 1791–1867, England

Faraday received very little by way of education early in life but, when he started reading many books, he became interested in science. In 1813 he was made Davy's assistant at the Royal Institution and in 1826 succeeded him as Professor of chemistry; in 1833 he was made director of the chemistry laboratory.

His published works contain no algebraic or chemical formulae. He maintained that the volume of scientific literature was too large and busied himself with the practical aspects and with lecturing.

He discovered that gases such as chlorine and carbon dioxide could be liquefied by the application of increased pressure (1823). He discovered benzene (1825). In 1833 he put forward the Laws of Electrolysis which gave a quantitative description of the phenomena discovered in his lifetime. He did many other experiments connecting and explaining the nature of electricity and magnetism.

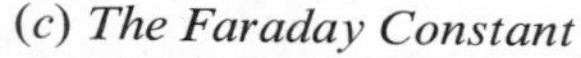

(*c*) *The Faraday Constant*

nF coulombs of electricity flow round the circuit when 1 mole of an 'n' valent ion are liberated. The faraday constant is the quantity of electricity carried by a mole of electrons, i.e. the quantity of electricity carried by 6×10^{23} electrons.

20.11 Ionic Reactions

Further evidence for the ionic theory comes from the fact that all solutions that would be expected to contain a given ion show many properties in common. Thus crystals and solutions of copper salts nearly all contain the hydrated copper(II) ion and are blue (or blue-green) in colour, and solutions of all acids show their characteristic behaviour because they all contain the ion H^+ (in some hydrated form).

Many precipitation reactions can be explained as the reactions between ions. All soluble sulphates, containing the ion SO_4^{2-}, form a white precipitate with all soluble barium compounds, which contain the ion Ba^{2+}.

Thus $BaCl_2 + H_2SO_4 \rightarrow BaSO_4\downarrow + 2HCl$

$$Ba(NO_3)_2 + Na_2SO_4 \rightarrow BaSO_4\downarrow + 2NaNO_3$$

$$Ba(OH)_2 + K_2SO_4 \rightarrow BaSO_4\downarrow + 2KOH$$

all simplify to

$$Ba^{2+} + SO_4^{2-} \rightarrow BaSO_4\downarrow$$

The explanation is similar for silver salts, all of which in solution give a precipitate of silver chloride when a soluble chloride is added.

$$Ag^+ + Cl^- \rightarrow AgCl\downarrow$$

The neutralization of an acid by an alkali can be represented by

$$H^+ + OH^- \rightarrow H_2O$$

but in this case the ionic equation is not in some cases as informative as is desirable.

The very different properties of ion and atom or molecule led to a long period before the existence of ions was fully accepted but it should now lead to the greater use of ionic equations to emphasize the essential features of a reaction. See also section 16.3.

20.12 The Applications of Electrolysis

(*a*) *The Refining of Metals*

The electrolysis of copper sulphate solution between an anode of impure copper and a cathode of pure copper is one of the most important processes to be considered (see sections 26.5 and 20.8). The thin sheets of copper used as cathodes in a factory may increase from 5 to 1000 kg in a fortnight. Nickel, lead and many other metals may be refined similarly.

(b) Electroplating

It may be advantageous to deposit a thin layer of one metal on top of another for purposes of protection or it may be desirable for decoration. Nickel and chromium, or zinc or tin, are deposited on iron to protect it from corrosion. Silver articles can be made tarnish-proof by a thin layer of the precious metal, rhodium.

Successful electroplating is not easy. The articles have to be thoroughly clean and free from grease; the other factors for successful electroplating are the concentration of the solution, the temperature, and a current according to the size of the article. If the conditions are not exactly right the metallic deposit may be rough or flaky. It is not always advantageous to use a solution of a simple salt; sometimes complex substances are used and very often mixtures of solutes. Even non-conductors of electricity can be electroplated by first coating them with a thin layer of graphite, e.g. plaster casts are coated with copper in this manner.

(c) The Preparation of Reactive Metals

The metals at the top of the electrochemical series are too reactive to be prepared in the laboratory by carbon or hydrogen reduction and react with water too readily to be prepared by electrolysis of an aqueous solution of one of their salts. Therefore sodium (see section 24.3), magnesium and calcium (see section 25.3) and aluminium (see section 27.3) are usually prepared by the electrolysis of fused salts or in non-aqueous solvents.

(d) The Preparation of Other Substances

Hydrogen and oxygen can be prepared by the electrolysis of water to which a small quantity of an alkali, acid or salt has been added (see sections 6.7 and 20.8). The electrolysis of concentrated sodium chloride solution yields chlorine, sodium hydroxide and hydrogen (see section 24.8) and can be made to yield sodium chlorate(I) and sodium chlorate(V) (see section 24.9). **N.B.** The manufacture of graphite, phosphorus and sometimes steel are **electrothermal** processes because the electricity is employed solely to generate heat.

20.13 The Movement of Ions

The movement of particles in a solution when a potential difference is applied can be shown. Strips of filter paper are soaked in sodium sulphate solution (to increase the conductance) and then small crystals of coloured substances, e.g. copper sulphate, potassium manganate(VII), etc., are put at the centres of the strips. A battery is then connected to the ends of the strip of paper and after a while some motion of the coloured material is apparent.

Another experiment is to soak the centre of a piece of filter paper in potassium nitrate solution and to put some silver nitrate at the end connected to the positive pole of the battery and some potassium chromate at the end connected to the negative pole. After a while a thin red line forms roughly mid-way along the strip where the silver ions meet the chromate ions and form a red precipitate of silver chromate.

$$2Ag^+ + CrO_4^{2-} \rightarrow Ag_2CrO_4\downarrow$$

Copper chromate will dissolve in dilute sulphuric acid and the dark solution can be put at the bottom of a U-tube; the movement of this solution by convection currents and diffusion can be reduced by putting glass wool plugs at each end of the curved portion of the tube. Then the arms of the tube are filled up simultaneously with dilute sulphuric acid and carbon electrodes attached to a battery inserted in the acid. After a while blue copper ions will be seen moving towards the electrode attached to the negative pole and yellow chromate ions towards the electrode attached to the positive pole.

20.14 A Conductimetric Titration

The movement of ions in solution gives that solution an electrical conductance and this can be assessed with a Wheatstone Bridge arrangement known as a conductance bridge.

Some barium hydroxide solution is put into a beaker and its conductance measured using platinum dipping electrodes. A magnetic stirrer is useful for agitating the solution. Then portions

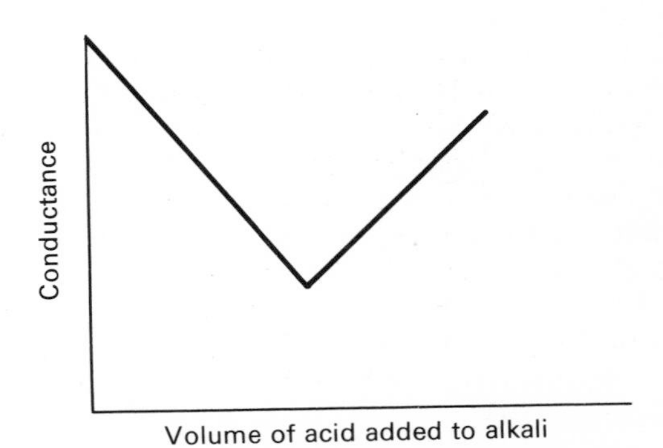

Figure 20.5
The result of a conductimetric titration

of a more concentrated solution of sulphuric acid are added and the conductance of the solution measured each time. The total volume of the solution should not vary too much as a result of the titration, hence the disparate concentrations of the reagents. No one reading is crucial, unlike the titrations done with indicators, and a graph (see figure 20.5) can be obtained from which the volume of acid and alkali needed for neutralization can be obtained.

$$Ba(OH)_2 + H_2SO_4 \rightarrow BaSO_4\downarrow + 2H_2O$$

or

$$Ba^{2+} + SO_4^{2-} \rightarrow BaSO_4\downarrow$$

20.15 Cells and Voltameters: A Summary

(a) A Cell ⊣⊢ *(several cells = a battery)*

Long thin line, positive pole, brown (red), cathode of cell.

Short thick line, negative pole, blue (black), anode of cell.

Conventional current flows from positive to negative in external circuit.

Electronic current flows in opposite direction.

At the negative pole in the battery the metal highest in the electrochemical series yields electrons, i.e. is oxidized.

(b) Wires

Brown, from positive pole of cell to anode of voltameter.

Blue, from negative pole of cell to cathode of voltameter.

(c) A Voltameter ⊣⊢

Cathode: cations attracted, reduction occurs here, gain of electrons by ions, metals deposited or hydrogen liberated.

Anode: anions attracted, oxidation occurs here, loss of electrons by ions, non-metals (often gases) liberated or metal anode may dissolve yielding cations and electrons.

(d) A Quick Test for a D.C. Circuit

This is very useful if the d.c. supply is from a plug in a socket leading away to a transformer or accumulators.

Connect the wires from the plug to two pieces of nichrome wire inserted through a piece of wood and dipping into dilute sulphuric acid: the blue wire should be connected to the nichrome wire where bubbles form, i.e. it comes from the negative pole of the cell.

CARE—do not let electrodes in any experiment touch as this will cause a short-circuit.

QUESTIONS 20

Electrochemistry

1 At which electrode, in electrolysis, does oxidation occur? What process is occurring during this oxidation? Quote an example of the practical application of this effect. [OC]

2 The terminals of a car battery are connected to carbon rods pressed down, one centimetre apart, on a piece of neutral litmus paper moistened with brine. Describe and explain any changes you would observe on the litmus paper. [OC]

3 A copper rod stands in a dilute solution of copper sulphate in a beaker. A zinc rod stands in a second beaker containing a dilute solution of zinc sulphate.
(*a*) Draw a labelled diagram of this apparatus, adding to it any items necessary to enable it to act as a voltaic cell.
(*b*) Show, on your diagram, the direction of flow of electrons when this cell is connected to an external circuit.
(*c*) Describe carefully all changes that will take place in each beaker while current flows. Explain these changes.
(*d*) Give two reasons why the flow of electrons will, after a time, diminish. [OC]

4 Suppose that you have to plate a graphite rod with copper. Describe, with the help of a labelled diagram and a brief description, how you would do this. State what changes occur during the operation. Calculate the mass of copper that would be deposited on the graphite during the passage of 500 coulombs. [OC]

5 An electric current is passed through two voltameters in series. One has silver electrodes and contains silver nitrate solution. The other has platinum electrodes and contains dilute sulphuric acid. If 4·32 g of silver are deposited in the first, what volume of hydrogen measured at 720 mmHg and 27°C is produced in the second? [S]

6 Calculate the volumes of hydrogen and oxygen (measured at s.t.p.) which would be liberated in the electrolysis of dilute sulphuric acid by the passage of 12 kC of electricity. What mass of water would be decomposed? [S]

7 Write ionic equations to show what happens at both the anode and the cathode, when an electric current is passed through a solution of copper sulphate (*a*) when platinum electrodes are used, (*b*) when copper electrodes are used. [S]

8 What mass of aluminium could be obtained by the complete decomposition of 25·5 g of aluminium oxide (Al_2O_3)? How many coulombs of electricity would be needed for this? [S]

9 (*a*) Explain why a solution of sugar is a non-conductor of electricity but a solution of sodium chloride is a good conductor. What is formed at each electrode when a current is passed through a solution of sodium chloride by means of carbon electrodes? Explain how you would demonstrate the presence of each product.
(*b*) A current of 2 A is passed for 2 hours through (i) a solution of copper sulphate (using copper electrodes) and (ii) a dilute solution of sulphuric acid (using platinum electrodes), the two solutions being connected in series. In (i), calculate the changes in mass of the cathode and anode, stating whether a gain or a loss. In (ii), calculate the volumes at s.t.p. of the gases liberated at the cathode and anode. Name the gases. [C]

10 (*a*) Dilute sulphuric acid is electrolyzed in a beaker, using copper electrodes. Answer the following questions about this electrolysis. (i) Give the formulae of all the ions present in the solution before electrolysis starts. (ii) Give the formulae of any new ion (or ions) which will be present in the solution after electrolysis has been taking place for a few minutes. (iii) Draw a simple diagram to indicate the direction of migration of the ions mentioned in (i) and (ii), and also the direction of flow of electrons in the circuit outside the beaker. (iv) What changes, if any, would you observe at the anode, at the cathode, and in the solution? (v) Explain the chemical change occurring at the anode.
(*b*) Draw a labelled diagram of the apparatus you would use to electroplate a small object with a coating of a metal such as nickel or silver. Do you think it would be possible to electroplate an object with magnesium? Give briefly the reasons for your answer. [C]

11 (*a*) How is copper purified electrically? State what happens to metallic impurities above and below copper in the electrochemical series.
(*b*) If two circuits were set up for the electrolysis of acidified water, one using copper electrodes and the other using platinum electrodes, what differences would be observed? Name the products at the separate electrodes in each case. If, in the case of the circuit using copper electrodes, the current was reversed after a short interval of time, what would be the immediate result at each electrode? [C]

12 What do you understand by the term *electrolyte*? Illustrate your answer by giving a brief description and explanation of the passage of an electric current through an aqueous solution of copper sulphate using platinum electrodes.

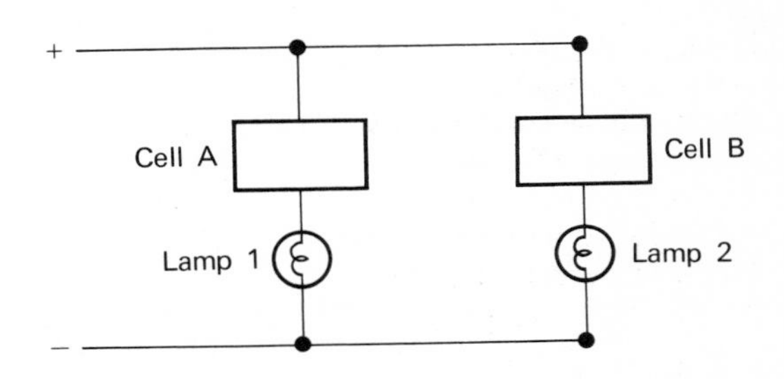

The diagram illustrates how an electric current could pass through 1 M solutions of ethanoic (acetic) acid (in cell A) and sulphuric acid (in cell B) in parallel. If the current flow is so arranged that the filament of lamp 1 just glows, what would you expect to observe at lamp 2? Briefly explain. Without altering the current setting, with what 1 M aqueous solution could you replace the ethanoic acid in cell A so that lamp 1 did not light at all? [O]

13 Describe a simple experiment to illustrate what is meant by *electrolysis*, giving a clearly labelled diagram of the apparatus. Briefly describe how electrolysis is used (*a*) to manufacture aluminium and (*b*) to purify copper.

14 Define the term *electrolyte*. Which of the following are electrolytes: zinc sulphate, copper, sodium hydroxide, cane sugar? A direct current is passed through a solution of sodium chloride, using carbon electrodes in a cell fitted with a diaphragm for keeping apart the products formed at the two electrodes. State the ions present in the solution, the names and polarities of the electrodes, and the products of the electrolysis. Write ionic equations for the reactions occurring at the electrodes. If the electrolysis is repeated, using a fresh solution of sodium chloride in a cell which allows the products to mix, what new compound is formed in solution? For what purpose is this compound used? [J]

15 A solution of potassium tetracyanoaurate(III) was electrolyzed using a gold anode and a gold cathode. During the electrolysis gold was deposited on the cathode. The electrodes were weighed before and after the experiment.

Results:

	Cathode	*Anode*
Initial mass	25·104 g	24·614 g
Final mass	25·596 g	24·122 g

Current flowing 0·20 A
Time of current flow 60 minutes

(*a*) Calculate the amount of electricity which flowed during the experiment. Give your answer in ampere-hours or coulombs.
(*b*) What mass of gold would have been deposited by 96 500 C of electricity?
(*c*) The relative atomic mass of gold is 197. How many coulombs of electricity are required to deposit 1 mol of gold?
(*d*) What does this experiment tell us about the number and kind of charges on one gold ion?
(*e*) On the basis of these results a pupil wrote the following conclusions. Which of these conclusions do you consider (i) justified from the results, (ii) not justified from the results?

A Potassium tetracyanoaurate(III) is a compound containing potassium and gold only.
B Potassium tetracyanoaurate(III) consists of molecules which change into ions when they are dissolved in water.
C For every atom of the gold anode which changes into an ion, an atom of gold is deposited on the cathode.
D Water is a good conductor of electricity.
E Gold electrodes are essential for the electrolysis of a solution of potassium tetracyanoaurate(III) in water.
F The formula of a chloride of gold is likely to be $AuCl_3$.

Indicate your answer by stating (i) or (ii) for each conclusion. [Nf]

16 Describe a simple fuel cell which uses hydrogen and oxygen, stating the reactions which give rise to the electricity. Give reasons why scientists and engineers consider that work leading to the development of a fuel cell for use in industry is worthwhile.

17 Describe the apparatus and the procedure to demonstrate that ions of a metal element *A*, in aqueous solution, each carry twice the charge as ions of the metal element *B*. State what measurements you would make and how you would use them in calculation (let *a* and *b* represent the relative atomic masses of the two metal elements).

18 The diagram (A) shows the arrangement used in an experiment to find the quantity of electricity required to liberate 1 mol of lead in electrolysis. Molten lead bromide was electrolyzed in a glass bowl, with carbon rods as electrodes, and the molten lead obtained was separated and weighed at the end of the experiment. The results of a series of experiments were not consistent and the amount of lead obtained per coulomb was less in the longer experiments than in those of shorter duration.

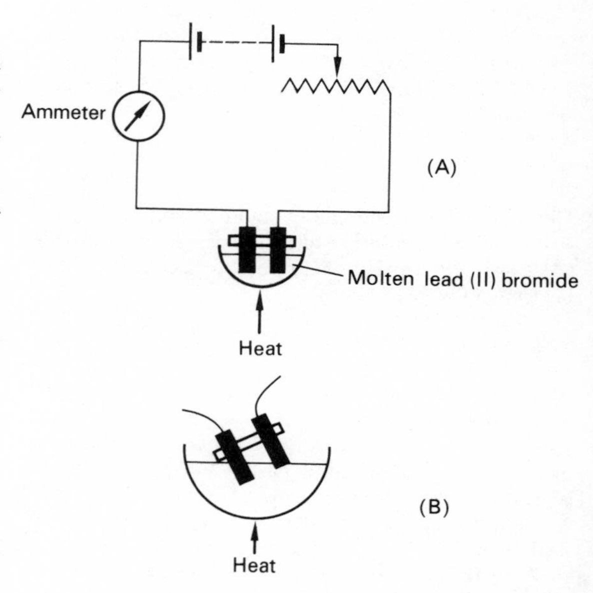

(*a*) In one of the shorter experiments the current was maintained at 3·25 A and the mass of lead obtained after exactly 10 minutes was 2·07 g. What value do these figures give for the quantity of electricity required to set free 1 mol of lead?
(*b*) One explanation suggested to account for the inconsistency of the results was that a pool of molten lead caused a *short-circuit* between the electrodes. (i) Explain why such a short-circuit would reduce the mass of lead obtained for a given quantity of electricity. (ii) What other indication would you expect there to be if a short-circuit occurred?
(*c*) Another suggestion was that the lead was attacked by the bromine which was also set free, so that lead bromide was reformed. Give a sketch of the apparatus, and a brief description, to show how you would attempt safely to find out whether bromine reacts with molten lead. (Boiling point of bromine = 59°C; melting point of lead = 327°C.)
(*d*) In an attempt to get more consistent results, the electrode assembly was tilted as shown in diagram (B). (i) Explain in what ways the tilting of the electrodes reduces the likelihood of errors arising from the two effects suggested above. (ii) Should the lower electrode in this arrangement be the positive or the negative, or does it not matter? Give reasons for your answer. [Nf]

19 Part of the list of standard redox potentials (against the standard hydrogen electrode) is given below:

Half cell	$E^\ominus$ (*volts*)
$Fe^{2+} \mid Fe$	$-0{\cdot}44$
$Sn^{2+} \mid Sn$	$-0{\cdot}14$
$Fe^{3+}, Fe^{2+} \mid Pt$	$+0{\cdot}77$
$Ag^{+} \mid Ag$	$+0{\cdot}80$
$Br_2, 2Br^{-} \mid Pt$	$+1{\cdot}09$
$(MnO_4^- + 8H^+), (Mn^{2+} + 4H_2O) \mid Pt$	$+1{\cdot}52$

(*a*) Describe carefully how you would attempt to make a cell with an e.m.f. of at least 0·5 volt, using electrodes selected from this list. You should make clear which is the positive and which is the negative terminal of the cell and state the e.m.f. you expect the cell to produce. What chemical reactions take place in your cell and where will they occur? If you found that the e.m.f. measured on a high resistance voltmeter was not the expected value, what explanation would you give?
(*b*) How would you try to convince a fellow pupil, who had not studied this topic, why it is that a table of $E^\ominus$ values should be useful in predicting a great many chemical reactions? You should, in fairness, also draw attention to any limitations which need to be placed upon your predictions. [Nf]

20 Describe the reactions in a lead accumulator when it is discharging. What happens when it is recharged? Why is it a bad policy to allow the concentration of sulphuric acid to fall too low?

21 Describe the apparatus you would use in order to set up the following cell which is to be used to measure the e.m.f. of an $Ag \mid Ag^+$ electrode against a standard hydrogen electrode.

$$Pt,H_2 \mid H^+ \mid Ag^+ \mid Ag \qquad E = +0{\cdot}80V$$

How would you measure the e.m.f. if a high resistance voltmeter was not available?

22 During the course of a titration, the pH changes and so does the conductance. The graphs representing these changes have been drawn below for the addition of sodium hydroxide solution to 25 cm^3 hydrochloric acid and compared by using the same axes for the

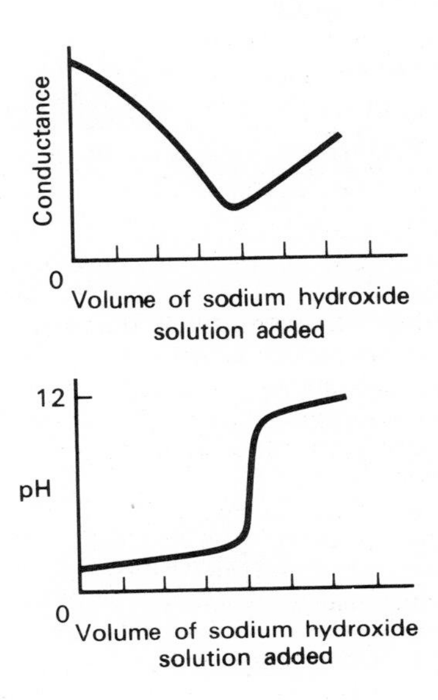

volume of alkali added. Describe how you would obtain one of these curves, drawing attention to the techniques needed to give reliable results. Explain why it is that these two curves have opposite slopes in the early part of the titration. [Nf]

21 Qualitative Analysis

21.1 Introduction

When faced with an unknown substance a chemist performs a logical set of tests to try to find out the nature and identity of the substance. The materials that are found in nature are usually far from simple and are often mixtures, not pure single substances. In elementary laboratory work it is unlikely that any substance will be set for analysis in which there is more than one metallic and one acidic radical.

It is important to record carefully all tests and observations made on the substance. If a gas is evolved it must be identified. It is sometimes useful to obtain the substance the unknown is thought to be and perform the same tests upon it. Using a centrifuge to separate a solid from a solution is faster than filtration but remember to counterbalance test tubes in a centrifuge and to keep the lid of the centrifuge closed whilst the rotor is in motion.

The analysis can be divided into three sections:

I. Preliminary tests—which are not always reliable and are done quickly:
(*a*) Appearance.
(*b*) Solubility.
(*c*) Action of heat.
(*d*) Heat with dry copper(II) oxide.

II. Tests for acidic radicals—which must be done very carefully; a negative result may suggest an element, oxide or hydroxide.
(*a*) Add dilute hydrochloric acid.
(*b*) Add concentrated sulphuric acid.
(*c*) Boil with excess concentrated sodium carbonate solution, centrifuge (filter), test the solution remaining.
(*d*) Some confirmatory tests.

III. Tests for metallic radicals—which again must be done carefully.
(*a*) Flame test (this is not always reliable).
(*b*) Add sodium hydroxide solution.
(*c*) Add a dilute solution of ammonia.
(*d*) Some confirmatory tests.

CAUTION:
These tests constitute a complete analysis of a substance at this stage: they are safe to perform upon the substances usually set for analysis at school. However, if you are doing experiments at home on a wider range of substances, the tests above, particularly II(*b*), must be done initially with very small quantities, or even omitted.

I PRELIMINARY TESTS

21.2 The Appearance of a Substance

This test, which can be very quickly dealt with, may give rise to the results shown opposite:

Colour	*Inference*
Blue	Cu^{2+} salts
Green	Ni^{2+}, Fe^{2+}, Cu^{2+} salts
Yellow	Fe^{3+} salts, PbO, SnS_2
Red or pink	Fe_2O_3, Pb_3O_4, Co^{2+} salts
Black	CuO, MnO_2, Fe_3O_4, PbS, CuS
Brown	PbO_2, SnS, Cu_2O
White	Absence of above

Nearly all the substances mentioned are compounds of transition metals (see section 26.1).

21.3 The Solubility of a Substance

Find out whether the substance is soluble in cold water or not: put one spatula load in a test tube, add 5 cm^3 water, shake well; observe whether the amount of solid diminishes.

At the end of the analysis consult section 17.9 to see whether there is any disagreement. If the substance is not soluble in cold water try hot water, then, in the same manner, try dilute and finally concentrated hydrochloric acid. If an acid is used in this test the results of the tests described in section 21.6 are obtained.

21.4 The Action of Heat

Place one spatula load of the substance in an ignition tube, heat gently and then strongly; have damp litmus paper, calcium hydroxide solution and a splint ready; watch and smell carefully. This test should be dealt with very quickly.

Result	*Inference*
White sublimate	Might be NH_4^+ salt
Decrepitation	Crystalline material
Characteristic smell, alkaline, white clouds with hydrogen chloride	NH_3 from NH_4^+ salt
Characteristic smell, acidic	SO_2 from SO_4^{2-} or SO_3^{2-} salt
Characteristic smell, acidic, fumes with ammonia	HCl from hydrated Cl^- salt
Characteristic smell, brown fumes, acidic, relights glowing splint	NO_2+O_2 from heavy metal nitrate
No smell, neutral, glowing splint reignites	O_2 from KNO_3 or $NaNO_3$
No smell, calcium hydroxide solution turns milky	CO_2 from CO_3^{2-} or HCO_3^- salt (exceptions Na_2CO_3, K_2CO_3)
No smell, calcium hydroxide solutions turns milky, some charring, slight flammability	$CO+CO_2$ from $C_2O_4^{2-}$ salt
Charring (due to formation of carbon)	$C_2O_4^{2-}$ or $C_2H_3O_2^-$ salt
Steam, colour change, change reversed (unless heated too strongly) by addition of water	Hydrated salt of transition metal
Temporary colour change: white ⇌ yellow	May be Zn^{2+} salt
yellow ⇌ red	PbO
red ⇌ black	Fe_2O_3

21.5 Heat with Dry Copper(II) Oxide

If the substance, by the previous test, has been shown to contain water of crystallization it should be heated gently to expel this before proceeding.

This test will show if the substance contains hydrogen or carbon, i.e. whether it is possibly wholly or partially organic. Heat one spatula load of the substance mixed with three spatula loads of dry copper(II) oxide in a dry test tube. If the substance is a reducing agent (most organic substances are) the residue in the test tube becomes red-brown (showing copper is formed) and carbon dioxide is evolved (proving carbon is present). Steam may condense on the cooler parts of the test tube (indicating hydrogen is present).

II TESTS FOR ACIDIC RADICALS

21.6 The Addition of Dilute Hydrochloric Acid

Add 2 cm^3 dilute hydrochloric acid to one spatula load of the solid in a test tube, and warm the substances if there is no reaction in the cold. The tests to be applied are the same as in section 21.4.

If one of these six results is obtained, then it is not necessary to perform the tests outlined in sections 21.7 and 21.8.

Result of test on gas evolved	*Inference*
Calcium hydroxide solution turns milky	CO_2 from CO_3^{2-} salt or $NaHCO_3$, or $KHCO_3$
Characteristic smell, acidic, potassium dichromate solution + dilute sulphuric acid (on paper) orange to very pale green	SO_2 from SO_3^{2-} salt
As previous result + white suspension	$SO_2 + S$ from $S_2O_3^{2-}$ salt
Characteristic smell, acidic, lead ethanoate paper turns black	H_2S from S^{2-} salt
Explosion on ignition	H_2. Metal high in electrochemical series
Characteristic smell, acidic, brown fumes particularly in upper part of test-tube	$NO_2 + NO$ from NO_2^- salt

21.7 The Addition of Concentrated Sulphuric Acid

There is no need to do this test if a result has been obtained in section 21.6. Put 1 cm^3 concentrated sulphuric acid on half a spatula load of the solid: if the substance reacted with dilute hydrochloric acid then it will react, probably vigorously, with concentrated sulphuric acid. If there is no reaction in the cold, the mixture may be warmed gently, **exercising great care.**

Result of test on gas evolved	*Inference*
Characteristic smell, acidic, fumes in air, white clouds with ammonia	HCl from Cl^- salt
Characteristic smell, acidic, red-brown, fumes in air, white clouds with ammonia	$HBr + Br_2$ from Br^- salt
Characteristic smell, acidic, violet-black, fumes in air, white clouds with ammonia (possibly hydrogen sulphide and sulphur dioxide as well from acid)	$HI + I_2$ from I^- salt
Characteristic smell, acidic, brown, oily drops in top of tube, on addition of copper clipping brown colour intensifies and solution becomes blue or green	$HNO_3 + NO_2$ from NO_3^- salt
No smell, calcium hydroxide solution turns milky, some charring, slightly flammable	$CO_2 + CO$ from $C_2O_4^{2-}$ salt

21.8 Tests on Sodium Carbonate Extract

These tests may be necessary to confirm results already obtained or to deal with difficult substances. In very simple cases the four tests below can be done on the original substance. Boil half a spatula load of the substance with 4 cm^3 saturated sodium carbonate solution or mix half a spatula load of the substance with two spatula loads of anhydrous sodium carbonate and 4 cm^3 distilled water. Centrifuge (or filter) and divide the alkaline solution into four portions. The absence of a precipitate suggests that the substance is a sodium, potassium or ammonium compound. The first step in each of the tests is to acidify the alkaline solution: the appropriate acid is added, with stirring, until all effervescence ceases.

Test	*Result*	*Inferences*
(*a*) Add dilute nitric acid, then silver nitrate solution	(i) White precipitate, soluble in ammonia solution	Cl^- salt
	(ii) Pale yellow precipitate, a little dissolves in ammonia solution	Br^- salt
	(iii) Yellow precipitate, insoluble in ammonia solution	I^- salt
(*b*) Add dilute sulphuric acid, then fresh iron(II) sulphate solution, finally concentrated sulphuric acid	Brown coloration	NO_3^- salt
(*c*) Add dilute hydrochloric acid, then barium chloride solution	White precipitate	SO_4^{2-} salt
(*d*) Add dilute sulphuric acid, then potassium manganate(VII) solution	(i) Goes colourless at room temperature	SO_3^{2-} salt
	(ii) Goes colourless on warming	$C_2O_4^{2-}$ salt

21.9 Confirmatory Tests for Anions

(a) Nitrates

Warm half a spatula load of the substance with 3 cm^3 sodium hydroxide solution in a test tube: if ammonia is released an ammonium salt is present (see section 21.11). Continue heating until no more ammonia is released. Then to the alkaline solution add half a spatula load of Devarda's alloy (Cu, Al, Zn) or coarse aluminium powder and warm the mixture. Hydrogen will be given off by the reaction of the alloy with the alkali: it is the production of an alkaline gas (ammonia) that is the crucial observation in the second stage to indicate that the substance is a nitrate (or a nitrite).

(b) Halides

Mix half a spatula load of the substance with one spatula load of manganese(IV) oxide and warm them in a test tube with 1 cm^3 concentrated sulphuric acid in a fume cupboard.

Colour of gas evolved	*Inference*
Pale yellow-green	Cl_2 from Cl^- salt
Red-brown	Br_2 from Br^- salt
Violet-black	I_2 from I^- salt

III TESTS FOR METALLIC RADICALS

21.10 Flame Test

Pour 2 cm^3 concentrated hydrochloric acid into a crucible or some other convenient vessel. Dip a 5 cm piece of nichrome wire held in tongs into the acid and then heat the wire in a non-luminous Bunsen flame: it should not cause any coloration of the flame unless it has been handled very much. When the wire is cleaned, dip it in the acid so that a particle of the unknown substance will adhere to it and then put it back in the outer flame of the burner. Observe the colour imparted. The test gives useful results with many substances that are colourless but it is not fully reliable. The wire can be used several times.

Colour	*Ion indicated*
Bright red	Li^+
Dark (brick) red or dull orange (usually contaminated with Na^+)	Ca^{2+}
Orange-yellow—persistent, masked by cobalt (blue) glass	Na^+
Yellow sparks	Fe^{2+}, Fe^{3+}, NH_4^+
Green, blue centre	Cu^{2+}
Light green, persistent	Ba^{2+}
Blue-white (dangerous fumes)	Pb^{2+}
Lilac—crimson through cobalt glass	K^+

21.11 The Addition of Sodium Hydroxide Solution

If the substance is suspected to be a metal from the results of the test described in section 21.6, then it should be dissolved in the minimum quantity of dilute nitric acid and then the tests outlined in sections 21.11(b), 21.12, and 21.13 may be performed.

(*a*) Warm half a spatula load of the substance with 2 cm^3 sodium hydroxide solution. If ammonia is evolved (characteristic smell, alkaline, white clouds with hydrogen chloride) then the substance is an ammonium salt and the tests described in sections 21.11(b), 21.12 and 21.13 may be omitted.

(*b*) Slowly add sodium hydroxide solution with stirring to 2 cm^3 solution of the substance until the sodium hydroxide is in excess.

Precipitate	*With excess alkali*	*Ion indicated*
White	Insoluble	Mg^{2+}, Ca^{2+}
White	Soluble	Zn^{2+}, Pb^{2+}, Al^{3+}, Sn^{2+}, Sn^{4+}
White, rapidly turning green then, slowly, brown	Insoluble	Fe^{2+}
White, slowly turning pale brown	Insoluble	Mn^{2+}
Red-brown	Insoluble	Fe^{3+}
Dark brown	Insoluble	Ag^+
Yellow	Insoluble	Hg^{2+}
Green	Insoluble	Ni^{2+}
Blue	Insoluble	Cu^{2+}, Co^{2+}
Black	Insoluble	Hg_2^{2+}
No precipitate	—	Na^+, K^+, NH_4^+, Ba^{2+}

21.12 The Addition of a Dilute Solution of Ammonia

Slowly add a dilute solution of ammonia (ammonium hydroxide) to 2 cm^3 solution of the substance, with stirring or shaking, until the ammonia solution is in excess.

Precipitate	*With excess alkali*	*Ion indicated*
White	Insoluble	Mg^{2+}, Al^{3+}, Pb^{2+}, Sn^{2+}, Sn^{4+} Temporarily hard water
White	Soluble	Zn^{2+}
White, rapidly turning green then, slowly, brown	Insoluble	Fe^{2+}
White, slowly turning pale brown	Insoluble	Mn^{2+}
Red-brown	Insoluble	Fe^{3+}
Dark brown	Soluble	Ag^{+}
Green	Soluble (blue)	Ni^{2+}
Blue (pale)	Soluble (deep blue)	Cu^{2+}
Blue	Insoluble	Co^{2+}
Black	Insoluble	Hg^{2+}, Hg_2^{2+}
No precipitate	—	Na^{+}, K^{+}, NH_4^{+}, Ca^{2+}, Ba^{2+}

21.13 The Addition of Hydrogen Sulphide

This test is part of systematic analysis, but in elementary work should be avoided if at all possible. Hydrogen sulphide is a dangerously poisonous gas and the test must be confined to a fume cupboard.

Some sulphides are precipitated under acidic conditions, others only under neutral or alkaline conditions. See section 31.14.

21.14 Some Confirmatory Tests for Cations

Make a small hole in a charcoal (carbon) block and mix the material which is displaced with the substance and replace it in the hole; a drop of water may be added to stop the powders blowing away. The powder is heated using a blowpipe and a luminous Bunsen flame. See figure 21.1.

If a copper compound is heated on a charcoal block, a red-brown residue of metallic copper is obtained. Copper carbonate is less stable to heat than, for example, copper sulphate and so mixing an unknown copper compound with anhydrous sodium carbonate may be helpful. The test should not be done with suspected lead compounds.

Other tests may be found in the chapter where the detailed chemistry of an element is discussed.

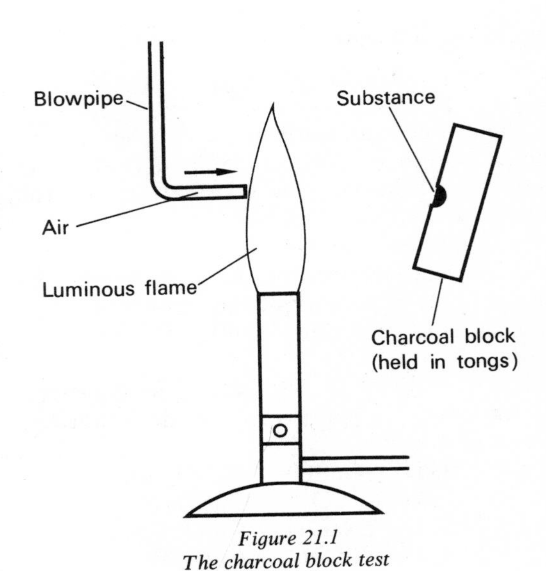

Figure 21.1
The charcoal block test

21.15 Conclusion

When writing up a qualitative analysis a three column table divided horizontally into three sections is a very convenient way of recording information:

Analysis of Substance X

Test	Result	Inference
1. Preliminary tests		
(a) Appearance	Blue	Cu^{2+}
(b) Solubility	Soluble in cold water	Nil (initially)
(c) Heat alone	Turns white, colour restored by water	Cu^{2+} hydrated
(d) Heat with dry copper oxide	Omitted	Not thought to be organic
2. Acidic radical tests		
(a) Dilute hydrochloric acid	—	Not CO_3^{2-} etc.
(b) Concentrated sulphuric acid	—	Not Cl^- etc.
(c) Boil with sodium carbonate solution, test filtrate		
(i) dilute nitric acid, silver nitrate solution	—	Not Cl^- etc.
(ii) dilute sulphuric acid, fresh iron(II) sulphate solution, concentrated sulphuric acid	—	Not NO_3^- etc.
(iii) dilute hydrochloric acid, barium chloride solution	White precipitate	SO_4^{2-}
(iv) dilute sulphuric acid, potassium manganate(VII) solution	Omitted	No suggestion of SO_3^{2-} or $C_2O_4^{2-}$
3. Metallic radical tests		
(a) Flame test	Green with blue centre	Cu^{2+}
(b) Dilute sodium hydroxide solution	Pale blue precipitate, insoluble in excess	Cu^{2+}
(c) Dilute solution of ammonia	Pale blue precipitate, soluble in excess giving deep blue solution	Cu^{2+}
(d) Charcoal block	Red-brown residue	Cu^{2+}

Conclusion: substance X is hydrated copper(II) sulphate.

The evidence all leads to the conclusion that the substance is hydrated copper(II) sulphate and this fits in with the observation that the substance is soluble in water.

It should be noted that negative as well as positive results are recorded because the elimination of many possibilities helps to narrow down the choices to be made at the end of the practical work.

The account of an analysis can be improved by recording the equations for the important tests, e.g.

2(c)(iii) $BaCl_2 + CuSO_4 \rightarrow BaSO_4 + CuCl_2$

or $Ba^{2+} + SO_4^{2-} \rightarrow BaSO_4\downarrow$

3(b) $CuSO_4 + 2NaOH \rightarrow Cu(OH)_2 + Na_2SO_4$

or $Cu^{2+} + 2OH^- \rightarrow Cu(OH)_2\downarrow$

QUESTIONS 21

Qualitative Analysis

1 A yellow metallic alloy is dissolved in nitric acid giving a blue solution, which on dilution and saturation with hydrogen sulphide gives a black precipitate. The filtrate from this precipitate, treated with excess of ammonia and hydrogen sulphide, gives a white precipitate. What is the alloy? Write equations for the above reactions, name the precipitates, and describe one experiment you would carry out with each precipitate to establish its identity. [O]

2 For each of the following pairs of substances, describe and explain one chemical test which, applied to each one of the pair, would distinguish between them. State what you would observe in each case: (*a*) potassium nitrate and potassium nitrite; (*b*) sodium carbonate and sodium hydrogencarbonate; (*c*) lead and zinc; (*d*) calcium carbonate and magnesium carbonate; (*e*) potassium chloride and potassium bromide. [O]

3 For each of the following pairs give one physical property and one chemical test by which the first may be distinguished from the second: (*a*) sulphur dioxide and carbon dioxide; (*b*) ammonium chloride and potassium sulphate; (*c*) iron(II) chloride and iron(III) chloride; (*d*) lead(II) oxide and copper(II) oxide. [J]

4 Explain how by means of suitable experiments and tests you would show that: (*a*) a given gas is dinitrogen oxide and not oxygen; (*b*) sea water contains sodium chloride; (*c*) one of the constituents of a fertilizer is potassium nitrate (it may be assumed that the others will not interfere in any tests performed). [J]

5 How would you distinguish by chemical experiments between the following pairs of substances: (*a*) aqueous solutions of hydrochloric and sulphuric acid; (*b*) oxygen and dinitrogen oxide; (*c*) carbon and manganese(IV) oxide; (*d*) anhydrous sodium carbonate and sodium hydrogencarbonate. State the chemical changes which take place in each case. [J]

6 For each substance, *A*, *B*, *C*, *D*, two tests are described below, together with the result of each test. Name each substance.
Liquid A. (i) When poured on to metallic copper a blue solution is formed and brown fumes are evolved. (ii) When boiled with sulphur, sulphuric acid is formed.
Solid B. (i) When water is added, a large amount of heat is evolved, it swells up and crumbles to a white powder. (ii) When moistened with hydrochloric acid and held on a nichrome wire in the Bunsen flame, the flame is coloured dark red.
Solution C. (i) When added to potassium iodide solution, iodine is liberated. (ii) Red cloth is bleached.
Solution D. (i) When added in excess to copper sulphate solution a deep blue solution is formed. (ii) When added to iron(III) chloride solution a red-brown precipitate is formed.

7 You are given four white powders known to be zinc sulphide, calcium carbonate, calcium chlorate(I) and sodium sulphite. Explain how, by adding hydrochloric acid to each in turn, you could distinguish between them. (If any gas is evolved, a test to prove its presence must be described.) [J]

8 Explain carefully how you would prove that: (*a*) the impurity in a sample of sodium hydroxide was a carbonate; (*b*) a liquid was nitric acid and not hydrochloric acid; (*c*) a given sample of tap water contains a chloride; (*d*) a given solid was sodium sulphate. [J]

9 You are provided with four black powders known to be copper(II) oxide, manganese(IV) oxide, triiron tetraoxide and iron(II) sulphide. Explain how by using hydrochloric acid you could distinguish between them. Say what happens in each case. [J]

10 Explain how by a chemical test you would distinguish between the two substances in each of the following pairs: (*a*) ammonium sulphate and sodium carbonate; (*b*) potassium chloride and barium nitrate; (*c*) solid sodium peroxide and anhydrous copper sulphate; (*d*) a solution of iron(II) chloride and a solution of iron(III) chloride. (Remember: (i) that only one chemical test is required for each pair of substances but you must state the results of that test on each substance of the pair; (ii) that the evolution of any invisible gas must be confirmed by a test.) [J]

11 Explain the following observations and state the name of the solid *X*: (*a*) a white crystalline solid *X* is left on the bench near to a hydrogen sulphide generator and it is noticed that the crystals become dark brown; (*b*) when the white crystalline solid is heated it gives off brown fumes which relight a glowing splint; (*c*) when the white solid is dissolved in water and dilute hydrochloric acid is added a white precipitate is formed. [J]

12 Explain the following observations and state the name of the solid V: (*a*) a white solid V when left in the air becomes a clear solution; (*b*) when dilute nitric acid and silver nitrate solution are added to this clear liquid a white precipitate is formed which dissolves when ammonia solution is added; (*c*) a nichrome wire when dipped into this clear liquid and held in a Bunsen burner produces a dark red coloration. [J]

13 From the tests described below name each of the given substances and explain the tests involved:
Substance A. (i) On warming with sodium hydroxide solution evolves a gas which turns moist red litmus paper blue. (ii) When dissolved in water and mixed with dilute nitric acid and silver nitrate solution a white precipitate is obtained.
Substance B. (i) When dissolved in water and mixed with a solution of ammonia a deep blue solution is produced. (ii) When dissolved in water and mixed with hydrochloric acid and barium chloride solution a white precipitate is formed.
Substance C. (i) When heated on a charcoal block a soft, malleable bead which marks paper is formed. (ii) When treated with sodium hydroxide solution and Devarda's alloy (or aluminium) ammonia is yielded.
Substance D. (i) When added to sodium carbonate a colourless gas is produced which turns calcium hydroxide solution milky. (ii) When added to copper turnings it reacts vigorously to give acidic brown fumes and a blue solution.

14 Four different solutions are provided. For each solution two tests are described below, together with the result of each test. Name each solution and explain the tests involved.
Solution A. (i) Addition of sodium hydroxide solution gives a green precipitate. (ii) Addition of dilute hydrochloric acid followed by a solution of barium chloride produces a white precipitate.
Solution B. (i) Addition of ammonia solution gives a red-brown precipitate. (ii) Addition of dilute nitric acid followed by a solution of silver nitrate gives a white precipitate.
Solution C. (i) On warming with sodium hydroxide solution a gas is evolved which turns red litmus paper blue. (ii) Addition of iron(II) sulphate solution followed by concentrated sulphuric acid produces a brown layer at the junction of the two liquids.
Solution D. (i) A nichrome wire dipped first into the solution and then held in a Bunsen flame gives an intense yellow flame. (ii) On adding dilute hydrochloric acid a gas is evolved which blackens lead ethanoate paper. [J]

15 Describe and explain how you would: (*a*) distinguish between magnesium oxide and zinc oxide; (*b*) test for the existence of a small amount of free iron in a sample of iron(II) sulphide; (*c*) determine whether a sample of town gas contains hydrogen sulphide; (*d*) distinguish between aqueous solutions of potassium chloride and potassium nitrate.

16 Describe the tests (one for each) which you would apply, and the results you would obtain when attempting to identify the following radicals: (*a*) nitrate; (*b*) chloride; (*c*) sulphate; (*d*) carbonate. Describe also two tests which would enable you to distinguish a solution of chlorine gas from one of hydrogen chloride.

17 Explain how to distinguish between the members of the following pairs, giving the results of one test applied to both of the substances concerned: (*a*) tap water and distilled water; (*b*) sodium nitrite and sodium nitrate; (*c*) copper(II) oxide and powdered charcoal; (*d*) sodium carbonate-10-water and sodium hydrogencarbonate; (*e*) carbon monoxide and carbon dioxide.

18 For each of the following pairs give one test to distinguish between the two substances, stating clearly the difference in behaviour which enables you to decide. (**N.B.** If any invisible gas is concerned, a test to prove its presence must be mentioned): (*a*) zinc sulphate and ammonium sulphate; (*b*) sodium nitrate and sodium bromide; (*c*) sodium chloride and sodium sulphate; (*d*) calcium carbonate and bleaching powder. [J]

19 Give two chemical tests in each case which would enable you to distinguish between: (*a*) zinc carbonate and barium carbonate; (*b*) copper nitrate and copper sulphate; (*c*) anhydrous sodium carbonate and sodium hydrogencarbonate; (*d*) mercury(II) oxide and trilead tetraoxide. [J]

20 Compound *A* is a green crystalline solid which, when heated, gives off steam, followed by dense white fumes which turn a moist potassium dichromate paper green. The residue is a red-brown powder *B* which dissolves in hydrochloric acid to form a yellow solution of *C*. Addition of sodium hydroxide solution to a solution of *A* gives a dirty green gelatinous precipitate *D*, but if hydrogen peroxide solution is added to an acidified solution of *A* and then excess of sodium hydroxide solution is added the resulting precipitate *E* is brown and gelatinous. Name the compounds *A* to *E* and explain all the changes which have been observed. [J]

21 Explain how to distinguish between the members of the following pairs, giving the results of one test applied to both of the substances concerned: (*a*) zinc sulphide and zinc carbonate; (*b*) hydrogen and carbon monoxide; (*c*) sodium nitrate and sodium sulphate; (*d*) zinc carbonate and lead carbonate; (*e*) iron(II) chloride and iron(III) chloride.

22 Describe one chemical test in each case by means of which the following pairs of substances may be distinguished from each other. Represent reactions by equations where possible. (*a*) Sodium carbonate and sodium sulphite. (*b*) Barium chloride and magnesium chloride. (*c*) Aluminium nitrate and zinc nitrate. (*d*) Iron(II) sulphate and iron(III) sulphate.

23 A commercial household cleaning-powder is labelled as 'softening the water and giving off bubbles of oxygen when used in washing'. Describe carefully how you would set about checking these two claims in the laboratory. Another commercial preparation claims to 'remove traces of hydrogen sulphide, from a sample of impure hydrogen'. Describe how you would test this claim chemically. [OC]

24 Explain and comment on the following observations:
(*a*) A strip of metal foil is heated in the air: it does not burn, but it changes colour and becomes black. It is then put into dilute sulphuric acid and warmed; the solution turns slightly blue. When excess ammonia solution is added, it turns a deep blue.
(*b*) (i) On adding a black powder to hydrogen peroxide solution a colourless gas is given off. This black powder itself does not change. (ii) The same powder gives off a yellow-green gas when warmed with concentrated hydrochloric acid. With excess acid the black powder is slowly dissolved.
(*c*) A colourless crystalline salt effervesces with dilute acid and gives an intense yellow colour in a Bunsen flame. When left on a watch glass in the air it gradually loses mass. [OC]

25 The human body has the following percentage composition by mass: oxygen 65·0, carbon 18·5, hydrogen 9·5, nitrogen 3·3, calcium 1·5, phosphorus 1·0. There are at least fifteen other elements in the remaining 1·2%. Is a human body a compound? Give reasons for your answer.

22 Volumetric Analysis

22.1 Introduction

Quantitative analysis of an unknown substance will follow on from qualitative analysis: once the substances present are identified their proportions can be found. Occasionally gravimetric analysis is employed in which the material is weighed, then chemical reactions carried out and finally a precipitate is obtained, washed, dried and weighed. More often volumetric analysis is to be preferred: initially a weighing may be made but thereafter all measurements are of the volumes of solutions. The latter method is quicker and will often yield an answer that is accurate enough for many purposes.

When a known mass of a substance is dissolved in water and then made up to a known volume of solution at a known temperature, a standard solution is obtained. The concentration of a standard solution can be expressed in many ways but the most useful to the chemistry student is in terms of mol/dm^3 (see next section) and to the laboratory assistant, is in grams per 1000 cm^3 (1 dm^3, 1 litre) of solution. The abbreviation M is often used for mol/dm^3 but it is not accepted internationally.

22.2 The Concentration of Solutions

A 1 M solution contains 1 mole, i.e. 1 relative molecular mass in grams (1 molar mass) of a substance, in 1000 cm^3 of solution.

If 1 dm^3 (1000 cm^3) of a c M solution is considered then c moles of the substance are present. If v cm^3 of a c M solution are considered then $\frac{vc}{1000}$ moles are taken.

e.g. in 25 cm^3 of a 1 M solution there are 0·025 moles of solute whereas in 25 cm^3 of a 0·1 M solution there are 0·0025 moles of solute.

In a mole of a substance there are 6×10^{23} molecules (the Avogadro constant, see section 7.11) so that when a concentration is expressed in terms of M the number of particles is really being considered.

If the equation for two reactants is known e.g.

$$\mathrm{a\,A + b\,B \rightarrow c\,C + d\,D}$$

then a and b, the number of molecules of substances A and B, must be in the same ratio as the experimentally determined numbers of moles of the substances, i.e.

$$\frac{v_A c_A}{v_B c_B} = \frac{\mathrm{a}}{\mathrm{b}}$$

Consider for example the reaction of hydrochloric acid with sodium hydroxide solution, the completion of which can easily be judged by using an indicator such as litmus: the equation is

$$HCl + NaOH \rightarrow NaCl + H_2O$$

so that the ratio a/b is 1/1. Another problem can also be solved: if the two volumes and the two concentrations in terms of M are known or found then a/b for a given reaction can be determined.

N.B. The relationship quoted above can only be applied to the volumes of solutions which react with one another, not to gases or solids.

For volumetric analysis, solutions which are 0·1 M are frequently employed. The process of adding one solution to another solution until an indicator shows that the reaction is complete is known as titration. Titrations at this stage are done usually with acids and alkalis. Volumetric analysis has many uses:

(*a*) to measure the concentration of a solution of a given substance (see section 6.12(d)),
(*b*) to measure the proportion of a substance in a mixture,
(*c*) to measure the solubility of a solute in a solvent (see section 6.12(d)),
(*d*) to measure the rate of a reaction (see section 18.9),
(*e*) to find the equation of a reaction or the formula of a substance,
(*f*) to prepare salts (see section 17.11).

22.3 Acids and Alkalis

Some of the common acids used in volumetric analysis are shown in the following table:

Acid	Formula	Basicity	Relative Molecular Mass
Hydrochloric	HCl	1	36·5
Nitric	HNO_3	1	63
Sulphuric	H_2SO_4	2	98
Phosphoric	H_3PO_4	3	98
Ethanoic (acetic)	CH_3COOH	1	60

The basicity of an acid is the number of hydrogen atoms in a molecule which are replaceable by metal atoms. In a given reaction not all the hydrogen atoms of an acid may be replaced so that acid and normal salts can be prepared. In the case of ethanoic acid only one of the four hydrogen atoms can be replaced by a metal atom, i.e. it is monobasic or monoprotic.

Some of the common alkalis used in volumetric analysis are shown in the following table:

Alkali	Formula	Acidity	Relative Molecular Mass
Sodium hydroxide	$NaOH$	1	40
Sodium carbonate	Na_2CO_3	2	106
Sodium carbonate-10-water	$Na_2CO_3,10H_2O$	2	286
Ammonia solution (Ammonium hydroxide)	NH_3	1	17
Sodium hydrogencarbonate	$NaHCO_3$	1	84

The carbonates of sodium are salts but their alkaline properties outweigh their acidic properties and they may be titrated with hydrochloric acid in the same manner as sodium hydroxide providing methyl orange is used as the indicator.

22.4 Apparatus and Preliminary Procedure

A weighing bottle may be used at the start of an analysis: it is often a plastic bottle which can easily be washed and then dried with paper tissue (or in a desiccator); it should have a lid. Into the bottle is put some very pure material known as a primary standard (commercially called an 'Analar' or 'Research' Reagent).

Many substances do not qualify as primary standards because they react with the atmospheric gases or lose mass by vaporization into the atmosphere. Substances such as sodium hydroxide and the three mineral acids (hydrochloric, nitric and sulphuric) are thus unsuitable. Anhydrous sodium carbonate is the most suitable compound available in the laboratory. It may react slightly with carbon dioxide and moisture giving sodium hydrogencarbonate but the change can be reversed by gentle heating and then the material stored in a desiccator until it is cool enough to weigh out.

The weighing bottle containing about 0·025 of a mole of the primary standard, is then weighed accurately. This proportion is taken for 250 cm^3 of a 0·1 M solution. The solid is then tipped out into a clean 250 cm^3 beaker containing about 50 cm^3 of distilled water. The bottle which still contains traces of the solid is reweighed accurately so that the amount taken is known. The solute in the beaker is stirred with a clean glass rod until it has all dissolved. If heating is employed to hasten dissolution then the solution must be cooled back to room temperature before proceeding.

A measuring flask is made to contain a certain volume of a liquid or solution at a stated temperature. A 250 cm^3 measuring flask (figure 22.1) and a funnel are rinsed with distilled water. The solution in the beaker is then poured down the glass rod into the funnel and thence into the measuring flask. The beaker, rod and funnel are rinsed twice with small portions of distilled water and the rinsings added to the solution in the flask. The solution in the flask is made up to 250 cm^3 (the meniscus of the solution should rest on the graduation mark) and then shaken to make it homogeneous.

A pipette is made to deliver a certain volume

of a liquid or solution at a stated temperature. A 25 cm^3 pipette is frequently used: it is rinsed with distilled water and then with the alkali. A 250 cm^3 conical flask should be rinsed with distilled water only. The pipette is used to transfer 25 cm^3 of the alkali into the conical flask. Two or three drops of indicator are then added. Pipette fillers are useful when the solutions involved are corrosive or poisonous.

A measuring cylinder is sometimes accurate enough and safer for elementary work.

An indicator is a substance which shows by its colour the nature of a solution, e.g. acidic, neutral or alkaline. The rules for choosing indicators depend on the extent of ionization of the acids and alkalis, i.e. their strength, and can be summarized as:

Acid	*Alkali*	*Indicator*
Strong	Strong	Any common one
Strong	Weak	Methyl orange
Weak	Strong	Phenolphthalein
Weak	Weak	None suitable

A burette is made to deliver a certain volume of a liquid or solution at a stated temperature. A 50 cm^3 burette is frequently used: it should be rinsed with distilled water and then with the acid to be standardized, not forgetting the portion below the tap. The burette is filled with the acid and the level adjusted so that the meniscus is on the scale; it does not have to be precisely at the zero mark each time. If the acid to be standardized is highly concentrated, it cannot be titrated directly and accurately but must be first diluted to about 0·1 M for a monobasic acid or 0·05 M for a dibasic acid.

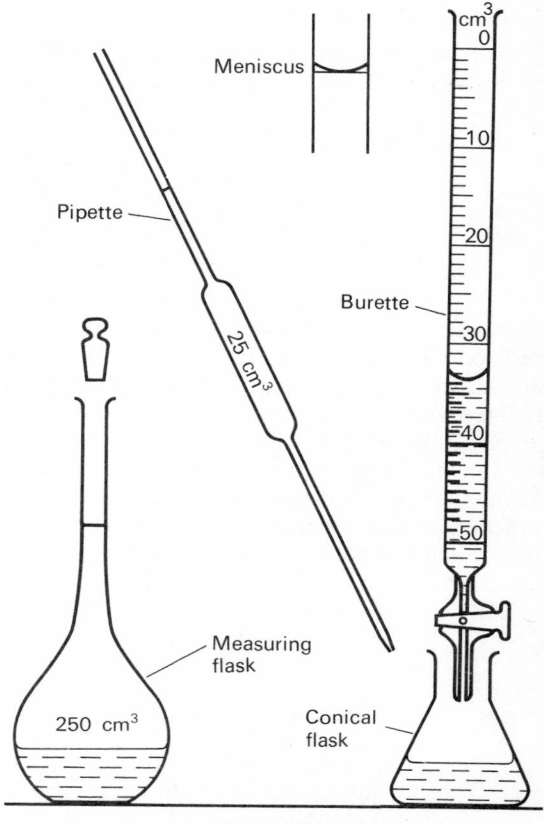

Figure 22.1
Apparatus used in volumetric analysis

22.5 The Titration

The acid is added from the burette until the indicator shows that the alkali in the conical flask has been neutralized. The flask should be gently swirled between additions from the burette. The colours shown by common indicators with various solutions are set out in the table at the foot of this page.

The first titration gives an approximate answer, thereafter accurate answers should be obtainable. The titration need only be done twice if two accurate titres are within 0·1 cm^3 of one another but three may be necessary for greater precision. (Initially a greater error may be acceptable). The average accurate titration value is used for the calculation.

If a class titrates an acid with a fixed quantity of alkali the results can be studied to see if they give a normal distribution curve (see section 4.13).

	Acidic solution	*Neutral solution*	*Alkaline solution*
Litmus	red	purple	blue
Methyl orange	red-pink	yellow	yellow
Screened methyl orange	red	pale blue	green
Methyl red	pink	yellow	yellow
Phenolphthalein	colourless	colourless	red

22.6 An Example of a Titration

The standardization of concentrated hydrochloric acid with anhydrous sodium carbonate.

Anhydrous sodium carbonate is a primary standard, so that if it is weighed out accurately and made into a known volume of solution, a standard solution is obtained. The mass to be taken for 250 cm^3 of 0·05 M solution is

$$106 \times \frac{250}{1000} \times 0{\cdot}05 = 1{\cdot}325 \text{ g}$$

Time need not be spent trying to attain precisely this mass: as long as the mass is between 1·2 and 1·5 g and is known accurately the experiment will give a satisfactory result.

The concentrated hydrochloric acid supplied to the laboratory is usually the saturated solution of hydrogen chloride in water, and it is much too concentrated to be titrated directly. In order that it is approximately 0·1 M it is diluted 100 times, i.e. 10 cm^3 to 1000 cm^3 of solution.

Results

Mass of bottle + anhydrous sodium carbonate = 6·190 g

Mass of bottle + traces of solid = 4·800 g

∴ Mass of anhydrous sodium carbonate = 1·390 g

Burette readings	*Titrations (in cm³)*			
	Rough	*1*	*2*	*3*
Finish	23·35	46·35	23·05	46·00
Start	0·10	23·35	0·15	23·05
Difference	23·25	23·00	22·90	22·95
∴ Average accurate titration = 22·95 cm³.				

Calculations

The 1·39 g of anhydrous sodium carbonate was dissolved to produce 250 cm^3 of solution. The number of grams of solid in 1000 cm^3 of such a solution would be

$$1{\cdot}39 \times 4$$
$$= 5.56$$

Thus the sodium carbonate solution has a concentration of

$$\frac{5{\cdot}52}{106}$$
$$= 0{\cdot}0525 \text{ M}$$

The equation for the reaction when methyl orange is used as the indicator is

$$Na_2CO_3 + 2HCl \rightarrow 2NaCl + H_2O + CO_2\uparrow$$

Using $\frac{v_A c_A}{v_B c_B} = \frac{a}{b}$

the concentration of the hydrochloric acid (c_B) is

$$\frac{25 \times 0{\cdot}0525}{22{\cdot}95 \times c_B} = \frac{1}{2}$$

$$\therefore c_B = \frac{25 \times 0{\cdot}0525 \times 2}{22{\cdot}95}$$

$$= 0{\cdot}114 \text{ M}$$

But the acid used for titration was obtained by diluting the saturated solution of hydrogen chloride in water 100 times. Thus the concentration of the saturated solution is 11·4 M.

The relative molecular mass of hydrochloric acid is 36·5, and so the number of grams per dm^3 (litre) of the solution of hydrogen chloride is

$$36{\cdot}5 \times 11{\cdot}4$$
$$= 416$$

N.B. The relationship

Concentration in mol/dm^3

$$= \frac{\text{Mass of solute in 1000 cm}^3\text{ of solution}}{\text{Molar mass of solute}}$$

is often used in calculations: it follows directly from the definition.

22.7 Volumetric Calculations

(*a*) *A solution containing 6 g/dm³ of glacial ethanoic acid was titrated with 0·1 M sodium hydroxide using phenolphthalein as the indicator: 25 cm³ of the diluted acid react with 25 cm³ of the alkali. What is the concentration of the diluted acid and the relative molecular mass of glacial ethanoic acid (it may be considered to be anhydrous for this calculation)?*

$$CH_3COOH + NaOH \rightarrow CH_3COONa + H_2O$$

By the equation derived in section 22.2 the ratio of acid to alkali is

$$\frac{25 \times c_A}{25 \times 0{\cdot}1} = \frac{1}{1}$$

hence c_A, the concentration of the diluted acid, is also 0·1 M. By the definition of concentration as rephrased in the previous section, the relative molecular mass of the (glacial) ethanoic acid is

$$\frac{6}{0{\cdot}1}, \text{ i.e. } 60$$

(b) A crystalline form of sodium carbonate used in industry was made into solution (6·2 g/dm^3) and titrated with 0·125 M hydrochloric acid using methyl orange as the indicator: 20 cm^3 of the acid were equivalent to 25 cm^3 of the alkali. What is the extent of water of hydration in the sodium carbonate crystals?

$$Na_2CO_3 + 2HCl \rightarrow 2NaCl + H_2O + CO_2$$

By the equation the ratio of acid to alkali is

$$\frac{20 \times 0{\cdot}125}{25 \times c_B} = \frac{2}{1}$$

hence c_B the concentration of the alkali is 0·05 M. Thus the relative molecular mass of the crystals is $\frac{6{\cdot}2}{0{\cdot}05}$, i.e. 124.

The relative molecular mass of anhydrous sodium carbonate is 106 so the relative molecular mass due to the water of crystallization is 18, i.e. there is one molecule of water for each of the carbonate.

(c) A sample of sodium carbonate solution (10·6 g/dm^3) was titrated with 0·1 M hydrochloric acid using phenolphthalein as the indicator: 25 cm^3 of the alkali was found to have reacted with the same volume of the acid when the indicator showed the end of the reaction. However, when methyl orange was used as the indicator double the quantity of acid was found to be necessary. What are the equations for the reactions involved?

The concentration of the sodium carbonate solution is $\frac{10{\cdot}6}{106}$, i.e. 0·1 M. Thus the molecular ratio of acid to alkali in the first titration is $\frac{25 \times 0{\cdot}1}{25 \times 0{\cdot}1}$, i.e. 1:1. Thus the equation for the reaction when phenolphthalein is used as the indicator starts off

$$HCl + Na_2CO_3$$

(and the reaction can be shown by further analysis to yield sodium chloride and sodium hydrogencarbonate solution).

In the second titration the molecular ratio of acid to alkali is $\frac{50 \times 0{\cdot}1}{25 \times 0{\cdot}1}$, i.e. 2:1. Thus the equation for the reaction starts off

$$2HCl + Na_2CO_3$$

(and the reaction can be shown by further analysis to yield sodium chloride, water and carbon dioxide).

(d) A saturated solution of sodium hydrogencarbonate at 15°C was diluted ten times, i.e. 100 cm^3 were made up to 1 dm^3 of solution. Upon titration of 25 cm^3 of the diluted alkali with 0·1 M hydrochloric acid using methyl orange as the indicator the volume of acid required was found to be equal. What is the solubility of sodium hydrogencarbonate at 15°C?

$$NaHCO_3 + HCl \rightarrow NaCl + H_2O + CO_2$$

The molecular ratio of acid to alkali is 1:1 and as the volumes of solutions are also equal the molar concentrations must be. The diluted solution of the alkali is 0·1 M and the original concentration is 1 M.

The relative molecular mass of the alkali is 84 so the concentration of the saturated solution is 84 g/dm^3.

(e) 0·3 g of magnesium was dissolved in 50 cm^3 of 1 M hydrochloric acid. After the reaction ceased it was found that 20 cm^3 of 1·25 M sodium hydroxide solution were needed to react with the excess acid. If the relative atomic mass of magnesium is 24 what is its valency?

$$HCl + NaOH \rightarrow NaCl + H_2O$$

Let v_A (cm^3) be the volume of 1 M acid which is in excess after the metal has dissolved.

Then $$\frac{v_A \times 1}{20 \times 1{\cdot}25} = \frac{1}{1}$$

i.e. $$v_A = 25$$

Thus the volume of 1 M acid that has reacted is also 25 cm^3. Let the equation for the reaction of metal and acid be

$$M + nH^+ \rightarrow M^{n+} + 0{\cdot}5n\ H_2$$

where n is the valency of magnesium.

i.e. 24 g of Mg $\equiv n$ dm^3 1 M HCl

$$\therefore 0{\cdot}3 \text{ g Mg} \equiv \frac{0{\cdot}3n}{24} \text{ dm}^3 \text{ 1 M HCl}$$

$$= \frac{0{\cdot}3 \times 1000n}{24} \text{ cm}^3 \text{ 1 M HCl}$$

By the titration this volume has been found to be 25 cm^3 hence

$$n = \frac{24 \times 25}{0{\cdot}3 \times 1000}$$

$$= 2$$

i.e. the valency of magnesium is two.

QUESTIONS 22

Volumetric Analysis

1 If 5·3 g of anhydrous sodium carbonate are dissolved in water and just neutralized with 0·5M hydrochloric acid: (*a*) What volume of acid will be required? (*b*) What will be the volume of carbon dioxide evolved measured at 26°C and 736 mmHg pressure? [J]

2 If 5·0 g of calcium carbonate were allowed to react with 50 cm^3 of 1 M hydrochloric acid, until there was no further reaction, what mass would remain undissolved? What volume of carbon dioxide (measured at s.t.p.) would be produced? [J]

3 (*a*) Find the mass of pure potassium carbonate in 500 cm^3 of 0·1M solution. (*b*) 9·2 g of a sample of impure potassium carbonate were dissolved in water, and the solution was made up to 1000 cm^3 with water. 20·0 cm^3 of this solution required 24·0 cm^3 of 0·1M hydrochloric acid for neutralization. Calculate (i) the concentration of the potassium carbonate solution; (ii) the percentage purity of the sample of potassium carbonate. [J]

4 A solution containing 14·3 g/dm^3 of crystalline sodium carbonate was found on titration to be exactly 0·05 M. From these figures determine the number of molecules of water of crystallization per molecule of the crystal. [J]

5 (*a*) Find the mass of pure sulphuric acid contained in 100 cm^3 of exactly 0·25 M solution. (*b*) Calculate the volume of 1 M sodium hydroxide solution which must be added to 100 cm^3 of 0·25M sulphuric acid in order to convert the acid into: (i) sodium hydrogensulphate; (ii) sodium sulphate. [J]

6 In an experiment it was found that after 0·225 g of a trivalent metal had been completely dissolved in 50 cm^3 of 1M hydrochloric acid, 25·0 cm^3 of a 1M sodium hydroxide solution were required to complete the neutralization. What is the relative atomic mass of the metal? [J]

7 (*a*) Write down the equation which represents the action of sodium on water. (*b*) Assuming the above reaction to be complete, what volume of gas measured at s.t.p. would be formed by 9·20 g of sodium? (*c*) What volume of 0·5M sulphuric acid would be needed to neutralize the resulting solution? (*d*) On carefully evaporating this neutralized solution to dryness, what mass of anhydrous salt would be obtained? [J]

8 Define *acid salt*. In a titration experiment it was found that 25 cm^3 of a sodium hydroxide solution required 15 cm^3 of 0·5M solution of sulphuric acid to give the neutral point. (*a*) What is the concentration in mol/dm^3 of the sodium hydroxide solution? (*b*) What is the concentration of the sodium hydroxide solution in g/dm^3? (*c*) Explain briefly how from these solutions you would obtain a pure, dry, crystalline specimen of sodium hydrogensulphate. [J]

9 Calcite is a pure crystalline form of calcium carbonate. A crystal of calcite weighing 5·50 g was allowed to stand in 100 cm^3 of a solution of hydrochloric acid until no further reaction took place. The crystal was removed, washed and dried, and then weighed 3·00 g. (*a*) Write down the equation for the reaction. (*b*) Calculate the concentration in mol/dm^3 of the hydrochloric acid. (*c*) Calculate the volume of carbon dioxide, measured at s.t.p., liberated during the reaction. [J]

10 (*a*) What mass of potassium hydroxide is formed when 1·95 g of potassium are allowed to react with excess of water? (*b*) If the solution is made up to 500 cm^3 with water what is its concentration in mol/dm^3? (*c*) What volume of 0·1M hydrochloric acid would be needed to neutralize this solution? (*d*) Explain the result which would be obtained if 1000 cm^3 of 0·05M sulphuric acid were added to the solution obtained in (*b*) and then the solution evaporated to obtain crystals? [J]

11 Describe how to perform the following laboratory operations, stating with reasons the precautions necessary to secure accuracy: (*a*) transfer exactly 25 cm^3 of a solution of sodium hydroxide to a clean beaker, being given a clean but wet pipette; (*b*) set up and fill a clean but wet burette with a given solution of sulphuric acid; (*c*) titrate this acid against the sodium hydroxide solution, naming the indicator. [J]

12 You are given 0·5M solution of sodium carbonate. (*a*) What is the mass of sodium carbonate in 100 cm^3 of solution? (*b*) Explain how you would obtain from this solution dry crystals of sodium carbonate. Why should you not evaporate to dryness? (*c*) What volume of 0·5M sulphuric acid would you add to 100 cm^3 of the sodium carbonate solution in order to obtain a solution of: (i) sodium sulphate; (ii) sodium hydrogensulphate? [J]

13 Define *molar solution*. What mass of sulphuric acid is needed to make 1000 cm^3 of 0·5M solution? What volume of this sulphuric acid would be needed to neutralize 6·40 g of sodium hydroxide? Name the product, and calculate what mass of it would be obtained, if the above neutral solution were evaporated to dryness. [J]

14 (*a*) What mass of sodium hydroxide is formed when 1·15 g of sodium are allowed to react with excess of water? (*b*) If the resulting solution is made up to 100 cm^3 with water, what is its concentration in mol/dm^3?

(*c*) If 3·25 g of zinc are completely dissolved in 150 cm^3 of 1M hydrochloric acid: (i) what will be the volume of hydrogen evolved, measured at s.t.p.; (ii) what volume of 1M sodium hydroxide will be required to neutralize the excess of acid? [J]

15 (*a*) What is a 1M solution of an acid? (*b*) What volume of 1M hydrochloric acid is required to neutralize 50 cm^3 of 0·2M sodium hydroxide? (*c*) How many grams of anhydrous sodium carbonate are required for the preparation of 100 cm^3 of 0·25M sodium carbonate solution? [J]

16 In an experiment 1000 cm^3 of 0·5M sulphuric acid are mixed with 500 cm^3 of 0·5M sodium carbonate solution, the reaction being represented by the equation

$$2H_2SO_4 + Na_2CO_3 \rightarrow 2NaHSO_4 + H_2O + CO_2 \uparrow$$

The mixture is then carefully evaporated to dryness. (*a*) What is the volume of carbon dioxide evolved, measured at s.t.p.? (*b*) Name the solid remaining after the evaporation and calculate the mass of it. [J]

17 Define 1*M solution.* What mass of each substance is required to make 1000 cm^3 of a 1M solution of (*a*) nitric acid; (*b*) potassium hydroxide? If 10·0 g of sodium hydroxide are mixed with 500 cm^3 of 1M hydrochloric acid, which of these two solutions (*a*) or (*b*) would you use in order to neutralize the mixture, and what volume of this solution would it be necessary to add? [J]

18 A solution contains 23·0 g/dm^3 of pure potassium carbonate. It was found that 100 cm^3 of this solution were neutralized by 33·3 cm^3 of 1M hydrochloric acid solution. Find the concentration in mol/dm^3 of the potassium carbonate solution (*a*) from the mass used; (*b*) assuming the equation, from the titration. Describe carefully how you would now proceed to make 1000 cm^3 of exactly 0·05M potassium carbonate starting with the pure solid. Why should you not add 1000 cm^3 of water?

19 What mass of sulphuric acid is contained in 1000 cm^3 of 0·5M solution? Write down the equations: (*a*) for the action of dilute sulphuric acid on zinc; (*b*) for the neutralization of sodium hydroxide by dilute sulphuric acid. Calculate: (i) the volume of hydrogen evolved (at s.t.p.) when 1000 cm^3 of 0·5M sulphuric acid reacts completely with excess of zinc; (ii) the volume of 0·5M sulphuric acid required to neutralize 20·0 g of sodium hydroxide.

20 0·56 g of metallic oxide (MO) were dissolved in 250 cm^3 of 0·1M hydrochloric acid. To neutralize the excess of acid 50 cm^3 of 0·1M sodium hydroxide were required. Calculate the relative atomic mass of the metal.

The Elements and their Compounds

23 Hydrogen

23.1 Introduction

Hydrogen is an element which is difficult to classify; most periodic tables put hydrogen with helium in a special position at the top. It is like the elements in Group I, the alkali metals, because it forms a singly charged positive ion and burns in air forming a stable oxide. Also it has similarities to the elements in Group VII, the halogens: it is a diatomic gas and forms many covalent compounds.

23.2 Occurrence

The proportion of hydrogen in the atmosphere and in the earth's crust is negligible. Water, which covers three quarters of the globe contains 11·1% by mass of hydrogen. Hydrogen is also extracted from petroleum and natural gas, both of which are hydrocarbons. Hydrogen was discovered by Cavendish in 1766.

23.3 Manufacture

(a) Steam reforming of natural gas

Natural gas consists mainly of methane (CH_4); an alternative is to use the gases which are available from a refinery (petroleum or naphtha feedstock). The preheated gas is passed with excess steam at 20 atmospheres over a nickel-chromium catalyst at 900°C.

$$CH_4 + H_2O \rightleftharpoons CO + 3H_2 \quad \Delta H = -206\text{kJ}$$

The reaction is reversible but the yield is very good. Taking heptane (C_7H_{16}) as a typical constituent of refinery gases:

$$C_7H_{16} + 7H_2O \rightarrow 15H_2 + 7CO$$

The second stage is often called the water-gas shift reaction: the gases are passed, still at 20 atmospheres, with excess steam over an iron(III) oxide catalyst at 500°C.

$$CO + H_2O \rightleftharpoons CO_2 + H_2 \quad \Delta H = -40\text{kJ}$$

The hydrogen previously made is unaffected.

The third stage is to purify the hydrogen by eliminating the unchanged carbon monoxide, the carbon dioxide and steam. The carbon dioxide is dissolved in warm potassium carbonate solution.

$$K_2CO_3 + H_2O + CO_2 \rightleftharpoons 2KHCO_3$$

or

$$CO_3^{2-} + H_2O + CO_2 \rightleftharpoons 2HCO_3^-$$

The solvent is regenerated by steam heating and by lowering the pressure. Carbon dioxide is a useful by-product. The second step in the purification is to remove unchanged carbon monoxide (0·3%). This is done by 'methanation' which is the reverse of the initial reaction, a very small proportion of hydrogen being lost: a nickel catalyst is again employed, this time at about 300°C. The final step is to remove any steam: this is done by compression of the gases which causes the steam to liquefy. The hydrogen produced is over 99% pure.

(b) Electrolysis

Hydrogen is produced as a by-product in the electrolysis of sodium chloride (see section 24.8) and some is made by the electrolysis of potassium hydroxide solution (similar to that of sodium hydroxide, see section 20.8).

The United Kingdom production is about one third of a million tonnes each year: most of it is immediately converted into compounds.

23.4 Laboratory Preparation

(a) Dilute acid and metals

Most reactive metals will give hydrogen with dilute hydrochloric or sulphuric acid but there are several drawbacks: sodium is likely to explode, iron gives a very impure gas and magnesium is rather expensive. In the laboratory zinc is generally used, but if the metal is very pure the reaction will be extremely slow in the cold, and so a few drops of copper sulphate are usually added to speed it up.

$$Zn + 2HCl \rightarrow ZnCl_2 + H_2\uparrow$$
$$Zn + H_2SO_4 \rightarrow ZnSO_4 + H_2\uparrow$$

or

$$Zn + 2H^+ \rightarrow Zn^{2+} + H_2\uparrow$$

The copper sulphate is not considered to be a catalyst because it cannot be recovered at the end of the reaction. By the reaction

$$Zn + Cu^{2+} \rightarrow Cu\downarrow + Zn^{2+}$$

a miniature voltaic cell is set up. At first air

mixed with hydrogen is displaced from the apparatus and this can be allowed to escape if there are no flames in the neighbourhood. Before proceeding with any experiment a sample of the gas produced should be collected in a test tube and ignited with a splint: if the gas explodes all the air has not been expelled from the apparatus, but if the gas burns quietly it is safe to collect the gas and/or proceed with further experiments.

Figures 23.1 and 23.2 illustrate two ways of preparing the gas. It may be collected over water. If it is required dry, concentrated sulphuric acid may be used and then the gas, at least for many purposes, collected by upward delivery. Kipp's apparatus is very convenient if an intermittent supply of gas is required: when the tap is open acid flows from the top compartment directly into the bottom compartment and then into the middle compartment where the reaction with the zinc occurs; when the tap is shut the reaction continues for a short while but then the increased pressure of hydrogen forces the acid back into the bottom and thence to the top compartment.

The other ways of preparing hydrogen are much less important.

(b) Water and metals

The very reactive (electropositive) metals such as sodium and calcium react with cold water. Magnesium and iron, on heating, react with steam (see sections 25.5 and 26.7).

(c) Alkalis and metals

Aluminium (and a few other metals) will dissolve in warm concentrated sodium hydroxide.

$$2Al + 6NaOH + 6H_2O \rightarrow \underset{\text{Sodium aluminate}}{2Na_3Al(OH)_6} + 3H_2\uparrow$$

or

$$2Al + 6OH^- + 6H_2O \rightarrow 2Al(OH)_6^{3-} + 3H_2\uparrow$$

(d) Electrolysis

Most of the common acids, alkalis and salts in dilute aqueous solution yield hydrogen (at the cathode) upon electrolysis (see sections 6.7 and 20.8).

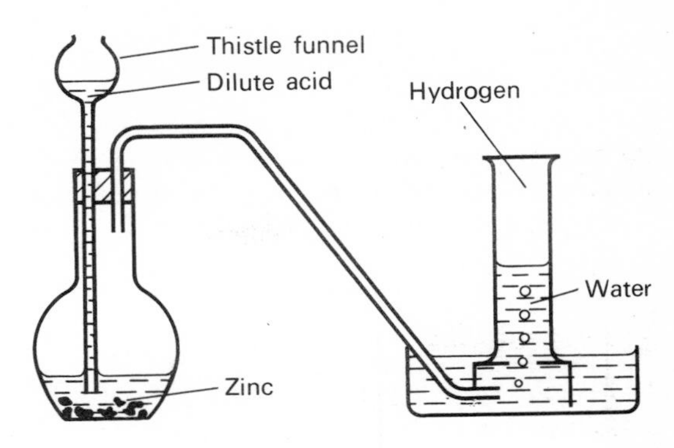

Figure 23.1
The preparation of hydrogen

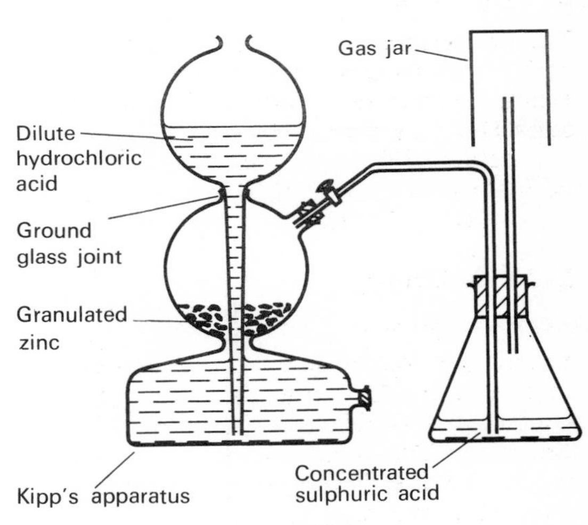

Figure 23.2
The preparation of dry hydrogen

23.5 Physical Properties

Hydrogen has an atomic number of one and as the commonest isotope contains no neutrons the relative atomic mass is one. There is one electron associated with the nucleus. The gas has a density of 0·09 kg/m^3 at s.t.p.: it is the gas with the lowest density and it diffuses through solid materials such as platinum. It is almost insoluble in water. It is colourless, odourless and tasteless; it is not poisonous, but it does not support life. The boiling point of liquid hydrogen is −253°C and it is very difficult to liquefy the gas. The standard electrode potential of hydrogen is, by convention, zero (see section 15.5).

Experiments Depending on the Low Density of Hydrogen

(*a*) It will flow upwards: a gas jar of hydrogen is brought up to one of air as shown in figure 23.3. After a few seconds each gas jar is tested with a lighted splint: both give explosions.

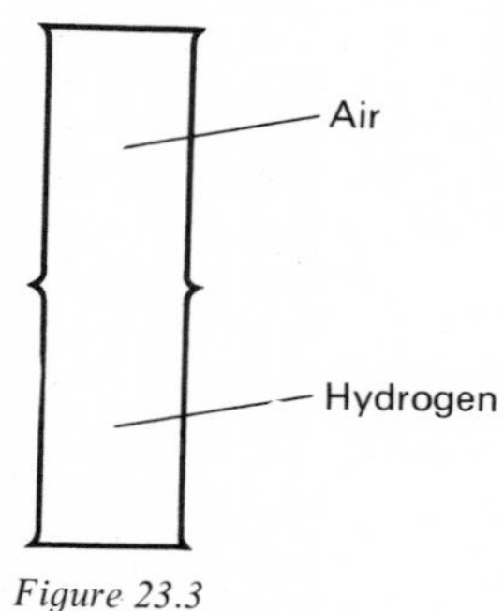

Figure 23.3
The density of hydrogen (a)

(*b*) An inverted beaker is fixed to one arm of a balance as shown in figure 23.4 and counter-poised with weights. The beaker is filled with hydrogen from a Kipp's apparatus and the pointer of the balance observed.

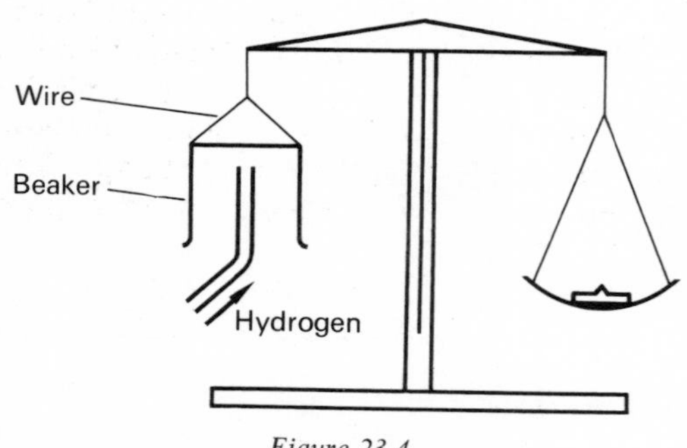

Figure 23.4
The density of hydrogen (b)

(*c*) The hydrogen supplied from a Kipp's apparatus is washed with water to remove volatilized acid and passed through a thistle funnel or clay pipe into a soap or detergent solution; bubbles filled with hydrogen rise to the ceiling. They burn with a mild explosion if touched with a lighted splint.

(*d*) Using the apparatus shown in figure 23.5 a balloon can be filled with hydrogen. The balloon should be softened by blowing it up by mouth before attaching it to the flask. When released it rises to the ceiling, but it descends after an hour or two because hydrogen diffuses out through the rubber.

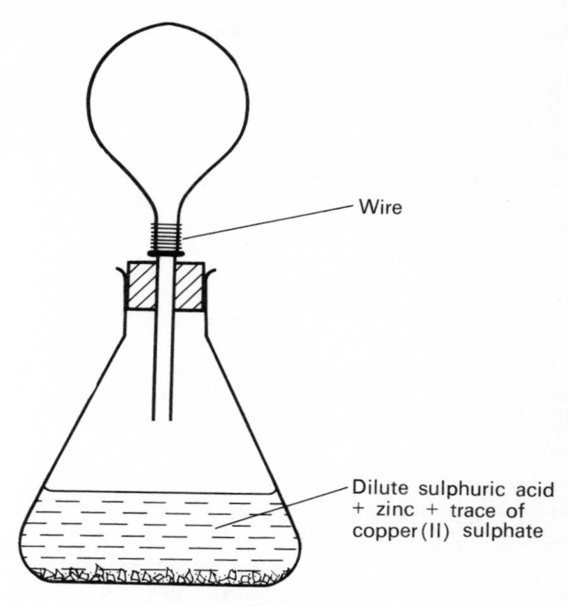

Figure 23.5
Filling a balloon with hydrogen

Experiment to Show that Hydrogen Diffuses Rapidly

The molecules of a gas are in rapid motion, so that a gas tends to spread throughout any vessel in which it is placed and through porous materials. See experiment (*d*) in section 7.8. Graham's Law (see section 7.9) gives a quantitative description of the rate of diffusion.

23.6 Chemical Properties

(*a*) *Stability to Heat*

Hydrogen is stable to heat up to 2000°C.

$$H_2 \rightleftharpoons 2H \qquad \Delta H = +435 \text{ kJ}$$

Atomic hydrogen, as produced by passing molecular hydrogen through an ozonizer (see section 31.4) or an electric discharge, is a very good reducing agent and is also employed in blowpipes to yield temperatures of over 4500°C, e.g. for flame cutting copper.

(*b*) *With Group I and II Elements*

Metals such as sodium and calcium combine on gentle heating with hydrogen giving salt-like crystalline solids which are hydrolyzed by water and are good reducing agents.

$$2Na + H_2 \rightarrow 2NaH\ [Na^+H^-]$$
$$Ca + H_2 \rightarrow CaH_2\ [Ca^{2+}(H^-)_2]$$

and $$NaH + H_2O \rightarrow NaOH + H_2\uparrow$$
$$CaH_2 + 2H_2O \rightarrow Ca(OH)_2 + 2H_2\uparrow$$
or $$H^- + H_2O \rightarrow OH^- + H_2\uparrow$$

(c) With Group V Elements

In the Haber process hydrogen combines with nitrogen at 350°C and 350 atmospheres pressure to give ammonia.

$$N_2 + 3H_2 \rightleftharpoons 2NH_3 \quad \Delta H = -92\ kJ$$

(d) With Group VI Elements

Hydrogen burns very easily when ignited in air or oxygen to give water vapour which may condense. Heating to 600°C suffices instead of a spark.

$$2H_2(g) + O_2(g) \rightarrow 2H_2O(l) \quad \Delta H = -568\ kJ$$

Pure hydrogen burns quietly with a pale blue flame, a temperature of 2800°C being attainable. There is a wide range of explosive mixtures of air with hydrogen, and before hydrogen is ignited upon emission from any apparatus the issuing gas should be collected in a test tube and held to a flame. If it burns explosively air has yet to be displaced from the apparatus; if it burns quietly the main stream may be ignited.

A mixture of two volumes of hydrogen with one volume of oxygen in a poly(ethene) vessel or soap bubble upon ignition explodes violently: it is used as a non-toxic detonating gas.

However, although hydrogen burns so easily it is not a supporter of combustion. If a lighted splint is pushed firmly into a jar of hydrogen it is extinguished, although the hydrogen burns readily enough at the mouth of the jar where it is in contact with oxygen. For the synthesis of water see section 6.7(b).

(e) With Group VII Elements

Hydrogen combines slowly with chlorine in daylight but explosively in sunlight or upon ignition: as with oxygen, hydrogen burns smoothly in an atmosphere of the gas, i.e. if a jet of hydrogen burning in air is lowered into a gas jar of chlorine combustion continues.

$$H_2 + Cl_2 \rightarrow 2HCl \quad \Delta H = -184\ kJ$$

(f) As a Weak Reducing Agent

Not only does hydrogen combine directly with oxygen but it removes the oxygen from certain metallic oxides: this process is called reduction. If a small sample of lead(II) oxide (litharge, PbO) is heated in a stream of hydrogen, (see figure 23.6), there is a colour change from yellow to silver after a few minutes, which indicates that reduction has taken place. The steam formed is swept out of the tube by excess hydrogen.

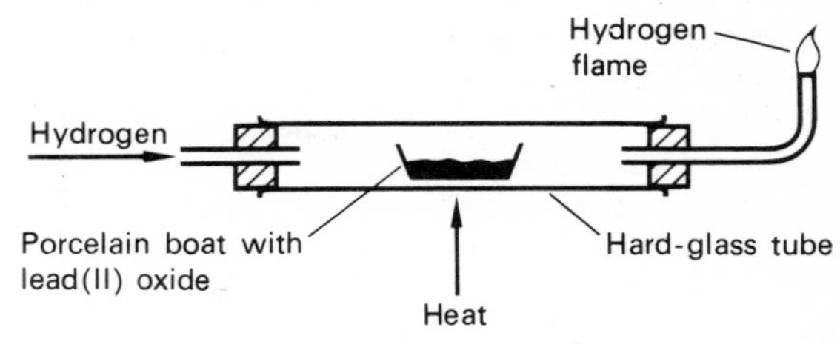

Figure 23.6
Reduction with hydrogen

The other oxides of lead or those of copper or of iron can be used, but not those of metals high in the electrochemical series such as calcium, sodium, magnesium, aluminium or zinc. See also the manufacture of margarine, section 23.7(d).

Hydrogen is a reducing agent towards nitrogen and chlorine—discussed previously—but it is an oxidizing agent towards sodium and calcium.

(g) Nascent Hydrogen

Molecular hydrogen does not always perform reactions which are attributed to nascent (freshly made) hydrogen, e.g. if hydrogen is bubbled into iron(III) chloride solution nothing happens, but if zinc and dilute hydrochloric acid are added to the solution there is a colour change from yellow to pale green, due to the production of iron(II) ions.

$$2Fe^{3+} + Zn \rightarrow 2Fe^{2+} + Zn^{2+}$$

Acid conditions facilitate this reduction.

A side reaction

$$Zn + 2H^+ \rightarrow Zn^{2+} + H_2\uparrow$$

yields hydrogen.

23.7 Uses of Hydrogen

(a) Haber and Ostwald Processes

Over 70% of the hydrogen made is converted into ammonia and 33% of the ammonia to nitric acid. For the uses of these two substances see sections 30.13 and 30.21 respectively.

(b) Methanol and nylon 66

About 25% of the hydrogen produced is reacted with carbon monoxide over a zinc catalyst at 300°C, under a pressure of 300 atmospheres to yield methanol.

$$CO + 2H_2 \rightarrow CH_3OH \quad \Delta H = -109\ kJ$$

(c) Refinery Uses
About 12% of the hydrogen made is used in refineries, e.g. for the hydrogenation of sulphur compounds which enables them to be removed from petroleum.

(d) Hardening of Oils
The supply of oils (whale, cotton, peanut, etc.) in the world is used to supplement the supply of fats: the process was devised by Sabatier and Senderens in 1902 (see section 29.15). 1% of uses.

(e) Rockets
Apollo 11, propelled by a Saturn V rocket, took off for the moon in 1969 powered by the combustion of hydrogen in oxygen, 2000 tonnes of these substances in liquid form providing the motive power.

(f) Low Temperature Research
Hydrogen has a very low boiling point and properties of materials when cooled by the liquid can be investigated.

QUESTIONS 23

Hydrogen

1 Describe the preparation and collection of hydrogen in the laboratory. Describe an experiment that will show hydrogen is a reducing agent. (A diagram is not required). Name two other gases that are reducing agents and for each gas name a different chemical that can be reduced by it. Give two uses for hydrogen other than in balloons. [C]

2 Describe a laboratory method for the preparation of hydrogen. Give two uses for hydrogen. Describe three experiments (one in each case) to show (*a*) that the gas has a low density; (*b*) that when it burns water is formed; (*c*) that it is a reducing agent. [J]

3 How is hydrogen produced on a large scale? Why is its manufacture important? How, and under what conditions, does hydrogen react with: (*a*) copper(II) oxide; (*b*) sulphur; (*c*) chlorine? [OC]

4 Explain why a saucepan full of cold water becomes wet on the outside when it is first placed on a gas ring. Would it become wet on an electric hot plate?

5 Explain briefly how hydrogen can be liberated, in sufficient quantity to fill a few gas jars, from: (*a*) dilute sulphuric acid; (*b*) water. Describe fully an experiment to determine the mass of oxygen which combines with 1 g of hydrogen to form water. Give a diagram of the apparatus you would use. [O]

6 Name two metals which react with cold water to give hydrogen and give equations for the reactions. Describe experiments in which hydrogen is (*a*) a reducing agent; (*b*) made to combine directly with chlorine. State and explain two common uses of hydrogen. What do you understand by *nascent hydrogen*?

24 Group I: Lithium, Sodium and Potassium

24.1 Introduction

The Group I elements (lithium, sodium, potassium, rubidium and caesium and francium) are known as the alkali metals because they form alkalis when dissolved in water. Of these elements only sodium is considered in detail; some reference will be made to lithium and potassium but rubidium and caesium are not important here. The six elements form a good example of a group or family of elements: what is said of sodium is generally true of the other elements in this column of the periodic table.

The alkali metals are near the top of the electrochemical series: they are strongly electropositive. They are powerful reducing agents: the outermost electron is easier to remove from caesium than from lithium. These metals usually show an electrovalency of one. Their compounds all give positive flame tests: when heated in a Bunsen flame on a nichrome wire which has been dipped into concentrated hydrochloric acid, lithium colours the flame bright red, sodium an intense yellow and potassium a lilac hue (red through cobalt blue glass). The melting points, boiling points, densities and hardnesses are low for metals, but the chemical properties are typical.

Most of the salts of these metals are soluble in water; they are very stable to heat. The free elements are too reactive to occur naturally and they are found in the form of their salts.

24.2 Occurrence

(a) Chlorides

Sodium chloride (common salt) is found in sea water—three quarters of the 3·5% by mass of dissolved solids—and underground in dried up sea beds (rock salt). In hot countries (even Southern France) sea water is evaporated to obtain crystals of salt. The rock salt may be mined using drills and cutters or it may be extracted as a solution in water.

Potassium chloride occurs as a simple salt known as sylvine (KCl): important deposits in Yorkshire are now being exploited commercially. These are known in industry as potash, a most confusing designation. The deposits of carnallite (a double salt $KCl,MgCl_2,6H_2O$) at Stassfurt in Germany have been known to exist for a long time.

(b) Carbonates

Sodium sesquicarbonate ($Na_2CO_3,NaHCO_3,2H_2O$) is found in East Africa. Potassium carbonate occurs in the ashes of burnt plants.

(c) Silicates

Sodium aluminosilicate and its lithium and potassium counterparts constitute many rocks. These 'weather' or react with water and carbon dioxide, and soluble salts pass into the soil and sea.

(d) Other important sodium salts are sodium nitrate ($NaNO_3$, Chile saltpetre) and sodium borate ($Na_2B_4O_7,10H_2O$, borax, found in California, U.S.A.).

I THE ELEMENTS

24.3 The Manufacture of Sodium

Davy in 1807 succeeded in electrolyzing fused sodium and potassium hydroxides and these processes (developed by Castner) were the basis of the production of the metals for many years.

Sodium is nowadays made in the Downs' cell (see figure 24.2): the melting point of pure sodium chloride is 800°C but by adding calcium chloride a liquid can be obtained for electrolysis at 600°C. Copper connections are made to the

Figure 24.1. The occurrence and production centres of chemicals from sodium chloride and potassium chloride in the British Isles

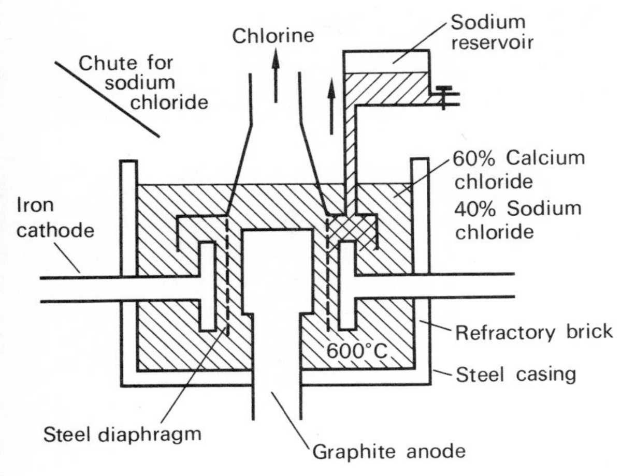

Figure 24.2
The manufacture of sodium—the Downs' Cell

electrodes: the cathode consists of two semi-circular pieces of iron. The cell is about 2·5 m high and 1·5 m in diameter, and it consumes up to 30 000 amperes at 6·7 volts. To start the cell working graphite blocks are put across the electrodes and, on becoming red hot, they melt the electrolyte. After removing these blocks the collecting assembly is lowered into place. The reactions at the electrodes are:

Cathode: $Na^+ + e^- \rightarrow Na$
Anode: $2Cl^- \rightarrow Cl_2\uparrow + 2e^-$

The world production of sodium is nearly 200 000 tonnes each year.

Potassium and lithium are produced in a similar manner but on a much smaller scale.

24.4 Physical Properties

	Lithium	*Sodium*	*Potassium*
Atomic number	3	11	19
Electronic structure	2,1	2,8,1	2,8,8,1
Relative atomic mass	6·941	22·99	39·10
Density (kg/m^3)	530	970	860
Melting point (°C)	186	98	63
Boiling point (°C)	1350	880	760
Crystal structure	They are all body centred cubic		
Standard electrode potential (volts)	−3·02	−2·71	−2·92

The alkali metals are soft, silvery, lustrous solids but they soon tarnish in air.

24.5 Chemical Properties

The alkali metals are so reactive that they have to be kept under oil in the laboratory; even this procedure destroys their lustrous surface. Their reactions usually follow the same course but their vigour increases from lithium to sodium and from sodium to potassium.

(*a*) *With air or oxygen*

If heated in air the metal burns with an intense yellow flame yielding a little sodium oxide (not an important substance)

$$4Na + O_2 \rightarrow 2Na_2O \quad [(Na^+)_2O^{2-}]$$

and much sodium peroxide, the latter being a pale yellow solid (see below).

$$2Na + O_2 \rightarrow Na_2O_2 \quad [(Na^+)_2O_2^{2-}]$$

A pure product is obtained when the air is free from moisture and carbon dioxide.

Sodium Chloride

South Durham and North Yorkshire: to Billingham and Wilton.

Cheshire: 80% of the U.K. supply from Middlewich, Northwich and Winsford.
Also sent to Chesterfield, Ellesmere Port, Runcorn, Sandbach and Widnes.

Furness (Lancashire): not worked.

Fylde (Lancashire): to Fleetwood.

Isle of Man: not worked.

Larne: Not used for the production of chemicals.

Somerset: not worked.

Staffordshire and Worcestershire: mainly for the non-electrolytic manufacture of chemicals from salt.
To Baglan Bay, Newport (Mon.), Stafford and Stoke Prior.

The Winsford mine (Cheshire) is the only one in Britain in which solid sodium chloride is hewn out. Most of the salt is obtained by sending down water to the beds (about 300 m below ground) and pumping the resulting brine to the surface; most of the brine is employed as a solution and the remainder is evaporated to yield the crystals.

The 1973 output of sodium chloride in the United Kingdom was 8·5 million tonnes; of this, 68% was obtained and used as brine, 19% was obtained as brine and evaporated to yield the solid, and 13% was hewn out as the solid.

The electrolysis of fused sodium chloride yields sodium and chlorine (Billingham and Ellesmere Port). The electrolysis of sodium chloride solution yields sodium hydroxide, hydrogen and chlorine (Billingham, Cheshire and Fleetwood).

Sodium carbonate is manufactured at Runcorn and Widnes.

Potassium Chloride

The United Kingdom potassium chloride mine-workings are at Loftus (near Whitby). Hot water is sent down 1200 m and the solution pumped back. The annual output is expected to be over a million tonnes by the mid-1970's, thus eliminating the need for costly imports.

If a fresh piece of sodium is left out in the air, a series of reactions takes place which may be shown diagrammatically as follows:

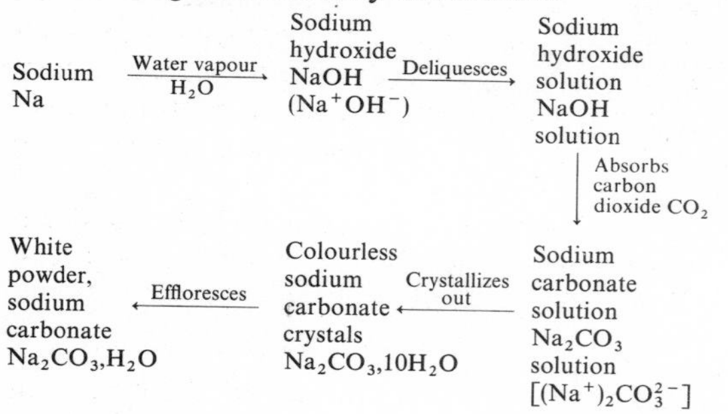

Sodium Na —Water vapour H_2O→ Sodium hydroxide NaOH (Na^+OH^-) —Deliquesces→ Sodium hydroxide solution NaOH solution

↓ Absorbs carbon dioxide CO_2

Sodium carbonate solution Na_2CO_3 solution $[(Na^+)_2CO_3^{2-}]$ —Crystallizes out→ Colourless sodium carbonate crystals $Na_2CO_3,10H_2O$ —Effloresces→ White powder, sodium carbonate Na_2CO_3,H_2O

N.B. Potassium forms more superoxide than monoxide or peroxide:

$$K+O_2 \rightarrow KO_2 \quad [K^+O_2^-]$$

(*b*) *With water*

A small piece of sodium metal can be placed on water—a vigorous reaction ensues which may terminate in an explosion.

$$2Na+2H_2O \rightarrow 2NaOH+H_2\uparrow \quad [Na^+OH^-]$$

N.B. Lithium reacts very slowly with water compared to sodium and potassium, but the actual reaction is the same. The reactions with acids are dangerous as they proceed explosively.

Ethanol reacts similarly, yielding sodium ethanolate.

$$2Na+2C_2H_5OH \rightarrow 2C_2H_5ONa+H_2 \quad [C_2H_5O^-Na^+]$$

(*c*) *With hydrogen*

When gently heated (380°C) in a stream of hydrogen, sodium forms a hydride.

$$2Na+H_2 \rightarrow 2NaH \quad [Na^+H^-]$$

Sodium hydride is unstable to water but is a useful reducing agent.

(*d*) *With non-metals*

Sodium combines readily with the halogens (chlorine etc.), sulphur and phosphorus, e.g.

$$2Na + Cl_2 \rightarrow 2NaCl \quad [Na^+Cl^-]$$

24.6 Uses of Sodium

Sodium is used in the manufacture of sodium cyanide (NaCN, for gold and silver extraction, case hardening iron); of sodium peroxide (Na_2O_2, for bleaching); of titanium in Britain (it is a reducing agent for titanium(IV) chloride); and of sodamide ($NaNH_2$, for detonators).

If sodium is alloyed with lead and then treated with chloroethane (C_2H_5Cl), tetraethyllead is formed: this is added to petrol as an anti-knock agent. In turn dibromoethane ($C_2H_4Br_2$) must be added to petrol to remove the lead as volatile lead(II) bromide, otherwise the lead from the tetraethyllead would foul sparking plugs.

Sodium has many other uses as a reducing agent in organic chemistry. It is a good heat conductor (used as a coolant in the Dounreay fast reactor) and has been proposed as a material for electricity cables (suitably sheathed).

Sodium-vapour lamps are a familar feature of our streets, a little neon being included in the lamp as a starter gas.

II THE COMPOUNDS

24.7 Sodium Peroxide (Na_2O_2)

This is manufactured by heating sodium in air to 200°C and then in oxygen to 300°C: this process is performed on aluminium trays in iron tubes, the air being free from moisture and carbon dioxide. The reaction is exothermic and heating is stopped to keep it under control. The product, which contains small quantities of other oxides, is very stable to heat.

In the laboratory sodium peroxide can be used to prepare small quantities of oxygen by adding cold water.

$$2Na_2O_2+2H_2O \rightarrow 4NaOH+O_2\uparrow$$

or

$$2O_2^{2-}+2H_2O \rightarrow 4OH^-+O_2\uparrow$$

With ice cold water or cold dilute sulphuric or hydrochloric acids, hydrogen peroxide is produced.

$$Na_2O_2+2H_2O \rightarrow 2NaOH+H_2O_2$$

and

$$Na_2O_2+2HCl \rightarrow 2NaCl+H_2O_2$$

or

$$O_2^{2-}+2H^+ \rightarrow H_2O_2$$

These reactions are the basis upon which sodium peroxide's use as a bleaching agent depends.

It combines with carbon dioxide and in so doing produces oxygen. This last fact enables it to be used for the purification of air in submarines.

$$2Na_2O_2+2CO_2 \rightarrow 2Na_2CO_3+O_2\uparrow$$

or

$$2O_2^{2-}+2CO_2 \rightarrow 2CO_3^{2-}+O_2\uparrow$$

Sodium peroxide is a vigorous oxidizing agent and if some is sprinkled on cotton wool and then moistened it bursts into flame. More useful applications of sodium peroxide's oxidizing power are evident in the manufacture of dyes and other organic compounds.

24.8 The Manufacture of Sodium Hydroxide (NaOH)

Small quantities of sodium hydroxide can be made by adding sodium to distilled water but the important methods are the two electrolytic and the lime-soda processes.

(a) The mercury cathode cell

The process devised by Castner, Kellner and Solvay is based on the electrolysis of sodium chloride solution: the plant is also referred to as the mercury cathode brine cell and soda cell (see figure 24.3). In the upper cell, the mercury cathode brine cell, the chlorine is formed at the anode and bubbles off.

$$2Cl^- \rightarrow Cl_2\uparrow + 2e^-$$

The sodium formed at the cathode dissolves in the mercury producing an amalgam (an alloy of a metal with mercury).

$$2Na^+ + 2e^- \rightarrow 2Na$$

N.B. On the basis of the electrochemical series it might be suggested that hydrogen ions from the water would be discharged, but a mercury cathode is employed here not a platinum one, and sodium ions are discharged rather than hydrogen ions.

In the lower cell, the soda cell, the sodium amalgam reacts with the water, this being facilitated by the iron (or graphite) grids.

$$2Na + 2H_2O \rightarrow 2Na^+ + 2OH^- + H_2\uparrow$$

The process gives a concentrated solution of sodium hydroxide, only some of which is evaporated to yield the solid; it is economic to transport the solution for industrial purposes. The capital costs of the plant are high (mercury is expensive) but the product is very pure. The brine cell may be up to 2 m wide, 0·3 m high and 20 m long; the base is slightly inclined to promote the flow of mercury; a current of up to 180 000 amperes is used at 4·3 volts. 27% of the sodium hydroxide made in the world (total about 20 m tonnes) is made by this process; it is the commonest method in the U.K.

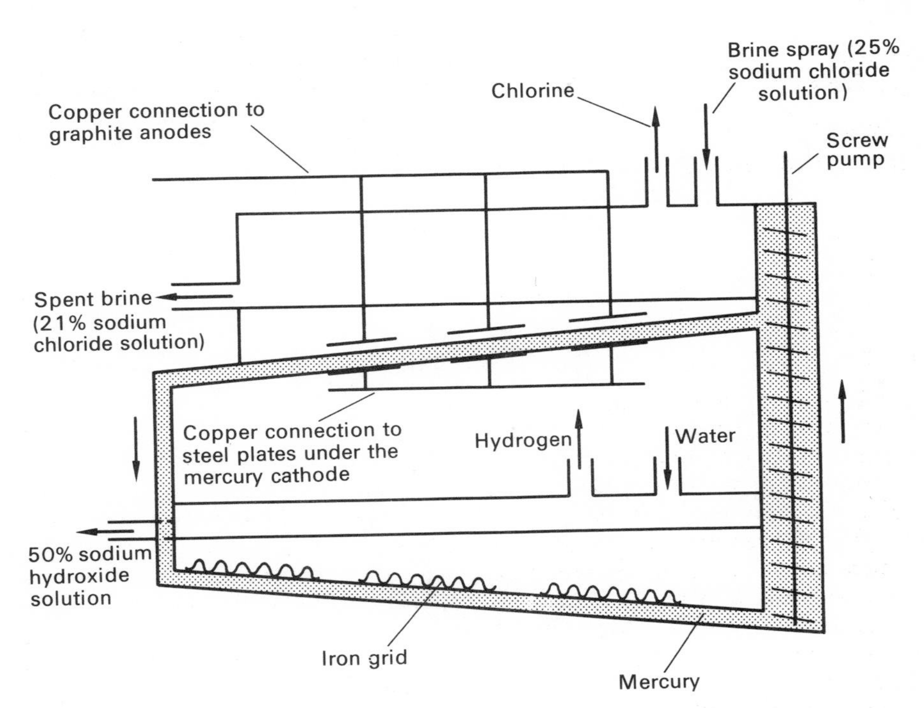

Figure 24.3
The manufacture of sodium hydroxide—the mercury cathode brine cell and soda cell

(b) The Gibbs' diaphragm cell

In the U.S.A. diaphragm cells are preferred, being cheaper to set up, but the product requires purification before use. The brine sprayed in percolates out through the asbestos and is partly electrolyzed before draining to the base of the cast iron vessel (see figure 24.4). At the cathode hydrogen is evolved:

$$2H^+ + 2e^- \rightarrow H_2\uparrow$$

leaving OH^-, hence more water ionizes. At the anode:

$$2Cl^- \rightarrow Cl_2\uparrow + 2e^-$$

The diaphragm keeps the chlorine away from the products at the cathode and enables the cathode section of the cell to be drained.

The catholyte (the solution draining from the cathode, in this case sodium hydroxide and sodium chloride) is evaporated until it contains about 50% sodium hydroxide, in which sodium chloride is almost insoluble and so the latter precipitates out. Such a cell will consume 1000 amperes at 3·2 volts and contains about 30 dm^3 of electrolyte. Diaphragm cells account for about 40% of the world production of sodium hydroxide; their use is popular in the U.S.A. and to an increasing extent in the U.K.

(c) The lime-soda process

The non-electrolytic process has been known for hundreds of years: it was put forward on an industrial basis by Gossage. A 10% solution of

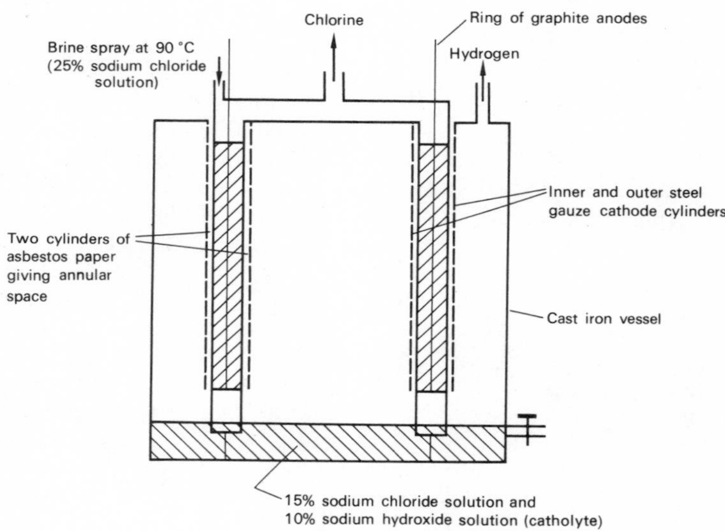

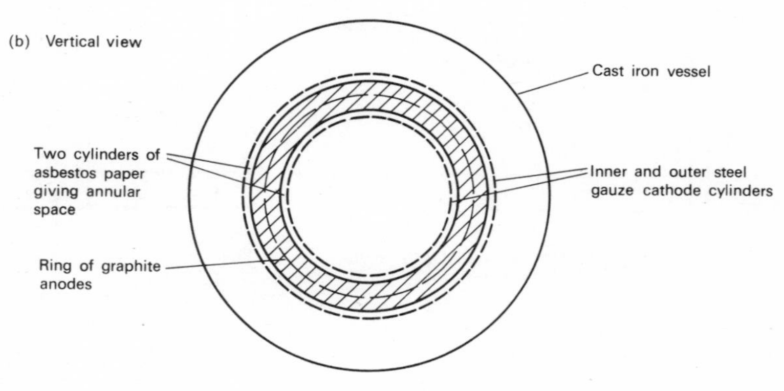

Figure 24.4
The manufacture of sodium hydroxide—the Gibbs' diaphragm cell

sodium carbonate is boiled with an excess of a suspension of calcium hydroxide (milk of lime).

$$Na_2CO_3 + Ca(OH)_2 \rightleftharpoons CaCO_3\downarrow + 2NaOH$$

The reaction is governed by the relative insolubilities of calcium hydroxide and carbonate: a 90% conversion is achieved. The solution of sodium hydroxide is filtered off and then concentrated in iron vessels. 33% of the world production is by this method; it is frequently used to remake sodium hydroxide which, during an industrial process has been converted to sodium carbonate.

24.9 Properties of Sodium Hydroxide

Sodium hydroxide is a white translucent solid which melts at 320°C. It is deliquescent in air, extremely soluble in water but not very soluble in ethanol, potassium hydroxide is more soluble and such solutions are used in organic chemistry. It is very stable to heat (boiling point 1400°C).

(a) With air

On exposure to air, sodium hydroxide absorbs moisture to form a saturated solution but this then reacts with carbon dioxide from the air to form sodium carbonate. This is the crust that is seen around the stoppers of sodium hydroxide bottles which have been left standing for some time in the laboratory.

(b) With metals

Molten sodium hydroxide attacks all metals except iron and nickel. Metals such as aluminium and zinc that form amphoteric compounds dissolve readily in hot concentrated sodium hydroxide forming solutions of sodium aluminate and zincate respectively.

$$2Al + 6NaOH + 6H_2O \rightarrow 2Na_3Al(OH)_6 + 3H_2\uparrow$$

or

$$2Al + 6OH^- + 6H_2O \rightarrow 2Al(OH)_6^{3-} + 3H_2\uparrow$$

$$Zn + 2NaOH + 2H_2O \rightarrow Na_2Zn(OH)_4 + H_2\uparrow$$

or

$$Zn + 2OH^- + 2H_2O \rightarrow Zn(OH)_4^{2-} + H_2\uparrow$$

(c) With non-metals

Non-metals react with sodium hydroxide in a more complicated manner: with chlorine cold dilute sodium hydroxide solution reacts to give sodium chlorate(I), etc.

$$Cl_2 + 2NaOH \rightarrow NaOCl + NaCl + H_2O$$

or $Cl_2 + 2OH^- \rightarrow OCl^- + Cl^- + H_2O$

but a hot concentrated solution reacts to give sodium chlorate(V), etc.

$$3Cl_2 + 6NaOH \rightarrow NaClO_3 + 5NaCl + 3H_2O$$

or $3Cl_2 + 6OH^- \rightarrow ClO_3^- + 5Cl^- + 3H_2O$

(see section 32.6(f)).

(d) As an alkali

Sodium hydroxide solution is a typical alkali: it will turn red litmus blue, react with acidic oxides, neutralize acids and give ammonia gas when heated with an ammonium salt:

$$NaOH + HCl \rightarrow NaCl + H_2O$$

or $OH^- + H^+ \rightarrow H_2O$

$$NaOH + NH_4Cl \rightarrow NaCl + H_2O + NH_3\uparrow$$

or $OH^- + NH_4^+ \rightarrow H_2O + NH_3\uparrow$

Sodium carbonate is not very soluble in sodium hydroxide, but potassium carbonate is fairly soluble in potassium hydroxide and so 'caustic potash' is used in preference to 'caustic soda' for the absorption of carbon dioxide (in the purification of air); otherwise precipitation will occur which blocks the apparatus.

$$2NaOH + CO_2 \rightarrow Na_2CO_3 + H_2O$$

or $2OH^- + CO_2 \rightarrow CO_3^{2-} + H_2O$

followed by

$$Na_2CO_3 + CO_2 + H_2O \rightarrow 2NaHCO_3$$

or $CO_3^{2-} + CO_2 + H_2O \rightarrow 2HCO_3^-$

Silicon dioxide (a constituent of glass) reacts similarly to carbon dioxide, forming sodium silicate (Na_2SiO_3). Because of this, plastic stoppers or rubber bungs are used as tops for sodium hydroxide bottles, instead of glass stoppers which would react with the alkali. If sodium hydroxide is used in a burette, the latter must be thoroughly washed out immediately after use.

(e) With metallic salts

With many metallic salts in solution sodium hydroxide will react giving a gelatinous precipitate of the metallic hydroxide:

(i) $CuSO_4 + 2NaOH \rightarrow Cu(OH)_2\downarrow + Na_2SO_4$ (Blue)

or $Cu^{2+} + 2OH^- \rightarrow Cu(OH)_2\downarrow$

(ii) $FeSO_4 + 2NaOH \rightarrow Fe(OH)_2\downarrow + Na_2SO_4$ (Usually green)

or $Fe^{2+} + 2OH^- \rightarrow Fe(OH)_2\downarrow$

(iii) $FeCl_3 + 3NaOH \rightarrow Fe(OH)_3\downarrow + 3NaCl$ (Red-brown)

or $Fe^{3+} + 3OH^- \rightarrow Fe(OH)_3\downarrow$

If the hydroxide is amphoteric the gelatinous precipitate will dissolve in excess alkali to form a colourless solution (in these two cases):

(iv) $ZnSO_4 + 2NaOH \rightarrow Zn(OH)_2\downarrow + Na_2SO_4$ (White)

or $Zn^{2+} + 2OH^- \rightarrow Zn(OH)_2\downarrow$
followed by

$$Zn(OH)_2 + 2NaOH \rightarrow Na_2Zn(OH)_4$$

or

$$Zn(OH)_2 + 2OH^- \rightarrow Zn(OH)_4^{2-}$$

(v)

$$Al_2(SO_4)_3 + 6NaOH \rightarrow \underset{\text{White}}{2Al(OH)_3\downarrow} + 3Na_2SO_4$$

or

$$Al^{3+} + 3OH^- \rightarrow Al(OH)_3\downarrow$$

followed by

$$Al(OH)_3 + 3NaOH \rightarrow Na_3Al(OH)_6$$

or

$$Al(OH)_3 + 3OH^- \rightarrow Al(OH)_6^{3-}$$

24.10 Uses of Sodium Hydroxide

About 25% of the total sodium hydroxide made in the United Kingdom (1·2 million tonnes) is used in the manufacture of rayon (artificial silk). Another 25% is used in the preparation of other sodium compounds such as the chlorate(I), the chlorate(V) and the carbonate. 13% is used in the manufacture of soap (section 29.15) and 10% in paper-making. Other uses include the mercerization of cotton, refining petroleum, and the purification of bauxite (impure aluminium oxide).

24.11 The Manufacture of Sodium Carbonate

Some sodium carbonate is manufactured by passing carbon dioxide into sodium hydroxide (see above), but most is made by the Solvay process, also known as the Fresnel or ammonia-soda process. Annual production is over 20 m tonnes (U.K. about 1·3 m tonnes).

The Solvay process depends for its efficient operation upon the fact that sodium hydrogen-carbonate is not very soluble in sodium chloride solution (brine). Ammonia is the vital intermediary permitting the conversion,

$$CaCO_3 + 2NaCl \rightarrow Na_2CO_3 + CaCl_2$$

which cannot be achieved directly.

Brine is saturated with ammonia by passing ammonia up a large tower down which the brine is sprayed. The ammoniacal brine (NaCl $+ NH_3$ in water) is then sprayed down second and third towers up which carbon dioxide is passed at 3 atmospheres pressure. Exothermic reactions occur, so the towers must be cooled.

$$NH_3 + H_2O + CO_2 \rightarrow NH_4HCO_3$$

followed by

$$NaCl + NH_4HCO_3 \rightarrow NaHCO_3\downarrow + NH_4Cl$$

or $Na^+ + HCO_3^- \rightarrow NaHCO_3\downarrow$

The sodium hydrogencarbonate is filtered off and, after rinsing, heated to yield anhydrous sodium carbonate (soda ash) and carbon dioxide which is recycled.

$$2NaHCO_3 \rightarrow Na_2CO_3 + H_2O + CO_2\uparrow$$

The initial supply of carbon dioxide is obtained by heating calcium carbonate in a kiln.

$$\underset{\text{Limestone}}{CaCO_3} \rightarrow \underset{\text{Quicklime}}{CaO} + CO_2\uparrow$$

Water is added to the residue yielding slaked lime.

$$CaO + H_2O \rightarrow Ca(OH)_2$$

and this then heated with the ammonium chloride solution to regain the ammonia

$$Ca(OH)_2 + 2NH_4Cl \rightarrow CaCl_2 + 2H_2O + 2NH_3\uparrow$$

The sodium carbonate obtained is very pure; it may be recrystallized from water to give a hydrate, if required.

The whole process is an excellent example of industrial economics. A small proportion of ammonia is lost from the plant and is continually replaced. Some uses have been found for the calcium chloride, but much of it is allowed to drain underground.

Potassium hydrogencarbonate is too soluble to be precipitated in any modification of the ammonia soda process, but can be prepared by another route from potassium chloride and hydrated magnesium carbonate.

24.12 Properties of Sodium Carbonate

Sodium carbonate is familiar in the home as large colourless crystals of washing soda ($Na_2CO_3,10H_2O$). Hydrated forms on heating yield the anhydrous substance, a white powder which is stable to heat. The decahydrate effloresces on standing in air, yielding the monohydrate.

N.B. The dihydrate of potassium carbonate ($K_2CO_3,2H_2O$) is a deliquescent solid.

The anhydrous substance has a low density and it tends to cake on dissolving in water. Industrialists prefer to use the monohydrate (Na_2CO_3,H_2O) which does not have these disadvantages. The decahydrate contains nearly twice as much water by mass as it does anhydrous sodium carbonate. Solutions of sodium carbonate (any form) are slightly alkaline to litmus because the carbonate ions react with the hydrogen ions of the water leaving hydroxide ions.

$$H^+ + CO_3^{2-} \rightarrow HCO_3^-$$

$$HCO_3^- + H^+ \rightarrow H_2CO_3$$

or $Na_2CO_3 + 2H_2O \rightarrow H_2CO_3 + 2NaOH$

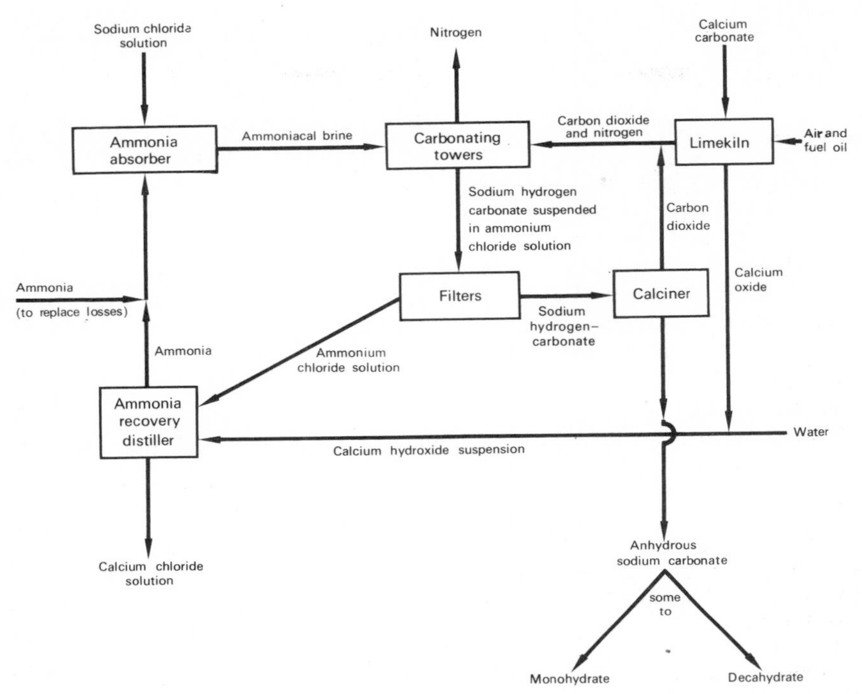

Figure 24.5
The manufacture of sodium carbonate—the Solvay or ammonia soda process

Sodium hydroxide is a strong alkali (highly ionized) whereas carbonic acid is only a weak acid (partly ionized). The excess of hydroxide ions over hydrogen ions makes the solution alkaline: it may be referred to as a mild alkali (as opposed to sodium hydroxide, a caustic alkali).

If sodium carbonate solution is added to a solution of a metallic salt a precipitate is formed (except in the case of potassium and ammonium which form soluble carbonates):

Ca^{2+} Sr^{2+} Ba^{2+}	yield the carbonate
Cu^{2+} Zn^{2+} Mg^{2+} Fe^{2+} Pb^{2+}	yield the basic carbonate
Al^{3+} Fe^{3+} Cr^{3+}	yield the hydroxide

These reactions are examples of ionic association (double decomposition),

e.g. $Ca^{2+} + CO_3^{2-} \rightarrow CaCO_3\downarrow$

24.13 Uses of Sodium Carbonate

35% of the sodium carbonate produced is in turn used for the manufacture of glass, 10% is used for making soap and in soap powders, another 10% is used in paper making, while 17% is used in washing and bleaching textiles. It is thus a chemical which makes a large but hidden contribution to modern life. Other industrial uses include metal refining (adding to molten iron to remove sulphur), making sodium hydroxide by the Gossage process, water softening (for temporary and permanent hardness), in agricultural sprays, for the manufacture of other sodium compounds (sodium borate or borax, $Na_2B_4O_7,10H_2O$ and sodium silicate or water-glass, Na_2SiO_3). In the laboratory sodium carbonate is very useful in qualitative and volumetric analysis.

Anhydrous potassium carbonate is used in the manufacture of glass and soft soaps and for drying alcohols.

24.14 Sodium Hydrogencarbonate ($NaHCO_3$)

This is also known as sodium bicarbonate.

In the laboratory preparation, excess carbon dioxide is bubbled through a saturated solution of sodium carbonate. The white precipitate of sodium hydrogencarbonate is filtered off and dried without heating.

$$Na_2CO_3 + H_2O + CO_2 \rightarrow 2NaHCO_3$$

If heated:

$$2NaHCO_3 \rightarrow Na_2CO_3 + H_2O\uparrow + CO_2\uparrow$$

It is used in baking-powder and in certain health salts. Baking powder contains sodium hydrogencarbonate and tartaric acid crystals.

When this is moistened the acid acts on the carbonate, liberating carbon dioxide, which helps the dough to 'rise' during the baking.

24.15 Sodium Chloride (NaCl)

Some of the uses of sodium chloride have already been illustrated: the manufacture of sodium, sodium hydroxide, hydrogen, chlorine, hydrogen chloride and hydrochloric acid, sodium chlorate(I), sodium chlorate(V), sodium carbonate and sodium hydrogencarbonate. Sodium chloride is used for preserving food and is an essential article of diet: the average person consumes 35 tonnes of food in his lifetime, a small proportion of which must be salt.

When sodium chloride is added to water the freezing point decreases; the lowest temperature that can be reached is $-23°C$. Use is made of this in winter: salt is sprinkled on ice and snow; when the heat generated by friction of footsteps and car tyres causes the formation of liquid water a solution is formed which does not freeze under the usual conditions encountered in the United Kingdom. Sodium chloride is stable in air but the commercial product often contains a little magnesium chloride which is deliquescent, and inhibits the free running properties of the crystals: a little sodium phosphate is added to react giving magnesium phosphate which is not deliquescent. Table salt may also contain a small proportion of potassium iodide: an iodide is another essential article of diet, finding its way to the thyroid gland in the neck.

24.16 Sodium Nitrate ($NaNO_3$)

Sodium nitrate is usually made by reacting sodium carbonate with nitric acid and evaporating the solution to yield the crystals.

$$Na_2CO_3 + 2HNO_3 \rightarrow 2NaNO_3 + H_2O\uparrow + CO_2\uparrow$$

It is used in flares, as a fertilizer and for manufacturing potassium nitrate and sodium nitrite.

24.17 Sodium Sulphate ($Na_2SO_4, 10H_2O$)

This is also known as Glauber's Salt; it effloresces to the anhydrous substance. It was manufactured by heating concentrated sulphuric acid with sodium chloride to 800°C in a reverberatory furnace.

$$2NaCl + H_2SO_4 \rightarrow Na_2SO_4 + 2HCl\uparrow$$

This was the first stage of the now obsolete Leblanc process for manufacturing sodium carbonate. Nowadays, sodium sulphate is made by neutralization.

Sodium sulphate shows a break in its solubility curve at 32.4°C, because below that temperature the solid in equilibrium with the solution is the decahydrate, and above that temperature it is the anhydrous substance. The anhydrous substance is useful as a drying agent for organic substances and for manufacturing glass.

24.18 Sodium and Potassium Salts

Like the sodium salts usually seen in the laboratory all the potassium salts encountered are soluble in water. Sodium salts are more abundant and cheaper than potassium salts and so the latter are only chosen when there are distinct advantages, e.g. in volumetric analysis, potassium manganate(VII) is better than sodium manganate(VII), because the former is not deliquescent.

QUESTIONS 24

Group I: Lithium, Sodium and Potassium

1 Describe the preparation of crystals of sodium sulphate (hydrated), starting from sodium, sulphur, air, water and any suitable catalyst. [OC]

2 How would you prepare in the laboratory a dry sample of sodium hydroxide, starting from sodium carbonate solution? Describe and explain fully what you would observe if some solid sodium hydroxide were left exposed to the air, on a watch glass, for a long period of time. [C]

3 Describe the manufacture of sodium hydroxide by an electrolytic process. Name any other products formed in the process and give two uses for each product named. How would you prepare a solid specimen of sodium hydrogencarbonate from sodium carbonate? Draw a diagram of the complete apparatus you would use. Give one method for distinguishing between sodium carbonate and sodium hydrogencarbonate. [C]

4 How would you distinguish between (*a*) sodium carbonate and sodium hydrogencarbonate, (*b*) sodium sulphate and sodium hydrogensulphate, (*c*) sodium sulphite and sodium thiosulphate? Suggest one reagent or experiment for application to each pair of substances. [O]

5 (*a*) The raw materials for the manufacture of sodium carbonate are calcium carbonate and sodium chloride. These substances do not give sodium carbonate directly, and the process depends upon the fact that sodium hydrogencarbonate is insoluble in brine saturated with ammonia and carbon dioxide. In the light of this information and without detailed reference to a technical plant, give an account of the chemical reactions which finally produce sodium carbonate crystals.
(*b*) Assuming no processing loss, what mass of sodium carbonate (anhydrous) is obtainable from 585 tonnes of sodium chloride?
(*c*) Show whether it is cheaper to buy anhydrous sodium carbonate at 47 p/kg or the hydrated crystals ($10H_2O$) at 33 p/kg for making a 1 M solution. [O]

6 Describe an electrolytic process by which sodium hydroxide is manufactured and name all the products. (A simple diagram is required.) Why is sodium hydroxide called *caustic soda*, and what do you understand by the statement that *sodium hydroxide is an alkali*? Give two uses of sodium hydroxide. [CO]

7 Describe what you would see, and explain the changes taking place, when: (*a*) a freshly cut piece of sodium is left exposed to the air for a long time; (*b*) a freshly cut piece of sodium is carefully dropped into a beaker of water containing some red litmus; (*c*) a piece of burning sodium is lowered into a jar of chlorine.

8 Give equations for the reactions which take place, and describe the colour changes, if any, when sodium hydroxide solution is added to solutions of: (*a*) copper sulphate; (*b*) magnesium sulphate; (*c*) iron(II) sulphate; (*d*) iron(III) chloride; (*e*) zinc sulphate; (*f*) warm ammonium chloride solution.

9 Name four gases which react with sodium hydroxide solution. In each case give the equation which represents the reaction and name the product. [J]

10 Describe briefly a method of manufacturing: (*a*) sodium hydroxide; (*b*) sodium carbonate; (*c*) sodium; starting in each case from sodium chloride. (In describing an electrolysis a labelled sketch should show the names of the electrolyte and the container material, the names, polarities and materials of the electrodes and where the product is obtained. A fully-labelled sketch of this kind requires no other written description.) [J]

11 Describe the appearance of sodium carbonate-10-water. State and explain what is seen to happen when the crystals are: (*a*) exposed to the atmosphere for some time; (*b*) heated in a test tube; (*c*) added to dilute hydrochloric acid. Starting with the crystals how would you prepare a sample of solid sodium hydroxide as free as possible from the carbonate? [J]

12 How would you distinguish by experiment between sodium carbonate and sodium hydrogencarbonate, both being in a finely powdered state? State what happens when a solution of sodium carbonate is: (*a*) boiled with excess of calcium hydroxide; (*b*) added to an aqueous solution of nitric acid; (*c*) added to a temporarily hard water. [J]

13 Explain the modern commercial method for the production of anhydrous sodium carbonate. Write equations for the essential reactions on which this process is based. Starting from sodium carbonate, explain in full detail how you would prepare a pure, dry, crystalline specimen of sodium sulphate. [J]

25 Group II: Magnesium and Calcium (Hard Water)

25.1 Introduction

The Group II elements (beryllium, magnesium, calcium, strontium, barium and radium) are known as the alkaline earth elements because some of their oxides and hydroxides have been known as earthy alkaline materials for a very long time. The six metals form a good example of a group or family of elements. Here only magnesium and calcium are considered.

The alkaline earths are amongst the elements at the top of the electrochemical series: they are strongly electropositive and usually show an electrovalency of two. They are powerful reducing agents. The compounds of most of them give positive flame tests: calcium a dark red, strontium a bright red and barium a pale green.

Compared with the compounds of the elements in Group I fewer are soluble in water and they are less stable to heat. The free elements are too reactive to occur naturally and they appear in the form of their salts.

25.2 Occurrence

(a) Halides

Carnallite ($KCl,MgCl_2,6H_2O$) is found in the Stassfurt salt deposits. Fluorspar (CaF_2) is mined in England. Magnesium ions occur to the extent of 0·13% by mass in sea water from which the chloride may be crystallized out.

(b) Carbonates

The separate carbonates magnesite ($MgCO_3$) and limestone ($CaCO_3$, also known as marble, chalk, etc.) as well as the double carbonate, dolomite, ($MgCO_3,CaCO_3$) occur widely in many countries.

(c) Phosphates

Rock phosphate ($Ca_3(PO_4)_2$), found also in bones, and fluorapatite ($CaF_2,3Ca_3(PO_4)_2$) are important in the manufacture of phosphorus (see section 30.4) but 85% are used as fertilizers.

(d) Sulphates

Epsom salts ($MgSO_4$, $7H_2O$) is found in some mineral springs and salt deposits. Anhydrite ($CaSO_4$) and gypsum ($CaSO_4,2H_2O$) are also important. Figure 25.3 is a map showing the occurrence and mining of calcium compounds in the United Kingdom.

I THE ELEMENTS

25.3 Manufacture

The manufacture of **magnesium** starts with dolomite which, on heating to 1200°C, gives dolime.

$$MgCO_3,CaCO_3 \rightarrow MgO,CaO + 2CO_2\uparrow$$

This compound is then made into a slurry with sea water.

$$\underset{\text{From sea water}}{MgCl_2} + MgO,CaO + 2H_2O \rightarrow 2Mg(OH)_2\downarrow + CaCl_2$$

or $$Mg^{2+} + 2OH^- \rightarrow Mg(OH)_2\downarrow$$

If limestone is used instead of dolomite only about half as much magnesium hydroxide is formed.

The precipitate of magnesium hydroxide is filtered off and heated to 1600°C.

$$Mg(OH)_2 \rightarrow MgO + H_2O\uparrow$$

The oxide is important enough for the process to stop here in many cases but some is converted into the metal.

If the magnesium oxide is heated with coke to 1300°C in a current of chlorine the chloride is formed.

$$MgO + C + Cl_2 \rightarrow MgCl_2 + CO\uparrow$$

The magnesium chloride melts at 950°C but, by adding sodium, potassium and calcium chlorides together with calcium fluoride as a flux, electrolysis of the molten mass can be carried out at 750°C to yield magnesium metal. At the graphite anode chlorine is released.

$$2Cl^- \rightarrow Cl_2\uparrow + 2e^-$$

and it can then be used to prepare more magnesium chloride. At the iron cathode magnesium is released.

$$Mg^{2+} + 2e^- \rightarrow Mg$$

It collects as droplets which rise to the surface and are periodically removed (figure 25.1).

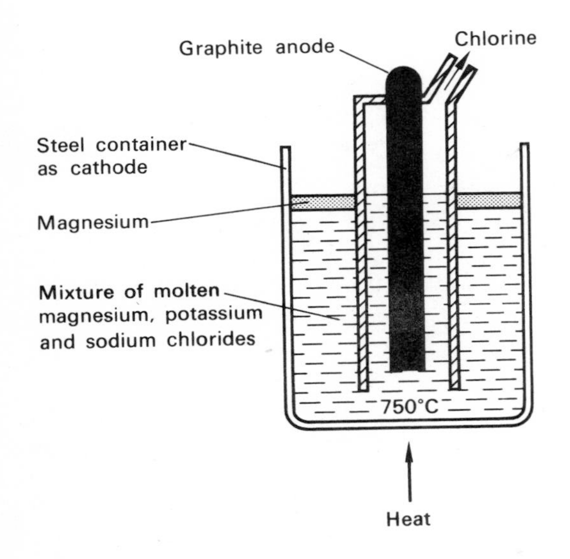

Figure 25.1
The manufacture of magnesium

An alternative method of preparing magnesium which uses less electricity—expensive in the United Kingdom—is the Pidgeon process in which calcined dolomite is heated in an electric furnace with ferro-silicon as a reducing agent (the manufacture of ferro-silicon, 25% iron, 75% silicon, itself involves the use of electricity but this can be done in countries where it is cheaper). The dolomite is heated as before and then mixed with ferro-silicon and calcium fluoride, which acts as a flux and a catalyst, before being loaded into cylindrical steel retorts. The retorts are evacuated to 0·0001 atmospheres and heated to 1200°C.

$$2(MgO,CaO) + Si \rightarrow \underset{\text{Calcium ortho-silicate}}{Ca_2SiO_4} + 2Mg\uparrow$$

The magnesium vapour is condensed at the cooler end of the retort. Under similar circumstances carbon can also be used as the reducing agent for pure magnesium oxide.

Calcium can be produced by aluminium or silicon reduction of the oxide but is usually prepared by electrolysis: at the temperature of the molten electrolyte (750°C) the calcium formed on the iron cathode by the reaction

$$Ca^{2+} + 2e^- \rightarrow Ca$$

solidifies (m.p. 850°C) and so the cathode is gradually withdrawn leaving a stick of the metal (see figure 25.2). The electrolyte is fused calcium chloride, to which a little calcium fluoride is added; chlorine is released at the graphite anode. Calcium chloride is readily available as the waste product of the Solvay process for sodium carbonate (see section 24.11).

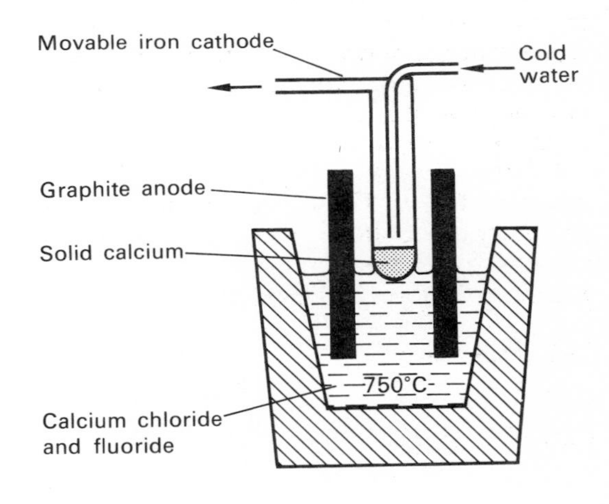

Figure 25.2
The manufacture of calcium

The primary production of **magnesium** in the world in 1973 was about 240 000 tonnes, of which the U.S.A. contributed 47%, the U.S.S.R. 22%, Norway 15% and Japan 5%; the U.K. used about 8500 tonnes including recovered scrap.

The extraction of gypsum, anhydrite, dolomite, limestone and chalk in the United Kingdom alone amounts to over 100 million tonnes every year, but the production of calcium metal is negligible.

△ Limestone ($CaCO_3$)

▲ Marble ($CaCO_3$)

▲ Chalk ($CaCO_3$)

○ Anhydrite ($CaSO_4$)

◑ Gypsum ($CaSO_4 . 2H_2O$)

■ Fluorspar (CaF_2)

Figure 25.3
The occurrence and mining of calcium compounds in the British Isles

Chalk (and cement factories)	
Chiltern Hills East:	Chinnor, Leighton Buzzard, Luton and Tring
East Anglian Ridge:	Cambridge, Cherry Hinton, Grays (2) and Ipswich
Lincoln Wolds:	Gainsborough, Stamford and South Ferriby
Norfolk Edge:	—
North Downs:	Northfleet, Rochester and Swanscombe Sittingbourne, Snodland and Strood
Salisbury Plain, White Horse Hills, Western Downs and Purbeck Downs:	Westbury
South Downs:	Lewes (2) and Steyning
Yorkshire Wolds:	North Ferriby
Marble	Clifden, Galway and Purbeck
Gypsum	
Northern:	Kirkby Thore, Langwathby, Long Meg, Long Marton and Sherburn-in-Elmet
Midlands:	Aston-upon-Trent, Barton-in-Fabis, Bunny, Cropwell Bishop, East Leake, Fauld, Gotham, Hawton, Kingston-upon-Soar, Newark-upon-Trent, Orston, Stanton-in-Staveley, Staunton-in-the-Vale, Stowe-Nine-Churches and Tutbury
South:	Brightling, Mountfield and Robertsbridge
Eire:	An Uaimh and Kingscourt
Anhydrite	
Eire:	As gypsum
England:	Whitehaven
Fluorspar	
Derbyshire:	Eyam (Ladywash, Sallet Hole and Longstone Edge), Matlock (Millclose and Clay Cross) and Youlgreave (2)
Durham:	Allenheads, Blackdene, Groverake, Whiteheaps, and Weardale, all near Stanhope.
Limestone (and cement factories)	
Derbyshire, especially Buxton:	Buxton, Hope and Stoke-on-Trent
North Yorkshire Moors:	—
Somerset:	Cheddar
Wales:	Barry (2) and Mold
These four are the most important; less important sources are:	
Cotswold Hills:	—
Glasgow–Edinburgh belt:	Clydebank, Coatbridge and Dunbar
Lake District:	—
Lincoln Edge:	—
Northampton Uplands:	Kidlington, Rugby and Southam
Northern Pennines:	Clitheroe and Stanhope
Northumberland and Durham:	—
Others:	Plymouth
Ulster:	Belfast and Cookstown
Eire:	Numerous quarries

25.4 Physical Properties

	Magnesium	*Calcium*
Atomic number	12	20
Electronic structure	2,8,2	2,8,8,2
Relative atomic mass	24·30	40·08
Density (kg/m^3)	1700	1600
Melting point (°C)	650	850
Boiling point (°C)	1100	1440
Crystal structure	Close packed hexagonal	Face centred cubic
Standard electrode potential (volts)	−2·37	−2·87

They are silvery white metals which are lustrous when freshly cut.

Magnesium is fairly ductile but the tensile strength is low unless the metal is alloyed. Calcium turnings are brittle.

25.5 Chemical Properties

(a) With Air

Both metals on exposure to air quickly become tarnished with a film of oxide which protects them from further attack. They burn with very bright flames when ignited.

$$2Mg + O_2 \rightarrow 2MgO$$
White light

$$2Ca + O_2 \rightarrow 2CaO$$
Dark red flame

Magnesium powder and ribbon burn very readily when melted and they may give a false impression of the metal: in the form of sheets and bars the metal can be safely cut with an oxy-acetylene torch. Magnesium also forms some magnesium nitride (Mg_3N_2). These three substances are ionic: $Mg^{2+}O^{2-}$, $Ca^{2+}O^{2-}$ and $(Mg^{2+})_3(N^{3-})_2$. The formation of magnesium nitride accounts for the production of ammonia when the residue from the combustion is moistened with water and heated.

$$Mg_3N_2 + 3H_2O \rightarrow 3MgO + 2NH_3\uparrow$$

(b) With Water

Magnesium decomposes boiling water or, more vigorously, steam.

$$Mg + H_2O \rightarrow MgO + H_2\uparrow$$

Calcium decomposes cold water. It is higher in the electrochemical series than magnesium and its hydroxide is more soluble; therefore the product of the reaction diffuses away into the water, although it may be precipitated after a while.

$$Ca + 2H_2O \rightarrow Ca(OH)_2 + H_2\uparrow$$

(c) With Acids

Magnesium dissolves in dilute solutions of the three mineral acids.

$$Mg + 2HCl \rightarrow MgCl_2 + H_2\uparrow$$

or

$$Mg + 2H^+ \rightarrow Mg^{2+} + H_2\uparrow$$

With nitric acid, however, the production of oxides of nitrogen is also likely. As can be predicted from a comparison of their behaviour with water, calcium should only be added to very dilute solutions of acids, if at all.

(d) With Hydrogen

Calcium reacts on gentle heating with hydrogen to give calcium hydride, also known as hydrolith.

$$Ca + H_2 \rightarrow CaH_2$$

or

$$Ca + H_2 \rightarrow Ca^{2+}(H^-)_2$$

This substance can be used as a transportable source of hydrogen for no heavy steel cylinder is involved.

$$CaH_2 + 2H_2O \rightarrow Ca(OH)_2 + 2H_2\uparrow$$

(e) With Other Oxidizing Agents

Burning magnesium will continue to burn in several gases, e.g. the oxides of carbon, nitrogen and sulphur, usually reducing them to their elements.

$$2Mg + CO_2 \rightarrow 2MgO + C$$

25.6 Uses of Magnesium and Calcium

Both metals are used as reducing agents: magnesium in the production of titanium and uranium from their halides and calcium in thorium production and as a deoxidizer for steel castings. Magnesium mixed with potassium chlorate(V) is used in photography as a flashlight powder.

Magnesium alloys have high strength but low density: they are used in the form of castings and wrought products for aircraft and in canning uranium for nuclear power stations. Magnesium can be used for the cathodic protection of iron pipelines: the magnesium undergoes preferential corrosion and no rust appears.

A cell has been developed which has magnesium and silver chloride as its plates; the electrolyte can be sea water. Such cells have a high power to mass ratio, and are useful for life jackets and emergency radio beacons.

Calcium is also useful for removing the water from rectified spirit after distillation has yielded a mixture of 96% ethanol and 4% water.

Racing car wheels.

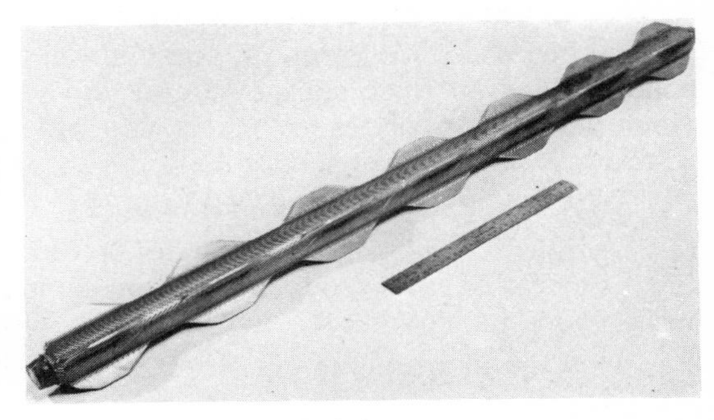

Nuclear fuel element can.

Television camera tripod

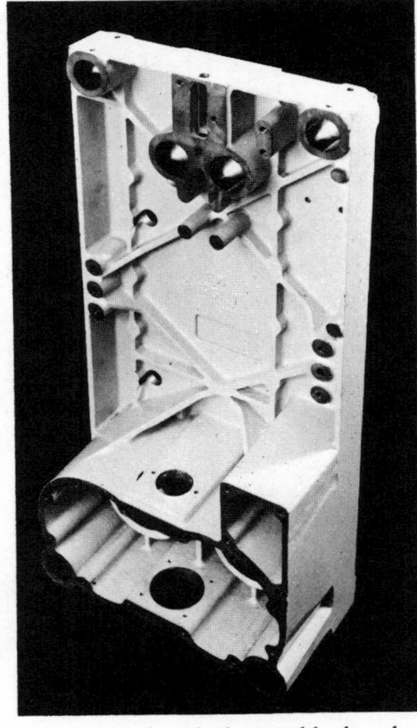

Main frame for telephone cable threader

Figure 25.4
Uses of magnesium

25.7 Calcium Dicarbide (CaC_2)

Commercial calcium dicarbide is an almost black substance which is prepared by heating calcium oxide and coke in an electric furnace to 2500°C.

$$CaO + 3C \rightarrow CaC_2 + CO\uparrow$$

The molten dicarbide is run off and allowed to solidify. It is an ionic substance and is white when pure. It hydrolyzes, i.e. reacts with water, to produce ethyne (acetylene).

$$CaC_2 + 2H_2O \rightarrow Ca(OH)_2 + C_2H_2\uparrow$$

(See section 29.11.) It is used also in desulphurizing iron and in the production of melamine resins (non-chip crockery).

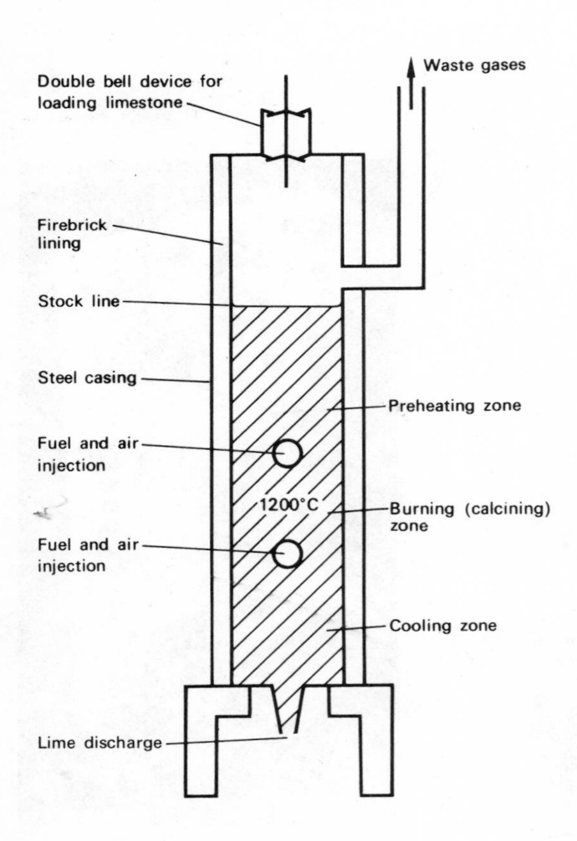

Figure 25.5
A limekiln

25.8 The Oxides

The oxides are white solids formed when the metals are heated or burnt in a plentiful supply of air. They are also formed on heating the hydroxide, carbonate or nitrate.

$$Mg(OH)_2 \rightarrow MgO + H_2O\uparrow$$

Commercially magnesium oxide is made from sea water (see section 25.3) or by heating the carbonate. Calcium oxide (quicklime) is made by heating the carbonate (limestone) in a limekiln to 1200°C (figure 25.5) by burning fuel oil, producer gas or coal.

$$CaCO_3 \rightarrow CaO + CO_2\uparrow$$

It is difficult to accomplish the latter reaction quantitatively under laboratory conditions. The carbon dioxide is swept out of the kiln by the nitrogen of the air, the oxygen of which enables the fuel to burn.

They are basic oxides which after strong heating become rather unreactive: they are not affected by hydrogen, carbon and other reducing agents under laboratory conditions. Magnesium oxide is hardly affected by water but calcium oxide, especially when freshly prepared, reacts vigorously to give calcium hydroxide (slaked lime).

$$CaO + H_2O \rightarrow Ca(OH)_2$$

If a lump of quicklime has water dripped on to it, it swells and disintegrates, evolving clouds of steam, and eventually yielding a dry white powder.

Because of their low reactivities the oxides, and the double oxide dolime already mentioned, are used in lining furnaces. At high temperatures they react with non-metal oxides to give the appropriate salts and are therefore used in the manufacture of iron and the conversion of iron into steel.

25.9 The Hydroxides

Magnesium hydroxide is usually prepared in the laboratory by adding a soluble hydroxide to a solution of a magnesium salt: the white base is precipitated.

$$MgSO_4 + 2NaOH \rightarrow Mg(OH)_2\downarrow + Na_2SO_4$$

or $$Mg^{2+} + 2OH^- \rightarrow Mg(OH)_2\downarrow$$

It is used as 'Milk of Magnesia' to neutralize the acidity which is a cause of stomach upsets.

The preparation of calcium hydroxide has

been discussed above. It is slightly soluble in water (1·7 g/dm^3 at 15°C), the solution being known as lime water. The solution is the best reagent to use to detect carbon dioxide: it rapidly gives a white suspension (goes 'milky') which clears so slowly to give a colourless solution if the passage of carbon dioxide is prolonged that the second stage is rarely seen in the laboratory.

$$Ca(OH)_2 + CO_2 \rightarrow CaCO_3\downarrow + H_2O$$
$$CaCO_3 + H_2O + CO_2 \rightarrow Ca(HCO_3)_2$$

If chlorine is passed over moist calcium hydroxide, bleaching powder, a complex compound is produced: it contains about 36% of chlorine available for bleaching purposes upon acidification.

Calcium hydroxide when mixed with sand and water in the form of a thick paste yields mortar which slowly hardens on standing. The water evaporates and the calcium hydroxide gradually reacts with the carbon dioxide in the air to form calcium carbonate.

Calcium hydroxide is used for industrial effluent treatment (it neutralizes waste acids), softening water, sewage treatment, leather treatment and sugar production. It is also used in the manufacture of sodium hydroxide, to counteract acidity in soils, and in reclaiming the ammonia in the Solvay process for sodium carbonate.

25.10 The Chlorides

Magnesium chloride can be crystallized out from sea water and its importance from this aspect has already been discussed. It can be prepared in the anhydrous form by heating the metal in a current of chlorine.

$$Mg + Cl_2 \rightarrow MgCl_2$$

Large amounts of calcium chloride are available as a waste product of the Solvay process for sodium carbonate. By the usual solution methods crystals of the hexahydrates of both salts can be obtained. Magnesium chloride-6-water decomposes on heating leaving the oxide but the calcium salt can be satisfactorily dehydrated and hence regenerated when the anhydrous substance is used as a drying agent. Both chlorides are very deliquescent.

Magnesium chloride is used for lubricating the cotton threads in the spinning industry and in dentistry as a constituent of cement. Calcium chloride is used in the manufacture of calcium and as a drying agent (not for ammonia or ethanol).

25.11 The Carbonates

Magnesium carbonate is formed as a white precipitate when sodium hydrogencarbonate solution is added to a solution of a magnesium salt.

$$MgSO_4 + 2NaHCO_3 \rightarrow MgCO_3\downarrow + Na_2SO_4 + H_2O + CO_2\uparrow$$

or $$Mg^{2+} + CO_3^{2-} \rightarrow MgCO_3\downarrow$$

If sodium carbonate is used a basic carbonate ($Mg(OH)_2,4MgCO_3,4H_2O$) is obtained.

Calcium carbonate, if needed in a purer form than that usually found in nature, can also be obtained by precipitation, but the choice of reagents is not so crucial.

$$CaCl_2 + (NH_4)_2CO_3 \rightarrow CaCO_3\downarrow + 2NH_4Cl$$

or $$Ca^{2+} + CO_3^{2-} \rightarrow CaCO_3\downarrow$$

The slow dissolution of calcium carbonate in rain water yields calcium hydrogencarbonate solution, temporarily hard water.

$$CaCO_3 + H_2O + CO_2 \rightarrow Ca(HCO_3)_2$$

Most of the magnesium carbonate is used with asbestos (itself a complex magnesium compound) for covering steam pipes. 35% of the calcium carbonate mined is used in making Portland cement—a British invention: it is roasted with clay at 1500°C for five hours in a horizontal rotary kiln internally heated by burning powdered coal in air. Cement consists of calcium silicate and calcium aluminate which when mixed with water and sand (and sometimes gravel) gives concrete.

25.12 The Sulphates

Magnesium sulphate, Epsom salts ($MgSO_4$, $7H_2O$), is prepared by the usual solution methods: it is very soluble in water. It is used in medicine and in leather and silk manufacturing.

Calcium sulphate, also found in nature (see section 25.2) is easily prepared in the laboratory by precipitation.

$$CaCl_2 + H_2SO_4 + 2H_2O \rightarrow \underset{\text{Gypsum}}{CaSO_4,2H_2O}\downarrow + 2HCl$$

or

$$Ca^{2+} + SO_4^{2-} + 2H_2O \rightarrow CaSO_4,2H_2O\downarrow$$

If gypsum is gently heated (to 120°C) it loses some of its water of crystallization to yield 'Plaster of Paris'.

$$2(CaSO_4,2H_2O) \rightarrow (CaSO_4)_2,H_2O + 3H_2O\uparrow$$
$$\text{or } CaSO_4,\tfrac{1}{2}H_2O$$

On adding some water to this hemi-hydrate and leaving the slurry it sets to a hard white mass, expanding slightly as it does so. Thus it can be used in moulds to give an accurate reproduction of a shape; it is also used to immobilize limbs which have been broken. Strong heating of gypsum yields anhydrite which does not combine with, but is slightly soluble in, water, which it makes permanently hard.

A small proportion of gypsum is included in cement to control the rate of setting, and it is also used as a fertilizer and as blackboard 'chalk'. Anhydrite is used to manufacture sulphur dioxide for sulphuric acid (see section 31.17) and in paints used for white lines on roads.

III HARD AND SOFT WATER

25.13 Water Supplies

The sun causes the evaporation of water, leaving dissolved salts behind. Eventually the clouds of water vapour in the atmosphere cool and rain falls back to the surface of the earth. Rain water is practically pure water containing only a small proportion of dissolved carbon dioxide and other materials derived from the atmosphere. Even so rain water would not be a good direct source of drinking water: it does not contain any calcium compounds necessary for health and it is tasteless. Sea water can only be drunk in very small quantities without causing sickness.

It is expensive to distil sea water but this has to be done in some parts of the world: British firms are competitive suppliers of equipment to do this. Distilled water suffers from the same disadvantages as rain water if it is used for drinking. The usual sources of drinking water are springs, rivers and lakes. During its course through the atmosphere and the earth's crust water takes up:

(*a*) carbon dioxide, sulphur dioxide, nitrogen dioxide and other gases soluble in water, salt spray near the sea, dust, radioactive materials, etc.;
(*b*) calcium sulphate as gypsum or anhydrite which is slightly soluble in water (about 2 g/dm^3) and silicon dioxide which is also slightly soluble in water and may cause trouble in high pressure boilers;
(*c*) materials from the rocks which are soluble in the slightly acidic rain water, e.g. calcium carbonate (limestone or chalk)

$$CaCO_3 + H_2O + CO_2 \rightarrow Ca(HCO_3)_2$$

(*d*) agricultural waste, including some fertilizers;
(*e*) domestic waste, including sewage;
(*f*) industrial waste;
(*g*) bacteria.

The water engineer has to consider all these factors in order that he can supply drinking water which has almost no taste and no smell, is clear and colourless and is at a fairly constant temperature. The presence of some dissolved calcium compounds is advantageous even though they make the water hard.

A hard water is one which will not lather readily with soap.

Soap ($C_{17}H_{35}COONa$) is the sodium salt of stearic (octadecanoic) acid; when it is put into a hard water it forms calcium stearate which, being insoluble, forms a scum (suspension).

Sodium stearate + Calcium sulphate → Calcium stearate ↓ + Sodium sulphate

A lather will form if an excess of soap is used but this is wasteful.

At the water works the water is stood to allow solid particles to settle: it may be necessary to add substances such as aluminium sulphate to assist in this process. Then it is filtered through sand and gravel to remove bacteria, fine particles, colour and algae. The water is then sterilized by passing in about one part in a million of chlorine.

In some areas sodium fluoride is present in the water already but in others one part in a million is added because it reduces dental decay in adolescence.

The best drinking water contains 50–120 parts in a million of calcium compounds (assessed as calcium carbonate). The determination of the degree of hardness can be carried out by titrating the water with a standard soap solution.

Experiment to determine the Hardness of Water

A burette is filled with standard soap solution made by dissolving soap in ethanol and water. It is added to 25 cm^3 of distilled water in 0·5 cm^3 portions until on shaking a permanent lather is obtained, i.e. one that lasts for three minutes. Distilled water is very soft and so very little soap should be required. Then the titration is repeated with 25 cm^3 of the water under examination. Much more soap will probably be required before a permanent lather is seen. By subtracting the first titre from the second the amount of soap needed to destroy the hardness is found: the conversion factor to give the hardness of the water may be quoted with the recipe for the soap solution or may have to be found by an experiment using a standard hard water.

25.14 Types of Hardness

The hardness of water is usually caused by calcium and magnesium salts in solution. There are two types of hardness:

(a) Temporary Hardness

This is caused by the hydrogen carbonates. It is said to be temporary because it is destroyed by heating when the compound decomposes precipitating the carbonate which henceforth cannot react with soap.

$$Ca(HCO_3)_2 \rightarrow CaCO_3\downarrow + H_2O + CO_2\uparrow$$

(b) Permanent Hardness

This is usually caused by the sulphates, but may be caused by the chlorides. These salts are not affected by boiling their solutions and so the hardness is said to be permanent.

25.15 The Advantages and Disadvantages

(a) The Advantages

(i) Calcium is needed by the body to build bones and teeth (mainly calcium phosphate) and hard water is a source of the necessary calcium salts (but it is not known how important it is).

(ii) Hard water coats lead pipes with a thin film of insoluble lead sulphate. Soft water slowly dissolves lead pipes which, because lead is a cumulative poison, may result in death. Nowadays many pipes are of copper or polythene so this advantage of a hard water is not always relevant.

(iii) Hard water has a better taste than pure water which is insipid.

(iv) Hard water is needed for brewing beer, e.g. at Burton-on-Trent.

(b) The Disadvantages

(i) Hard water causes a waste of soap in the home and in laundries. Soap consists of the sodium salts of one or more organic acids and is made by boiling sodium hydroxide with oils and fats yielding substances such as sodium stearate ($C_{17}H_{35}COONa$)—see section 29.15. The hydrocarbon chain of these substances will dissolve in the grease that usually causes the adhesion of dirt to a surface and the carboxyl group dissolves in the water: thus the soap molecule breaks down the boundary between grease and water releasing the dirt which passes into suspension in the water. The presence of calcium salts in the water causes the precipitation of the calcium salt of the organic acid and this is seen as a scum (suspension) which may leave marks on clothes or baths.

(ii) Temporary hardness in water causes a nuisance in the home and in industry if the water has to be heated. The formation of a precipitate of calcium carbonate by the reaction

$$Ca(HCO_3)_2 \rightarrow CaCO_3\downarrow + H_2O + CO_2\uparrow$$

gives fur in kettles or scale in boilers and pipes. Thereafter the heat supplied has to pass through a layer of calcium carbonate (ever increasing in thickness) as well as the metal container to reach the water being heated. The fur in kettles can be removed by using ethanoic acid (vinegar). The scale in pipes may lead to blockages and hence explosions: industry usually softens the water before use.

(iii) Hard water spoils the finish of fabrics, causes off-shades in dyeing and prevents successful tanning of leather.

Despite these disadvantages, drinking water usually contains some dissolved calcium compounds: London 300 parts in a million and Manchester 15 parts in a million. 60% of the United Kingdom's water supplies are classified as hard, 45% of them being very hard.

25.16 Detergents

During the last thirty years an increasing proportion (now about 70%) of the market for cleaning materials has been taken by detergents. They are better than soap at breaking down the interface between water and grease, thus releasing the dirt. Like most organic chemicals the starting point in their manufacture is petroleum: one third of the detergents are liquids and two thirds are powders. They are superior to soap

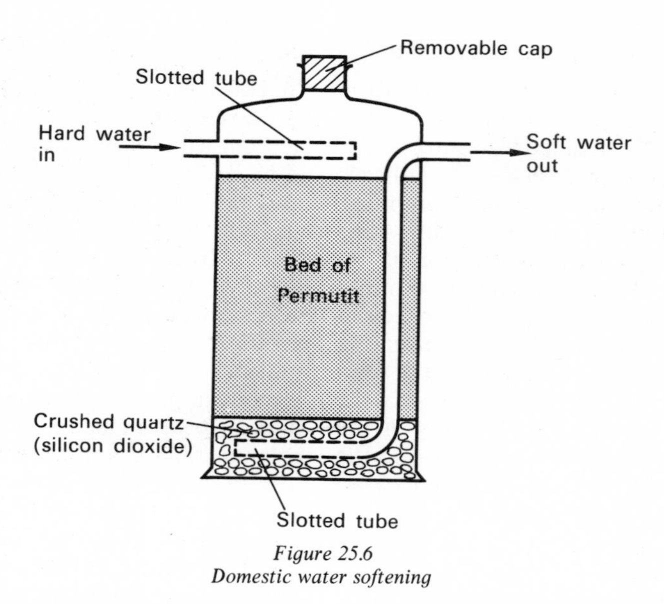

Figure 25.6
Domestic water softening

in coping with hard water because the calcium ions are enveloped in a soluble cage—giving a clathrate compound—and no scum is deposited (see section 29.15).

25.17 Water Softening

To soften water it is necessary to remove the calcium and magnesium salts which cause hardness, though this may not be done completely for reasons which are obvious from section 25.15. Usually this is done by precipitating the calcium in some insoluble form which therefore cannot cause hardness.

(*a*) **Temporary Hardness only**

(i) *Boiling and Filtering*

This is too expensive to employ on a large scale.

(ii) *Addition of Calcium Hydroxide* (*Clark's Process*)

When calcium hydroxide is added to temporarily hard water, calcium carbonate is precipitated.

$$Ca(HCO_3)_2 + Ca(OH)_2 \rightarrow 2CaCO_3\downarrow + 2H_2O$$

If too much calcium hydroxide is added the water will be made hard again because it is a slightly soluble calcium compound: careful analysis must be made to determine the right quantity needed.

(*b*) **Both Types of Hardness**

(i) *Distillation*

In some parts of the world there is no alternative to the distillation of sea water, but in most temperate regions other sources of water, which may or may not be hard, are cheaper.

(ii) *Addition of Sodium Carbonate*

Sodium carbonate will react with any calcium salts present to precipitate them as carbonates: a hydrated salt functions most efficiently.

$$Ca(HCO_3)_2 + Na_2CO_3 \rightarrow CaCO_3\downarrow + 2NaHCO_3$$

$$CaSO_4 + Na_2CO_3 \rightarrow CaCO_3\downarrow + Na_2SO_4$$

or $$Ca^{2+} + CO_3^{2-} \rightarrow CaCO_3\downarrow$$

The method is employed more often at home than in industry; it leaves sodium compounds in dilute solution but these do not affect soap. Bath salts are coloured and scented crystals of sodium carbonate.

(iii) *Ion Exchange Materials*

Natural minerals of the zeolite family (greensand and sodium aluminosilicate) are sold under trade names for softening water. The zeolite consists of a giant anion (the aluminosilicate) which is fixed and insoluble, and sodium cations which have a moderate mobility in the crystal. Nowadays more reliable resins can be made in the laboratory. The hard water is allowed to percolate slowly through the column of the zeolite and the water exchanges its calcium ions for the sodium ions of the resin. A solution of sodium hydrogencarbonate and/or sodium sulphate emerges from the bottom of the column: this solution is a soft water (see figure 25.6).

Sodium alumino-silicate		Calcium sulphate		Calcium alumino-silicate		Sodium sulphate
(Insoluble)	+	(Hard water)	→	(Insoluble)	+	(Solution)

Eventually all the sodium ions have been displaced from the column and it must be regenerated to continue its job. This is done by pouring in a concentrated solution of sodium chloride and as a result of a second exchange reaction a solution of calcium chloride is obtained (which is thrown away):

Calcium alumino-silicate		Sodium chloride		Sodium alumino-silicate		Calcium chloride
(Insoluble)	+	(Solution)	→	(Insoluble)	+	(Solution)

The column should then be washed out to eliminate any remaining sodium or calcium chloride solution before it is used again. The method is employed both in the home and in industry.

Recently synthetic resins (often made starting from petroleum) have extended the scope of this method. A cation exchange resin in its hydrogen form is used to replace calcium ions by hydrogen ions (regenerated by dilute hydrochloric acid) and an anion exchange resin in its hydroxide form is used to replace sulphate and hydrogen carbonate ions by hydroxide ions (regenerated by dilute sodium hydroxide solution). These resins can be used together to demineralize water. Most of the hydrogen and hydroxide ions combine to form undissociated water,

$$H^+ + OH^- \rightarrow H_2O$$

the only impurities being organic (covalent) materials.

(iv) *Addition of Sodium Metaphosphate*

This material sold under the name 'Calgon' softens a hard water by rendering the calcium ions non-reactive towards stearate ions (by forming a clathrate compound).

25.18 Stalactites and Stalagmites

Limestone regions are often noted for their remarkable caves, e.g. Cheddar in Somerset (figure 25.7), Han-sur-Lesse in Belgium and Postojna in Yugoslavia. Temporarily hard water, a solution of calcium hydrogencarbonate, may drip from the roof of the cave depositing some calcium carbonate as it does so, giving a hollow stalactite.

$$Ca(HCO_3)_2 \rightarrow CaCO_3\downarrow + H_2O + CO_2$$

Some may also be precipitated by evaporation

Figure 25.7
Cheddar cave— stalactites, stalagmites and pillars

and decomposition where the solution lands on the floor of the cave, giving a stalagmite. The rate of growth of the deposits may vary considerably but is usually very slow. The growths may unite to give a pillar (figure 25.8) and slight variations in the water flow and the action of air currents may produce fantastic shapes.

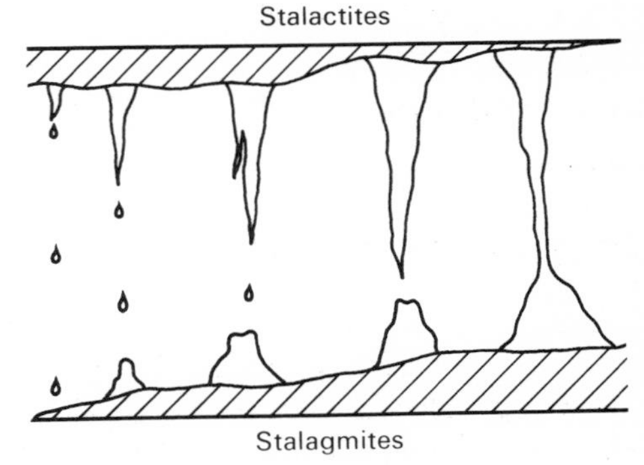

Figure 25.8
The growth of stalactites and stalagmites

Certain natural springs, e.g. Matlock in Derbyshire, may produce such a hard water that a rapid deposit of calcium carbonate forms upon objects placed in them and they become fossilized in a few months.

QUESTIONS 25

Group II: Magnesium and Calcium (Hard Water)

1 How is calcium oxide (quicklime) obtained commercially? What happens when (*a*) a small quantity of water is added to calcium oxide, (*b*) much water is added to the product of (*a*)? Describe how you would prepare a dry specimen of calcium sulphate from calcium oxide. Mention two important commerical compounds of calcium, other than calcium oxide, and state one use of each. [OC]

2 A specimen of water was found to exhibit both temporary and permanent hardness. After boiling, or adding to it a small quantity of calcium hydroxide, the water was found to be less hard than at first. On adding sodium carbonate in excess to some of the original water it was found to be softened completely.
(*a*) What is meant by *hardness* of water?
(*b*) What do you understand by (i) temporary hardness, (ii) permanent hardness?
(*c*) Give the name of one compound which causes temporary hardness and one which causes permanent hardness.
(*d*) Explain why the water was not as hard after boiling.
(*e*) Explain why the water was not as hard after adding a small quantity of calcium hydroxide.
(*f*) Explain why all hardness was removed when excess sodium carbonate was added to the water.
(*g*) With the aid of a diagram, explain how you could obtain chemically pure water from some of the original specimens. [A]

3 What is meant by the term *hard water*? Explain briefly how the chemicals which cause (*a*) permanent hardness, (*b*) temporary hardness get into natural water. Describe and explain what happens when (*c*) calcium hydroxide, (*d*) hydrated sodium carbonate, (*e*) an ion exchanger such as Zerolit, are added to water containing both temporary and permanent hardness. [S]

4 Describe simply how calcium oxide is manufactured from calcium carbonate. Give three ways by which carbon dioxide is added to the air and give two uses for carbon dioxide (excluding photosynthesis). [C]

5 (*a*) Describe and explain what happens when: (i) metallic calcium is dropped into water, (ii) magnesium ribbon is heated in steam. State two respects in which the action of iron on steam differs from the reaction between magnesium and steam.
(*b*) Outline (with equations and the conditions of reaction but without technical details) how hydrogen is manufactured from naphtha and steam.
(*c*) How could you reconvert hydrogen into water without burning the hydrogen? [C]

6 Name the two forms of hardness that may be present in water and, in each case, name and give the formula of one compound that, by its presence, causes the water to be hard. Explain the action of sodium carbonate in softening hard water. [C]

7 In an experiment to measure the volume of carbon dioxide released from a given mass of calcium carbonate, 580 cm^3 carbon dioxide were collected at 17°C and 760 mmHg pressure; what mass of calcium carbonate was used? What is the difference between gypsum and plaster of Paris (they are both forms of calcium sulphate)? Give a brief explanation of the use of the latter substance. [O]

8 Calcium belongs to Group II of the periodic table; state whether electrons are gained or lost when a calcalcium atom becomes a calcium ion and give the number of electrons per atom concerned. Mention two reactions as a result of which calcium ions are formed from calcium atoms. Calcium is a very reactive metal. What method can you suggest for converting calcium ions into calcium atoms? (No description of a process or naming of materials is required.) Describe how (*a*) calcium oxide, (*b*) calcium hydroxide, can be obtained from calcium carbonate, and mention two large-scale uses of calcium carbonate. [O]

9 Give the names of three forms of calcium carbonate which occur naturally. Starting from calcium carbonate describe briefly how you could prepare (*a*) calcium hydroxide, (*b*) calcium sulphate. Describe and explain a commercial use of each of these substances. [S]

10 (*a*) Calcium metal is often encountered as calcium turnings, white in colour but becoming darker on scraping with a penknife. (i) What is the white layer that the penknife removes? (ii) When the pure metal is dropped into water a reaction occurs and a gas is evolved. Write an equation for this reaction. (iii) The liquid remaining from (ii) is treated with sodium carbonate solution and a white precipitate is formed. What new substance would be in the filtrate if this precipitate were filtered off? (iv) The precipitate obtained in (iii) can also be obtained by adding ammonium carbonate solution to calcium chloride solution. Write an ionic equation for the formation of this precipitate. (v) The substance precipitated in (iii) and (iv) is used in the blast furnace for the production of iron from iron ore. Name the compounds into which it decomposes in the blast furnace, and show how each of the products of this decomposition is utilized in the manufacture of iron by this process. (vi) Write equations for two of the chemical reactions occurring in (v).
(*b*) Why should rain water slowly dissolve calcium carbonate? [NI]

11 Describe what you would observe and explain the changes taking place, when: (*a*) a piece of clean calcium is carefully dropped into some water containing red litmus; (*b*) magnesium is heated to redness in a current of nitrogen and the substance formed was decomposed by water; (*c*) calcium carbonate is roasted for a long time, the product allowed to cool and cold water slowly added to it; (*d*) dilute hydrochloric acid is added to calcium chlorate(I) and a piece of moist blue litmus paper is held in the gas given off; (*e*) water is slowly dropped on to calcium carbide.

12 What are the common names and the formulae for calcium carbonate, calcium oxide and calcium hydroxide? Starting with the carbonate, how could you prepare samples of the oxide and hydroxide? Describe and explain what happens when the gas liberated by the action of an acid on the carbonate is bubbled through a solution of the hydroxide for a long time. Give two commercial uses for the hydroxide and illustrate your answer throughout by equations. [J]

13 Name three common forms of calcium carbonate. Starting from calcium carbonate how would you prepare: (*a*) gas jars of carbon dioxide; (*b*) calcium hydroxide; (*c*) pure, dry calcium sulphate? How is it that the percentage of carbon dioxide in the air remains constant? [J]

14 Given some magnesium turnings, describe in detail how you would prepare pure dry samples of: (*a*) magnesium sulphate crystals ($MgSO_4,7H_2O$); (*b*) magnesium carbonate. [O]

15 Starting from calcium carbonate, explain how you would obtain: (*a*) calcium hydroxide; (*b*) a solution of calcium nitrate; (*c*) a pure, dry specimen of calcium sulphate. Why is the presence of calcium sulphate in certain natural waters considered to be objectionable? How may such waters be made more suitable for domestic use? (Distillation will not be accepted as an answer to this question.) [J]

16 Name two naturally occurring compounds of calcium and give their formulae. What calcium compound causes: (*a*) the permanent hardness of water; (*b*) the temporary hardness of water? Given a supply of calcium carbonate, explain with all necessary experimental detail how you would obtain from it a pure dry sample of calcium oxide. (The action of heat on calcium carbonate will not be accepted in answer to this question.) [J]

17 If you were provided with three white powders known to be calcium carbonate, calcium hydroxide and calcium oxide, describe tests by which you would identify them. What are: (*a*) mortar (*b*) cement? What happens when mortar sets? [O]

18 If you were provided with some calcium carbonate, describe carefully each stage in the process you would use to prepare a specimen of pure dry precipitated calcium carbonate. How would you use this precipitate to prepare a specimen of hard water which contains both temporary and permanent hardness? Write equations (no description is required) to show what happens when: (*a*) calcium hydroxide is added to temporarily hard water; (*b*) soap solution is added to permanently hard water.

19 Outline the manufacture of magnesium. Give two physical and two chemical reasons for classifying magnesium as a metal. Give experimental details of: (*a*) one reaction in which magnesium removes oxygen from an oxide; (*b*) one reaction in which it displaces a metal from a salt.

20 Why is water sometimes described as being hard? Name one substance causing temporary hardness of water and another causing permanent hardness. Give one method for softening: (*a*) temporarily hard water; (*b*) permanently hard water; (*c*) water with both varieties of hardness. Illustrate your answer with equations. [J]

21 Describe and explain a process (other than distillation) for softening water that is both permanently and temporarily hard. How would you find out which of two specimens of water is the harder? [OC]

22 Given distilled water, carbon dioxide and calcium oxide, how would you make a sample of temporarily hard water? What is the chemical reaction that occurs when soap solution is added gradually to temporarily hard water, and what change is visible?

23 Explain: (*a*) why distilled water is used in ships' boilers; (*b*) why bath salts are made of sodium carbonate; (*c*) why stalactites and stalagmites are often found in caves in limestone districts.

24 Describe how you would obtain from some muddy salt water: (*a*) a specimen of sodium chloride; (*b*) some pure water. Explain why tap water often produces: (*c*) a curd with soap solution; (*d*) a scale when boiled; (*e*) a precipitate with sodium carbonate.

25 In order to determine the relative degree of hardness of a specimen of tap water, a standard soap solution was prepared. It was found that, to give a permanent lather, 25 cm^3 of distilled water needed 0·6 cm^3, 25 cm^3 of the tap water needed 10·6 cm^3 and 25 cm^3 of the tap water after boiling needed 6·6 cm^3 of the soap solution. Calculate the percentage of the total hardness that is only temporary.

26 The Transition Elements: Iron, Copper and Zinc

26.1 General Introduction

Mendeléeff originally conferred the title 'the transition elements (metals)' upon the nine elements in Group VIII of his periodic table but nowadays the term refers to about thirty elements (if the inner transition elements are included there are about sixty). The main feature is that the penultimate shell of electrons plays an important part in determining their physical and chemical properties. The elements are more similar than successive members in a short period (e.g. sodium, magnesium, aluminium . . .) but not as similar as successive members in group I (lithium, sodium, potassium . . .). The manner in which these elements occur in nature and the methods used to extract them follow similar patterns.

The most important **physical** properties of the transition elements are as follows.

(*a*) They are lustrous white metals (except copper and gold). Pure iron is white but most forms seen are grey.
(*b*) The density of the first series, scandium to zinc, is approximately 7000 kg/m^3.
(*c*) They have high melting points and boiling points.
(*d*) The cations derived from these elements are small, often hydrated and often coloured (exception: $Zn(H_2O)_6^{2+}$ is colourless).
(*e*) Most of them appear just above hydrogen in the electrochemical series (exceptions include copper).
(*f*) They are good conductors of heat and electricity.
(*g*) Most of them form alloys with iron.
(*h*) They are ductile, of high tensile strength and malleable: to the layman they are better examples of metals than the alkali metals.
(*i*) They are often used as catalysts.

The most important **chemical** properties of the transition elements are as follows.

(*a*) Variable valency is an important aspect of their chemistry: redox reactions are common.
(*b*) Many of them form films of the oxide, hydroxide or carbonate on the surface when exposed to air: some are tenacious like zinc carbonate, others are porous and flake off like iron(III) hydroxide.
(*c*) Their oxides are usually reducible with carbon.
(*d*) The oxide EO is usually basic, but E_2O_3 is often amphoteric and the higher oxides are usually acidic.
(*e*) They are usually oxidized by steam when red-hot (exceptions include copper).
(*f*) They usually dissolve readily in dilute hydrochloric acid (exceptions include copper).
(*g*) They absorb hydrogen but do not form simple hydrides.
(*h*) Their chlorides react slightly with water (hydrolysis) so that the solutions are acidic to litmus. Hydrated chlorides can be crystallized from aqueous solutions; anhydrous chlorides are usually made by synthesis.
(*i*) Most of the sulphates and nitrates of these elements are also hydrolyzed on addition to water.

Only three transition elements, iron, copper and zinc, are considered here: they are very important metals and are used to illustrate the general properties outlined above. Other transition elements such as chromium, manganese, cobalt, nickel, etc. have been included in the qualitative analysis scheme in Chapter 21. Metals other than iron and steel are often referred to as the non-ferrous metals.

26.2 Occurrence

Iron and copper have been known for about 5000 years. Zinc, as an individual element, has only been known since about 1700 A.D. but it has been employed in alloys for 3000 years.

Copper is the only one of the three metals to occur as an element; besides this rare occurrence in North America there are over 360 known minerals but the copper content is usually very small.

	Iron	*Copper*	*Zinc*
Oxide	Fe_2O_3 haematite Fe_3O_4 magnetite $Fe_2O_3,1{\cdot}5H_2O$ limonite	Cu_2O cuprite, ruby ore CuO tenorite	ZnO zincite
Sulphide	FeS_2 iron pyrites, marcasite	Cu_2S chalcocite $CuFeS_2$ copper pyrites, chalcopyrites Cu_5FeS_4 bornite	ZnS zinc blende
Carbonate	$FeCO_3$ ironstone, chalybite, spathic iron ore	$CuCO_3,Cu(OH)_2$ malachite $Cu(OH)_2,2CuCO_3$ azurite	$ZnCO_3$ smithsonite

The commonest compounds found as minerals are tabulated above.

Iron pyrites is used for the manufacture of sulphur dioxide (see section 31.17) but not for the manufacture of iron because the sulphur makes the metal brittle.

Iron is abundant in the earth's crust and it is found in all parts of the world. Copper and zinc are comparatively rare. The chief suppliers of copper minerals are the U.S.A., Zambia, Chile and Canada; of zinc minerals Canada, the U.S.A. and Australia.

I THE ELEMENTS

26.3 The Manufacture of Pig Iron

The United Kingdom ores, mined mostly in the Humber-Severn belt by opencast methods, consist of iron(II) carbonate, known as ironstone. The ore, with a little coal, is given a preliminary roasting (sintering) by gas heating to 1200°C.

$$6FeCO_3+O_2 \rightarrow 2Fe_3O_4+4CO_2$$

Volatile oxides such as water, carbon dioxide, sulphur dioxide and arsenic(III) oxide are driven off at this stage and granules of material suitable for the blast furnace are obtained. About 25% of the United Kingdom supply of pig iron comes from home ores which only contain about 28% iron. Imported ores (chiefly from Sweden, Canada, Mauretania and Venezuela) are richer in iron—about 61%—and are usually the oxides, which only require granulation before being reduced.

Into the top of the blast furnace (see figures 26.1(a) and (b)) is charged a mixture of the sintered ore and high grade metallurgical coke together with a little calcium carbonate (limestone) to remove earthy impurities. Hot air (750–1000°C at 2 atmospheres pressure) is blown in near to the base of the furnace through tuyères (jets) at supersonic speeds and as the air moves upwards the charge moves downwards. The coke is burnt to carbon monoxide via the dioxide

$$C+O_2 \rightarrow CO_2$$
$$CO_2+C \rightleftharpoons 2CO$$

and the carbon monoxide reduces the iron oxide to iron which then melts and runs down to the bottom of the furnace. Excess carbon monoxide is generated because the key reaction is reversible.

$$Fe_3O_4+4CO \rightleftharpoons 3Fe+4CO_2$$

The calcium carbonate decomposes to the oxide which reacts with the silicon dioxide (earthy impurities).

$$CaCO_3 \rightarrow CaO+CO_2\uparrow$$
$$CaO+SiO_2 \rightarrow CaSiO_3$$

The molten calcium silicate (the slag) does not mix with the iron and, being less dense, forms a layer on top of the iron at the bottom of the furnace.

Periodically the molten iron and slag are tapped off from the base of the furnace: the clay plugs are drilled out and red-hot fluids gush forth. The iron is run either into large ladles to be taken directly to the steel maker or into iron moulds which are then sprayed with water to cool the slabs of cast iron (also known

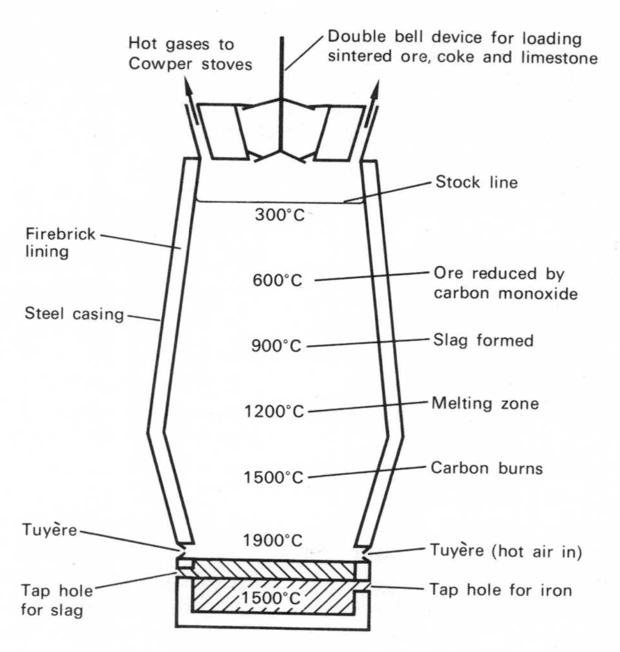

Figure 26.1(a)
Cross section of a blast furnace

Figure 26.1(b)
A blast furnace in South Wales

Figure 26.1(c)
Tapping a blast furnace

as pig iron). This cast iron contains nearly 4% carbon and smaller proportions of silicon, sulphur, phosphorus and manganese. The slag is run into a pit where on cooling it solidifies: it is then broken up and used for road making, cement manufacture and for building blocks.

The hot gases rushing out of the top of the furnace go through a network of pipes (which may double the height of the furnace) before being passed into dust catchers. They are then burnt in Cowper stoves to heat the ingoing air. The percentage composition of the gases is nitrogen 60, carbon monoxide 24, carbon dioxide 12 and hydrogen 4; two of these are inflammable.

Recent variations to increase productivity have been to enrich the air to 25% oxygen and to inject steam, powdered calcium oxide, powdered coal or heavy fuel oil through the tuyères.

The whole process is an excellent example of continuous flow but an account of the chemistry gives little indication of the excitement of the tasks involved (for which films or a visit will be helpful) or of the magnitude of the production. At a typical modern steelworks 3·5 tonnes of ironstone give 2·3 tonnes of sintered ore, to which 0·7 tonne of coke and 0·1 tonne of limestone are added and then 4·5 tonnes of hot air are blown in (about 3200 m^3); the yield of iron is about 1 tonne, but this is accompanied by 1·25 tonnes of slag. The furnace produces about 350 tonnes between each four hour tapping and thus in a year produces about 750 000 tonnes of pig iron.

The furnace is made of firebrick and has an outer casing of thin steel for strength. The lining wears out after about three years but it is about five weeks before the furnace is cool enough to enter for maintenance work!

26.4 The Conversion of Iron into Steel

In 1973 the United Kingdom produced about 17 m tonnes of new iron and reclaimed about 19 m tonnes of scrap, producing 26·7 m tonnes of steel. About 3·5 m tonnes of iron and 0·5 m tonnes of steel were cast into objects used directly. Cast iron is brittle but has considerable compression strength: it is used in castings for cars, machinery, pipes, manhole covers and gas stoves.

The world production of steel in 1973 was 696 m tonnes, of which the U.S.S.R. produced 20%, the U.S.A. 19%, Japan 17%, West Germany 7%, and the U.K. 4%. The map (figure 26.3) shows the location of the iron and steel industry in the British Isles. The process of steelmaking is one of reducing or even eliminating the impurities in the cast or molten iron and then adding materials to give iron the desired properties.

The silicon content of the molten iron is reduced while still in the ladle by oxygen injection (pre-refining)—a process which causes the evolution of brown clouds of iron(III) oxide.

The **Bessemer** process (1856) was invented in England, and, whilst it proved to be very useful to European manufacturers, British pig iron which contained a higher proportion of phosphorus, could not be used until Gilchrist and

Fume–collecting hood
Refractory lining
Water–cooled oxygen lance
Taphole
Converter rotates about external axle for pouring
Slag
Steel shell of converter
Molten iron being converted into steel

Figure 26.2
The manufacture of steel in the Basic Oxygen Convertor

Figure 26.3
The iron and steel industry in the British Isles

Thomas (1879), substituted a basic lining (calcined dolomite or dolime, MgO, CaO) for the original acidic lining (silica SiO_2). The chemistry of the conversion is the same as in the open hearth furnace, but the carbon is usually eliminated during the oxygen blow and then replaced to the required extent at the same time as the alloying elements are added. The original converter allowed air to be blown in underneath the molten pig iron: in the L.D. converter (Linz and Donawitz are in Austria), oxygen is blown in by a lance at the top (see figure 26.2). The L.D. converter is now referred to as the basic oxygen converter (B.O.C.) and it produces 48% of the U.K. steel.

In the B.O.C. process, the furnace is tipped to the horizontal position, molten pig iron is poured in and some scrap iron is added; the furnace then resumes the vertical position and a water-cooled copper lance blows oxygen at about 15 atmospheres pressure on to the surface of the iron. As in the Bessemer process, the oxidation of the non-metals proceeds, giving an excellent firework display, the heat of the oxidation maintaining the charge above melting point. Impurities which do not escape as gaseous oxides react with the calcium carbonate (added after the blow) to give a slag. An alloy of iron, manganese and carbon (ferro-manganese) is then added: the manganese acts as a deoxidizer, manganese(II) oxide passing into the slag, and the carbon converts the iron into steel. The B.O.C. process is very fast, about 300 tonnes of steel being produced every 40 minutes.

The B.O.C. and open hearth processes are used chiefly to produce carbon steel: a steel is a form of iron containing up to 1·7% carbon. The carbon is mostly present as a compound (cementite, Fe_3C). These steels can be classified by the percentage of carbon as follows.

(*a*) Low carbon steels (i) up to 0·07, (ii) 0·07 to 0·15.
These soft steels are used for wires, rivets and sheets.

(*b*) Mild steels 0·15 to 0·25.
These are used for general engineering purposes.

(*c*) Medium carbon steels 0·2 to 0·5.
These are used for springs.

(*d*) High carbon steels 0·5 to 1·4.
These are used for hammers and chisels.

In the United Kingdom until the early 1970's, most of the steel was made in oxygen blown **Siemens-Martin open hearth furnaces** (invented in 1851); now about 32% is produced by this method. The lining of the furnace is often made of basic materials such as calcined dolomite (MgO, CaO); the charge consists of molten iron together with up to 50% scrap iron. The furnace is oil-fired and oxygen lances project through the side to expedite production: about 300 tonnes of steel can be produced in 6–8 hours at a temperature of 1650°C. Most of the carbon and

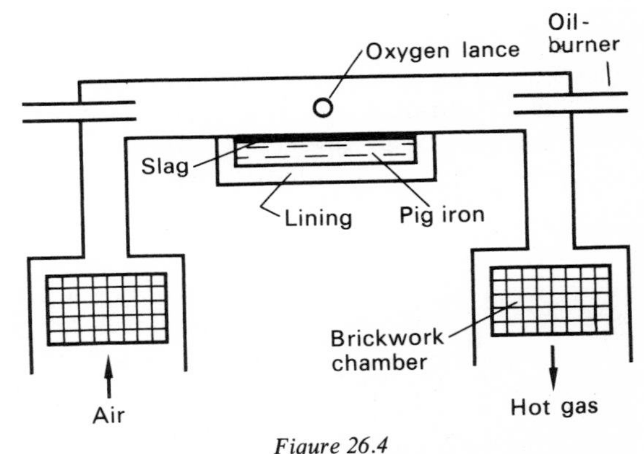

Figure 26.4
The manufacture of steel—the Siemens-Martin open-hearth furnace

---

Haematite Ores (Iron(III) oxide)
Cumberland and Glamorgan.

Jurassic Ores (Iron(II) carbonate and some aluminosilicate)
Leicestershire, Lincolnshire, Northamptonshire and Rutland.

Major Import Terminals for Ores
Immingham, Port Talbot and Redcar (Hunterston projected).

Blast Furnaces (only)
Dagenham and Workington (to be steel later).

Steelworks (only)
Barrow, Cork, Glengarnock, Panteg, Sheerness, Sheffield and Stocksbridge.

Blast Furnaces and Steelworks
Bilston, Brymbo (Wrexham), Cardiff (East Moors, until 1976), Consett, Corby, Glasgow (and Motherwell, including Ravenscraig), Port Talbot (Margam-Abbey), Middlesbrough, Llanwern (Newport, Spencer Works), Scunthorpe (Anchor, Appleby-Frodingham, Redbourn, Normanby Park), Rotherham (Park Gate) and Templeborough, Skelton (until 1975) and Shotton (until 1979).

Tinplate Works (only)
Trostre and Velindre.

Blast Furnaces, Steelworks and Tinplate Works
Ebbw Vale (to be tinplate only from 1976).

any silicon left is oxidized by the flame. The tap hole is at the back of the furnace, which may be tipped to pour out the steel and slag into ladles. The air for the combustion of the oil passes, on entering, through a chamber of hot bricks and, on leaving, through a second chamber of open brickwork; when the bricks in the second chamber are hot, the flow of air is reversed (see figure 26.4). Haematite (Fe_2O_3) is added to remove the remainder of the carbon in the molten iron. Limestone is added to assist in the removal of impurities, forming a slag (e.g. calcium phosphate) with any phosphorus present.

Samples of the metal and slag are taken at intervals and analyzed by a mass spectrometer, the results being available in a matter of minutes. The final addition of alloying elements and deoxidizers (manganese and aluminium) is made to the ladle while tapping. The molten steel is then poured into ingot moulds and taken away to the rolling mills or forging shops.

Electric furnaces produce an increasing proportion (now 20%) of the steel made in the United Kingdom. In the arc furnace massive graphite electrodes enter through the roof and a large current, up to 40 000 A, passes through the charge heating it up to 3400°C. An oxygen blast can be used to conserve electricity; scrap iron can be processed, limestone being added to remove impurities as a slag. Electric furnaces are used mainly to produce alloy steels, i.e. those steels in which nickel, chromium, molybdenum, manganese, vanadium, tungsten and cobalt are added to the iron and carbon mixture —stainless steel was invented by Brearley in the United Kingdom in 1913 (74% iron, 18% nickel and 8% chromium). High grade carbon steels for cutting tools are also made in electric furnaces.

The steel can be further improved by subjecting it to heat treatment. If heated in molten potassium cyanide (KCN) or by burning methane the steel is case hardened—the surface is hardened but the interior is relatively soft and thus it is suitable for gear wheels. Annealing steel—to take out strains after fabrication of objects—is carried out by heating to 750°C in an inert atmosphere and then allowing the objects to cool slowly in the furnace: it makes the material ductile and tough. This type of steel can be used to make chains.

Quenching of steel, or sudden cooling from red heat, can be done using oil or water: it gives a hard and brittle steel which is then tempered, i.e. reheated and then allowed to cool, thus reducing the hardness but making the product tougher. The temperature for reheating is often gauged by the colour of the oxide film—see the table on the next page.

Figure 26.5
An electric arc furnace

Colour of Oxide Film	*Temperature (°C)*	*Use of the Steel*
Pale straw	220	Razor blades, hammers
Medium straw	235	Penknives, scribers
Dark straw	245	Dies, scissors
Brown	255	Punches, tappers
Light purple	270	Knives, drills
Dark purple	290	Screwdrivers
Blue	300	Springs, saws

Figure 26.6
Hot and cold processing of steel

26.5 The Manufacture of Copper and Zinc

The extraction of **copper** from its sulphide ores is a complex process especially if the ore contains Iron sulphide; it is very important, particularly for electrical purposes, to remove all iron and sulphur from copper. After reduction the copper is refined by electrolysis. The impure copper is the anode of the cell, the cathode is a thin sheet of pure copper and the electrolyte is an acidified (with dilute sulphuric acid) solution of copper sulphate. The net result of the process is the transfer of copper from the anode to the cathode; impurities such as gold and silver in the copper fall to the bottom of the cell (the anode sludge) whilst impurities such as iron pass into and remain in solution. See section 20.12.

Anode: $Cu \rightarrow Cu^{2+} + 2e^-$
Cathode: $Cu^{2+} + 2e^- \rightarrow Cu$

For general engineering purposes the copper is pure enough (99·99%) but for electrical purposes a further stage may be necessary. The world mine production of copper in 1973 was 7·51 million tonnes of which the U.S.A. produced 20%, U.S.S.R. 15%, Canada 11% and Chile 10%. The United Kingdom consumed 699 000 tonnes of which 22% was recovered scrap; imports were mainly from Zambia and Canada.

As in the case of copper ores the first stage in the extraction of **zinc** is the removal of as much earthy material as possible: this is usually done by oil flotation—the crushed ore is mixed with a little pine oil, put into water, and air blown through the suspension. The zinc sulphide floats to the surface and is skimmed off.

On a world wide basis electrolysis is the most important method of reducing zinc ores to the metal. The zinc sulphide is heated in air to

Chambesi pit.

Casting wire bars.

Electrolytic refining.

Cathode refining furnace.

Figure 26.7
Copper

convert it into the oxide, the sulphur dioxide evolved being processed to give sulphuric acid (see section 31.20).

$$2ZnS + 3O_2 \rightarrow 2ZnO + 2SO_2$$

The zinc oxide is dissolved in dilute sulphuric acid.

$$ZnO + H_2SO_4 \rightarrow ZnSO_4 + H_2O$$

The zinc sulphate solution, containing a little more acid and some gum, is then electrolyzed between lead anodes and aluminium cathodes at 35°C.

Anode: $4OH^- \rightarrow 2H_2O + O_2\uparrow + 4e^-$
Cathode: $Zn^{2+} + 2e^- \rightarrow Zn$

The zinc stripped off the cathodes is 99·95% pure.

The most important method in the United Kingdom – one which has been adopted on a very large scale elsewhere in the last decade— is the Imperial Smelting process which is capable of dealing with zinc and lead ores at the same time (the two often occur together). The ore is concentrated by oil flotation and sintered by the combustion of fuel oil so that the material remains porous during distillation. The zinc sulphide is converted into zinc oxide as above and the roasted concentrates (zinc oxide, lead(II) oxide) together with metallurgical coke are fed into the top of the blast furnace (see figure 26.8). Hot air is blown in near the base through tuyères and the coke burns to give, eventually, carbon monoxide.

$$2C + O_2 \rightarrow 2CO$$

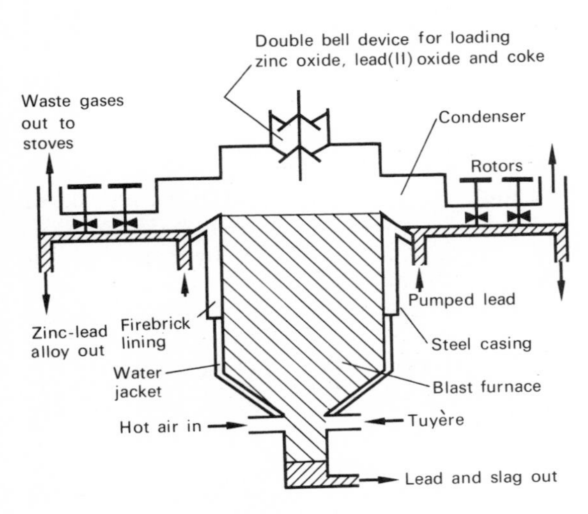

Figure 26.8
The Imperial Smelting zinc-lead blast furnace

The carbon monoxide reduces the metal oxides to the elements.

$$ZnO + CO \rightarrow Zn\uparrow + CO_2\uparrow$$
$$PbO + CO \rightarrow Pb + CO_2\uparrow$$

The molten lead and any slag formed run out of the base of the furnace and being immiscible are easily separated. The hot gases leaving the top of the furnace contain about 6% zinc, 11% carbon dioxide and 18% carbon monoxide, the balance being nitrogen. The zinc is condensed out by a shower of molten lead thrown up by the rotors: on cooling the zinc and lead, being immiscible, separate and the lead is returned to the condensers. The remaining gases after washing are burnt to heat the air and coke used by the blast furnace. The zinc separated and cast is 98·5% pure; fractional distillation can increase the purity to 99·99%.

The world mine production of zinc in 1973 was 5·84 million tonnes, of which Canada produced 23%, the U.S.S.R. 11%, Australia 8%, the U.S.A. 8% and Peru 7%. The United Kingdom consumed 391 000 tonnes of which about 20% was recovered scrap.

26.6 Physical Properties

See table overleaf.

Iron loses its magnetic properties at a temperature of 770°C (its Curie temperature). At 880°C the iron crystal becomes face centred cubic but at 1388°C it reverts to body centred cubic: these allotropic (enantiotropic, see section 8.4) transitions are harder to investigate in the laboratory than those of the non-metals. Pure iron is soft, malleable and ductile, the addition of carbon affecting mainly the first named property and heat treatment of the alloy the second and third.

Copper is malleable (sheets can be made 4 μm thick) and ductile (1 g can be extruded to give 200 m of wire). Only silver is a better conductor of heat and electricity than copper, but it is too expensive for general use.

Zinc is malleable in the range 100–150°C but it is otherwise slightly brittle; it is fairly ductile.

26.7 Chemical Properties

(a) With Oxygen

Iron, especially in powder or wire form, burns when heated in air, sparks being seen.

$$3Fe + 2O_2 \rightarrow \underset{\text{Triiron tetraoxide}}{Fe_3O_4}$$

Copper on heating forms copper(II) oxide

	Iron	Copper	Zinc
Atomic number	26	29	30
Electronic structure	2, 8, 14, 2	2, 8, 18, 1	2, 8, 18, 2
Relative atomic mass	55·85	63·55	65·38
Density (kg/m^3)	7900	8960	7100
Melting point (°C)	1535	1080	419
Boiling point (°C)	2800	2580	907
Crystal structure	Body centred cubic	Face centred cubic	Distorted hexagonal close packed
Standard electrode potential (E^{2+}, volts)	−0·44	+0·34	−0·76
Colour	White (if pure)	Red-brown	Blue-white

(CuO) on the surface only—this is not a good method of carrying out the conversion.

Zinc burns, especially if finely divided, with a blue-green flame to give white clouds of zinc oxide

$$2Zn + O_2 \rightarrow 2ZnO$$

(b) With Water

Hot iron reacts with steam to give hydrogen, this reaction being utilized by industry.

$$3Fe + 4H_2O \rightleftharpoons Fe_3O_4 + 4H_2\uparrow$$

The reaction is reversible (see section 19.2).

Copper is not attacked, even when hot, by water or steam.

Hot zinc reacts reversibly with steam.

$$Zn + H_2O \rightleftharpoons ZnO + H_2\uparrow$$

(c) With Moist Air

Iron in moist air—the humidity must be sufficiently high to cause a film of liquid on the surface—reacts to give iron(II) hydroxide which is converted rapidly to iron(III) hydroxide (rust) which forms as a red-brown, porous, flaky deposit on the surface. Oxygen from the atmosphere dissolves in the water on the surface of the iron. The presence of electrolytes accelerates the process of rusting: iron under compression or in the presence of metals lower in the electrochemical series behaves as an anode, electrons being released.

$$Fe \rightarrow Fe^{2+} + 2e^-$$

Iron under tension or with a metal higher than it in the electrochemical series behaves as the cathode: electrons are taken up by the water containing oxygen.

$$2H_2O + O_2 + 4e^- \rightarrow 4OH^-$$

Then in quick succession:

$$Fe^{2+} + 2OH^- \rightarrow Fe(OH)_2\downarrow$$
$$4Fe(OH)_2 + 2H_2O + O_2 \rightarrow 4Fe(OH)_3\downarrow$$

Rust is sometimes considered to be hydrated iron(III) oxide (Fe_2O_3, $3H_2O$). In 1971 it was estimated that this corrosion of iron objects costs £1365 million annually in the United Kingdom alone by way of protective measures and replacement.

Copper in moist air, over a long period of time, becomes coated with a green deposit (a patina) which is very noticeable where copper is used for roofing: it frequently consists of the basic sulphate $CuSO_4,3Cu(OH)_2$.

Zinc in moist air becomes coated with a thin tenacious film of a very basic oxide so that a further reaction occurs giving zinc carbonate; this too is very tenacious and is so thin that the object is referred to as being tarnished rather than corroded.

(d) With Acids

Iron and impure zinc dissolve readily in dilute hydrochloric or sulphuric acid.

$$Fe + 2HCl \rightarrow FeCl_2 + H_2\uparrow$$
$$Zn + H_2SO_4 \rightarrow ZnSO_4 + H_2\uparrow$$

or, e.g.

$$Fe + 2H^+ \rightarrow Fe^{2+} + H_2\uparrow$$

Pure zinc reacts slowly and the addition of a few drops of copper(II) sulphate solution is advantageous: it does not act as a catalyst, for it cannot be recovered (see section 23.4(a)).

Copper, which is below hydrogen in the electrochemical series, does not dissolve in either acid unless air is bubbled through the mixture: the reaction is still slow.

$$2Cu + 2H_2SO_4 + O_2 \rightarrow 2CuSO_4 + 2H_2O$$

This reaction is used commercially to make copper(II) sulphate.

Copper, zinc and iron all react with dilute nitric acid: the equations should be regarded as approximations indicating the principal products.

$$4Fe + 10HNO_3 \rightarrow 4Fe(NO_3)_2 + NH_4NO_3 + 3H_2O$$

The iron(II) nitrate readily oxidizes to iron(III) nitrate.

Zinc may react like iron with very dilute nitric acid.

$$4Zn + 10HNO_3 \rightarrow 4Zn(NO_3)_2 + NH_4NO_3 + 3H_2O$$

or

$$4Zn + 10HNO_3 \rightarrow 4Zn(NO_3)_2 + 5H_2O + \underset{\text{Dinitrogen oxide}}{N_2O}\uparrow$$

Zinc and copper react with dilute nitric acid in the same manner:

$$3Zn + 8HNO_3 \rightarrow 3Zn(NO_3)_2 + 4H_2O + \underset{\text{Nitrogen oxide}}{2NO}\uparrow$$

Concentrated hydrochloric acid usually reacts in the same manner, but more vigorously than the dilute acid if there is a reaction. Hot concentrated sulphuric acid is a more vigorous oxidizing agent than dilute sulphuric acid and reacts with all three metals:

$$2Fe + 6H_2SO_4 \rightarrow Fe_2(SO_4)_3 + 6H_2O + 3SO_2\uparrow$$

$$Cu + 2H_2SO_4 \rightarrow CuSO_4 + 2H_2O + SO_2\uparrow$$

$$(Zn) \qquad (ZnSO_4)$$

Concentrated nitric acid renders sheets of iron passive, i.e. they rapidly become covered with a thin layer of triiron tetraoxide (Fe_3O_4). This phenomenon of passivity renders the iron unreactive, e.g. on addition to copper sulphate solution no displacement occurs (see section 15.6(d)). The passivity can be destroyed by scratching the iron. On the other hand concentrated nitric acid reacts vigorously with copper and zinc, brown clouds of nitrogen dioxide being evolved:

$$Cu + 4HNO_3 \rightarrow Cu(NO_3)_2 + 2H_2O + \underset{\text{Nitrogen dioxide}}{2NO_2}\uparrow$$

$$(Zn) \qquad (Zn(NO_3)_2)$$

(*e*) *With Alkalis*

Only zinc, being amphoteric, reacts with alkalis: hot concentrated sodium hydroxide solution causes the evolution of hydrogen.

$$Zn + 2NaOH + 2H_2O \rightarrow Na_2Zn(OH)_4 + H_2\uparrow$$

(*f*) *With Non-metals*

The two most important reactions are that iron on heating with sulphur yields iron(II) sulphide (FeS) and with chlorine iron(III) chloride ($FeCl_3$) (see section 26.13).

There are similar reactions with copper (giving copper(II) sulphide and copper(II) chloride) and with zinc (giving zinc sulphide and zinc chloride). The reactions with sulphur can be used to illustrate the relative reactivity of the metals.

26.8 Uses of Iron, Copper and Zinc

The large scale use of **iron** started in about 1709 when Darby discovered that it was better to smelt iron ores with coke rather than with coal (the supply of charcoal from wood was running out): the production of pig iron in the U.K. has since increased ten thousand fold. The production of iron is much greater than that of all the other metals put together and it would be hard to imagine a world without it. The main uses of cast iron and steel have already been mentioned. Large quantities of sheet steel are coated with tin or zinc for the prevention of rusting before being used. The minor uses of iron include nuts, bolts and screws; it is also utilized in tyres and in ball-point pens and as a catalyst, e.g. in the Haber process for the production of ammonia (see section 30.9).

The major proportion (50%) of metallic copper is for applications in electrical engineering, principally in the form of wire and cable for generator, motor and transformer windings and conductors of all kinds. Printed circuits, though very important, hardly consume much copper. Tube for plumbing and heating services account for a possible further 15%. Some 35% is used in the manufacture of copper alloys (see table overleaf) for applications in all branches of industry.

One of the main uses of **zinc** (about 26% in the United Kingdom but 35% in the world) is in galvanizing, i.e. putting a protective coating on steel: this can be done directly by dipping the steel in molten zinc, rolling off the excess, or by electrolysis, the steel being the cathode. The steel is then used for constructional purposes such as pylons or tanks. The zinc corrodes in preference to the iron if the article is scratched (see section 15.6). Although galvanized steel can be used for water tanks it cannot unfortunately be used for food containers because juices in food are often acidic enough to dissolve the zinc so tin has to be used instead (at about twelve times the price), section 15.6(e). About 20% of the zinc made in the United Kingdom is converted into 'Mazak', an alloy (zinc 96%, aluminium 4%, traces of magnesium or copper) used for zips, radiators and many other objects made by die casting; about 28% is used in

Name	% copper approx.	Principal metal added	Typical uses
Brasses	60–70	Zinc	Propellors, cartridges, lamp caps, water fittings
Bronzes	80–90	Tin	Chemical plant, bearings, springs, architectural work
Cupro-nickels	60–80	Nickel	Condensers for power stations and ships
Nickel Silvers	65–70	Nickel	Tableware, contact springs
'Silver' 'Bronze'	75 95·5	25% nickel 3% tin + 1·5% zinc	U.K. coinage
—	98	Beryllium	Springs
—	99	Cadmium	Overhead lines for railway electrification

making brass; small proportions of zinc are sometimes included in bronze. The remaining 24% of zinc is used in many ways: for roofing, as the cathodes in dry cells, in lithographic plates for printing, as a reducing agent and in silver and gold manufacture.

Figure 26.9
Uses of zinc

Lifting out a panel that has been dipped in molten zinc.

Pylon of galvanized iron (cable is of steel cored aluminium)

II THE COMPOUNDS

26.9 Compounds of Monovalent Copper

(a) Copper(I) Oxide

This is prepared in the laboratory by warming Fehling's solution with a reducing sugar such as glucose or dextrose. **N.B.** If Fehling's solution is supplied in two parts then equal volumes of those solutions are taken for the test; solution A contains copper(II) sulphate (0·3 M) and solution B contains potassium sodium tartrate-4-water (Rochelle salt, 1·2 M) and sodium hydroxide (2·5 M). The aldehyde group (—CHO) of the sugar is converted into a carboxylic group (—COOH) whilst the copper is reduced from the divalent to the monovalent state: a reddish precipitate of copper(I) oxide is soon formed. It is a basic precipitate which is used in rectifiers and in manufacturing red glass.

(b) Copper(I) Chloride

This is prepared by boiling copper turnings with copper(II) chloride in the presence of concentrated hydrochloric acid (a mixture of copper and copper(II) oxide with the acid may also be used). The dark solution, on decanting into cold distilled water, yields a white precipitate of copper(I) chloride.

$$CuCl_2 + Cu \rightarrow 2CuCl\downarrow$$

When dissolved in aqueous ammonia the substance is a good absorbent for carbon monoxide (see section 23.3).

(c) Copper(I) Iodide

When potassium iodide is added to copper(II) sulphate, copper(I) iodide is precipitated and iodine is released—this will be precipitated also unless there is an excess of potassium iodide.

$$2CuSO_4 + 4KI \rightarrow \underset{\text{White}}{2CuI\downarrow} + \underset{\text{Brown or black}}{I_2} + 2K_2SO_4$$

or $$2Cu^{2+} + 4I^- \rightarrow 2CuI\downarrow + I_2$$

26.10 The Oxides of the Divalent Metals

Iron(II) compounds are often referred to as ferrous salts and copper(II) compounds as cupric (or just copper) salts. No designation is necessary in the case of zinc because it only forms one series of salts.

Iron(II) oxide is a black powder prepared by heating iron(III) oxide to 300°C in a stream of hydrogen.

$$Fe_2O_3 + H_2 \rightarrow 2FeO + H_2O\uparrow$$

It is oxidized by air spontaneously to reform the higher oxide. It will dissolve in dilute acids to yield the appropriate salts.

The best ways of preparing **copper(II) oxide** are by heating the hydroxide, carbonate or nitrate.

$$Cu(OH)_2 \rightarrow CuO + H_2O\uparrow$$
$$CuCO_3 \rightarrow CuO + CO_2\uparrow$$
$$2Cu(NO_3)_2 \rightarrow 2CuO + 4NO_2\uparrow + O_2\uparrow$$

It is a black hygroscopic powder which with dilute acids yields the appropriate salts. It is a useful oxidizing agent when hot:

$$CuO + H_2 \rightarrow Cu + H_2O\uparrow$$
$$3CuO + 2NH_3 \rightarrow 3Cu + 3H_2O\uparrow + N_2\uparrow$$
$$2CuO + C \rightarrow 2Cu + CO_2\uparrow$$

The oxidation of carbon and hydrogen still occurs even if they are constituents of an organic compound.

Zinc oxide is prepared in the laboratory by methods similar to those employed for producing copper(II) oxide. It can also be made by burning the metal or its sulphide in air.

$$2Zn + O_2 \rightarrow 2ZnO$$
$$2ZnS + 3O_2 \rightarrow 2ZnO + 2SO_2\uparrow$$

It is a white powder which becomes yellow on heating; the change is reversed on cooling. It absorbs water vapour and carbon dioxide from the atmosphere. It is amphoteric, dissolving in acids to give the appropriate zinc salts and in alkalis to yield zincates:

$$ZnO + 2HCl \rightarrow ZnCl_2 + H_2O$$
$$ZnO + 2NaOH + H_2O \rightarrow Na_2Zn(OH)_4$$

Zinc oxide is reduced by carbon under industrial conditions to the metal. Paints containing zinc oxide do not have as good a covering power as lead paint (see section 28.19(b)) but they have the advantage that they do not tarnish in industrial atmospheres. It is also used in 'zinc' ointments, in rubber processing and in spinning baths for rayon.

26.11 The Hydroxides of the Divalent Metals

These are gelatinous precipitates obtained on adding sodium hydroxide solution or aqueous ammonia to a solution of the appropriate metal salt:

e.g. $$Fe^{2+} + 2OH^- \rightarrow Fe(OH)_2\downarrow$$

Iron(II) hydroxide is white when highly pure but it rapidly becomes (if it is not already so) a dirty green colour. It is a basic precipitate that on standing slowly oxidizes to iron(III) hydroxide.

Copper(II) hydroxide is a pale blue basic substance but it dissolves in excess aqueous ammonia to give a deep blue solution.

$$Cu(OH)_2 + 4NH_3 \rightarrow \underset{\text{Tetraamminecopper(II) hydroxide}}{Cu(NH_3)_4(OH)_2}$$

This deep blue solution will dissolve cellulose (blotting paper, filter paper) giving a viscous solution that on squirting into dilute acid re-precipitates the cellulose as a fibre (cuprammonium rayon), see sections 14.4 and 29.18.

Zinc hydroxide is a white amphoteric substance which will dissolve in acids and alkalis:

$$Zn(OH)_2 + 2HCl \rightarrow ZnCl_2 + 2H_2O$$
$$Zn(OH)_2 + 2NaOH \rightarrow Na_2Zn(OH)_4$$
$$Zn(OH)_2 + 4NH_3 \rightarrow Zn(NH_3)_4(OH)_2$$

26.12 The Salts of the Divalent Metals

(a) *Iron(II) Salts*

Iron(II) sulphate (green vitriol) is prepared by dissolving iron in dilute sulphuric acid and crystallizing out the salt ($FeSO_4,7H_2O$) without causing too much oxidation to occur.

$$Fe + H_2SO_4 \rightarrow FeSO_4 + H_2\uparrow$$

On heating the crystals first lose their water of crystallization becoming pale yellow in colour, then extensive decomposition occurs giving a red or black residue of iron(III) oxide.

$$2FeSO_4 \rightarrow Fe_2O_3 + SO_3\uparrow + SO_2\uparrow$$

It is less susceptible to oxidation as a solid than when in solution: the crystals are gently shaken with distilled water to rinse them, then with a further portion of water to obtain a solution when the latter is required. Iron(II) sulphate solution absorbs nitrogen oxide to form an addition compound having the formula $FeSO_4,NO$: this is the compound responsible for the brown ring in the common test for nitrates. The salt is used as a weed destroyer and in the manufacture of ink.

Iron(II) chloride crystals ($FeCl_2,4H_2O$) can be obtained by dissolving the metal in hydrochloric acid; the anhydrous substance can be obtained by passing hydrogen chloride over the heated metal.

$$Fe + 2HCl \rightarrow FeCl_2 + H_2\uparrow$$

Although iron(II) carbonate occurs in nature it is not of much interest in the laboratory.

(b) *Copper(II) Salts*

Copper(II) sulphate (blue vitriol, $CuSO_4,5H_2O$) is most easily obtained by dissolving the oxide, hydroxide or carbonate in warm dilute sulphuric acid and evaporating the solution until crystals are obtained. On heating these crystals lose their water of crystallization and become white: the reversal of this colour change is the usual test for free water. On strong heating the crystals will decompose further:

$$CuSO_4 \rightarrow CuO + SO_3\uparrow$$

Copper sulphate is used on a large scale in vineyards as a fungicide, e.g. in Bordeaux mixture, calcium hydroxide, sulphur and copper sulphate are used.

Copper(II) chloride crystals ($CuCl_2,2H_2O$) can be made by dissolving the oxide in dilute hydrochloric acid; the anhydrous substance is made by passing chlorine over the heated metal.

$$Cu + Cl_2 \rightarrow CuCl_2$$

Copper(II) carbonate is easily made by precipitation:

$$Cu^{2+} + CO_3^{2-} \rightarrow \underset{\text{Green}}{CuCO_3\downarrow}$$

It is unstable to heat but can be used as a green pigment; it dissolves readily in acids. The precipitate is a salt of variable composition.

(c) *Zinc Salts*

Zinc sulphate (white vitriol) can be made by dissolving the metal in dilute sulphuric acid. On evaporation crystals ($ZnSO_4,7H_2O$) are obtained. Zinc sulphate solution when added to barium sulphide solution yields a mixture known as lithopone.

$$ZnSO_4 + BaS \rightarrow ZnS\downarrow + BaSO_4\downarrow$$

The precipitate is a white pigment used in the manufacture of paints and as a filler in rubber and linoleum.

Zinc chloride crystals ($ZnCl_2, H_2O$) are prepared by the standard methods of preparing salts; the anhydrous substance is obtained by heating the metal in a stream of chlorine or hydrogen chloride.

Zinc carbonate is a basic salt made by precipitation: it is a white powder used in calamine ointments. It is unstable to heat and dissolves readily in acids but more slowly in alkalis.

26.13 Compounds of Trivalent Iron

(a) *Iron(III) Oxide*

This occurs naturally but is prepared in the laboratory as a red powder by heating iron(II) sulphate or iron(III) hydroxide. It is insoluble in water and alkalis but dissolves slowly in acids yielding the appropriate salts. It is used as a polishing powder (jeweller's rouge) and as a pigment in cosmetics.

The compound, triiron tetraoxide (Fe_3O_4), made by heating iron in air or steam, behaves as though it were a compound of iron(II) oxide and iron(III) oxide. It is magnetic and has a high melting point; it is used in the electrodes for arc-lights.

(b) Iron(III) Hydroxide

This is formed as a red-brown gelatinous precipitate when an alkaline solution is added to a solution of an iron(III) salt.

$$Fe^{3+} + 3OH^- \rightarrow Fe(OH)_3\downarrow$$

The precipitate will dissolve in solutions of strong acids yielding the appropriate salts.

(c) Iron(III) Chloride

If iron(III) hydroxide is dissolved in hydrochloric acid and the solution evaporated yellow crystals ($FeCl_3,6H_2O$) are obtained. These crystals react slightly with water in the cold: hydrolysis, as this is called, makes the solution acidic. On heating the reaction occurs to a greater extent:

$$FeCl_3 + 3H_2O \rightleftharpoons Fe(OH)_3 + 3HCl$$

The iron(III) hydroxide is present as a colloid: the particles are so fine that they cannot be seen directly (a light beam at right angles to the viewer shows a misty solution) nor immediately (but they may appear as the solution 'ages', see section 7.1). A solution of iron(III) chloride keeps best if a little hydrochloric acid is initially present: this solution is used for etching copper to make printed circuits.

Anhydrous iron(III) chloride cannot be made

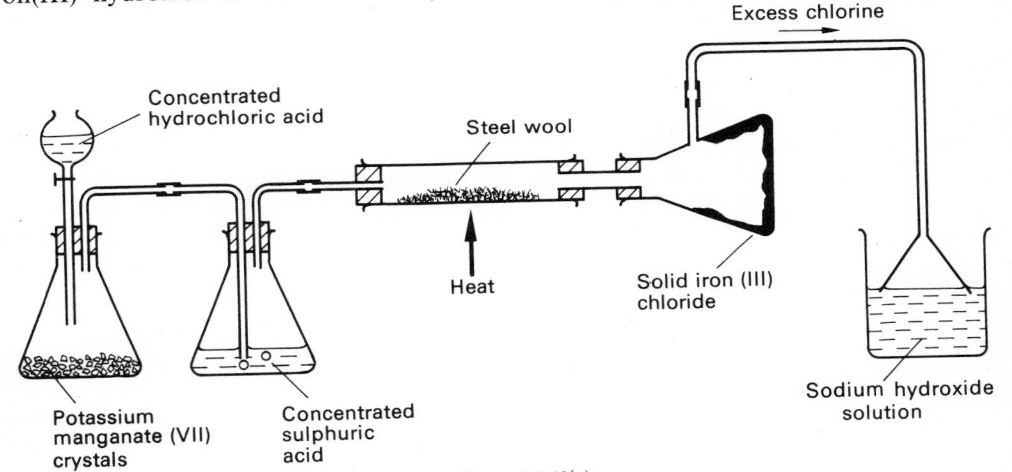

Figure 26.10(a)
The preparation of iron(III) chloride

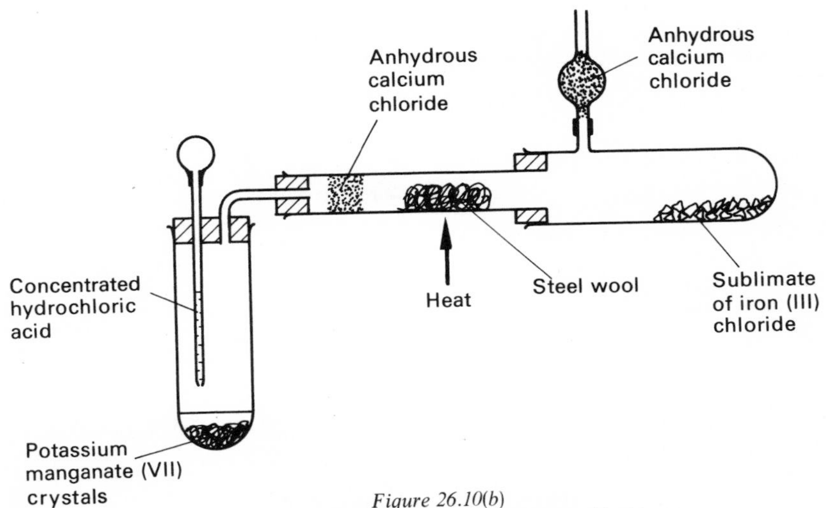

Figure 26.10(b)
The small scale preparation of iron(III) chloride

by dehydration of the hydrated salt because of decomposition: it is prepared by synthesis instead (see figures 26.10(a) and (b)). The anhydrous substance is a useful catalyst in organic chemistry.

$$2Fe + 3Cl_2 \rightarrow 2FeCl_3$$

(d) Iron(III) Sulphate

Iron(III) sulphate is prepared by heating concentrated sulphuric acid with iron(II) sulphate crystals for a long time.

$$2FeSO_4{,}7H_2O + 2H_2SO_4 \rightarrow Fe_2(SO_4)_3 + 16H_2O + SO_2\uparrow$$

26.14 Iron(II) and Iron(III) Salts

Iron shows the variable valency, typical of transition elements, much more readily than copper; zinc is not a transition element in this respect.

Iron(II) salts slowly oxidize in the air to form iron(III) salts but a rapid conversion is achieved by warming the solution with a few drops of concentrated nitric acid in the presence of dilute sulphuric acid.

$$6FeSO_4 + 3H_2SO_4 + 2HNO_3 \rightarrow 3Fe_2(SO_4)_3 + 4H_2O + 2NO\uparrow$$

or

$$3Fe^{2+} + 4H^+ + NO_3^- \rightarrow 3Fe^{3+} + 2H_2O + NO\uparrow$$

Oxidation can also be achieved by chlorine and hydrogen peroxide.

Iron(III) salts can be reduced to iron(II) salts by passing hydrogen sulphide into the solution.

$$2FeCl_3 + H_2S \rightarrow S\downarrow + 2FeCl_2 + 2HCl$$

or

$$2Fe^{3+} + H_2S \rightarrow S\downarrow + 2Fe^{2+} + 2H^+$$

The best tests for iron(II) and iron(III) salts at this stage are by using sodium or potassium hydroxides or aqueous ammonia to obtain the coloured precipitate of the appropriate hydroxide.

$$Fe^{2+} + 2OH^- \rightarrow Fe(OH)_2\downarrow$$

White or dirty green

$$Fe^{3+} + 3OH^- \rightarrow Fe(OH)_3\downarrow$$

Red-brown

An iron(II) salt is very likely to be contaminated by an iron(III) salt because of atmospheric oxidation but the reverse is not true.

QUESTIONS 26

The Transition Elements: Iron, Copper and Zinc

1 In producing pig iron, a mixture of iron ore, coke and calcium carbonate is fed into a blast furnace. Answer the following questions about this process:
(*a*) Give the common name and the chemical name of one iron ore used.
(*b*) How is the furnace heated?
(*c*) Why is coke added, and what happens to it?
(*d*) How is the iron ore converted to iron?
(*e*) Why is limestone added, and what happens to it?
(*f*) Give equations for three important reactions taking place in the furnace.
(*g*) Name two other elements usually present in pig iron.
(*h*) Why is this type of method used for producing pig iron but not for producing aluminium? [A]

2 Name an ore of zinc and describe how the metal is extracted from it. What happens when (*a*) zinc oxide is heated, (*b*) sodium hydroxide solution is slowly added to aqueous zinc sulphate until it is definitely in excess, (*c*) zinc is placed in copper sulphate solution? [OC]

3 Name the substances put in a blast furnace for obtaining pig iron. State the main reactions which occur in the furnace and give equations for them. Describe two experiments to show the relationship between iron and copper in the electrochemical series. [OC]

4 Give three reasons (two chemical and one physical) why both magnesium and copper are classified as metals. How would you prepare pure specimens of (*a*) copper(II) oxide from metallic copper, (*b*) iron(II) sulphate crystals from iron? Describe how to obtain a solution of iron(III) sulphate starting from iron(II) sulphate crystals. [C]

5 Describe the extraction of iron from haematite (Fe_2O_3) by means of the blast furnace. Name three impurities that are removed from cast iron in the manufacture of steel and give two methods by which the quality of steel can be varied. Describe how to find the relative atomic mass of iron using copper(II) sulphate, assuming that you know the relative atomic mass of copper and the equation of the reaction. Name the type of reaction involved in your experiment. [C]

6 Give two conditions necessary for iron to rust. Give also two ways of preventing the formation of rust. [S]

7 Give the name of one important ore of iron, and state briefly how the metal may be obtained from it. Describe the reaction of iron with (*a*) chlorine, (*b*) copper(II) sulphate solution. Calculate the mass of sulphur which combines with 3·5 g of iron, and the

volume of hydrogen sulphide (measured at s.t.p.) which would be evolved if the whole of the product were treated with excess of hydrochloric acid. [S]

8 A, B and C are three metals. When the oxide of A was mixed with the powdered metal B and heated, a vigorous reaction took place, and from the resulting mixture some metallic A and the oxide of B were isolated. Samples of each element were placed in test-tubes containing dilute hydrochloric acid. There was vigorous effervescence from B but no reaction from the other two metals. When this experiment was repeated using dilute nitric acid, all three metals caused effervescence. The smell of nitrogen dioxide was apparent in each case. When C was added to a solution containing ions of A, a reaction took place in which A was deposited. Answer the following:
(*a*) What is the descending order of reactivity of the three metals?
(*b*) What gas would be given off when B reacted with dilute hydrochloric acid?
(*c*) Does the smell of nitrogen dioxide from each test-tube necessarily mean that this gas is a reaction product in each case? Explain your answer.
(*d*) Which two metal oxides would be most easily reduced? If reduction were carried out by passing water gas over the heated oxide, what gaseous products would be formed? [Sc]

9 Explain how zinc is manufactured from its naturally occurring sulphide, zinc blende. State two large scale uses of metallic zinc. Giving all essential practical details, describe how you would obtain from zinc foil (*a*) crystals of zinc sulphate, (*b*) a solution of sodium zincate. Quote two facts which justify the statement that zinc is chemically more active than copper. [W]

10 Name a common ore of zinc and describe how the metal is obtained from it on the industrial scale. Quote two chemical properties of zinc which show that it is a metallic element. State and explain all that would be observed in the following experiments:
(*a*) Sodium hydroxide solution is added, in turn, to separate solutions of zinc sulphate and copper(II) sulphate until an excess of the alkali is present.
(*b*) Zinc foil is left for some time in a solution of lead nitrate.
(*c*) Granulated zinc is added to a solution of iron(III) chloride which has been mixed with an equal volume of dilute hydrochloric acid. [W]

11 Describe how, starting with iron filings, you would in turn prepare in the laboratory reasonably pure specimens of: (*a*) a solution of iron(II) sulphate, (*b*) crystalline iron(III) chloride, (*c*) iron(III) hydroxide, (*d*) iron(III) oxide.

12 What raw materials are used in the production of iron? State briefly the role of each, giving equations where appropriate. A certain iron ore is found to contain 60% of iron(III) oxide (Fe_2O_3) and no other iron-containing substance. Calculate how many tonnes of ore are required to produce one tonne of iron, assuming the process to be 100% efficient.

13 Given a dilute solution of copper(II) chloride, how could you obtain from it (*a*) water, (*b*) dilute hydrochloric acid, (*c*) dry copper(II) oxide? Anhydrous copper(II) sulphate turns blue when water is added to it. State, with a reason, whether you consider this to be a physical or a chemical change. [J]

14 Draw a fully labelled diagram of the apparatus you would use to prepare anhydrous iron(III) chloride if you were given some manganese(IV) oxide and iron wire, and allowed to use any of the usual laboratory reagents and apparatus. Describe the appearance of the product so formed and say what would happen to a portion of it if it were (*a*) heated in the bottom of a test tube, (*b*) placed on a watch glass and left exposed to the air. What chemical test would you apply to distinguish between iron(III) chloride and iron(II) chloride solutions? The result of the test should be given for each solution. [J]

15 50 cm^3 of a solution of iron(III) chloride *A* containing 0·1 mol/dm^3 Fe^{3+} were heated until nearly boiling and then a solution of tin(II) chloride *B* (containing 1 mol/dm^3 Sn^{2+}) was added slowly; when 2·5 cm^3 of the latter solution had been added the yellow colour of the iron(III) solution had just been removed. At this stage it may be assumed that all the iron(III) chloride has been converted to iron(II) and all the tin(II) has been converted to an ion with a higher formal charge.
(*a*) How many moles of iron(III) were present in the 50 cm^3 of solution *A*?
(*b*) How many moles of tin(II) were present in the 2·5 cm^3 of solution *B*?
(*c*) How many ions of Fe^{3+} have reacted with one ion of Sn^{2+}?
(*d*) What is the formula of the tin ion which has been formed?
(*e*) Write an equation for the reaction.
(*f*) What alteration in structure takes place when Fe^{3+} changes to Fe^{2+}?
(*g*) The reaction does not seem to go at room temperature. Why does heating to near boiling cause it to take place?
(*h*) If some of solution *B* is electrolyzed with a current of 1 A for 160 minutes (i.e. 2·67 hours), how many moles of tin would you expect to be deposited at the cathode?
(*i*) How do you explain the fact that when solution *A* is electrolyzed it loses its colour but there is no immediate deposition of iron on the cathode? [Nf]

16 Describe fully how iron is obtained from its ores. Explain how you would: (*a*) convert a solution of an iron(III) salt into one of an iron(II) salt, and then prove by a chemical test that it was an iron(II) salt; (*b*) convert a solution of an iron(II) salt into one of an iron(III) salt, and then prove by a chemical test that it was an iron(III) salt. [J]

17 Starting with metallic iron in each case, how would you prepare reasonably pure specimens of: (*a*) iron(II) sulphate; (*b*) iron(III) oxide? Describe these two substances and their properties.

18 Give a fully labelled diagram to show how you would prepare and collect a sample of anhydrous iron(III) chloride. State what changes would have to be made if you wished to use the same apparatus to prepare anhydrous iron(II) chloride. Give one test in each case to distinguish between solutions of (*a*) iron(III) sulphate and iron(II) sulphate, (*b*) iron(III) sulphate and iron(III) chloride. The result of the test must be stated for both of the solutions to which it is applied. [J]

19 Starting from metallic copper, outline three methods by which copper(II) oxide may be prepared. (Heating the metal in air is not suitable for this experiment, and will not be accepted.) State briefly how you would obtain metallic copper from the copper(II) oxide. Mention two commercial uses of copper. [J]

20 Outline a method for the preparation of a sample of crystalline copper(II) sulphate, starting from metallic copper. How could you recover copper from copper(II) sulphate by a chemical method? [J]

21 Describe briefly how you would prepare: (*a*) copper(II) sulphide from copper(II) sulphate; (*b*) a solution of copper(II) nitrate from copper(II) chloride; (*c*) copper(II) oxide from copper(II) nitrate; (*d*) metallic copper from copper(II) sulphate. What happens when a solution of ammonia is slowly added to a solution of copper(II) sulphate until there is no further change? [J]

22 Starting with copper turnings, describe how you would prepare a small sample of: (*a*) copper(II) oxide; (*b*) copper(II) chloride; (*c*) copper(II) hydroxide. Calculate the theoretical mass of copper(II) sulphate crystals ($CuSO_4,5H_2O$) which could be prepared from 6·4 g of copper.

23 (*a*) Describe how you would obtain, by electrolysis, pure copper from a plate of copper containing iron as an impurity. What would happen to the iron?
(*b*) Describe three tests to show that zinc is a more electropositive metal than copper. [O]

24 Given some metallic copper, how would you prepare: (*a*) copper(II) nitrate crystals; (*b*) copper(II) oxide from the nitrate; (*c*) copper(II) sulphate crystals from the oxide. Give equations for the reactions you describe. If you used 10 g of copper in (*a*) and the whole of this was eventually turned into copper(II) sulphate crystals ($CuSO_4,5H_2O$), what mass of crystals, to the nearest gram, would you obtain? 2·88 g of copper(I) oxide when reduced leaves 2·56 g of copper. Calculate the number of atoms of copper to one of oxygen in copper(I) oxide. [L]

25 What is the chief ore of zinc? Outline the manufacture of zinc from this ore. How would you prepare zinc hydroxide from the metal? How does the hydroxide behave with: (*a*) hydrochloric acid; (*b*) sodium hydroxide solution? [OC]

26 Describe briefly how zinc is obtained from a zinc ore. What is the action of zinc on: (*a*) dilute sulphuric acid; (*b*) a solution of copper(II) sulphate solution; (*c*) an acidified solution of iron(III) chloride; (*d*) sodium hydroxide solution?

27 Triiron tetraoxide or magnetic oxide of iron (Fe_3O_4) is a 'mixed' (or compound) oxide, i.e. it behaves as though it is a mixture in fixed proportions of iron(II) oxide (FeO) and iron(III) oxide (Fe_2O_3). Describe how you would prepare from this oxide: (*a*) a solution containing iron(II) chloride but no iron(III) chloride, (*b*) a solution containing iron(III) chloride but no iron(II) chloride. Describe chemical tests by means of which you would prove that the solutions obtained did contain the required salts. [L]

27 Group III: Aluminium

27.1 Introduction

Often aluminium is considered after sodium and magnesium and this is useful because it emphasizes the decline in metallic character of the elements (also exhibited in compounds) as one goes from left to right across a series in the periodic table. Aluminium is fairly high in the electrochemical series when this is determined by electrical experiments but when chemical experiments are performed in order to establish the relative reactivities of the elements the coherent oxide film on the surface leads to a distinct lack of reactivity (compare calcium).

Aluminium is trivalent: the electrovalent character of its compounds is not so strong as in Groups I and II, e.g. the chloride behaves as a covalent compound particularly on heating, but the atoms do not achieve noble gas electronic structures when there is simple covalency.

The solubility of compounds is usually lower than in previous groups (the solubility decreases from sodium to magnesium to aluminium hydroxides). The salts are less stable to heat than those in previous groups (the stability decreases from sodium to magnesium to aluminium nitrates). **N.B.** The carbonate does not exist.

27.2 Occurrence

Aluminium is the most abundant metal in the earth's crust (8%) and the third most abundant element. Commercially the most common form is bauxite, the hydrated oxide ($Al_2O_3,2H_2O$) which is formed by the weathering of many rocks. It is mined chiefly in the U.S.A., West Africa and France (at Les Baux where it was originally recognized). Its composition is approximately: Al_2O_3 55%, H_2O 25%, Fe_2O_3 15%, SiO_2 3%, TiO_2 2%. The sources are often far from the reduction plants.

Cryolite, sodium hexafluoroaluminate(III) (Na_3AlF_6) is found in Greenland; much is made synthetically.

The widespread forms of aluminium are kaolin (china clay), feldspar (sodium aluminosilicate, a common rock), and corundum, sapphire and ruby (all forms of aluminium oxide). Aluminium is not extracted from these compounds.

Davy (1807) first recognized aluminium in the form of its oxide but the metal was not cheap enough to be really useful until the middle of the twentieth century.

27.3 The Manufacture of Aluminium

This involves two main stages:

(*a*) The separation of pure aluminium oxide from the ore (Bayer Process).

(*b*) The electrolytic reduction of aluminium oxide to metallic aluminium (Hall-Héroult method).

(*a*) *Separation of the Pure Oxide by the Bayer Process*

The bauxite is first crushed, washed, and screened to remove loose earth. It is then subjected to the Bayer Process to make pure aluminium oxide.

(i) It is digested by grinding it to powder and dissolving it in a hot solution of sodium hydroxide under pressure. The aluminium oxide is dissolved, being amphoteric; the remaining oxides are filtered off and discarded.

$$Al_2O_3 + 6NaOH + 3H_2O \rightarrow \underset{\text{Soluble sodium aluminate}}{2Na_3Al(OH)_6}$$

or $$Al_2O_3 + 6OH^- + 3H_2O \rightarrow 2Al(OH)_6^{3-}$$

(ii) The solution is pumped into tall tanks and seeded with aluminium hydroxide crystals as it cools down. As the solution is stirred, crystals are precipitated.

$$Na_3Al(OH)_6 \rightarrow Al(OH)_3\downarrow + 3NaOH$$

or $$Al(OH)_6^{3-} \rightarrow Al(OH)_3\downarrow + 3OH^-$$

Though this reaction is a reversible one, operating conditions are such that it proceeds in the direction shown. The sodium hydroxide is returned to the cycle and used again.

(iii) The crystals of aluminium hydroxide are filtered off, washed and calcined at about 1100°C to drive off water.

$$2Al(OH)_3 \rightarrow Al_2O_3 + 3H_2O\uparrow$$

The aluminium oxide so produced is a white powder which will not reabsorb atmospheric moisture. It is known as alumina.

(*b*) *Electrolytic Reduction of the Pure Oxide (Alumina) in the Hall-Héroult Cell* (*Figure 27.1*)

Attempts were made to obtain the metal by reducing the oxide with coke, but this cannot be done, as aluminium oxide is a very stable compound. This fact delayed the large scale commercial preparation of the metal until 1886, when Hall in America and Héroult in France, working independently, discovered that alumina (pure Al_2O_3) would dissolve in molten cryolite (Na_3AlF_6) at a working temperature of about 1000°C, and whilst in the solution it could be decomposed by electrolysis. The consumption of electricity is high.

The reduction cell is lined with a layer of carbon, which acts as the cathode of the cell. The cell operates at 970°C. Pure aluminium is continuously deposited on the bottom of the cell, and is siphoned out periodically.

In modern plants only one carbon (graphite) block is used, and aluminium, of purity greater than 99%, is obtained.

During the electrolysis, complicated reactions take place. The net effect is

$$\underset{\text{Cathode}}{2Al_2O_3 \rightarrow 4Al} + \underset{\text{Anode}}{3O_2\uparrow}$$

Thus the cryolite remains unchanged, and fresh supplies of alumina are added as required. The oxygen given off attacks the carbon anode and burns it away to carbon monoxide and dioxide.

The following figures show the actual amount of basic materials required to make 1 tonne of aluminium:

Bauxite	4 tonnes	give 2 tonnes alumina
Sodium hydroxide	0·15 tonne	
Cryolite	Very small consumption—whilst essential to the reduction process, it is not consumed in the electrochemical reaction	
Anode	0·6 tonne	
Electricity	18 000 kilowatt-hours	

The cell is about 5 m by 3 m and 0·5 m deep; it consumes up to 125 000 amperes at 5·5 volts.

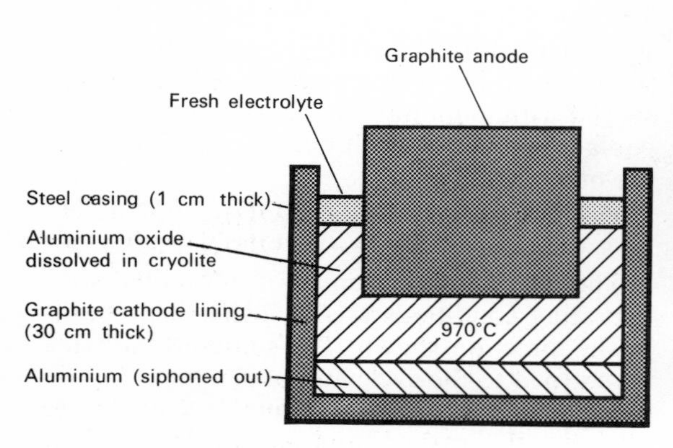

Figure 27.1

The manufacture of aluminium—the Hall-Héroult Cell

Extracting molten aluminium by suction from the cell.

Preparing to recharge the electrolytic reduction cell.

Figure 27.2
The production of aluminium at Fort William

The world production of aluminium in 1973 was 12·7 m tonnes of which the U.S.A. contributed 32%, the U.S.S.R. 15%, Japan 8% and Canada 7%. The United Kingdom consumption was 668 000 tonnes of which 26% was recovered scrap and 38% was primary production, the balance being imported.

27.4 Physical Properties

Aluminium has an atomic number of 13 and electronic structure of 2,8,3; its relative atomic mass is 26·98. The crystal is face centred cubic and has a density of 2700 kg/m^3; it melts at 660°C and boils at 2330°C. The metal is a blue-white colour and it is malleable and ductile enough to be rolled into thin leaf or drawn into wire. It is a good conductor of heat and electricity. Its standard electrode potential is $-1{\cdot}66$ volts.

27.5 Chemical Properties

(a) With Air

Aluminium remains unaltered in dry air; in moist air a superficial film of oxide protects the metal from further action. The statue of Eros in Piccadilly Circus, London, has resisted atmospheric corrosion since 1893; the supporting leg is solid aluminium, the rest of the body being made up of hollow aluminium castings.

The oxide film can be loosened by rubbing with mercury or mercury(II) chloride which forms an amalgam. An oxide film may be put on deliberately by electrolysis: this film can be made to incorporate dyes, e.g. for colourful kettles, milk bottle tops, etc.

Aluminium, except in finely divided form, does not burn in air at any temperature. At very high temperature the finely divided metal undergoes an extremely slow reaction, forming the oxide (Al_2O_3) and the nitride (AlN).

(b) With Acids

The metal dissolves slowly in cold dilute hydrochloric or sulphuric acid, but quickly in concentrated hydrochloric acid or warm dilute sulphuric acid.

$$2Al + 6HCl \rightarrow 2AlCl_3 + 3H_2\uparrow$$

or

$$2Al + 6H^+ \rightarrow 2Al^{3+} + 3H_2\uparrow$$

Sulphuric acid, if hot and concentrated, attacks it, forming aluminium sulphate and sulphur dioxide.

$$2Al + 6H_2SO_4 \rightarrow Al_2(SO_4)_3 + 6H_2O + 3SO_2\uparrow$$

Nitric acid of all concentrations has little or no effect on the metal due to the formation of a protective layer of aluminium oxide: the metal is rendered 'passive'. A 'furred' kettle may be cleaned with nitric or ethanoic acid.

(c) With Alkalis

The metal is rapidly attacked by caustic alkalis, hydrogen being evolved and the corresponding aluminate left in solution.

$$6NaOH + 2Al + 6H_2O \rightarrow 2Na_3Al(OH)_6 + 3H_2\uparrow$$

or

$$6OH^- + 2Al + 6H_2O \rightarrow 2Al(OH)_6^{3-} + 3H_2\uparrow$$

Thus caustic abrasives must not be used for cleaning aluminium pans.

(d) With Iron(III) Oxide

Because of its great affinity for oxygen, aluminium can reduce, at high temperatures, certain metallic oxides (namely iron(III) oxide, Fe_2O_3, chromium(III) oxide, Cr_2O_3, manganese(IV) oxide, MnO_2), e.g.

$$Fe_2O_3 + 2Al \rightarrow Al_2O_3 + 2Fe \quad \Delta H = -850\ kJ$$

i.e. a tremendous amount of heat is given out during this reaction. The reactions are examples of the Goldschmidt process for preparing metals. The mixture of aluminium and iron(III) oxide is known as 'Thermit' and it can be used for welding cracks in iron. The mixture and a magnesium primer are placed in the crack, and the reaction initiated by lighting the magnesium. Then, during the reduction of the iron oxide by the aluminium, the heat is so great that it melts the iron, which fills the crack and forms a solid mass with the rest of the iron on cooling.

27.6 Uses of Aluminium

It is not surprising that a metal which
(*a*) has a density about a third that of steel;
(*b*) is a good conductor of heat and electricity;
(*c*) is resistant to corrosion—especially to sea-water;
(*d*) is non-magnetic;
(*e*) is non-toxic;
(*f*) is capable of being converted into high-strength light alloys with other metals such as magnesium, manganese, and copper should be used to a large and varied extent. In terms of tonnage it is second to, but a long way behind, iron. The following are some typical uses.

(*a*) High strength alloys for air, marine, and road transport; structural engineering, e.g. H30 (Al 97·25%, Mg 1%, Si 1% and Mn 0·75%).

Canberra.

Warehouses.

Concorde.

Delivery vans.

Curtain walling and glazing.

Queen Elizabeth 2.

Trident.

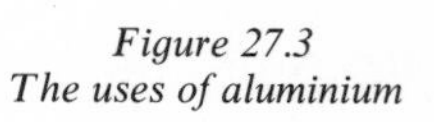
Figure 27.3
The uses of aluminium

(*b*) In the manufacture of electric kettles, cooking utensils, food and chemical plants, and foil for packaging.
(*c*) An electrical conductor for cables (with a steel core for strength) and busbars.
(*d*) As a paint to protect iron and its alloys from rusting.
(*e*) For silvering large telescope mirrors, because of its lustre and resistance to corrosion.
(*f*) In steel manufacture as a reducing agent (desulphurizer).

27.7 Aluminium Compounds

(*a*) *Aluminium Oxide* (Al_2O_3)
The impure oxide, bauxite, is mined on a large scale (73 m tonnes in 1973); much of it is merely purified and less than half is used to produce aluminium. In the laboratory the pure oxide is made by heating the hydroxide or the sulphate. It is a white powder, infusible (bauxite bricks for furnaces) and amphoteric (unless strongly heated); it is insoluble in water.

(*b*) *Aluminium Hydroxide* ($Al(OH)_3$)
This white precipitate is formed when ammonia or an alkali is added to a solution of an aluminium salt. If sodium hydroxide is used, the precipitate which is first formed redissolves in excess sodium hydroxide; it is amphoteric.

$$AlCl_3+3NaOH \rightarrow Al(OH)_3\downarrow+3NaCl$$

or $$Al^{3+}+3OH^- \rightarrow Al(OH)_3\downarrow$$

$$Al(OH)_3+3NaOH \rightarrow Na_3Al(OH)_6$$

or $$Al(OH)_3+3OH^- \rightarrow Al(OH)_6^{3-}$$

Sodium carbonate is a weak base and it has been shown why it is alkaline in solution (see section 24.12): on adding sodium carbonate solution to a solution of an aluminium salt the hydroxide ions combine with the aluminium ions as shown above and no aluminium carbonate is produced.

(*c*) *Aluminium Chloride* ($AlCl_3$)

If prepared by dissolving aluminium in hydrochloric acid and evaporating the solution to the point of crystallization, the hexahydrate ($AlCl_3$, $6H_2O$) is deposited. The anhydrous crystals cannot be obtained by this process because on heating hydrolysis takes place.

$$2AlCl_3,6H_2O \rightarrow Al_2O_3+6HCl\uparrow+9H_2O\uparrow$$

Anhydrous aluminium chloride is therefore prepared by passing a current of dry chlorine over a heated mixture of alumina and charcoal.

$$Al_2O_3+3C+3Cl_2 \rightarrow 2AlCl_3+3CO\uparrow$$

Or, aluminium may be heated in a current of chlorine or hydrogen chloride.

$$2Al+3Cl_2 \rightarrow 2AlCl_3$$

$$2Al+6HCl \rightarrow 2AlCl_3+3H_2\uparrow$$

The chloride is volatile and is collected in a manner similar to that used in the collection of iron(III) chloride (see figure 26.10). Anhydrous aluminium chloride fumes strongly in moist air because of hydrolysis.

$$AlCl_3+3H_2O \rightarrow Al(OH)_3+3HCl\uparrow$$

Aluminium chloride is used as a catalyst in organic chemistry and in the manufacture of synthetic lubricating oils.

(*d*) *Aluminium Sulphate* ($Al_2(SO_4)_3$)
This is made by dissolving the metal or the oxide in hot concentrated sulphuric acid. Crystals of aluminium sulphate-18-water ($Al_2(SO_4)_3$, $18H_2O$) separate out.

The salt is very soluble in water and has an acid reaction due to hydrolysis. Hence it is sometimes used instead of sulphuric acid in fire extinguishers. It is acid by hydrolysis because the reaction

$$Al^{3+}+3OH^- \rightarrow \underset{\text{colloidal}}{Al(OH)_3}$$

leaves hydrogen ions in solution together with the sulphate ions.

On heating, it undergoes thermal decomposition, giving sulphur trioxide.

$$Al_2(SO_4)_3 \rightarrow Al_2O_3+3SO_3\uparrow$$

Aluminium sulphate is used in the purification of sewage, as a mordant in dyeing, in waterproofing cloth, and in sizing paper.

(*e*) *Aluminium Potassium Sulphate-12-Water* ($K^+Al^{3+}(SO_4^{2-})_2,12H_2O$)

It is otherwise known as alum and frequently given the double formula $K_2SO_4,Al_2(SO_4)_3$, $24H_2O$ because it can be obtained by evaporating a solution containing the separate salts in equi-molar proportions. There is a series of compounds of this general formula: potassium can be replaced by sodium or ammonium, and aluminium by chromium or iron(III). These alums form well defined octahedral crystals (see figure 8.1). Alum itself is used as a mordant in dyeing and in the leather industry, and in styptic pencils to stop bleeding.

QUESTIONS 27

Group III: Aluminium

1 Name an ore of aluminium and describe the extraction of the metal from it. Describe very briefly how you would prepare aluminium hydroxide from the metal and say how it reacts with (*a*) dilute hydrochloric acid, (*b*) sodium hydroxide solution. Give two important uses of aluminium. [OC]

2 Describe how aluminium is obtained industrially from pure aluminium oxide. Give three physical properties of aluminium which render the metal specially useful, illustrating each property by one example. Give the reactions of aluminium with (*a*) oxygen, (*b*) chlorine, (*c*) iron(III) oxide, stating the necessary conditions and giving equations. [A]

3 Aluminium forms the compounds $AlCl_3$ and Al_2S_3 with chlorine and sulphur. These compounds react with water as indicated by the following equations

$$AlCl_3 + 3H_2O \rightarrow Al(OH)_3 + 3HCl\uparrow$$
$$Al_2S_3 + 6H_2O \rightarrow 2Al(OH)_3 + 3H_2S\uparrow$$

(*a*) Write formulae for the compounds you would expect aluminium to form with nitrogen and with carbon.
(*b*) Write equations for the reactions you would expect when the compounds in (*a*) are treated with water.
(*c*) Suggest simple qualitative experiments by which you would attempt to confirm your expectations in (*b*). [C]

4 Describe the physical and chemical properties of aluminium and explain how these determine the uses made of the metal.

5 From what substance is aluminium obtained on a commercial scale? Describe the process by which the metal is isolated, and point out the ways in which this process differs from the electrolysis of a solution such process differs from the electrolysis of a solution of sodium chloride. What advantages have aluminium alloys over the pure metal for industrial purposes? [O]

28 Group IV: Carbon, Silicon, Tin and Lead

28.1 Introduction

This is the first group usually considered in moderate detail in the periodic table in which both non-metals and metals are clearly present: the transition from one to the other as the group is descended is of great interest and importance. That is not to imply that the properties of carbon are completely non-metallic (its crystalline forms are lustrous; one allotrope conducts electricity) nor that tin and lead are entirely metallic (they possess amphoteric oxides and have volatile chlorides).

Tin and lead are just above hydrogen in the electrochemical series; carbon and silicon do not feature in it. All four elements are fairly unreactive compared with sodium and chlorine. In many compounds the elements show a covalency of four: hydrides, oxides, sulphides and halides. Tetravalent lead compounds are often oxidizing agents. There are also many compounds in which the elements show a covalency or electrovalency of two; the stability of these rises from carbon to lead—carbon monoxide (covalent) and tin(II) chloride (electrovalent) are well known reducing agents. In several compounds the elements appear as complex ions, e.g. CO_3^{2-} carbonates, $Pb(OH)_3^-$ plumbate(II)'s.

The stability of many compounds to heat is quite low. Many of the compounds of these elements are insoluble in water; the only compounds of lead that are fairly soluble are the nitrate and the ethanoate.

The ability of carbon atoms to link together in chains makes carbon the key element in many large molecules. The importance of carbon is out of all proportion to its abundance. All living things are made up of carbon compounds, and it is, apparently, the unique properties of carbon which make life possible. There are more carbon compounds known than those of all the other elements put together: the number is now over four million, and hundreds more are added every year as chemists synthesize new dyes, drugs, plastics and fabrics. So many and so important are carbon compounds that they are usually studied as a separate branch of chemistry, called organic chemistry (see chapter 29). Here the element itself and a few of its simpler compounds are studied, as is frequently done, under the heading of inorganic chemistry which is the chemistry of all the other elements and their compounds.

28.2 Occurrence

Some **carbon** occurs in an elemental form as diamond and graphite, the latter looking like lead sulphide—hence the term pencil 'lead' even though no lead is present. Compounds of carbon are more common: coal (a complex mixture of compounds of carbon, hydrogen and oxygen), natural gas (methane, CH_4), petroleum (a complex mixture of compounds mainly of carbon and hydrogen with a little sulphur, etc.), many carbonates (sodium, magnesium, calcium, zinc, etc.) and carbon dioxide (0·03% in air).

Silicon is the second most abundant element in the earth's crust (27%). The free element is difficult to obtain and is unimportant chemically but its compounds include such common substances as sand and glass.

Tin has only one naturally occurring compound: cassiterite or tinstone, SnO_2.

The chief source of **lead** is the sulphide, galena, PbS; but the carbonate (cerussite) and the sulphate (anglesite) occur to smaller extents.

28.3 Manufacture

Half the **diamonds** used are those found naturally in the 'blue earth' of volcanic origin; they were apparently formed when carbon was subjected to high temperature and pressure. To separate them, the earth is mixed with water and then poured down a greasy sloping surface to which the diamonds sink and adhere. Most of the **graphite** used is manufactured. In the Acheson process coke is ground, binders added and the mixture baked at 1200°C; the carbon is finally turned into graphite by passing an electric current through to raise the temperature to 2500°C.

Coke (98% carbon) is made by the destructive distillation of coal, in the absence of air, at temperatures ranging from 600 to 1200°C depending upon whether an organic substance or coke is the main object (see section 29.4).

Wood charcoal is made by the destructive distillation (pyrolysis) of hard wood, that is by heating it to 350°C out of contact with air so that it cannot burn (figure 28.1). A brown tar containing phenol, and a watery layer containing ethanoic acid, methanol and propanone collects in the second boiling tube. The wood gas obtained in the last tube is inflammable. The charcoal is left in the first tube in the form of a black porous solid.

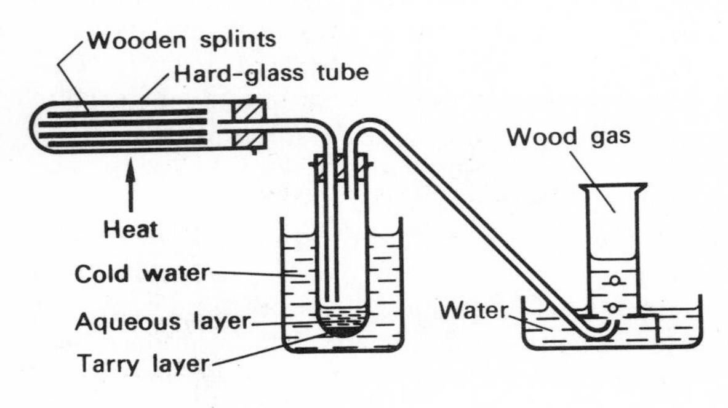

Figure 28.1
The preparation of wood charcoal

Before coal was common, the charcoal burner was responsible for the destruction of much of our woodland. He made piles of wood and let them burn with insufficient air for complete combustion. Then, charcoal was a valuable fuel; now, it owes its importance to its remarkable power of adsorbing gases.

Animal charcoal (boneblack) is made by heating bones out of contact with air: it consists of 10% carbon thinly spread out on calcium phosphate.

Carbon black (lampblack) is formed when petroleum, refinery tail gases or natural gas (methane) is burnt in a limited supply of air so that only the hydrogen is oxidized.

Hydrocarbon + Oxygen → Carbon + Steam

It is a fairly pure form of carbon unlike soot, which is formed by the incomplete combustion of coal.

Sugar charcoal can be made by the pyrolysis of sugar or by adding sugar to concentrated sulphuric acid; in both cases the carbon must be thoroughly washed with water and dried. It is a very pure form of carbon.

$$C_{12}H_{22}O_{11} \rightarrow 12C + 11H_2O$$

Pure **silicon** is very difficult to make; fortunately in its main use, transistors, a little goes a long way. It is only made on a small scale and that is beyond the scope of this book.

The manufacture of **tin** starts with washing the ore, tin(IV) oxide, to remove earthy matter. It is then roasted in a current of air to remove arsenic and sulphur impurities as their volatile oxides. The concentrated material is then mixed with anthracite and calcium carbonate and heated in a horizontal reverberatory furnace or a vertical blast furnace to 1300°C.

$$SnO_2 + 2C \rightarrow Sn + 2CO\uparrow$$

In a reverberatory furnace the hot gases are reflected or reverberated from the low roof on to the ores to be heated. The tin is purified by liquation, that is it is heated on a sloping surface and the molten metal runs down the slope, the dross (impurities such as copper and iron oxides) being left behind. It may be stirred while molten with poles of green birch wood: this prevents oxidation of the hot metal and at the same time other impurities form a scum on the surface and are removed.

The world mine production of tin in 1973 was 221 000 tonnes of which Malaysia contributed 33%, Bolivia 13%, China 10% and Indonesia 10%. The United Kingdom consumed 18 000 tonnes, of which 11% was recovered scrap.

The first step in the extraction of **lead** is the concentration of the ore by oil flotation: the crushed ore is added to water and a little pine oil. On blowing in air the lead sulphide rises to the surface and can be skimmed off while the earthy material sinks. On sintering the purified sulphide with silicon dioxide, the sulphur burns off

$$2PbS + 2SiO_2 + 3O_2 \rightarrow 2PbSiO_3 + 2SO_2\uparrow$$

The lead silicate is then heated with coke and calcium oxide in a blast furnace, together with some iron to remove any lead sulphide remaining.

$$PbSiO_3 + CaO + C \rightarrow CaSiO_3 + Pb + CO\uparrow$$

The lead is then desilvered by stirring the molten metal at 800°C with zinc; the silver is 300 times more soluble in zinc than in lead and distillation in a vacuum is used to separate the silver from the zinc (Parkes' process). For an alternative process, see section 26.5.

The world mine production of lead in 1973 was 3·56 m tonnes of which the U.S.A. contributed 16%, the U.S.S.R. 14%, Australia 11%, Canada 11% and Peru 6%. The United Kingdom consumption was 364 000 tonnes, of which 22% was recovered scrap.

28.4 Physical Properties

	Carbon	*Silicon*	*Tin*	*Lead*
Atomic number	6	14	50	82
Electronic structure	2,4	2,8,4	2,8,18,18,4	2,8,18,32,18,4
Relative atomic mass	12·01	28·09	118·7	207·2
Density (kg/m^3)	2250	2400	7300	11 300
Melting point (°C)	sublimes	1410	232	327
Boiling point (°C)	at 3700	2500	2270	1750
Crystal structure	hexagonal	tetrahedral	tetragonal	face centred cubic
Standard electrode potential (volts)	—	—	−0·14	−0·13
Colour	black	brown	silver-white	silver-white

All the elements are lustrous but lead soon tarnishes when exposed to the air. Carbon filaments are used in lamps but, like tin and lead wire, they normally have a low tensile strength. A recent development is the production of carbon fibres or whiskers (from an organic compound where long chains of carbon atoms already exist, see section 29.10) in which the near perfection of crystalline structure confers tremendous strength: a steel rod four times as thick is not as strong! Tin and lead are very soft and malleable.

In the table, the properties of carbon are those of graphite, the stable allotrope; those of tin are of white tin, the allotrope which is stable from 13 to 232°C.

28.5 The Allotropy of Carbon

The element carbon can exist in many forms, but there are only two in which the arrangement of carbon atoms is fundamentally different: these are diamond and graphite. The crystals of diamond and graphite look different (figure 28.2(a)). X-ray crystallography reveals that in diamond the atoms form a three-dimensional network (figure 28.2(b)) while in graphite they are arranged in definite layers (figure 28.2(c)). The other forms of carbon (e.g. charcoal, lampblack, coke) are often said to be amorphous (without shape) but X-ray analysis has shown that they consist of very small fragments of graphite crystals, i.e. they are micro-graphitic. The physical properties of diamond and graphite are given in the table overleaf.

The allotropy of carbon is monotropy, i.e. there is no transition temperature between the allotropes below the melting point at atmospheric pressure. Graphite is more stable than diamond at room temperature; compare rhombic and monoclinic sulphur (see section 31.6), but there is little likelihood of a transformation from diamond to graphite taking place as the

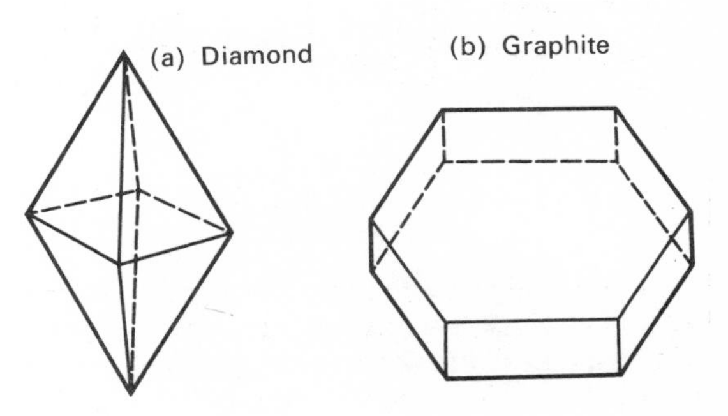

Figure 28.2(a)
Diamond and graphite crystals (as seen)

Property	*Diamond*	*Graphite*
Hardness	Hardest naturally occurring substance	Soft; shears easily; feels greasy; marks paper
Towards light	Transparent; refractive index 2·4. Colourless; lustrous	Opaque Grey-black; metallic lustre
Density (kg/m^3)	3520	2250
Thermal and electrical conductivity	Low	High

Figure 28.2(b)
The structure of diamond

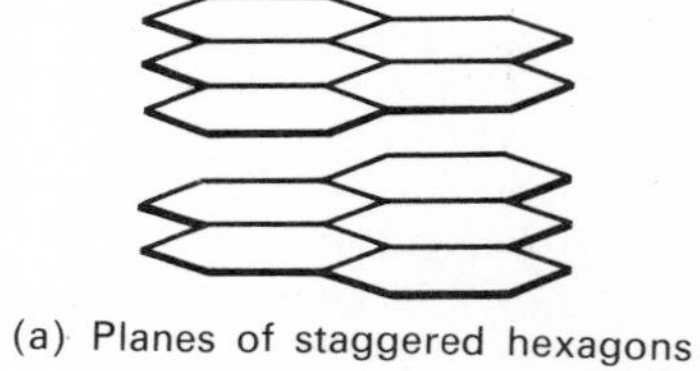

(a) Planes of staggered hexagons viewed at 45°

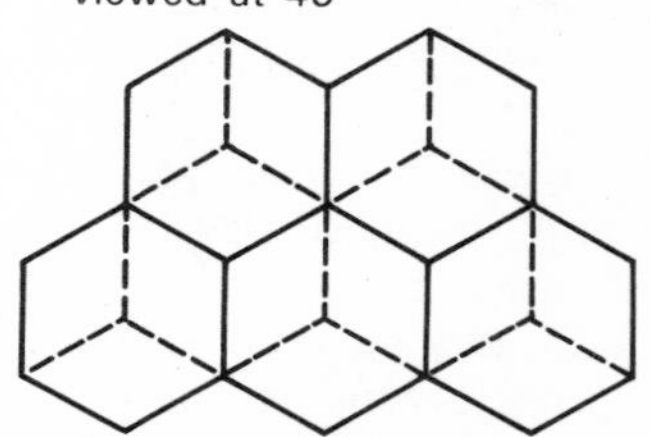

(b) Vertical view of crystal

(c) Horizontal view of crystal

Figure 28.2(c)
The structure of graphite

activation energy is very high (see section 18.5). This stability of graphite seems contradictory to its softness and combustibility.

Experiments with the Charcoals

(*a*) A piece of wood charcoal, which usually floats on water, is weighted with some copper wire so that it is immersed in some water in a beaker. The water is boiled for several minutes then the wire removed: the charcoal now does not float as the air previously contained in its pores has been expelled.

(*b*) A piece of wood charcoal is heated strongly and then buried in sand in a crucible while it cools. One drop of liquid bromine is put in a gas jar and shaken to speed the volatilization of the bromine. The wood charcoal is dropped in: the red gas is rapidly adsorbed, the colour vanishing. This property makes wood charcoal useful in gas masks.

N.B. Charcoal does not adsorb carbon monoxide and so breathing apparatus must be worn to combat some fires.

(*c*) Some water coloured with litmus and containing animal charcoal is boiled: on filtering, a colourless filtrate is obtained.

28.6 Chemical Properties

(*a*) *With Air*

Carbon burns on heating to give the monoxide (poisonous) or the dioxide, depending upon the availability of oxygen.

$$2C+O_2 \rightarrow 2CO$$
$$C+O_2 \rightarrow CO_2$$

In a coke fire (figure 28.3) the main air draught comes in at the bottom. Here the carbon burns to give carbon dioxide.

$$C+O_2 \rightarrow CO_2$$

As the carbon dioxide passes up through the red-hot carbon it is partly reduced to carbon monoxide.

$$CO_2+C \rightarrow 2CO$$

If there is a good supply of air, this carbon monoxide burns at the surface of the fire, with its characteristic blue flame. This flame is often seen on the top of the coke fire.

$$2CO+O_2 \rightarrow 2CO_2$$

It is wasteful, as well as possibly dangerous, not to burn the escaping carbon monoxide. The air supply must be good both at the bottom and at the top of the fire, and the temperature at the surface of the fire must be kept high enough to ignite the gas.

The charcoals take fire about 400°C, graphite at 700°C, but diamond not until 900°C. If the conditions are carefully controlled this gives a method of demonstrating that diamond, graphite and sugar charcoal all consist of pure carbon. A known mass of the substance is heated in a stream of dry oxygen, and the carbon dioxide formed is absorbed in weighed potash bulbs (figure 9.2). After cooling, the mass of carbon used and the mass of carbon dioxide formed can be found. It is found that a given mass of each of the substances gives the same mass of carbon dioxide.

Silicon burns on heating to form silicon dioxide.

$$Si+O_2 \rightarrow SiO_2$$

Tin is stable even in moist air at room temperature but burns on heating to give tin(IV) oxide; **lead** is stable in dry air but oxidizes on heating giving lead(II) oxide.

$$Sn+O_2 \rightarrow SnO_2$$
$$2Pb+O_2 \rightarrow 2PbO$$

The tarnishing of lead in moist air is caused

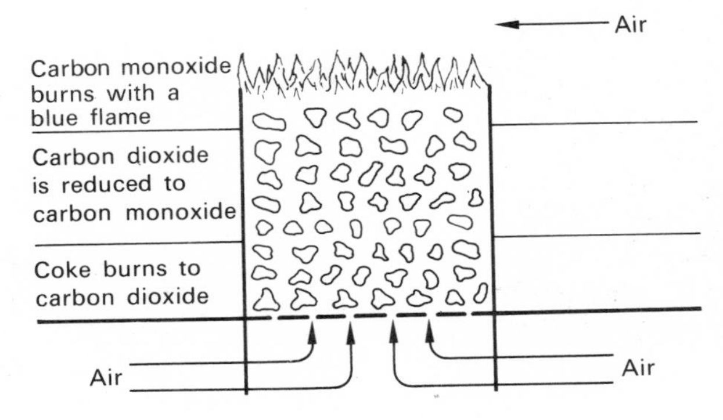

Figure 28.3
The coke fire

by the formation of a thin film of the oxide, hydroxide and carbonate which prevents the remaining metal from being attacked.

(b) With Water

Steam will react with white hot coke to give water gas (see section 6.4).

$$C + H_2O \rightarrow H_2 + CO$$

The other elements in Group IV do no react with pure water at room temperature, but lead dissolves slowly in soft water because its hydroxide is slightly soluble; in hard water a film of lead sulphate or carbonate forms and soon prevents the remainder of the metal dissolving. Because of this, drinking water passing through lead pipes is often artificially hardened to some extent to prevent lead poisoning.

(c) With Acids

Carbon is not attacked except by hot concentrated sulphuric and nitric acids which are vigorous oxidizing agents.

$$C + 2H_2SO_4 \rightarrow CO_2\uparrow + 2H_2O\uparrow + 2SO_2\uparrow$$
$$C + 4HNO_3 \rightarrow CO_2\uparrow + 2H_2O\uparrow + 4NO_2\uparrow$$

Tin dissolves slowly in dilute hydrochloric acid (but rapidly in the concentrated acid).

$$Sn + 2HCl \rightarrow SnCl_2 + H_2\uparrow$$

or $$Sn + 2H^+ \rightarrow Sn^{2+} + H_2\uparrow$$

The reaction of tin with dilute sulphuric acid is very slow but with the hot concentrated acid a more vigorous reaction occurs.

$$Sn + 4H_2SO_4 \rightarrow Sn(SO_4)_2 + 4H_2O + 2SO_2\uparrow$$

Dilute nitric acid slowly dissolves tin by a complex reaction with no gases being evolved; with the concentrated acid, however, a white precipitate forms which upon heating leaves a residue of anhydrous tin(IV) oxide.

$$Sn + 4HNO_3 \rightarrow SnO_2\downarrow + 2H_2O + 4NO_2\uparrow$$

Lead does not react with dilute hydrochloric or sulphuric acids except momentarily while a film of lead chloride or sulphate forms on the surface. Hot concentrated sulphuric acid attacks lead.

$$Pb + 2H_2SO_4 \rightarrow PbSO_4 + 2H_2O + SO_2\uparrow$$

Warm dilute nitric acid and cold concentrated nitric acid both attack lead.

$$\underset{\text{Dilute}}{3Pb + 8HNO_3} \rightarrow 3Pb(NO_3)_2 + 4H_2O + 2NO\uparrow$$

or

$$3Pb + 8H^+ + 2NO_3^- \rightarrow 3Pb^{2+} + 4H_2O + 2NO\uparrow$$

$$\underset{\text{Concentrated}}{Pb + 4HNO_3} \rightarrow Pb(NO_3)_2 + 2H_2O + 2NO_2\uparrow$$

or

$$Pb + 4H^+ + 2NO_3^- \rightarrow Pb^{2+} + 2H_2O + 2NO_2\uparrow$$

(d) With Alkalis

Hot concentrated sodium hydroxide solution attacks all the Group IV elements except carbon: hydrogen is released and the element is found as part of the anion (similar to aluminium—see section 27.5(c)).

(e) As Reducing Agents

Carbon is used in the laboratory, e.g. a carbon (charcoal) block, and in industry, e.g. in a blast furnace, to convert easily reducible metal oxides to the metals.

$$2CuO + C \rightarrow 2Cu + CO_2\uparrow$$
$$2PbO + C \rightarrow 2Pb + CO_2\uparrow$$

28.7 Uses

The uses of **carbon** are many and are best considered in terms of the separate forms. Less than 10% of the **diamonds** obtained are used as gemstones—only the clear colourless ones can be used in jewellery and they have to be skilfully cut using other diamonds for the purpose. Most diamonds are used for industrial purposes: as teeth in rock-cutting saws, dies (for drawing metal wires) and as bits in mining drills. About half the **graphite** manufactured is used for refractory crucibles and foundry moulds. Its structure makes it suitable as a lubricant for the bearings of small dynamos and motors, and it is also employed in suspension in oil or water. Its use as the 'lead' in pencils is also based on its structure: it is mixed with clay to harden it and baked at 1500°C. It is used in electrodes and as the central pole in dry batteries. It is able to slow neutrons, i.e. act as a moderator, in nuclear piles.

Wood charcoal is used to adsorb gases and volatile liquids: coconut charcoal is very efficient because it has a large surface area but even that is no use against carbon monoxide. Activated charcoal—charcoal which has been heated in steam—may have a surface area of up to 1000 m^2/g. Such charcoals are useful for removing the last traces of gas from vacuum tubes and benzene from coal gas. **Animal charcoal** is used as an adsorbent of dyes, e.g. in converting brown sugar to white. 90% of the **carbon black** manufactured is used in hardening rubber tyres; the remainder is used in making printer's ink, carbon paper, black shoe polish and hardening the plastic for records.

Silicon is used in transistors and in the manufacture of silicones.

Tin is used to coat mild steel plate after the iron has been derusted by pickling it in dilute

Lead bricks surrounding a radioactivity experiment.

Cables in Preston Power Station.

Lead crystal glassware.

Roofing on Daylesford House, Glos.

Chloride batteries for emergency lighting.

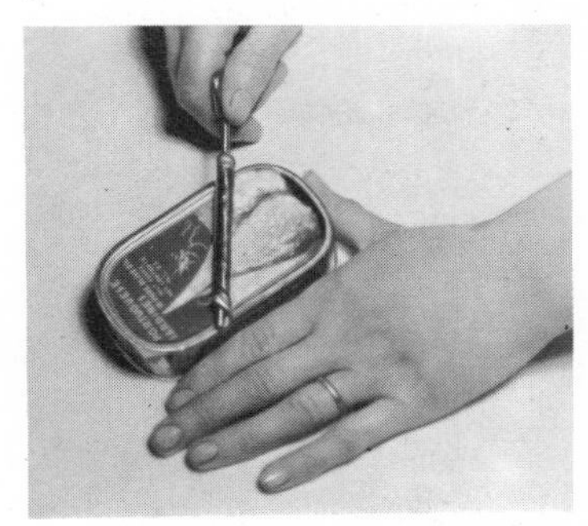

Lead solder seals a sardine tin.

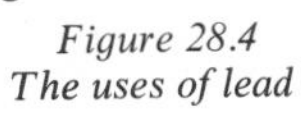

Figure 28.4
The uses of lead

sulphuric acid: the tin is usually put on by electrolytic methods. The tin coating (only 100 nm thick) prevents the iron rusting and is stable to the organic acids found in foods. The familiar tin can was first patented in the United Kingdom in 1810; nowadays it is calculated that we consume 2 million out of our 30 million tonnes of food every year out of 1000 million cans. Tinplate and tinning use 51% of the U.K. tin supply. 32% of the tin is used in alloys:

Soft solder	66% Sn + 34% Pb
Plumber's solder	30% Sn + 70% Pb
Bronze	15% Sn + 85% Cu
Type metal	8% Sn + 76% Pb + 16% Sb

A recent use for molten tin is as the undersurface for making float glass—another British invention. About 8% of the tin is used in manufacturing compounds and 1% as wrought tin.

Some 29% of the U.K. consumption of lead is based on its property of being a soft metal: cable covers (resistant to corrosion), water and gas pipes, damp-proof courses for buildings and roofing. 25% of the lead is converted into its compounds, chiefly the oxides. 15% is used in accumulators where it is alloyed with antimony (10% by mass) to harden it. 15% is converted into tetraethyllead for addition to petrol (see section 24.6). 9% is converted into other alloys, including solder mentioned above. Amongst the other uses are shielding blocks for radioactivity experiments.

II THE MONOXIDES

28.8 Preparation

Carbon monoxide is produced whenever carbon or a carbon compound is burnt in insufficient air for complete combustion, but this reaction and the process in which carbon dioxide is reduced by passing it over red-hot charcoal in a combustion tube are difficult to control in the laboratory.

$$2C + O_2 \rightarrow 2CO$$

$$\text{Hydrocarbon} + \text{Oxygen} \rightarrow \text{Carbon monoxide} + \text{Steam}$$

$$CO_2 + C \rightarrow 2CO$$

A better method of preparation is to drip concentrated sulphuric acid on to sodium methanoate crystals or methanoic acid. The concentrated sulphuric acid acts as a dehydrating agent.

$$HCOONa + H_2SO_4 \rightarrow NaHSO_4 + \underset{\text{Methanoic acid}}{HCOOH}$$

$$HCOOH \rightarrow \underset{\text{To sulphuric acid}}{H_2O} + CO\uparrow$$

The reaction is vigorous and it is advisable to

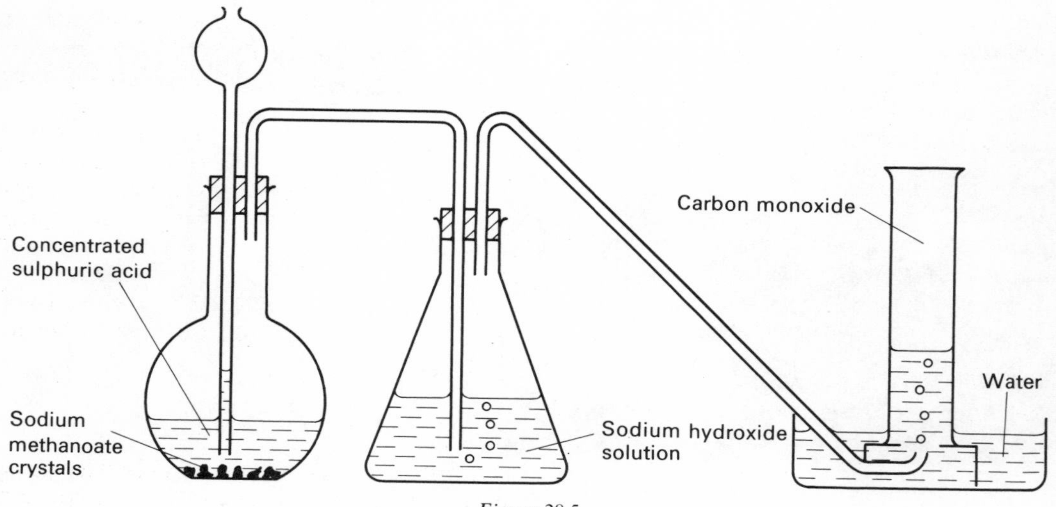

Figure 28.5
The preparation of carbon monoxide

wash the vapours evolved with sodium hydroxide solution to remove carbon dioxide and sulphur dioxide (see figure 28.5).

An alternative method is to put concentrated sulphuric acid on sodium ethanedioate or ethanedioic acid crystals. In this case heat and the removal of the carbon dioxide are essential.

$$Na_2C_2O_4 + 2H_2SO_4 \rightarrow 2NaHSO_4 + \underset{\text{Ethanedioic acid}}{H_2C_2O_4}$$

$$H_2C_2O_4 \rightarrow H_2O + CO\uparrow + CO_2\uparrow$$

The concentrated sulphuric acid is employed as a dehydrating agent, i.e. a substance which removes water from another substance. The colourless, odourless gas is usually collected over water as it is almost insoluble; it has approximately the same density as air.

Lead(II) oxide (litharge) is made by heating lead for a long time in air at a temperature of about 850°C.

$$2Pb + O_2 \rightarrow 2PbO$$

It can also be made by heating lead nitrate, carbonate or hydroxide, or any other oxide of lead above 600°C.

$$2Pb(NO_3)_2 \rightarrow 2PbO + 4NO_2\uparrow + O_2\uparrow$$
$$PbCO_3 \rightarrow PbO + CO_2\uparrow$$
$$Pb(OH)_2 \rightarrow PbO + H_2O\uparrow$$

The yellow colour of the litharge slowly changes on standing to a pale red.

28.9 Chemical Properties of the Monoxides

(a) With Air

Carbon monoxide burns upon ignition with a blue flame.

$$2CO + O_2 \rightarrow 2CO_2 \qquad \Delta H = -283\ \text{kJ}$$

It is less likely to explode with air than hydrogen is under laboratory conditions. Before combustion it does not affect calcium hydroxide solution but afterwards the gas, being now carbon dioxide, does turn calcium hydroxide solution milky.

(b) With Water

Under industrial conditions carbon monoxide reacts with steam (see section 23.3(a)) undergoing the water-gas shift reaction.

$$CO + H_2O \rightarrow CO_2 + H_2$$

(c) As Redox Agents

Carbon monoxide will reduce iron, copper and lead oxides at high temperatures:

$$Fe_3O_4 + 4CO \rightarrow 3Fe + 4CO_2\uparrow$$
$$CuO + CO \rightarrow Cu + CO_2\uparrow$$
$$PbO + CO \rightarrow Pb + CO_2\uparrow$$

These reactions are used in industry. Carbon monoxide can also be an oxidizing agent (see section 23.7):

$$CO + 2H_2 \rightarrow CH_3OH$$

Lead(II) oxide usually behaves as an oxidizing agent, as above with carbon monoxide and also on heating with various non-metals:

$$PbO + H_2 \rightarrow Pb + H_2O\uparrow$$
$$2PbO + C \rightarrow 2Pb + CO_2\uparrow$$

(d) With Acids and Bases

Carbon monoxide is neutral in aqueous solution but it will react with fused or a concentrated aqueous solution (150°C, at 7 atmospheres) of sodium hydroxide to yield sodium methanoate.

$$NaOH + CO \rightarrow HCOONa$$

Lead(II) oxide is amphoteric: it will react with acids, especially nitric, yielding lead salts,

$$PbO + 2HNO_3 \rightarrow Pb(NO_3)_2 + H_2O$$

or

$$PbO + 2H^+ \rightarrow Pb^{2+} + H_2O$$

and with hot sodium hydroxide solution yielding sodium plumbate(II).

$$PbO + NaOH + H_2O \rightarrow NaPb(OH)_3$$

(Known as 'Doctor' solution, this is used in refining mineral oil.)

N.B. Carbon monoxide is very poisonous and it is particularly dangerous because of its lack of smell. Oxygen is usually carried to all parts of the body from the lungs by the red colouring matter of the blood (haemoglobin). This forms a loose compound, oxyhaemoglobin, which gives up its oxygen to the tissues which need it. Carbon monoxide, however, reacts with haemoglobin to give a much more stable compound (carboxyhaemoglobin), which is cherry-pink. This prevents the haemoglobin from acting as an oxygen-carrier, and death eventually occurs from asphyxia.

Carbon monoxide is present in coal gas (up to 3% by volume), and is responsible for its poisonous properties. It is formed whenever carbon or carbon compounds are incompletely burnt, and so is present in the exhaust fumes of a car, and is the 'after-damp' produced in a mine explosion. Death occasionally occurs when a car engine has been run for some time in a closed garage.

28.10 Uses of the Monoxides

Carbon monoxide, in the form of water gas and producer gas, is used as a fuel and is concerned in the manufacture of hydrogen and methanol.

As a reducing agent it is generated in a blast furnace for the reduction of oxides of iron to the metal. Carbon monoxide is used in the Mond process of refining nickel in Wales: it reacts with warm nickel to give volatile nickel carbonyl, $Ni(CO)_4$, which is decomposed on small hot nickel pellets to yield more of the pure metal.

Lead(II) oxide is used in the manufacture of lead glass and as an accelerator in the vulcanization of rubber.

III THE DIOXIDES

28.11 Preparation of the Dioxides

Carbon dioxide is formed whenever carbon or any of its compounds is completely burnt in air; it is also present in various natural gases associated with coal mines, oil wells and volcanoes. In the laboratory the gas is made by the action of an acid on a carbonate: dilute hydrochloric acid is added to calcium carbonate, in the form of marble chips. A Kipp's apparatus (figure 28.6(a)) may be used instead of that illustrated in figure 28.6(b).

$$CaCO_3 + 2HCl \rightarrow CaCl_2 + H_2O + CO_2\uparrow$$

or $$CaCO_3 + 2H^+ \rightarrow Ca^{2+} + H_2O + CO_2\uparrow$$

N.B. Sulphuric acid is not suitable because it coats the marble chips with a layer of insoluble calcium sulphate which will stop the reaction after a few seconds.

The gas can easily be collected by downward delivery (upward displacement of air): it is much denser than air. If required pure and dry it may be rinsed with water to remove hydrogen chloride and then dried by concentrated sulphuric acid. It may be collected over water but there is some loss because it is fairly soluble.

Carbon dioxide may also be made by heating a carbonate (except those of sodium and potassium) or a hydrogencarbonate (only sodium and potassium form solid hydrogencarbonates):

$$ZnCO_3 \rightarrow ZnO + CO_2\uparrow$$

$$2NaHCO_3 \rightarrow Na_2CO_3 + H_2O\uparrow + CO_2\uparrow$$

Commercially, carbon dioxide is obtained by heating calcium carbonate in a limekiln (see section 25.8). Other commercial processes which yield carbon dioxide as a by-product are the manufacture of hydrogen (see section 23.3(a)) and the fermentation of sugars and starches to yield ethanol, catalyzed by yeast.

$$\underset{\text{Glucose}}{C_6H_{12}O_6} \rightarrow \underset{\text{Ethanol}}{2C_2H_5OH} + 2CO_2\uparrow$$

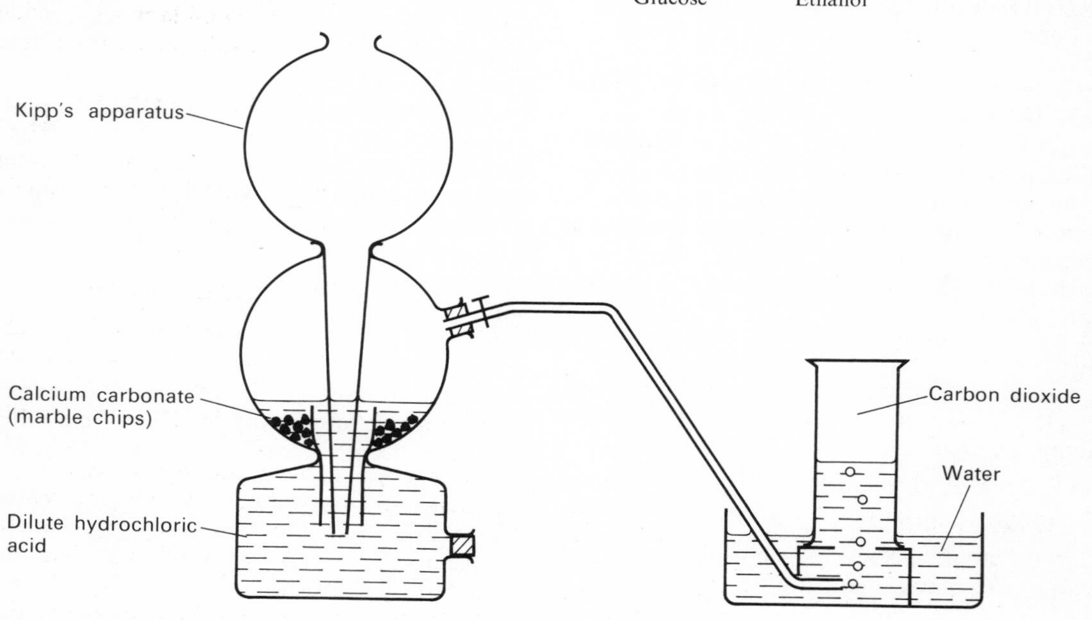

Figure 28.6(a)
Kipp's apparatus used for the preparation of carbon dioxide

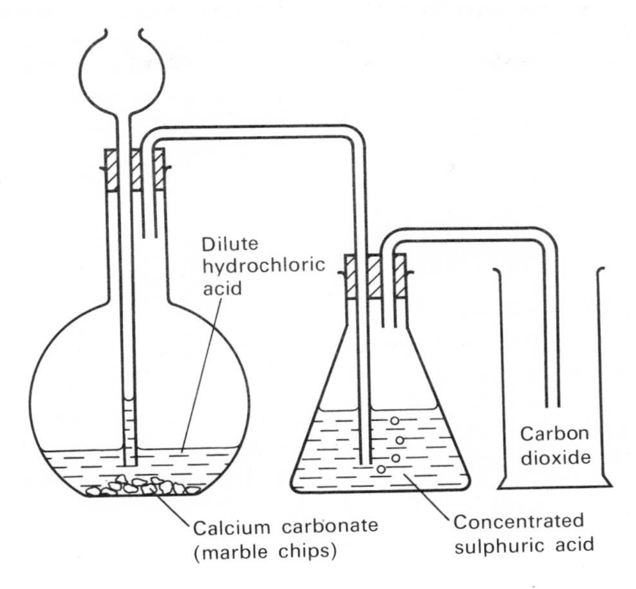

Figure 28.6(b)
The preparation of dry carbon dioxide

Silicon dioxide (silica) is an important mineral, occurring in several different forms.

(*a*) Quartz is the purest form of silica, and is often found as beautiful transparent crystals (rock crystal). Sometimes it is coloured by traces of impurity, and it is then prized as a gem stone. Amethyst is a purple variety.

(*b*) Sand consists of tiny quartz particles which have been broken and rounded by weathering and friction. Silver sand is practically pure silica, but ordinary yellow sand owes its colour to traces of iron(III) oxide, which may be removed by boiling with concentrated hydrochloric acid. Sandstone is made up of sand granules bonded together with iron(III) oxide and clay.

(*c*) Flint is a hard, impure, and non-crystalline form.

(*d*) Agate, opal, jaspar, and other semi-precious stones are amorphous forms of silica coloured by various impurities.

(*e*) Kieselguhr is a porous rock made up of the accumulated skeletons of microscopic sea plants called diatoms. It was originally used to absorb the explosive liquid nitroglycerine, thereby making dynamite. This is much safer to handle than the liquid explosive. Nowadays it finds its main uses in industrial filtration and as an abrasive.

Tin(IV) oxide occurs naturally; it can be made by heating tin in air but is more easily made by allowing tin to react with concentrated nitric acid and then heating the residue to give the white anhydrous powder.

Lead(IV) oxide is a brown-black solid made by adding trilead tetraoxide to warm dilute nitric acid.

$$Pb_3O_4 + 4HNO_3 \rightarrow PbO_2\downarrow + 2Pb(NO_3)_2 + 2H_2O$$

The precipitate is filtered off, rinsed with water and then dried by very gentle heating (to 100°C)—it decomposes at about 300°C.

$$2PbO_2 \rightarrow 2PbO + O_2\uparrow$$

28.12 Physical Properties of the Dioxides

Carbon dioxide is a colourless gas with a slight smell and taste (soda water). It is fairly soluble in water: one volume of gas dissolves in one volume of water at room temperature and pressure. Under higher pressure more will dissolve, and often with suitable colouring and flavouring the solution is sold as 'mineral' water. It is a dense gas and it is easily liquefied by increasing the pressure providing that the temperature is below 31°C. The liquid only exists at pressures greater than atmospheric; the solid sublimes at −78°C and used as a refrigerant.

Experiments to Show that Carbon Dioxide is Denser than Air

(*a*) A trough is filled with carbon dioxide from a Kipp's apparatus. A few bubbles are blown with soap solution and shaken into the trough. They remain, apparently suspended in mid-air, floating on the dense layer of carbon dioxide.

(*b*) A beaker is placed on a balance and counterpoised with weights. Some carbon dioxide is poured in from a gas jar. The beaker immediately sinks.

As a result of its high density, carbon dioxide formed naturally may collect in valleys, wells, or caves. It is not poisonous, but it is suffocating, so that it may be very dangerous.

Silicon dioxide at first sight is very different from carbon dioxide. Although both are covalent molecules carbon dioxide is of a simple molecular structure, CO_2, whereas silicon dioxide is a macromolecule, a giant structure, the empirical formula being SiO_2 (see figure 28.7).

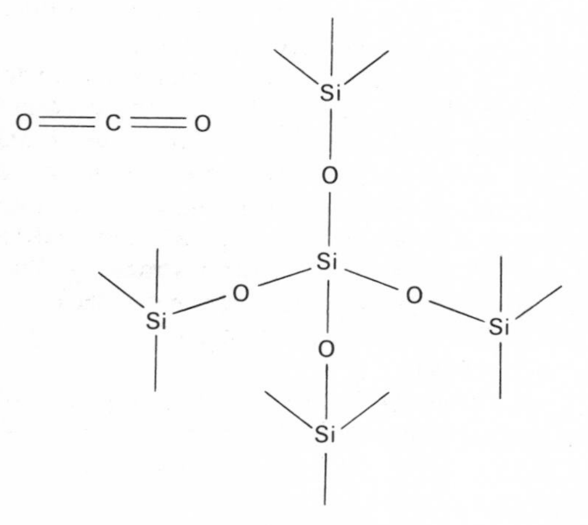

Figure 28.7
The structures of carbon dioxide and silicon dioxide

28.13 Chemical Properties of the Dioxides

(*a*) *With Acids and Bases*

Carbon dioxide dissolves in water giving a weakly acidic solution (carbonic acid) which turns litmus a pale red.

$$H_2O+CO_2 \rightleftharpoons 2H^+ + CO_3^{2-}$$

The acid is very unstable: it cannot be isolated, and on boiling the solution it decomposes. If carbon dioxide is bubbled into sodium hydroxide solution, sodium carbonate is formed first but the reaction cannot be accurately stopped at that stage and continues to the formation of the hydrogencarbonate (which, if the solution is concentrated, will be precipitated).

$$2NaOH+CO_2 \rightarrow Na_2CO_3+H_2O$$
$$Na_2CO_3+H_2O+CO_2 \rightarrow 2NaHCO_3$$

Potassium hydroxide solution is capable of absorbing even more carbon dioxide and is preferred if the expense is justified. A similar reaction occurs with calcium hydroxide solution (lime water) but the formation of the carbonate is obvious because it is insoluble whereas the hydrogencarbonate is soluble.

$$Ca(OH)_2+CO_2 \rightarrow CaCO_3\downarrow+H_2O$$
$$CaCO_3+H_2O+CO_2 \rightarrow Ca(HCO_3)_2$$

The first of these two reactions is much faster than the second and there is no danger of mistaking carbon dioxide when the lime-water test is used. On boiling a solution of calcium hydrogencarbonate (the solid cannot be isolated) it decomposes.

$$Ca(HCO_3)_2 \rightarrow CaCO_3+H_2O\uparrow+CO_2\uparrow$$

Calcium hydrogencarbonate is the cause of temporary hardness in water (see section 25.14).

Silicon dioxide with about 4% of water in it is known as silica gel, and this is a useful drying agent which is coloured by anhydrous cobalt chloride (see section 6.4). Silicon dioxide dissolves in alkalis to give silicates (e.g. Na_2SiO_3, sodium silicate, and $CaSiO_3$, calcium silicate, frequently formed as a slag in metallurgical processes).

Tin(IV) oxide and lead(IV) oxide are amphoteric.

(*b*) *As Oxidizing Agents*

Carbon dioxide does not burn or support the combustion of a splint, taper or candle; however, burning magnesium continues to burn in the gas.

$$2Mg+CO_2 \rightarrow 2MgO+C$$

Amid many crackling noises a white and a black powder are deposited.

Silicon dioxide reacts with magnesium when the two are heated strongly but the silicon produced is not very pure. Silicon dioxide also reacts with carbon in an electric furnace (2000°C) to produce the important abrasive silicon carbide (carborundum).

$$SiO_2+3C \rightarrow SiC+2CO\uparrow$$

Lead(IV) oxide is a strong oxidizing agent; it will convert hot concentrated hydrochloric acid to chlorine.

$$PbO_2+4HCl \rightarrow PbCl_2+2H_2O+Cl_2\uparrow$$

It undergoes an additive reaction when warmed with sulphur dioxide.

$$PbO_2 + SO_2 \rightarrow PbSO_4$$

Like the other oxides of lead, lead(IV) oxide will yield the metal when heated on a charcoal block.

$$PbO_2 + C \rightarrow Pb + CO_2\uparrow$$

It loses oxygen readily on heating to give lead(II) oxide.

$$2PbO_2 \rightarrow 2PbO + O_2$$

(*c*) *As a Non-volatile Oxide*

Silicon dioxide is only very weakly acidic, but it is capable of driving off much more strongly acidic oxides from their salts, because it is much less volatile (b.p. 2200°C) than the other oxides so that, at a high temperature, they are removed. For instance, in the manufacture of phosphorus (see section 30.4) calcium phosphate is heated with sand.

$$2Ca_3(PO_4)_2 + 6SiO_2 \rightarrow 6CaSiO_3 + P_4O_{10}\uparrow$$

In the method for manufacturing sulphuric acid from anhydrite, calcium sulphate is heated with sand and coke.

$$CaSO_4 + SiO_2 + C \rightarrow CaSiO_3 + SO_2\uparrow + CO\uparrow$$

Similarly, sodium silicate may be made by heating silicon dioxide with sodium carbonate.

$$Na_2CO_3 + SiO_2 \rightarrow Na_2SiO_3 + CO_2\uparrow$$

28.14 Uses of the Dioxides

About 90% of the **carbon dioxide** manufactured is used in making mineral waters and beer.

Another important use is in fire extinguishers: as carbon dioxide is a non-supporter of combustion and a very dense gas, it will form a blanket which will exclude air from a fire and so extinguish it. Sometimes liquid carbon dioxide is used, but many hand extinguishers contain a carbonate solution (often sodium hydrogencarbonate) and a strong acid. By inverting the extinguisher or tapping the knob the carbonate and acid are brought into contact, and carbon dioxide and water are forced out under pressure (figure 28.8). Most of the extinction is actually done by the water.

Carbon dioxide is used in refrigeration as 'dry ice' for keeping ice cream cold and for shrink fitting. It is very convenient because (*a*) it sublimes (solid directly to gas) not leaving a messy fluid, (*b*) it gives a much lower temperature than ordinary ice; (*c*) it lasts much longer than ordinary ice.

Fourthly, carbon dioxide is used as the heat transmission agent in nuclear power stations and in manufacturing urea (used in making plastics and as a fertilizer).

Silicon dioxide (silica, sand) also has many uses: glass making (see section 28.15), mortar (see section 25.9) and cement (see section 25.11), sodium silicate (see section 28.15), phosphorus manufacture (see section 30.4) and sulphuric acid manufacture (see section 31.20).

At high temperature quartz melts and forms a glass which can be used for making crucibles and tubing, or drawn out into fine but strong threads. Silica has a very low coefficient of expansion, so that articles made from fused silica can be subjected to great and sudden changes of temperature without cracking. They are also very resistant to chemical attack.

Tin(IV) oxide is used in white opaque glass and glazed bricks.

Lead(IV) oxide is used for packing into the positive plates of accumulators (see section 20.4).

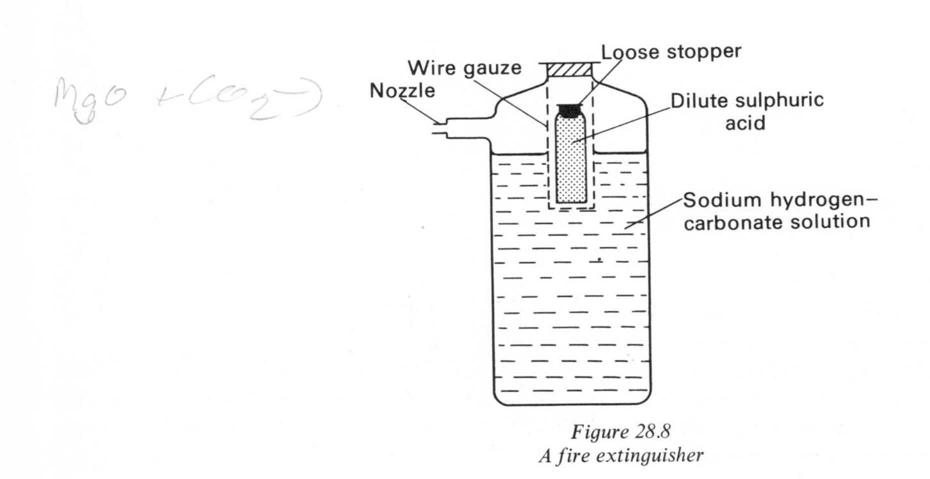

Figure 28.8
A fire extinguisher

IV OTHER COMPOUNDS

28.15 Acids and Hydroxides

The compound $C(OH)_2$ does not exist but the isomer HCOOH is the well known methanoic (formic) acid.

Silicic acid (H_4SiO_4) is not as important as sodium silicate (Na_2SiO_3) which is made commercially by heating sand with concentrated sodium hydroxide solution for some hours under a pressure of five atmospheres.

$$SiO_2 + 2NaOH \rightarrow Na_2SiO_3 + H_2O$$

This gives a concentrated, viscous solution of the silicate. Another method is to heat powdered silica with sodium carbonate.

$$SiO_2 + Na_2CO_3 \rightarrow Na_2SiO_3 + CO_2\uparrow$$

The glass-like solid formed can be dissolved in water under pressure.

It is usually sold as a syrupy solution under the name of water glass, and is used for preserving eggs. It covers the porous shell with a silicate layer which prevents air from entering. It is also used in paper making, in television tubes, in foundry moulds, for metal protection and in detergents (it keeps the dirt in suspension). If a solution of sodium silicate is acidified, a gelatinous precipitate of hydrated silica is obtained. On filtering and strong heating, pure silicon dioxide is obtained.

The most commonly used glasses are mixtures of silicates, made by fusing sand with various metallic oxides or carbonates. Ordinary soft glass (window glass) is made by heating a mixture of silicon dioxide (sand), sodium carbonate, and calcium carbonate to a temperature of about 1500°C. It consists of a mixture of sodium silicate and calcium silicate.

$$Na_2CO_3 + SiO_2 \rightarrow Na_2SiO_3 + CO_2\uparrow$$
$$CaCO_3 + SiO_2 \rightarrow CaSiO_3 + CO_2\uparrow$$

If the sand is not very pure, the glass will be coloured. Iron is the most serious impurity, and makes the glass green. The colour can be removed by adding a trace of selenium. This soda-lime glass softens readily, and so may be cast or blown into various shapes, or rolled into sheets at a red heat.

A harder and more resistant glass may be made by using potassium carbonate instead of sodium carbonate. This may be used for combustion tubing. If some lead(II) oxide is added to the hard-glass melt, flint glass is obtained, containing lead silicate.

$$PbO + SiO_2 \rightarrow PbSiO_3$$

This has a high refractive index, and is used for making lenses and prisms, and also for 'cut-glass' ornaments.

Coloured glasses are made by adding traces of metallic oxides to ordinary glass. Thus cobalt oxide gives blue glass and nickel and chromium oxides give green glass.

'Pyrex' glass contains boron oxide (B_2O_3) as well as silica. It has a low coefficient of expansion, and so can be safely used for making cooking-vessels which can stand changes of temperature without cracking.

The most remarkable property of a glass is the fact that, when heated, it does not melt at a definite temperature, but instead becomes gradually more and more plastic. Hence it can be blown or moulded easily while it is semi-fluid. Moreover, on cooling it does not crystallize or become solid at a definite freezing point. It is considered to be a super-cooled liquid and not a true solid. At room temperature it has a very high viscosity, and this gradually decreases as the temperature rises. The molecules are not arranged in definite patterns as in a true solid, but retain the random positions they have in a liquid.

If molten glass is cooled too rapidly, internal strains are set up and it is likely to crack. So glass articles are always cooled slowly (annealed) unless they are made from one of the types of glass which have low coefficients of expansion. When the large glass mirror for the 5 m telescope at Mount Palomar was cast, it was cooled so slowly that the process lasted for six months.

[The easiest way of making a glass in the laboratory is to use borax ($Na_2B_4O_7,10H_2O$) which on heating on a platinum loop yields $NaBO_2$ in the form of a polymer with a structure similar to that of normal glass. By dissolving in it very small quantities of transition element compounds coloured glasses can often be obtained. This was formerly used in analysis—the borax bead test.]

Tin salts in solution with an alkali yield a white precipitate, the nature of which is similar to lead(II) hydroxide.

Lead(II) hydroxide is easily prepared by precipitation: sodium hydroxide or ammonia solution is added to lead ethanoate or nitrate solution.

$$Pb(NO_3)_2 + 2NaOH \rightarrow Pb(OH)_2\downarrow + 2NaNO_3$$
or $$Pb^{2+} + 2OH^- \rightarrow Pb(OH)_2\downarrow$$
The white precipitate is soluble in excess sodium hydroxide or in nitric acid—it is amphoteric.

28.16 The Sulphides

Carbon disulphide is prepared by heating coke in an electric furnace and passing in sulphur vapour.

$$C + 2S \rightarrow CS_2$$

It is a colourless liquid with a characteristic smell; it is very volatile (b.p. 46°C) and being poisonous must be treated with great care. It is immiscible with water but it is a very good solvent for octaatomic sulphur, white phosphorus and fats; industry uses most of it to make viscose rayon and cellophane. It is very inflammable.

Tin(II) sulphide is formed as a brown precipitate when hydrogen sulphide is bubbled into a solution of a tin(II) salt.

$$SnCl_2 + H_2S \rightarrow SnS\downarrow + 2HCl$$
or $$Sn^{2+} + S^{2-} \rightarrow SnS\downarrow$$

Tin(IV) sulphide is formed as a yellow precipitate if hydrogen sulphide is bubbled into a solution of a tin(IV) salt.

$$SnCl_4 + 2H_2S \rightarrow SnS_2\downarrow + 4HCl$$
or $$Sn^{4+} + 2S^{2-} \rightarrow SnS_2\downarrow$$

The yellow salt is used as a pigment, e.g. mosaic gold.

Lead(II) sulphide is formed as a black precipitate, often with a metallic lustre, when hydrogen sulphide is bubbled into a solution of a lead(II) salt.

$$Pb(NO_3)_2 + H_2S \rightarrow PbS\downarrow + 2HNO_3$$
or $$Pb^{2+} + S^{2-} \rightarrow PbS\downarrow$$

It is the substance formed when lead paints tarnish in town air. The restoration of oil paintings is carried out by converting the lead sulphide into lead sulphate by oxidation with hydrogen peroxide dissolved in ethoxyethane.

$$PbS + 4H_2O_2 \rightarrow PbSO_4 + 4H_2O$$

28.17 The Chlorides

Carbon tetrachloride is manufactured by passing chlorine into warm carbon disulphide, the reaction being catalyzed by iron(III) chloride.

$$CS_2 + 3Cl_2 \rightarrow CCl_4 + S_2Cl_2$$

Fractional distillation separates the products (b.p. 77°C and 138°C respectively). Carbon tetrachloride is a good grease solvent ('Thawpit') and fire extinguishing fluid ('Pyrene'; useful in confined spaces, but there is a danger of poisonous products).

Tin(II) chloride can be made by passing dry hydrogen chloride over heated tin. The crystalline salt, which has two molecules of water of crystallization, can be made by dissolving tin in hydrochloric acid: it cannot be dehydrated by heating.

$$Sn + 2HCl \rightarrow SnCl_2 + H_2\uparrow$$

Tin(II) chloride is a strong reducing agent that can be easily converted into a tin(IV) salt by potassium manganate(VII), potassium dichromate, iron(III) salts, mercury(II) salts etc.

$$SnCl_2 + 2HgCl_2 \rightarrow Hg_2Cl_2\downarrow + SnCl_4$$

or

$$Sn^{2+} + 2Hg^{2+} + 2Cl^- \rightarrow Hg_2Cl_2\downarrow + Sn^{4+}$$

followed by

$$SnCl_2 + Hg_2Cl_2 \rightarrow 2Hg\downarrow + SnCl_4$$

or

$$Sn^{2+} + Hg_2Cl_2 \rightarrow 2Hg\downarrow + Sn^{4+} + 2Cl^-$$

Tin(IV) chloride can be prepared by synthesis; it is not very important.

$$Sn + 2Cl_2 \rightarrow SnCl_4$$

Lead(II) chloride can be prepared by precipitation.

$$Pb(NO_3)_2 + 2HCl \rightarrow PbCl_2\downarrow + 2HNO_3$$
or $$Pb^{2+} + 2Cl^- \rightarrow PbCl_2\downarrow$$

The white precipitate is fairly soluble in hot water.

28.18 The Silicones

In recent years remarkable new plastic materials have been prepared which contain long chains of silicon atoms alternating with oxygen atoms. The remaining valencies are usually occupied by methyl groups (CH_3), e.g.

```
  CH3    CH3    CH3
   |      |      |
—Si—O—Si—O—Si—
   |      |      |
  CH3    CH3    CH3
```

These silicones find their main applications as water-repellent liquids for proofing textiles, and as lubricants which retain their effectiveness over a wide range of temperatures. One silicone is used to prevent bread from sticking to the tin during baking; another can be moulded with the fingers like putty and yet will bounce from a hard surface!

28.19 Other Lead Compounds

(a) Trilead Tetraoxide (red lead)

If lead(II) oxide is carefully heated in air to a temperature between 400°C and 470°C it is converted into trilead tetraoxide (dilead(II) lead(IV) oxide).

$$6PbO + O_2 \rightarrow 2Pb_3O_4$$

It is an oxidizing agent when heated:

$$Pb_3O_4 + 2C \rightarrow 3Pb + 2CO_2\uparrow$$
$$Pb_3O_4 + 4H_2 \rightarrow 3Pb + 4H_2O\uparrow$$
$$Pb_3O_4 + 8HCl \rightarrow 3PbCl_2 + 4H_2O\uparrow + Cl_2\uparrow$$

It is used as a pigment in paints, as a priming paint because it inhibits rusting and in the manufacture of crystal glass.

(b) Lead Carbonate

A basic lead carbonate ($Pb(OH)_2,2PbCO_3$) is precipitated by adding sodium carbonate solution to a solution of a lead(II) salt. It is known as white lead and is used in lead paints, which have a good covering power but suffer from the disadvantages of tarnishing in town air (the white substance is converted to a black one, lead sulphide) and being poisonous.

(c) Lead Nitrate

When lead, lead(II) oxide or lead(II) carbonate are dissolved in dilute nitric acid and the solution evaporated to the point of crystallization, lead(II) nitrate is obtained. Because this is one of the few anhydrous heavy metal nitrates it is used to prepare nitrogen dioxide (see section 30.14).

(d) Lead Sulphate

When a soluble lead salt is added to a soluble sulphate, lead sulphate is formed as a white precipitate.

$$Pb(NO_3)_2 + H_2SO_4 \rightarrow PbSO_4\downarrow + 2HNO_3$$

or

$$Pb^{2+} + SO_4^{2-} \rightarrow PbSO_4\downarrow$$

QUESTIONS 28

Group IV: Carbon, Silicon, Tin and Lead

1 Two pieces of lead foil are dipped, side by side but not touching, into a beaker of fairly concentrated sulphuric acid. After some days' immersion the plates are coated with a white deposit. The two pieces of lead are now connected to a 6 V d.c. supply, and a current passes. After some hours, one piece of lead has lost its white deposit and bubbles of hydrogen are streaming off it. The other piece of lead now appears to be coated with a dark brown deposit, replacing the white deposit.
(*a*) Draw a labelled diagram, showing the apparatus when current is being passed through it.
(*b*) Suggest a name for the white deposit first formed on each plate.
(*c*) Suggest a name for the dark brown deposit. If some of this deposit could be scraped off, describe how you would attempt to identify it.
(*d*) Suggest any change that you would expect to occur in the liquid during the passage of the current.
[OC]

2 (*a*) How would you prepare and collect carbon monoxide gas free from carbon dioxide? How would you test a jar of this gas so as to prove that it was carbon monoxide and not another inflammable gas such as hydrogen?
(*b*) Give two ways in which carbon dioxide is formed on a large scale. State two uses of this gas. How do you explain the constant though small proportion of carbon dioxide in the atmosphere? [C]

3 (*a*) Describe how you would prepare a reasonably pure sample of lead(IV) oxide using lead(II) oxide as the starting material.
(*b*) What chemical changes take place and what observations could be made when to each of these oxides is added moderately concentrated hydrochloric acid, the mixtures boiled for a few minutes, and then allowed to cool?
(*c*) Indicate very briefly how you would attempt to obtain a small sample of lead from lead(II) oxide.
[O]

4 Write equations for three reactions, of different types, in which carbon dioxide is a product. For each reaction, give a brief statement, without diagrams, of the conditions required for it to occur. Describe and explain how you would obtain pure samples of (*a*) carbon, (*b*) carbon monoxide, from a supply of carbon dioxide.
[W]

5 Given lead carbonate, how would you prepare from it: (*a*) lead(II) oxide; (*b*) lead nitrate crystals; (*c*) a dry sample of lead sulphate? How could you prepare lead sulphate from lead(IV) oxide? Describe what you would observe in the course of this preparation.
[J]

6 Litharge is a yellow-red powder: how would you show that it contains both lead and oxygen?

How would you demonstrate that it is *a basic oxide*? Lead nitrate is soluble in water but lead carbonate is insoluble. Making use of these facts, describe in detail how you would make some lead carbonate from litharge.

7 What is meant by *allotropy*? Give the names of two allotropes of carbon and compare their physical properties. Describe with a diagram how you could show that the allotropes of carbon are all forms of the same element.

8 How is wood charcoal prepared? Give an account of its more important uses, explaining how these depend on its physical and chemical properties.

9 Describe a laboratory method for the preparation and collection of carbon dioxide. Describe what happens when carbon dioxide is passed into calcium hydroxide solution for a considerable time. Write equations for the changes which take place. Give two uses for carbon dioxide and state how it is stored. [J]

10 Explain briefly the importance of carbon dioxide to a living plant. How and under what conditions does carbon dioxide react with: (*a*) magnesium (*b*) carbon (*c*) sodium carbonate?

11 Describe how carbon monoxide can be prepared and collected in the laboratory. Give two physical and two chemical properties of carbon monoxide. What mass of carbon monoxide could be made from 100 g of carbon? [OC]

12 When crystalline ethanedioic acid is heated with concentrated sulphuric acid a reaction represented by the following equation takes place:

$$\begin{array}{l} COOH \\ | \\ COOH \end{array} \rightarrow CO_2\uparrow + CO\uparrow + H_2O\uparrow$$

How does the sulphuric acid act in this reaction? How, starting from ethanedioic acid, could you prepare: (*a*) carbon monoxide (*b*) carbon dioxide? When a piece of carbon is burnt (in a suitable apparatus) in excess of oxygen the volume of gas when cooled to the original temperature and pressure remains unchanged. The relative density of the gas is 22. Use this data to show that the formula of carbon dioxide is CO_2. [J]

13 Draw a diagram of the apparatus you would use to prepare a specimen of carbon monoxide from carbon dioxide, and describe how the experiment would be conducted. Describe a test to distinguish carbon monoxide from hydrogen. What reactions occur between carbon monoxide and: (*a*) blood (*b*) oxygen? In what region of a briskly burning coke fire would carbon monoxide be present? [O]

14 Name the materials often used in the laboratory to prepare carbon dioxide. Give the equation for the reaction. A blue flame is often seen above a coke fire. Explain fully how air and carbon produce this phenomenon. When a mixture of 50 cm^3 of oxygen and 50 cm^3 of carbon monoxide is sparked what volume of gas will remain: (*a*) after explosion, and (*b*) after treating the product with potassium hydroxide solution?

15 Why is carbon considered to be an element? Describe experiments (one in each case) to show that: (*a*) carbon dioxide contains carbon (*b*) carbon is a reducing agent (*c*) carbohydrates contain carbon. [O]

16 How would you demonstrate: (*a*) that exhaled air contains more carbon dioxide than inhaled air (*b*) that carbon dioxide can be converted into carbon monoxide (*c*) the composition of carbon dioxide by mass?

17 Show in tabular form how the properties of carbon monoxide compare with those of carbon dioxide. How would you prove that the formula of carbon monoxide is CO, knowing that the formula of carbon dioxide is CO_2?

18 Describe the preparation and collection of carbon monoxide in the laboratory. Show how you would obtain some carbon dioxide from a mixture of carbon dioxide and carbon monoxide. A mixture of 10 cm^3 carbon monoxide and 25 cm^3 air (containing 20% oxygen) was exploded. Find the volume of each residual gas, the temperature and the pressure remaining constant.

19 Illustrate, by means of a fully labelled diagram, the usual laboratory preparation of carbon dioxide, its conversion into carbon monoxide, and the collection of the latter free from carbon dioxide. Name three gaseous fuels which contain carbon monoxide. If 1000 cm^3 of carbon dioxide were completely converted into carbon monoxide, what volume of the latter, measured under the same conditions of temperature and pressure, would be obtained? [J]

20 Describe briefly how tin is obtained from its main ore. What would you see if a solution of tin(II) chloride were added slowly to excess to a solution of mercury(II) chloride? Give equations for the reactions taking place.

21 State briefly how metallic lead may be obtained from a naturally occurring compound. Starting from metallic lead, how would you prepare a solution of lead nitrate? How would you obtain from lead nitrate a specimen of: (*a*) lead chloride (*b*) lead(II) oxide (*c*) metallic lead? [J]

22 What is the action of heat on trilead tetraoxide? How would you prepare lead(IV) oxide from trilead tetraoxide? What is seen to happen when trilead tetraoxide is heated with moderately concentrated hydrochloric acid? What occurs when the gaseous product of this reaction is led into a jar containing damp neutral litmus paper? [J]

23 How and under what conditions does tin react with: (*a*) air (*b*) sodium hydroxide (*c*) hydrochloric acid (*d*) chlorine (*e*) nitric acid?

24 Describe in detail how, from lead nitrate, you would prepare, in a reasonable state of purity, a few grams of: (*a*) lead carbonate (*b*) metallic lead. [O]

25 How is trilead tetraoxide made from metallic lead? Name one important use of this oxide. Describe how you would prepare from trilead tetraoxide specimens of: (*a*) lead(II) sulphate (*b*) lead(IV) oxide (*c*) lead(II) carbonate.

26 How would you prepare a sample of lead from one of its oxides and convert this sample into lead sulphide? Lead paints blacken in the air. What causes this and how can the paints be renovated? [J]

29 The Organic Chemistry of Carbon

29.1 Introduction

Carbon is unique among the elements in the number and variety of compounds it can form. All living matter is composed of carbon compounds: it was once believed that a special 'vital' force was necessary for the production of organic compounds in living organisms, but although this idea was repudiated experimentally in 1828 by Wöhler, the title 'organic chemistry' still remains for the study of the majority of the compounds of carbon. Under the title of 'inorganic chemistry' the elements and a few simple compounds of carbon have already been studied.

In most of its compounds carbon exhibits a valency of four—the bonds are directed as from the centre to the four corners of a tetrahedron, though it is often drawn planar—and many carbon atoms may link up to form chains and rings (see figure 29.1). Organic compounds can be broadly classified as aliphatic and aromatic. The aliphatic compounds of high relative molecular mass are often fats; in aliphatic compounds the carbon atoms are linked in chains and occasionally rings. Aromatic compounds may have characteristic aromas; they are based on benzene in which six carbon atoms form a ring with distinctive properties and, compared with aliphatic compounds, remarkable stability.

The manufacture of organic chemicals has undergone a great transformation in the last twenty years and the order of importance of the sources is now:

1. Petroleum and natural gas (over 85% in the U.K., over 90% in the U.S.A.).
2. Coal.
3. Carbohydrates (fermentation process).
4. Wood, and animal and vegetable oils and fats.

For the location of the manufacturing centres for organic compounds in the British Isles see figure 1.9.

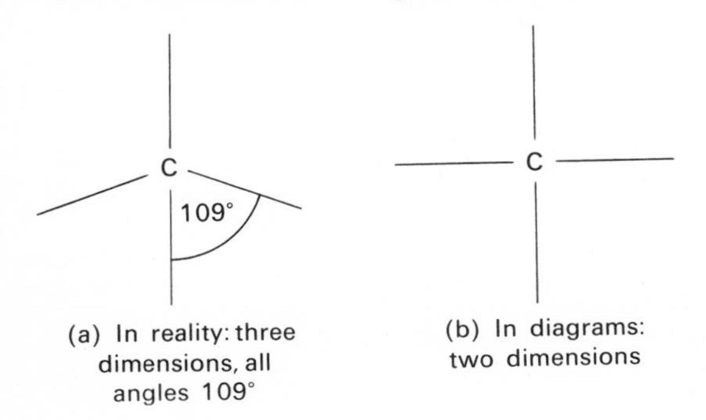

Figure 29.1
The valencies of carbon

29.2 Petroleum

(a) Occurrence and Production

Petroleum or crude oil is a dark viscous fluid which is found underground in many parts of the world: figure 29.2(a) shows the production statistics for this century and figure 29.2(b) shows the detailed statistics for 1973.

Oil probably was formed by the bacterial decomposition of marine animal and plant remains under pressure and possibly at high temperatures. It is a complex mixture of substances and consists of hydrocarbons (compounds of hydrogen and carbon) mixed chiefly with compounds of carbon, hydrogen and sulphur. The hydrocarbons may be straight chain molecules, chains with side arms or ring molecules based on and including benzene.

In the U.S.A. Drake (1859) drilled the first well to obtain crude oil successfully; production has approximately doubled every decade in this century.

(b) Processing

The first stage is a rough separation of the substances by *fractional distillation*. The crude oil is heated to 400°C by passing it in a pipe through a gas heated furnace and is then fed into a vertical cylinder or tower, known as a fractionating column, about one third of the way up. The tower is divided horizontally into many compartments and as the hot gases move upwards they bubble through liquids

Year	Millions of tonnes
1900	20
1910	44
1920	106
1930	206
1940	300
1950	539
1960	1091
1970	2349
1973	2837

Figure 29.2(a)
The rise in world production of crude oil and natural gas liquids

It can be seen that the production approximately doubles every decade. The U.S.A., despite its huge home production of petroleum (518 million tonnes), imported a further 313 million tonnes in 1973.

The world production of petroleum in 1973 was 2837 million tonnes which was made up as follows:

Area	%	*Country*	%
North America	22·8	U.S.A.	18·3
		Canada	3·6
Caribbean	6·9	Venezuela	6·3
South America	1·7	—	
Western Europe	0·8	—	
Middle East	36·8	Saudi Arabia	12·9
		Iran	10·3
		Kuwait	4·9
		Iraq	3·4
		Abu Dhabi	2·2
Africa	10·3	Libya	3·7
		Nigeria	3·5
		Algeria	1·8
South East Asia	2·9	Indonesia	2·3
Other Eastern hemisphere	17·8	U.S.S.R.	14·8
		Eastern Europe and China	

Figure 29.2(b)
The world production statistics for petroleum in 1973 (including natural gas liquids)

The total capacity of the U.K. refineries in 1973 was 142 million tonnes but the consumption was 113 million tonnes. In order of capacity in millions of tonnes they are:

Fawley (Hampshire)	19·0
Milford Haven (Pembrokeshire, Esso)	15·0
Isle of Grain (Kent)	10·9
Stanlow (Cheshire)	10·8
Killingholme (Lincolnshire)	10·0
Shell Haven (Essex)	10·0
Coryton (Essex)	9·0
Pembroke (Pembrokeshire)	9·0
Grangemouth (Stirling)	8·8
Llandarcy (Glamorgan)	8·3
Teesport (Yorkshire)	6·0
Billingham (Teesside)	5·0
Milford Haven (Pembrokeshire, Gulf)	5·0
South Killingholme (Lincolnshire)	4·5
Milford Haven (Pembrokeshire, Amoco)	4·0
Heysham (Lancashire)	2·2
Belfast (Ulster)	1·5
Ellesmere Port (Cheshire)	1·5
Eastham (Cheshire)	0·6
Kingsnorth (Kent)	0·3
Ardrossan (Ayrshire)	0·2
Dundee (Angus)	0·1

The principal inland U.K. oilfields produce oil on the scale of thousands, rather than millions, of tonnes. In 1973 production totalled 88 100 tonnes, the principal sources being at:

Gainsborough	30%
Kimmeridge	20%
Beckingham	15%
Bothamsall	14%
Egmanton	11%

Oil was found in 1970, in addition to natural gas, under the North Sea. The oil is being shipped to nearby refineries pending completion of pipelines. The pipelines with their expected dates of completion run from:

Efofisk field to Teesside	(1975)
Forties field to Cruden Bay, Peterhead then to Grangemouth	(1975)
Piper field to Flotta, Orkney	(1975)
Brent and Thistle fields to Sullom Voe, Shetland	(1976)

The Efofisk field is on the Norwegian side of the boundary line.

Meanwhile the U.K. imports nearly all its requirements of petroleum. The principal sources are:

Saudi Arabia	24%
Kuwait	18%
Iran	18%
Libya	10%
Nigeria	8%

A new refinery (capacity 6 million tonnes) is due to open on Canvey Island, Essex in mid-1975. The Amoco refinery at Milford Haven should be enlarged to 5·4 m tonnes by the beginning of 1976.

Ten million tonnes of the United Kingdom crude oil were used as chemical feedstock. Sulphur is an important by-product of refining petroleum, e.g. Fawley produces 30 thousand tonnes of sulphur every year.

Figure 29.3
The petroleum refineries in the British Isles

which are flowing down. Some gases escape from the top of the tower but most of the products are collected as liquids at various levels according to their boiling points: lowest at the top, highest at the bottom. From the top of the tower proceeding downwards:

Fraction	*Carbon atoms per molecule*	*Boiling point (°C)*	*Derivatives*
Gases and volatile liquids	1–4	less than 40	Fuel gases, solvents
Gasoline	5–10	40–190	Petrol, naphtha
Kerosine	11–14	190–260	Paraffin oil, jet fuel
Light fuel oils	15–19	260–330	Diesel oil, gas oil
Light lubricating oil	19–30	330–400	Medicinal paraffin, paraffin wax
Bitumen	over 30	above 400	Heavy fuel oils, asphalt

The second stage is the *catalytic cracking* of the components of the above fractions to yield products that are in demand. The fractional distillation does not yield the proportions of the products that the market requires: nearly half the demand is for gasoline and nearly one quarter for light and heavy fuel oils. Usually it is required to convert large molecules into small molecules and a catalyst such as aluminium oxide at a temperature of about 500°C is employed.

The third stage may be *reforming molecules*: a straight chain molecule may have been produced but a molecule with side chains might be required. An example of this process is the isomerization of butane to 2-methylpropane, which occurs in the presence of aluminium chloride. An alternative third stage is *polymerization*: a small molecule can be converted into a larger molecule having a precise multiple of the original formula and relative molecular mass (see section 29.9).

A map (figure 29.3) shows the location of the refineries in the British Isles.

Figure 29.4
The Shell Petroleum Refinery at Teesport

29.3 Coal

Coal is of vegetable origin: giant tree-like ferns flourished about three hundred million years ago in the Carboniferous Period. The remains of these ferns became covered by earth and under the influence of heat and pressure, in the absence of air, chemical changes occurred. Wood consists mainly of cellulose, a carbohydrate with an empirical formula of $C_6H_{10}O_5$. The progressive change from wood to anthracite is accompanied by an increase in the proportion of carbon.

Stage	*Percentage by Mass*		
	Carbon	*Oxygen*	*Hydrogen*
Wood	53	42	5
Peat-like material	60	34	6
Lignite—a soft brown coal	67	28	5
Bituminous (household) coal	88	6	6
Anthracite (hard coal)	94	3	3
For comparison: Petroleum (approx.)	85	0 (sulphur: 1)	14

Without coal the Industrial Revolution, which occurred in England in the eighteenth and nineteenth centuries, would have been impossible. The production and uses of coal in the United Kingdom are shown in figure 29.5. The proportion burnt in domestic fires has decreased sharply since the Clean Air Act of 1956 and will decrease still further. Unless coal is burnt under carefully controlled conditions it produces much smoke and soot with consequent pollution of the atmosphere—the carbon of the soot ought to have been burnt to produce heat. Coal is not as easy to transport as liquid or gaseous fuels and it is not so adaptable as electricity. Thus it is more sensible to burn it in power stations to produce electricity or to convert it into a multitude of products at a coal gas or coke works.

29.4 The Destructive Distillation of Coal

The destructive distillation (pyrolysis, carbonization, thermal degradation) of coal can be carried out at temperatures between 600°C (low yield of gas but useful smokeless solid fuel) and 1200°C (coke production for metallurgical processes and gas). The first commercial use of coal gas was by Murdoch in the United Kingdom in 1792 and in 1855 Bunsen devised his ubiquitous burner.

When coal is heated out of contact with air it gives a complex mixture of substances. In the laboratory this can be carried out in a manner similar to the destructive distillation of wood already described (see section 28.3 and figure 29.6). Very impure coal gas collects in the last tube: it has a characteristic smell and is inflammable. If lead ethanoate paper is placed inside a tube of the gas it turns black, signifying that hydrogen sulphide is present (see section 31.14). Brown coal tar and a watery liquid are left in the second tube. The watery liquid is ammoniacal liquor and can be recognized by its characteristic smell and its action on litmus paper. The black porous solid left in the first tube is coke.

In industry complete carbonization may take twelve hours or more, the coal being heated in a horizontal or vertical retort by burning producer gas. The coke is then pushed out of the retort by a ram and cooled by water. During the heating a mixture of gases is given off which has to be carefully purified to yield an acceptable coal gas.

(a) Tar

If it condensed in gas distribution pipes it would block them, so it is removed beforehand by condensation under cold water and processed to yield useful substances.

(b) Ammonia

If ammonia were not removed copper objects near the coal gas burners would rapidly corrode (see section 30.12); so it is dissolved out in water.

Figure 29.5
The location of the main coalfields in the British Isles

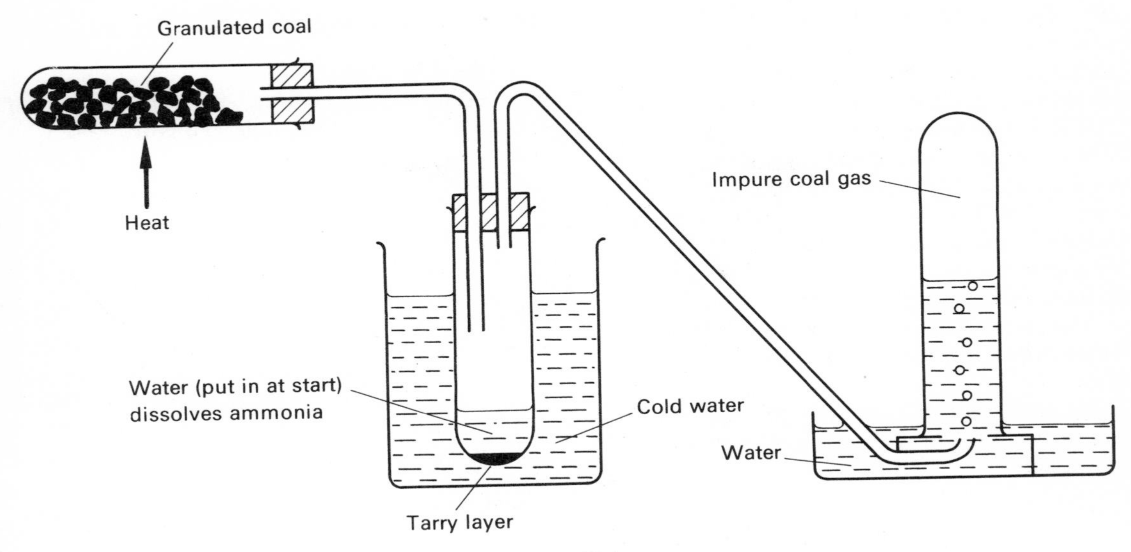

Figure 29.6
The destructive distillation of coal

(c) Hydrogen Sulphide

Hydrogen sulphide is removed by passing the gas over hydrated iron(III) oxide, a small proportion of air being admitted so that the overall reaction is:

$$2H_2S + O_2 \rightarrow 2S\downarrow + 2H_2O$$

When the proportion of sulphur in the 'spent' oxide reaches 50% the iron(III) oxide must be replaced. The hydrogen sulphide must be removed because it is poisonous and also on combustion it yields a poisonous and even more corrosive gas—sulphur dioxide.

(d) Benzene and Naphthalene

These two compounds are removed by washing the gas with anthracene oil, benzene because it is highly poisonous and naphthalene because it could condense in the gas pipes and block them.

(e) Moisture

Water vapour is removed by concentrated calcium chloride solution, otherwise in cold weather it would condense and block the pipes. The above processes leave a coal gas of approximate percentage composition by volume: hydrogen 48, carbon monoxide 15, methane 18, other hydrocarbons (propane, butane, ethene, benzene, ethane) 7, carbon dioxide 4 and nitrogen 8. The proportion of the poisonous gas, carbon monoxide, is much too high for safety and is reduced to about 3% by processes such as those described in section 23.3, the methane content simultaneously rising to about

---

The United Kingdom output of coal in 1973–74 (March) was 107 million tonnes of which 92% was deep mined (mines go as deep as 1100 metres), and 8% open cast. The output in Kent is markedly less than in other areas because there are only a few mines. The Oxfordshire coalfield (announced 1971) adds one third to the United Kingdom reserves but has yet to be exploited. The Irish coalfield in Leinster produced 75 thousand tonnes (1973).

In the U.K. the uses of coal (1973–74) are:

Use	%
Power stations	58
Coke ovens	16
Domestic	11
Industry	12
Miscellaneous	3

In the United Kingdom, electricity is generated using many fuels (1973): coal 65·7% (a tonne of coal gives over 2000 kW h), oil 24·5% (a tonne of oil produces about 4000 kW h), nuclear energy 7·3%, water power 1·5% and natural gas 1·0%. In 1973 sales of electricity amounted to 224000 GW h [G (giga) = 10^9].

30%. Tetrahydrothiophene (C_4H_8S) is added to coal gas to give it its characteristic odour: this serves as a warning if it is escaping from a pipe or appliance. Coal gas has a high calorific value: 500 B.T./ft^3 or approximately 450 kJ/22·4 dm^3. The hydrogen and carbon monoxide are known as lean gases (low calorific value), the methane and other hydrocarbons as rich gases (high calorific value) and the carbon dioxide and nitrogen as ballast gases (to give an overall density of half that of air).

The presence of the ethene causes the luminous flame when the airhole of a Bunsen burner is shut.

By carbonizing a tonne (1000 kg) of coal at about 1200°C the following products may be obtained:

700 kg coke
10 kg ammonium sulphate (from ammonia and sulphur)
40 dm^3 tar
14 dm^3 benzene
34 m^3 coal gas

The term 'coal gas' has become obsolete because of the change in pattern of production in the United Kingdom. The first significant step was in 1964 when tankers started bringing liquefied methane to Canvey Island from Algeria: from Canvey Island a natural gas grid is being constructed to cover the United Kingdom (see figure 29.7). The second step was in 1967 when methane (natural gas) supplies started to be piped in from wells in the North Sea: eventually all appliances will be converted to burn natural gas. Methane has a calorific value twice that of coal gas and needs twice the volume of air for complete combustion. Its burning velocity (34 cm/s) is a third that of coal gas and thus it is supplied at three times the pressure.

In 1973–74 the U.K. sales of gas amounted to 11 487 million therms ($1{\cdot}21 \times 10^{18}$ J) of which the domestic proportion was 44%, industrial 46% and commercial 10%. The production sources were: natural gas 92%, oil 6% and coal 2%. The vast majority of customers (77%) were receiving natural gas direct. The natural gas being obtained from the North Sea by pipelines:

West Sole Field to Easington;
Viking Field to Theddlethorpe St. Helen;
Indefatigable, Leman Bank, Deborah and Hewett Fields to Bacton.

Soon these pipelines should be joined by others from Frigg Field to St. Fergus, Peterhead and Rough Field to Easington.

29.5 Uses of the Products of Coal Distillation

(a) Coal gas

Gas is used for heating, temperatures from 40–2000°C being attainable. Thus it is used by the metal fabrication industries, in pottery kilns, for banana ripening, etc.

(b) Coke

This is a useful smokeless solid fuel because it contains up to 98% carbon and though difficult to ignite it burns well once the reaction has started (see section 28.6). It is used to manufacture several fuel gases.

Producer gas is made by passing air through hot coke (1200°C).

$$2C + \underbrace{O_2 + 4N_2}_{\text{From air}} \rightarrow 2CO + 4N_2 \quad \Delta H = -226 \text{ kJ}$$

It is best used straight away while it is still hot: it readily burns giving out still more heat, but its calorific value when cold is only one quarter that of coal gas. It is used to heat retorts for manufacturing coal gas and zinc.

Water gas is made by passing steam through white hot coke (1600°C).

$$C + H_2O \rightarrow CO + H_2 \quad \Delta H = +130 \text{ kJ}$$

The reaction is endothermic so the temperature of the coke rapidly falls: usually the temperature is maintained by making water gas alternately with producer gas on a 2·5 minute cycle. Both gases produced are inflammable, but the calorific value is only half that of coal gas. Water gas was used to manufacture hydrogen on a large scale but this process has been superceded by one based on naphtha (see section 23.3). Nowadays its principal use is for adding to coal gas in peak periods and for this purpose it must be carburetted, i.e. sprayed with paraffin oil at a high temperature to raise its calorific value.

Coke is prepared from coal especially by high temperature carbonization, which yields a hard form, for use as a reducing agent in the manufacture of iron in a blast furnace.

(c) Sulphur

The 'spent' iron(III) oxide is heated in air to give sulphur dioxide which is then converted into sulphuric acid and finally ammonium sulphate, a fertilizer.

(d) Ammonia

The impure solution of ammonia in water obtained is heated with calcium hydroxide and the ammonia driven off absorbed in dilute sulphuric acid yielding ammonium sulphate.

(*e*) *Coal tar*

Over 2 000 chemicals are, or have been, prepared from coal tar: these range from pitch (as binder in fuel briquettes and electrodes) and road tar to phenol (for disinfectants) and pyridine (for weedkillers). Distillation of the tar and chemical conversions are required.

(*f*) *Benzene*

Benzene and naphthalene may be found in coal tar but they may also be obtained by the special washing process. Benzene has many uses (see section 29.12); naphthalene is used for making resins and paints, dyestuffs and as a moth killer.

29.6 The Manufacture of Town gas from Petroleum

A minor source for the manufacture of town gas at the moment is petroleum. The off-cut from the refinery is called naphtha: it is rich in hydrocarbons such as heptane (C_7H_{16}) and octane (C_8H_{18}) but also contains sulphur compounds such as ethanethiol (C_2H_5SH). Once in operation the plant functions as follows.

The naphtha is vaporized, mixed with some of the final product (to supply hydrogen) and passed over zinc oxide. This removes any sulphur in the naphtha (desulphurization).

$$C_2H_5SH + H_2 + ZnO \rightarrow ZnS + \underset{\text{Ethane}}{C_2H_6}\uparrow + H_2O\uparrow$$

Any alkenes present, i.e. substances containing carbon to carbon double bonds, also react with hydrogen (hydrogenation).

$$\underset{\text{Alkene}}{>C{=}C<} + H_2 \rightarrow \underset{\text{Alkane}}{>CH{-}CH<}$$

The second stage is the steam reforming of the paraffins which is the same as in the manufacture of hydrogen (see section 23.3): more steam is added and the gases passed over a nickel-chromium catalyst at 750°C. The third stage is the water-gas shift reaction to remove carbon monoxide and this is followed by the removal of carbon dioxide using potassium carbonate solution.

$$K_2CO_3 + H_2O + CO_2 \rightarrow 2KHCO_3$$

The gas at this stage may consist of 70% hydrogen, 15% methane, 10% carbon dioxide and 5% carbon monoxide and has a calorific value too low for sale. It must be enriched by adding more methane (another refinery product) and its density adjusted to half that of air. Thus the gas which is sold may have the percentage composition: hydrogen 45, methane 36, carbon dioxide 16 and carbon monoxide 3.

29.7 The Properties of Organic Compounds

Organic compounds differ from inorganic compounds in many respects:

(*a*) only carbon shows such a remarkable ability to link atoms in chains and rings;
(*b*) there are more compounds of carbon than those of all the other elements put together;
(*c*) although organic compounds usually contain only a few other elements (often hydrogen, oxygen and nitrogen and less frequently sulphur, phosphorus, etc.), they have a tremendous range of properties;
(*d*) covalency is more common than electrovalency;
(*e*) isomerism is more important (see section 29.9);
(*f*) they are less stable to heat;
(*g*) except for the compounds of low relative molecular mass they are usually insoluble in water, though more soluble in other organic compounds;
(*h*) many are inflammable because of their high carbon and hydrogen content.

In order to rationalize the study of organic chemistry the compounds are divided into **homologous series** associated with **functional groups**. In a homologous series the members:

(*a*) have the same general formula, e.g. C_nH_{2n+2} for the paraffins, successive members differing by CH_2;
(*b*) are made by similar methods;
(*c*) have physical properties showing a steady gradation, e.g. the boiling point increases with relative molecular mass, thus for the paraffins the first few members are gases, the second batch are liquids and the remainder are solids at room temperature;
(*d*) have chemical properties showing a considerable measure of similarity;
(*e*) have a first member which may be slightly different from the other members.

Thus there are a number of parallels between the homologous series of compounds in organic chemistry and the groups of elements with their compounds in inorganic chemistry. The radicals of inorganic chemistry correspond to the functional groups of organic chemistry.

Gas Fields: (north to south) Frigg, *Rough, *West Sole, Ann, *Viking, *Indefatigable, Accumulation, *Leman Bank, *Deborah, Dottie and *Hewett
(* In production)

Terminals for North Sea Gas:
Easington (Yorkshire)
Theddlethorpe St. Helen (Mablethorpe)
Bacton
Terminal for Algerian supplies: Canvey Island

Figure 29.7
The natural gas grid in Britain

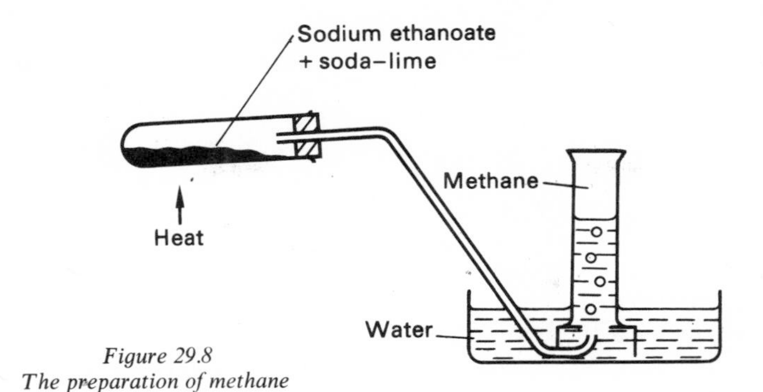

Figure 29.8
The preparation of methane

Some common functional groups are:

—CH_3	methyl	—OH	alcohol
—NH_2	amine	—Cl	chloride
>CO	ketone	—CHO	aldehyde
—COOH	carboxylic acid	>C=C<	alkene

Compounds are made up of units such as these, e.g. CH_3OH, methanol or methyl alcohol, for which the properties can be divided into those dependent upon the methyl group and those upon the hydroxyl group.

29.8 The Alkanes (Paraffins)

These are important constituents of petroleum; they have the general formula C_nH_{2n+2}. Methane (CH_4), the first member, is the main constituent of natural gas (90% of North Sea gas), of fire-damp in coal mines and of marsh gas from rotting vegetation. Ethane (C_2H_6), the second member, is found to a small extent in natural gas (3–5%).

In the laboratory methane can be made by Dumas' method: anhydrous sodium ethanoate is heated with soda-lime (sodium hydroxide and calcium hydroxide)—see figure 29.8.

$$CH_3COONa + NaOH \rightarrow Na_2CO_3 + CH_4\uparrow$$

Ethane can be made by Kolbe's method: the electrolysis of sodium ethanoate dissolved in ethanoic acid. At the anode ethane and carbon dioxide are evolved, the latter being eliminated by absorption in sodium hydroxide solution.

Methane has a boiling point of −160°C; it is a colourless, odourless, tasteless gas, with a relative density of 8. Like the other paraffins it is insoluble in water. The name paraffin is indicative of their lack of chemical reactivity: they are **saturated** compounds and, unless they disintegrate, they can only react by **substituting** atoms or radicals for the hydrogen attached to the carbon—often a slow process.

e.g. $$CH_4 + Cl_2 \rightarrow CH_3Cl + HCl$$

Thus two quick tests for the possible presence of a paraffin are that bromine water, and potassium manganate(VII) solution made alkaline with sodium carbonate solution (von Baeyer's reagent), are not affected.

The paraffins are inflammable; they burn in a free supply of air with hot non-luminous flames:

$$CH_4 + 2O_2 \rightarrow CO_2 + 2H_2O \quad \Delta H = -890\ \text{kJ/mol}$$

In a restricted supply of air carbon monoxide may be produced upon explosion and this

Figure 29.9
Methane, ethane and propane

'after-damp' makes rescue work after a coal mine disaster difficult. If the flame of burning methane impinges on a cold surface carbon black is deposited (see section 28.3); nowadays most is made using oil not methane. The liquid phase oxidation of paraffins yields ethanoic acid (see section 29.14). The steam reforming of methane and naphtha has already been dealt with (see sections 23.3 and 29.6).

Propane (C_3H_8) burns with the evolution of much heat. It is used as a solvent for extracting wax from lubricating oils to make candles and to cut metals. Butane (C_4H_{10}) may be more familiar as 'Calor' gas or 'Camping Gaz' because it is easily liquefied (b.p. 1°C) and transported in cylinders: it too burns with a hot flame and is dissolved in petrol for ease of starting internal combustion engines.

29.9 The Isomerism of the Paraffins

There is only one substance corresponding to each of the formulae CH_4, C_2H_6 and C_3H_8; usually represented in diagrams as

```
   H          H  H            H  H  H
   |          |  |            |  |  |
H—C—H    H—C—C—H      H—C—C—C—H
   |          |  |            |  |  |
   H          H  H            H  H  H
```

though in three dimensions the angles in between the bonds are 109°—an average value because rotation and vibration occur. However, in the case of butane there are two molecules each having the formula C_4H_{10} and in this case of isomerism the properties of the substances are very similar.

```
   H  H  H  H             H  H  H
   |  |  |  |             |  |  |
H—C—C—C—C—H        H—C—C—C—H
   |  |  |  |             |  |  |
   H  H  H  H             H  |  H
                            H-C-H
                              |
                              H
     Butane        2-Methylpropane (isobutane)
```

Isomers are substances having the same molecular formula but different structural formulae.

The naming of a paraffin is done in two parts —the first shows the number of carbon atoms: 1:meth-, 2:eth-, 3:prop-, 4:but- and the second, -ane, shows that the substance is a paraffin, an alkane, and is a saturated compound. When a complex paraffin is to be named the longest chain is first found and named. Then the side chains are named and their positions on the main chain are numbered so that the lowest set of numbers is obtained.

In the case of the pentanes there are three isomers

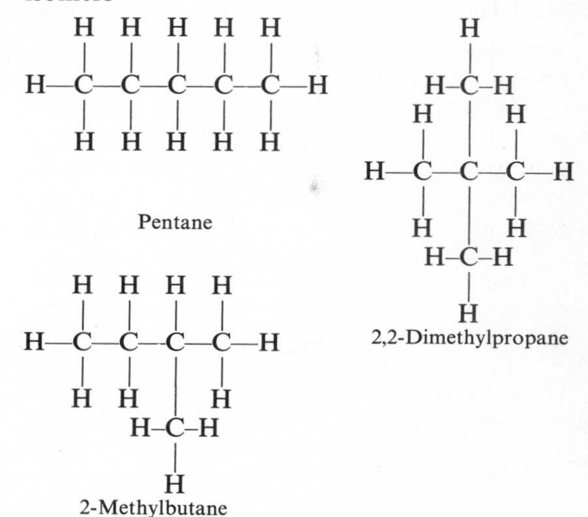

In accordance with the above rules the third isomer is called 2-methylbutane and not 3-methylbutane. Also, allowing for free rotation, there are only 3 isomers, although the first and third can be drawn in many different forms.

The numbers of isomers possible increases rapidly as the number of carbon atoms in the molecule increases:

Carbon atoms:	6	7	10	14	30
Number of isomers:	5	9	75	1858	4111846793

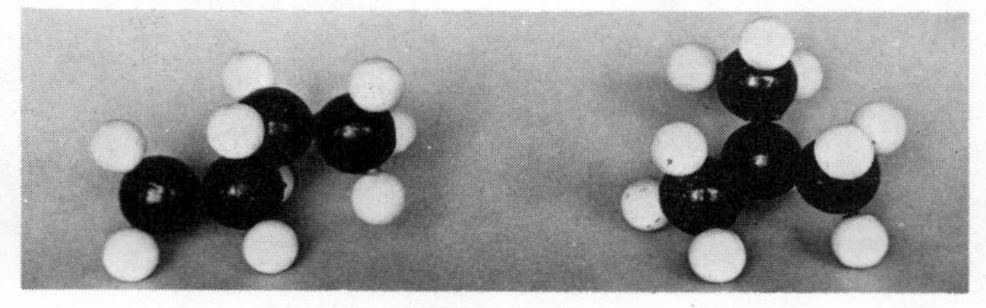

Figure 29.10
The two isomers of butane: butane and 2-methylpropane

Figure 29.11
The three isomers of pentane: pentane, 2-methylbutane and 2,2-dimethylpropane

29.10 The Alkenes (Olefines)

(a) Preparation

Ethene (ethylene) is the first member of the alkene series, in which members have the general formula C_nH_{2n}. Industry manufactures huge quantities of ethene from petroleum by catalytic cracking. In the laboratory the easiest preparation is to convert ethanol into ethene. This can be done by heating a mixture of two volumes of concentrated sulphuric acid with one volume of ethanol to 160–180°C; the addition of a little aluminium sulphate helps to reduce frothing. Because some sulphuric acid may be reduced, charring of the ethanol is evident (i.e. production of carbon); the gas evolved must be washed with an alkali to remove any sulphur dioxide or carbon dioxide (see figure 29.12). Concentrated phosphoric acid is sometimes preferred as an alternative to sulphuric acid.

$$C_2H_5OH \rightarrow C_2H_4\uparrow + H_2O$$

An alternative is to pass ethanol vapour over hot aluminium oxide.

(b) General Properties

Ethene has a boiling point of -103°C; it is a colourless gas with a slightly sweet smell and is almost insoluble in water; it has a relative density of 14. The general formula C_nH_{2n} means that all alkenes have the same empirical formula

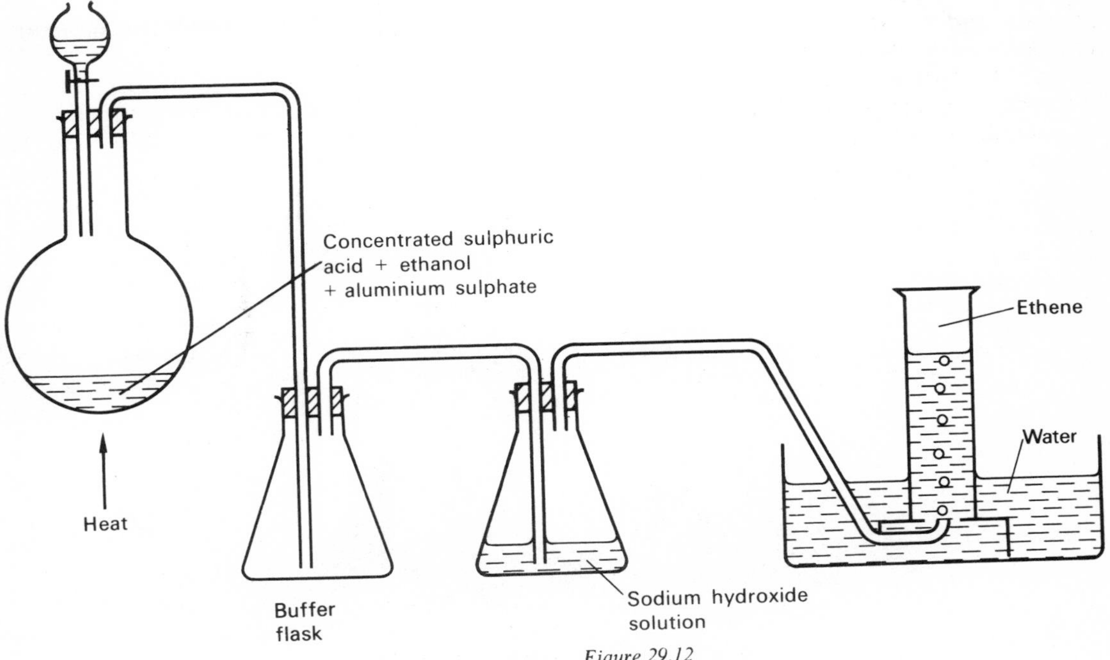

Figure 29.12
The preparation of ethene (ethylene)

(CH_2) and so differentiation between the members is by boiling point and relative density. The usual four-valency of carbon is indicated in these compounds by drawing a double bond between two of the carbon atoms

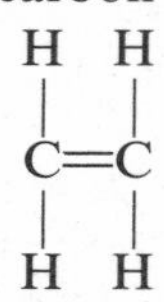

and it will be seen in this case using 'ball and spoke' models how this yields a flat molecule (see section 14.3 and figure 29.16). The double bond between the carbon atoms is not a sign of strength but of reactivity. The naming of the alkenes is also done in two parts: the first shows the number of carbon atoms (as for paraffins) and the second, -ene, shows that the substance has a double bond and hence it is an alkene (a number may have to be included in naming higher members to indicate the position of the double bond).

Ethene will burn in air with a hot luminous flame but an explosion is likely.

$$C_2H_4 + 3O_2 \rightarrow 2CO_2 + 2H_2O \qquad \Delta H = -1412 \text{ kJ}$$

In the presence of a silver catalyst, ethoxyethane is formed which, with water, gives ethane-1,2-diol $CH_2OH.CH_2OH$ (ethylene glycol, used as 'Bluecol' antifreeze for car radiators, for the manufacture of Terylene and for the manufacture of hydraulic fluids and solvents).

Ethene is an **unsaturated** compound: it will readily form compounds by **addition**. Thus two quick tests are that bromine water and potassium manganate(VII) solution made alkaline with sodium carbonate solution are both rapidly decolorized, the latter giving a brown precipitate of manganese(IV) oxide.

$$\underset{\text{(in bromine water)}}{C_2H_4 + \underset{\text{Bromic(I) acid}}{HOBr}} \rightarrow \underset{\text{2-Bromoethanol (Ethylene bromohydrin)}}{CH_2Br\ CH_2OH}$$

$$\begin{matrix} CH_2 \\ \| \\ CH_2 \end{matrix} \xrightarrow[\text{alkaline KMnO}_4]{\text{Oxidation with dilute}} \begin{matrix} CH_2OH \\ | \\ CH_2OH \end{matrix}$$

Ethene will also decolorize bromine gas giving 1,2-dibromoethane (ethylene dibromide) which is added to petrol to remove the lead of the tetraethyllead ($Pb(C_2H_5)_4$), itself added to prevent pre-ignition (knocking).

$$C_2H_4 + Br_2 \rightarrow CH_2BrCH_2Br$$

Hydrogen, in the presence of a nickel catalyst, adds on to ethene, giving ethane.

It is possible for ethene to undergo substitution reactions, e.g. at 350°C with chlorine it yields vinyl chloride, used in making plastics.

$$C_2H_4 + Cl_2 \rightarrow CH_2{=}CHCl + HCl$$

Ethene under 24 atmospheres pressure is absorbed by warm (65°C) concentrated sulphuric acid giving ethyl hydrogensulphate. If this is then diluted with water and distilled ethanol is produced, together with dilute sulphuric acid.

$$C_2H_4 + H_2SO_4 \rightarrow C_2H_5HSO_4$$
$$C_2H_5HSO_4 + H_2O \rightarrow C_2H_5OH + H_2SO_4$$

However, it is a costly and inconvenient business to reconcentrate the sulphuric acid and so the direct hydration process is now finding greater favour:

$$C_2H_4 + H_2O \rightarrow C_2H_5OH$$

This is carried out at 300°C in the vapour phase at a pressure of 67 atmospheres, the catalyst being phosphoric acid on a siliceous material such as 'Celite'. Under similar circumstances propene gives propan-2-ol.

$$CH_3CH{=}CH_2 + H_2O \rightarrow CH_3CHOHCH_3$$

(c) *Polymerization*

The largest use of ethene is in the manufacture of poly(ethene), (polythene, 'Alkathene'). Fawcett and Gibson (1933) in the United Kingdom, discovered that if a trace of oxygen was added to ethene gas at 200°C under 1500 atmospheres pressure a wax-like solid residue was produced. This low density form of poly(ethene), (920 kg/m³) softens at about 90°C; it has a relative

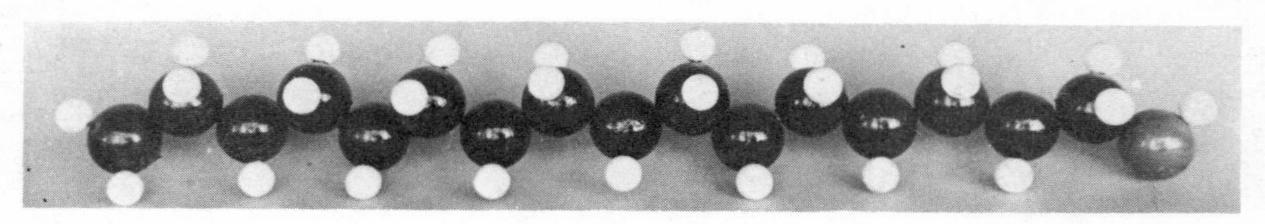

Figure 29.13
Poly(ethene) or polyethylene or polythene

Weather protection (perspex).

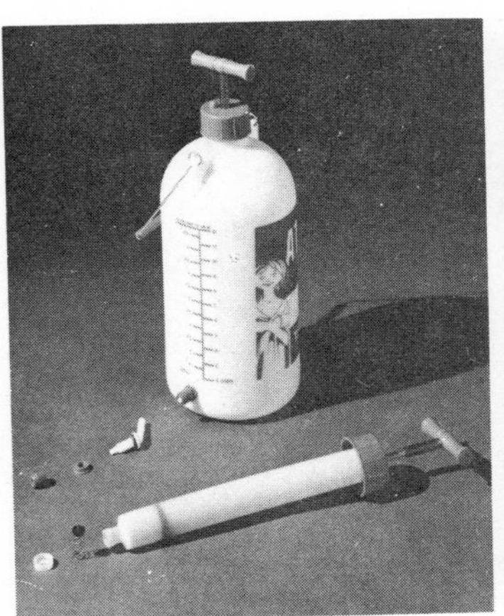

Car washer of poly(ethene) and poly(propene).

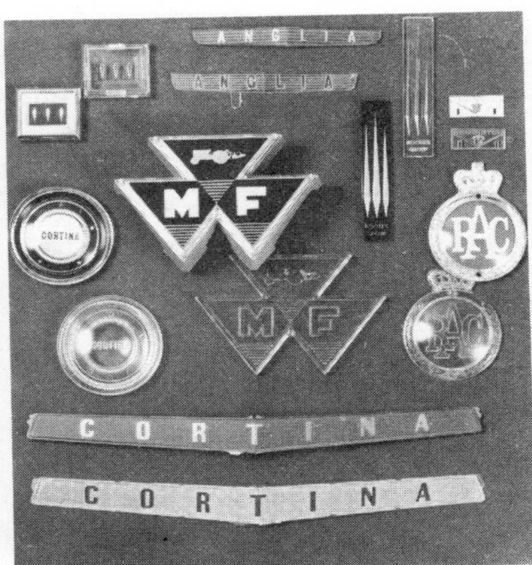

Insignia (acrylic).

Bottles of poly(vinyl chloride).

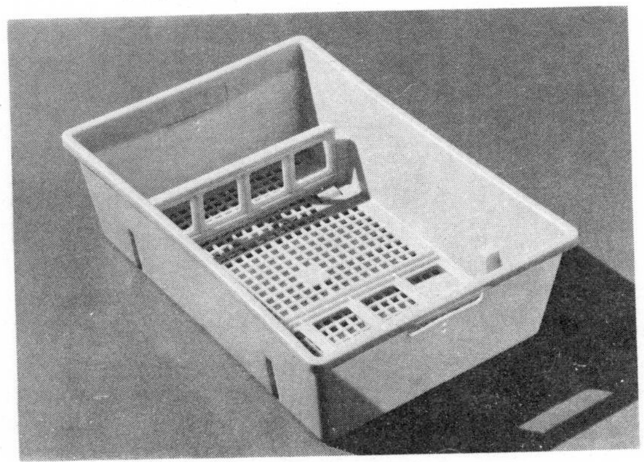

Box of poly(propene).

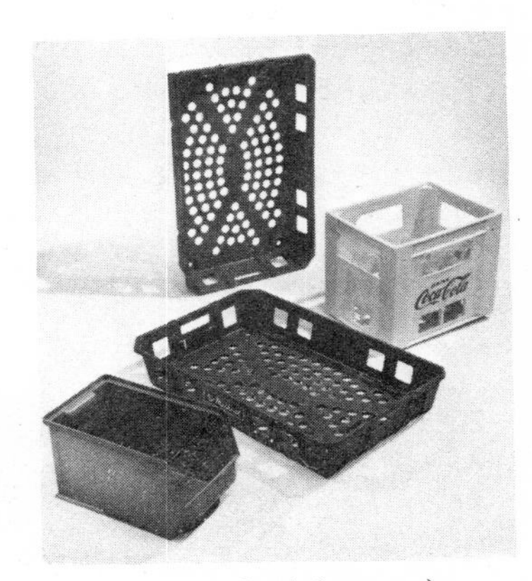

Boxes of poly(propene).

Figure 29.14
The wide range of plastic goods

molecular mass of about 30000; and although most of the molecules have linked in long chains by no means all of them have done so. It appears that the oxygen activates one end of the ethene molecule thus making it a radical which then attacks the next ethene molecule giving a larger radical until something happens which terminates the chain.

$$CH_2{=}CH_2 + CH_2{=}CH_2 + CH_2{=}CH_2$$

Ethene (monomers)

$$\rightarrow \begin{array}{cccccccccccc} & H & & H & & H & & H & & H & & H \\ & | & & | & & | & & | & & | & & | \\ - & C & - & C & - & C & - & C & - & C & - & C - \\ & | & & | & & | & & | & & | & & | \\ & H & & H & & H & & H & & H & & H \end{array}$$

Poly(ethene)—a polymer

The precise nature of the terminal groups is not clearly known. Ziegler (1953) in Germany discovered a low pressure method of polymerizing ethene and Natta (1954) in Italy used a similar process for propene (C_3H_6). At atmospheric pressure or slightly above (7 atmospheres) and at room temperature or slightly above (up to 170°C) using a complex catalyst of titanium and aluminium compounds dissolved in heptane (C_7H_{16}), polymerization readily takes place giving a high density (960 kg/m^3) polymer. This poly(ethene) softens at about 105°C (so it can be safely sterilized) and the molecules have fewer and shorter sidechains, hence the more tightly packed structure.

Vinyl chloride ($CH_2{=}CHCl$) is not usually made by the substitution reaction mentioned above but by a more complex process. It too will polymerize giving poly(vinyl chloride) (P.V.C.) used for floor tiles, pipes and synthetic fibres.

$$CH_2{=}CHCl + CH_2{=}CHCl \quad CH_2{=}CHCl$$

Vinyl chloride (monomers) or chloroethene

$$\rightarrow \begin{array}{cccccccccccc} & H & & H & & H & & H & & H & & H \\ & | & & | & & | & & | & & | & & | \\ - & C & - & C & - & C & - & C & - & C & - & C - \\ & | & & | & & | & & | & & | & & | \\ & H & & Cl & & H & & Cl & & H & & Cl \end{array}$$

Poly(vinyl chloride)—a polymer

The reaction of ethene with benzene and subsequent dehydrogenation yields styrene which is yet another substance with a double bond ($C_6H_5CH{=}CH_2$) and it may be polymerized alone to give poly(styrene) (used for heat and sound insulation, packing and cups) or with other substances (e.g. butadiene to give a synthetic rubber).

Propene, the second known member of the alkene series, has already been mentioned. On polymerization it gives the plastic with the lowest density (910 kg/m^3) but it has a softening point of 145°C. Nowadays it is the starting point for making acrylonitrile: it is passed with ammonia, air and steam at a pressure slightly above atmospheric over a catalyst such as bismuth phosphomolybdate or uranium at 450°C.

$$2CH_2{=}CH.CH_3 + 2NH_3 + 3O_2 \rightarrow 2CH_2{=}CHCN + 6H_2O$$

Acrylonitrile after polymerization is retailed as fibres such as 'Orlon', 'Acrilan' and 'Courtelle'. These fibres (or rayon) are the starting point for the manufacture of carbon fibres for bonding into metals and plastics for extra strength.

If propene reacts with benzene, it may then be converted into phenol (C_6H_5OH) and the useful solvent propanone (CH_3COCH_3). Propanone can be converted by using hydrogen cyanide, concentrated sulphuric acid and finally methanol into methyl methacrylate (2-methylpropenoate).

$$(CH_2{=}C(CH_3).COOCH_3).$$

This polymerizes in a manner similar to those of the substances described above to yield a substance usually known as Perspex (discovered in the United Kingdom in 1932) which is a better transmitter of light than glass.

Plastics and synthetic rubbers account for over half the world's production of organic chemicals (36 million tonnes a year) and the growth rate is tremendous. See section 29.20.

29.11 The Alkynes (Acetylenes)

The manufacture of calcium dicarbide and hence ethyne has already been discussed (see section 25.7). The laboratory preparation utilizes the manufactured calcium dicarbide in the same manner (see figure 29.15). The manufacture of

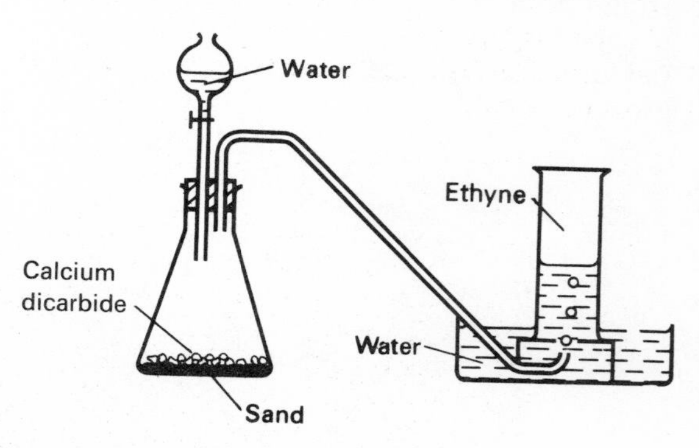

Figure 29.15
The preparation of ethyne (acetylene)

ethyne direct from hydrocarbons is increasing in importance.

The colourless gas thus prepared is impure and has a characteristic smell—it is poisonous. It has a density just less than that of air; it is moderately soluble in water but is usually collected over water. Ethyne is very soluble in propanone and is usually stored dissolved in it. Ethyne is the first member of the alkyne series which has a general formula of C_nH_{2n-2}. In order to maintain the four-valency of carbon a triple bond is drawn between the two carbon atoms giving a linear molecule H—C≡C—H (see figure 29.16). This triple bond is not a sign of stability or strength but of reactivity.

Mixtures of ethyne with oxygen or air can be dangerously explosive so **care** must be taken in performing experiments. On ignition (and using a full gas jar) ethyne burns with a hot, sooty, luminous flame. If mixed with the right amount of air or oxygen and ignited at a jet it burns with a very hot flame which is used for cutting and welding metals.

$$2C_2H_2 + 5O_2 \rightarrow 4CO_2 + 2H_2O \quad \Delta H = -1\,308 \text{ kJ}$$

Ethyne is an unsaturated compound; it will form compounds by addition of hydrogen (giving finally ethane), of halogens and of halogen hydrides.

Thus it decolorizes bromine water and alkaline potassium manganate(VII) in a similar manner to ethene but much more slowly. It can be converted into trichloroethene ($CHCl{=}CCl_2$) which is a useful dry cleaning solvent and into 2-chloro-1,3-butadiene which on polymerization gives an oil-resistant rubber (neoprene). A useful test to distinguish ethyne from ethene is to pass the gases into a solution of copper(I) chloride in aqueous ammonia when the former gives a red precipitate but the latter gives no result.

$$2CuCl + C_2H_2 \rightarrow \underset{\text{Copper(I) dicarbide}}{Cu_2C_2\downarrow} + 2HCl$$

29.12 Cyclohexane and Benzene

The compounds considered in detail so far have had their carbon atoms arranged in chains. However, there is a form of hexane which has its carbon atoms arranged in a regular hexagon and has two hydrogen atoms attached to each carbon; because it behaves as a paraffin, although it does not have the same general formula as the linear molecules, it is called cyclohexane (C_6H_{12}).

If hexane gas mixed with hydrogen is passed over a platinum catalyst at 550°C under a pressure of 20 atmospheres a molecular rearrangement occurs yielding benzene, which liquefies on cooling to normal temperatures, and more hydrogen which is a very useful by-product.

$$C_6H_{14} \rightarrow C_6H_6 + 4H_2$$

This way of manufacturing benzene is now more important than its extraction as one of the products of the carbonization of coal.

Benzene has a melting point of 5°C and a boiling point of 80°C; it has a pleasant but **most dangerous** odour; it is almost insoluble in water upon which it floats (density 880 kg/m³). The vapour is carcinogenic (causes cancer) and its use in school must be very carefully controlled or forbidden.

The chemical properties of benzene are very odd, for it does not show the unsaturation which would be expected of a compound with six carbon and six hydrogen atoms. Thus its structure is often represented by a regular (planar) hexagon with a circle inside to denote the difference from a saturated compound such as cyclohexane (which incidentally is not planar).

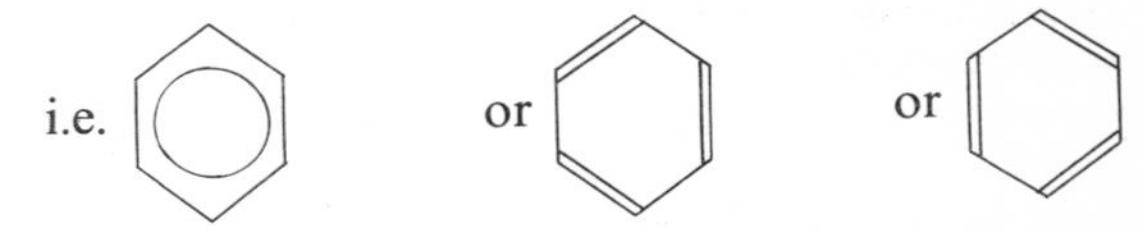

There is a carbon atom, with one hydrogen atom attached, at each corner. Benzene and other compounds which show a similar stability are

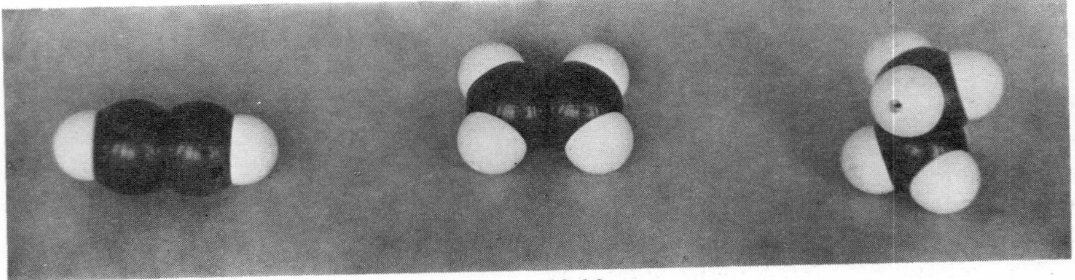

Figure 29.16
Ethyne, ethene and ethane

considered in a class by themselves, known as the aromatics.

Benzene burns with a very smoky, but nevertheless hot, flame unless there is a forced supply of oxygen: it is frequently included in petrol to improve its octane rating. It can be reduced to cyclohexane (used in making nylon). It is used to manufacture phenol (C_6H_5OH, for nylon and plastics), styrene (for plastics), aniline ($C_6H_5NH_2$, for dyes) and many other very important substances.

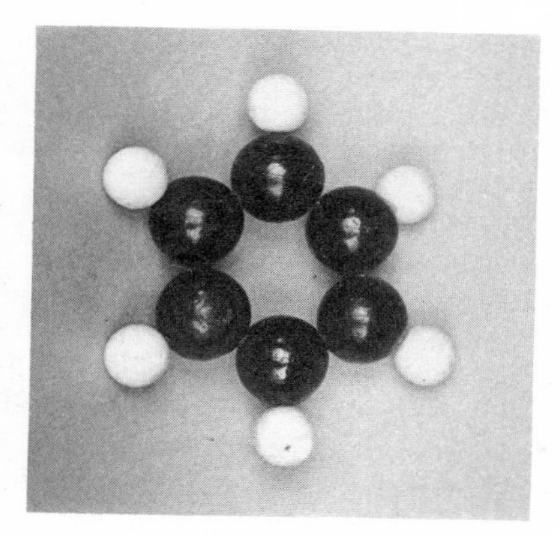

Figure 29.17
Benzene

29.13 Ethanol (Ethyl Alcohol)

The first member of the homologous series of alcohols, methanol (CH_3OH), is not as important as the second, ethanol (C_2H_5OH). The naming should be viewed in three parts: eth- which shows that there are two carbon atoms, -an- that all the bonds between carbon atoms are single (paraffinic) and -ol that the substance contains an -OH functional group. The alternative system (obsolescent but persistent) starts by considering a paraffin losing a hydrogen atom giving an alkyl radical, a monovalent constituent found in many compounds. Ethanol is commonly referred to just as alcohol.

The industrial process for preparing ethanol has already been mentioned in section 29.10. It is usually sold containing 5% of methanol (poisonous, causes blindness) to laboratories as industrial spirit and with methanol, a violet dye and other obnoxious substances to the general public as methylated spirits. The reason for these additions is to prevent it being used in drinks which carry a high rate of tax. Ethanol is the second most important solvent.

For the purpose of alcoholic beverages a carbohydrate is fermented (see section 29.18). The commonest starting point is sucrose or cane sugar ($C_{12}H_{22}O_{11}$). This is dissolved in water and added to brewers' yeast. This is a simple plant which grows in the sugar solution and contains enzymes: these are very specific catalysts found associated with living cells but able to act away from them. Two different enzymes catalyze the two reactions which occur:

$$\underset{\text{Sucrose}}{C_{12}H_{22}O_{11}} + H_2O \xrightarrow{\text{(Sucrase)}} \underbrace{\underset{\text{Glucose}}{C_6H_{12}O_6} + \underset{\text{Fructose}}{C_6H_{12}O_6}}_{\text{Isomeric substances}}$$

$$C_6H_{12}O_6 \xrightarrow{\text{Zymase complex}} 2C_2H_5OH + 2CO_2$$

(if glucose $\Delta H = -71$ kJ).

The temperature must be kept below 35°C and if the alcohol concentration rises too high the yeast will die. An air lock (water seal) should be fitted to the top of the vessel. It is easy to show that carbon dioxide is evolved.

Home Experiment: To Make Orange Wine

1 dm^3	Orange juice (not squash).
1·25 kg	Sugar (sucrose).
20 g	Brewers' or bakers' or wine-makers' granulated yeast.
20 g	Citric acid.
0·9 g	Potassium metabisulphite, $K_2S_2O_5$ (2 Campden tablets).
15 g	Yeast nutrient (8·5 g citric acid, 4 g ammonium sulphate, 2 g dipotassium hydrogenphosphate, 0·5 g magnesium sulphate).
50 cm^3	Wine finings e.g. processed isinglass.

The orange juice is put in a bottle of volume about 5 dm^3. 1 kg of the sugar is dissolved in the minimum quantity of hot water, the citric acid is added and then the solution poured into the orange juice. The bottle is filled up so that the liquid level is about 3 cm below the top. A fermentation lock must be fitted, e.g. a Bunsen

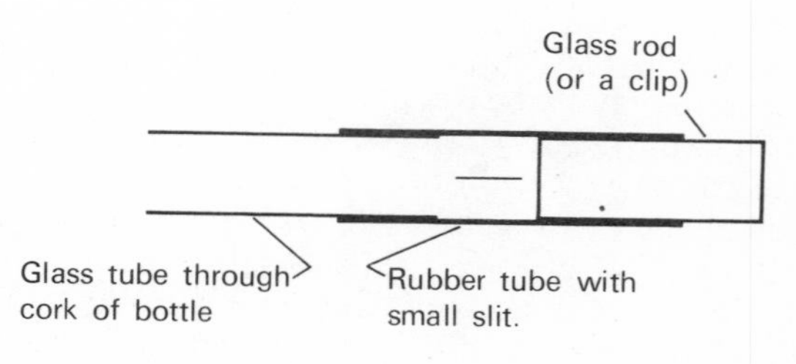

Figure 29.18
The Bunsen valve

valve (see figure 29.18). The solutions should be left to cool to about 22°C before the yeast and its nutrient are added.

The bottle is then placed in a warm position, e.g. near the hot water tank, to allow fermentation to proceed. After about 5–6 weeks no further bubbles should be seen and the cloudy solution can be siphoned away from the sediment. The solution should be left to settle for a further 1–2 weeks before adding a Campden tablet and the wine finings which will help to clarify the solution. The solution is left until it is clear and then siphoned away from any fresh sediment.

If a sweet wine is required 0·25 kg of sugar can be stirred into the cold solution. The bottles in which the wine is to be kept should then be sterilized with a Campden tablet. The wine can now be bottled, corked and drunk in moderation!

Beers and ciders contain by volume 5–10% of alcohol, table wines 10–15% and spirits 35–40%. Small quantities of alcohol produce a sensation of well-being and frequently make imbibers more talkative; large quantities in a short period produce unconsciousness, and moderate or large quantities over a long period produce bodily changes resulting in alcoholism, a degenerative disease ending in death. Alcoholic drinks have been socially acceptable for so long that it is sometimes hard to realize that it is really a drug with its attendant dangers. It is morally indefensible that anyone should drive a car when under the influence of a drug which is likely to make him reckless, and recent legislation making it an offence to drive with more than 80 mg of alcohol in 100 cm^3 of blood is an important factor in road safety.

The fermentation process is still economic for small scale production of ethanol. The laboratory product can be processed in the same way: after fermentation the solution is filtered and the filtrate put in a flask fitted with a fractionating column. On careful distillation a colourless liquid which boils at 78°C is obtained (rectified spirit): it contains 96% ethanol and 4% water. The water can be removed by adding calcium or its oxide and then the pure ethanol distilled off.

Ethanol can be used in low temperature thermometers because it melts at −117°C. It is completely miscible with water and is used alone or in such mixtures as a solvent.

Ethanol is readily inflammable, burning with a blue flame: it can be used as a fuel when camping and is included in some petrols.

$$C_2H_5OH + 3O_2 \rightarrow 2CO_2 + 3H_2O \qquad \Delta H = -1\,372 \text{ kJ}$$

It can be less drastically oxidized by warming with potassium dichromate (or manganate(VII)) solution and sulphuric acid: the colour change seen is from orange to very dark green (or purple to colourless with a brown suspension).

$$C_2H_5OH \rightarrow \underset{\text{Ethanal}}{CH_3CHO} + \underset{\text{To oxidizing agent}}{2H^+ + 2e^-}$$

Ethanal (acetaldehyde) has an apple-like odour, readily boils off (b.p. 21°C) but if the oxidation can be prolonged then ethanoic acid (vinegar) is produced.

$$CH_3CHO + H_2O \rightarrow CH_3COOH + 2H^+ + 2e^-$$

Atmospheric oxygen may cause the formation of ethanoic acid over a period of time: the wine is said to have gone sour.

The reaction of ethanol with concentrated sulphuric acid evolves much heat.

$$C_2H_5OH + H_2SO_4 \rightarrow \underset{\text{Ethyl hydrogensulphate}}{C_2H_5HSO_4} + H_2O$$

On boiling ethyl hydrogensulphate (an ester, see section 29.14) with water, ethene is formed (see section 29.10). See also sections 30.22 (with phosphorus chlorides), 24.5 (with sodium) and 29.14 (with ethanoic acid).

Other important alcohols are:

CH_2OHCH_2OH	ethane-1,2-diol
$CH_3CH_2CH_2OH$	propan-1-ol
$CH_2OHCHOHCH_2OH$	propane-1,2,3-triol
$CH_3CHOHCH_3$	propane-2-ol

The sugars and other carbohydrates contain many alcoholic —OH groups (see section 29.18).

29.14 Ethanoic Acid (Acetic Acid)

Ethanoic acid is the second member of the series of carboxylic acids, i.e. those containing the group of atoms $-C\overset{\displaystyle O}{\underset{\displaystyle O-H}{\lessgtr}}$, frequently written —COOH or $-CO_2H$. The first member of the homologous series is methanoic acid (formic acid, HCOOH) but this is not typical of the majority. Ethanoic acid can be produced in the laboratory by the oxidation of ethanol as described in the previous section. In industry similar processes are giving way to the liquid phase oxidation of butane under 60 atmospheres pressure at 175°C, a cobalt or manganese compound being used as the catalyst. The butane used is in a mixture with other hydrocarbons (from petroleum distillation) and the liquid produced on cooling is fractionally distilled yielding not only ethanoic acid but several other acids as well.

Pure ethanoic acid melts at 17°C (because it freezes in a cold room it is sometimes called glacial ethanoic acid) and boils at 118°C. It is a weak acid but is very corrosive to the skin. The dilute solution, about 5% in water, is known as vinegar.

As an acid it will react with magnesium to yield hydrogen.

$$Mg+2CH_3COOH \rightarrow (CH_3COO)_2Mg+H_2\uparrow$$

It will neutralize sodium hydroxide to give sodium ethanoate and water.

$$CH_3COOH+NaOH \rightarrow CH_3COONa+H_2O$$

If added to sodium carbonate, carbon dioxide is given off and sodium ethanoate is left in solution.

$$2CH_3COOH+Na_2CO_3 \rightarrow 2CH_3COONa+H_2O+CO_2\uparrow$$

The salts of some acids of high relative molecular mass are known as soap (see section 29.15).

The reaction of an acid—inorganic (see section 29.13) or organic—with an alcohol gives an ester and water:

$$CH_3COOH+C_2H_5OH \rightleftharpoons \underset{\text{Ethyl ethanoate}}{CH_3COOC_2H_5}+H_2O \quad \Delta H = 0\text{ kJ}$$

The reaction is catalyzed by hydrogen ions, the yield being enhanced by the removal of the water formed: thus concentrated sulphuric acid (a dehydrating agent) is added in a small proportion to perform both functions. The reaction has some similarity to the preparation of a salt: see the table below.

The main use of ethanoic acid in the U.K. is as cellulose ethanoate (34%, for synthetic fibres, plastics and packaging materials); 20% is used in preservatives in food (vinegar), tobacco, rubber and photographic processing; 17% as vinyl ethanoate for emulsion paints and adhesives; 12% in solvents and 10% for other fibres.

29.15 Oils, Fats, Soaps and Detergents

Several esters of glycerol occur in nature: three important fatty acids used in forming these esters are oleic acid ($C_{17}H_{33}COOH$) in olive oil, stearic acid ($C_{17}H_{35}COOH$) in beef or mutton fat and palmitic acid ($C_{15}H_{31}COOH$) in palm oil. At room temperature an oil is liquid and a fat is solid.

Fats are energy-producing foods, by oxidation (like that of carbohydrates): the body keeps most of its reserve supply of fuel in the form of fat and foodstuffs rich in fat are particularly important in cold climates. Butter, lard and suet contain fat in a solid or semi-solid form. Milk contains tiny fat particles in suspension (a colloid).

The supply of fats in the world does not meet the demand, except in Western Europe where supplies and economic considerations led recently to a stockpile of butter greater than the annual consumption in the United Kingdom! The difference in chemical nature between some oils and some fats is that the oils contain olefinic double bonds and are unsaturated whilst the corresponding fats are saturated.

The process for converting oils into fats was devised by Sabatier and Senderens in 1902. Under the influence of a nickel catalyst at 180°C and at a pressure of 5 atmospheres the hydrogen molecule adds on to the unsaturated organic compound giving a saturated compound which on cooling solidifies to a fat. The essential reaction is the hydrogenation of the olefinic double bond of the oil.

$$>C{=}C< +H_2 \rightarrow >CH{-}CH<$$

Margarine is made by mixing and vigorously stirring (beating) fat-free milk with oils, fats, salt and vitamins A and D until on standing a solid product is obtained. The word margarine means that the solid has a 'pearly' appearance. It is a cheap substitute for butter and has the advantage that it is of a consistent quality

	Organic	*Inorganic*
Reaction	Acid + Alcohol ⇌ Ester + Water	Acid + Base → Salt + Water
Name	Esterification	Neutralization
Speed	Takes time—hastened by heating	Instantaneous
Mechanism	– OH from organic acid, – H from alcohol form water	OH^- from base, H^+ from acid form water
Direction	Mostly two way, kinetic equilibrium established	Mostly one way, some salts react slightly with water (hydrolysis)
Yield of product	Variable, often about 50%	Nearly 100%

throughout the year. Margarine has 55% of the market (total about 0·978 m tonnes in the United Kingdom in 1973).

Soap is made from oils and fats by boiling them with sodium hydroxide solution (for a hard soap) or with potassium hydroxide solution (for a soft soap): the process is known as saponification.

Oil or Fat + Sodium hydroxide
Ester Alkali

→ Sodium salt of acid + Glycerol
Soap Alcohol

Sodium chloride is added to reduce the solubility of soap in the water present: the soap separates as a white, curdy mass which hardens as it cools. The glycerol can be extracted and purified by distillation. For the interaction of soap with hard water see section 25.13.

Soapless detergents are substitutes for soap: they are made by treating benzene derivatives (synthesized from petroleum distillates) with concentrated sulphuric acid; one of the first was 'Teepol' produced by Shell in 1942. It is a sulphate and is decomposed by bacteria, i.e. it is biologically soft. During the last decade many soapless detergents that are biologically hard have been made but their manufacture is being phased out so that foaming on rivers should soon become a thing of the past. The powder detergents are mixtures of these organic sulphates (or more often sulphonates) with sodium phosphate, sodium silicate, sodium carbonate and minor constituents such as bleaches and whitening agents. The advantages of soapless detergents over soap is that hard waters do not give precipitates with the former. On the other hand for those in rural areas who pipe their sewage into septic tanks detergents (and bleaches) are used sparingly. Synthetic detergents have captured 70% of the U.K. market for cleaning materials; the total used was 0·893 m tonnes in 1973.

29.16 Aminoethane and Aminoethanoic Acid

If bromoethane (ethyl bromide) is heated with a solution of ammonia in ethanol, aminoethane (ethylamine) is produced in the form of its salt ethylammonium bromide.

$$C_2H_5Br + NH_3 \rightarrow C_2H_5NH_3^+Br^-$$

On the addition of sodium hydroxide solution the salt decomposes releasing the free amine. There are alternative but more complicated methods of preparation which give a better yield. Many similarities exist between ammonia and aminoethane: both are substances with characteristic odours and a high solubility in water. Aminoethane is a stronger base than ammonia and it can be titrated with hydrochloric acid.

$$C_2H_5NH_2 + HCl \rightarrow C_2H_5NH_3^+Cl^-$$
In aqueous solution

It will precipitate metal hydroxides and gives a deep blue coloration finally with copper(II) salts. With organic acids (or their derivatives) compounds with the linkage —NH—CO— are formed: this configuration is known as the peptide linkage.

$$C_2H_5NH_2 + (CH_3CO)_2O$$
Ethanoic anhydride

$$\rightarrow C_2H_5NHCOCH_3 + CH_3COOH$$
N-ethyl ethanamide

Aminoethanoic acid (glycine, aminoacetic acid) has the formula CH_2NH_2COOH but when it is written H_2NCH_2COOH the fact that one end of the molecule is basic and the other acidic is emphasized. It is a crystalline solid which is very soluble in water: it exists as an internal salt $^+H_3NCH_2COO^-$. Like aminoethane it is a member of a homologous series and there are many substances related to it. The amino acids are the main constituents of proteins which are the structural materials of body tissue (the muscles, blood, skin, internal organs, hair and nails) and also of enzymes which are essential for metabolic processes in any living organism. Proteins are essential articles of diet because animals need a supply of those amino acids which they cannot synthesize. The linkage between amino acids is a peptide link and thus a protein is a polypeptide. The test for a peptide linkage is that, on the addition of sodium hydroxide solution and then a drop of copper(II) sulphate solution, a violet coloration is produced. On hydrolysis polypeptides produce a mixture of amino acids which are hard to separate; paper chromatography and, on a large scale, ion exchange are the best methods. X-ray crystallography has made it possible to unravel the structures of proteins such as haemoglobin (red cells in blood), myoglobin, ribonuclease and lysosyme, and also of DNA (deoxyribonucleic acid) and RNA (ribonucleic acid) which make the proteins.

29.17 Terylene and Nylon

The polymerizations discussed so far have been of the pattern:

$$nA \rightarrow A_n$$

Figure 29.19
The nylon plant at Wilton

the empirical formulae of the monomer and polymer being the same. A second type of polymerization is known as condensation polymerization: molecules of the polymer are produced with the elimination of a small molecule such as water:

$$nA + nB \rightarrow (AB)_n + nH_2O$$

The reaction of ethanoic acid with ethanol produces the simple ester, ethyl ethanoate, and water. If a dibasic acid (terephthalic acid,

$HOOCC_6H_4COOH$ or HOOC–⟨benzene ring⟩–COOH

is added to a dihydric alcohol (ethane-1,2-diol, ethylene glycol, $HOCH_2CH_2OH$) a polyester (Terylene) and water are produced.

$$\underset{\text{The monomers}}{HOOCC_6H_4COOH + HOCH_2CH_2OH}$$
$$\rightarrow \underset{\text{The dimer}}{HOOCC_6H_4COOCH_2CH_2OH} + H_2O\uparrow$$

By continuation of this process a polymer is produced: the acid, in the form of its dimethyl ester, is heated with ethane-1,2-diol to 180°C in an atmosphere of nitrogen. Terylene was discovered by Whinfield and Dickson in the United Kingdom in 1941. The molten polymer is extruded through fine nozzles and the fibres are then cold drawn to give them strength. It is easy to see where the word Terylene has come from; the fibre is also known as 'Dacron' and is used in tyres, belting, nets and clothing.

Nylon is another condensation polymer: it was discovered by Carothers in the U.S.A. in 1937 and contains peptide linkages. The most common form is nylon 66 which is made by heating hexanedioic acid (adipic acid, $HOOC(CH_2)_4COOH$) with hexane-1,6-diamine (hexamethylenediamine, $H_2N(CH_2)_6NH_2$) to give the dimer:

$$HOOC(CH_2)_4CONH(CH_2)_6NH_2 + H_2O\uparrow$$

By continuation of this process the polymer is

			Thousand tonnes
Natural Fibres			
	Vegetable:	Cotton	133
		Jute	67
		Hemp and Sisal	48
		Flax	26
	Animal:	Wool	137
		Silk	0·11
Man-made			
	Regenerated:	Viscose rayon	224
		Acetate rayon	52
		Glass	67
	Synthetic:	Polyamide	196
		Acrylic	136
		Polyester	120

Figure 29.20
The consumption of fibres in the United Kingdom in 1973 (by mass)

produced: there are about 2000 units in each molecule. As before, extrusion and cold drawing of the polymer give a fibre. The fibre is stronger, per unit mass, than any natural fibre. Nylon has a melting point of 263°C and a very low coefficient of friction. It is used for ropes, nets, tyres, belting, gear wheels, carpets and clothing.

Synthetic fibres have captured a large proportion of the market which used to be dominated by cotton and wool because they are superior in some respects such as crease resistance, water repellency and strength.

29.18 Carbohydrates

The carbohydrates are compounds of carbon, hydrogen and oxygen: in many of the ones studied in the last century the ratio of hydrogen atoms to oxygen atoms was 2:1, hence the name. The water is not present even as water of crystallization but a strong dehydrating agent such as concentrated sulphuric acid may be able to remove water leaving carbon.

The carbohydrates can be divided into the monosaccharides, e.g. glucose and fructose, which cannot be hydrolyzed to simpler sugars; the disaccharides, e.g. sucrose, which hydrolyze giving two molecules of monosaccharides, etc. In the polysaccharides, e.g. starch and cellulose, the number of saccharide residues is large but variable.

The carbohydrates are important items in our nutrition because on oxidation they release energy. The reactions involve many enzyme-catalyzed steps but overall they are:

$$\underset{\text{Glucose}}{C_6H_{12}O_6} + 6O_2 \rightarrow 6CO_2 + 6H_2O \qquad \Delta H = -2815\,\text{kJ}$$

$$\underset{\text{Sucrose}}{C_{12}H_{22}O_{11}} + 12O_2 \rightarrow 12CO_2 + 11H_2O \qquad \Delta H = -5650\,\text{kJ}$$

The energy released keeps our body temperature at 37°C, at which temperature our enzymes work most efficiently, despite the temperature of a room usually being about 17°C. They also enable the muscles to do work by causing the necessary chemical reactions and foster the synthesis of new body material, i.e. proteins.

The fermentation of starch, sucrose and glucose yields ethanol (see section 29.13).

(*a*) *Cellulose*

The fibrous parts of plants are composed of cellulose, a carbohydrate of complex structure, $(C_6H_{10}O_5)_n$. The molecules consist of chains of varying length (up to 8000 glucose units) which, when bundled together, give the plant its strength and flexibility. Cotton, jute, flax (linen), hemp and sisal are natural vegetable fibres. Wood is composed largely of cellulose and lignin (another polysaccharide derivative). Although there are many hydroxyl groups present the materials are insoluble in water because of their high relative molecular mass. The glucose units are linked in a different way than in starch.

Cotton can be mercerized (named after its inventor Mercer) by soaking it in sodium hydroxide solution: it then has a shiny surface. Cross and Bevan (1892) discovered that wood-pulp would dissolve in sodium hydroxide solution to yield a viscous solution which would react with carbon disulphide to give a bright yellow compound known as sodium cellulose xanthate. When this is extruded through fine circular holes into dilute sulphuric acid the cellulose of the wood is regenerated as fibres, known as viscose rayon (about 400 glucose units in each molecular chain). By using a slot instead of a hole a film can be obtained—cellophane. Cellulose nitrate is used as an explosive—gun cotton or cordite. Cellulose ethanoate (acetate) is used for fibres and for the backing of films (it is not inflammable).

In order to make paper, wood is debarked and then, in the form of chips, soaked in calcium hydrogensulphate(IV) solution to dissolve the non-cellulosic materials. The fibres left are beaten mechanically to reduce them to a very small size, mixed with china clay (twice as much used here as in making pottery), aluminium sulphate and other materials and then poured as a very dilute suspension (0·5%) on to a moving belt of bronze gauze. The water rapidly drains away, a process enhanced by rollers and steam heated cylinders, and a strip of paper is made at speeds up to 20 km/hour. Paper, like soap, still contains some water (about 8%): it would be too hard and brittle otherwise. The many uses of paper almost defy any attempt to list them but they include: books, newspapers, bags, boxes, cards for computers, paper for duplicators, copiers and writing, drying towels and handkerchiefs.

Cellulose cannot be used as a foodstuff by human beings, but herbivores (animals which feed on plant material) are able to digest it because they either possess an extra four-chambered stomach (the rumen) or a caecum. These structures contain bacteria which produce enzymes and thereby break down cellulose.

(b) Starch

Starch is another macromolecular substance composed of units of glucose linked by the elimination of water between molecules. It is found in most vegetables and cereals: the percentages by mass in some common foodstuffs are rice 80, oatmeal 70, white bread 45 and potatoes 20. Starch is enclosed in a protective coat of cellulose and so such foods need plenty of cooking to break open the granules containing it. Saliva and the duodenum contain an enzyme (amylase), which is able to catalyze the hydrolysis of starch to maltose, so digestion begins in the mouth and is completed to glucose in the small intestine where there is a second enzyme (maltase). Partial hydrolysis of starch gives the dextrins which are used as the gum on stamps and envelopes. If hydrolysis is catalyzed by dilute acids such as hydrochloric (the 0·0001 M acid is present in the stomach) glucose is produced.

$$(C_6H_{10}O_5)_n + nH_2O \rightarrow nC_6H_{12}O_6$$

Pure starch is a white powder which exists in two forms: one soluble, the other insoluble, in water. The test for soluble starch is to add iodine solution (dissolve the crystals in aqueous potassium iodide): a dark blue coloration is produced. A solution of starch in water is used in laundry work for giving materials rigidity.

(c) Sucrose

Two thirds of the sucrose used is obtained from sugar cane grown in tropical countries and one third from sugar beet grown in temperate climes. It is obtained by crystallization of an aqueous solution: this is not easy and is performed under reduced pressure. The fibrous part of the sugar-cane is used for making paper and boards; molasses, which are the non-crystallizable parts of the solution, are used as cattle feeds, syrup and treacle. Heating sucrose gives barley sugar and then caramel. Strong heating in the absence of air or in the presence of concentrated sulphuric acid (a dehydrating agent) yields sugar charcoal (carbon). See section 28.3.

$$C_{12}H_{22}O_{11} \rightarrow 12C + 11H_2O$$

Animal charcoal is used to decolorize the sugar initially obtained by crystallization.

Sugar, as it is usually referred to, is very soluble in water and has a sweet taste; the molecule consists of two parts—glucose and fructose (see below). The consumption of sugar in the U.K. has increased from about 2 kg per person in 1770 to about 55 kg per person in 1973: the consumption of sweet foods, especially between meals, leads to increased dental decay. Sucrose is a non-reducing sugar. It is the carbohydrate form in which most of the products of photosynthesis are transported from leaves to storage organs.

The body is unable to make use of sucrose directly: in the intestines an enzyme (sucrase) catalyzes its conversion into glucose and fructose.

$$C_{12}H_{22}O_{11} + H_2O \rightarrow \underbrace{C_6H_{12}O_6 + C_6H_{12}O_6}_{\text{Isomers}}$$

The same change can be brought about by warming sucrose with a dilute acid: the mixture obtained in these ways is called invert sugar. Fructose is the sweetest of sugars.

(d) Glucose

Glucose is found in the blood, and in honey and syrup. It is a very good source of energy as the body can absorb it without any preliminary digestion: a very ill patient can be fed by injecting the solution directly to his veins. It is also able to pass through animal and plant membranes, being energetically helped to do so by the cells there.

Glucose is made industrially by boiling starch with dilute hydrochloric acid: it will crystallize out from the resultant solution. It is not so sweet as sucrose.

In order to keep the concentration of glucose in the blood constant after a starch meal, the body converts the glucose into glycogen (animal starch) and stores it with the aid of a hormone (insulin) in the liver and muscles. Glycogen can be readily converted into glucose or lactic acid ($CH_3CHOHCOOH$), the latter occurring when the muscles are active but not receiving oxygen quickly enough. It is important to breathe deeply during and after exercise because in the second stage oxygen can be taken up by the lactic acid (see 'Kreb's cycle reactions' in a biology book).

Glucose and fructose can be distinguished from sucrose by their reaction with Fehling's solution (copper(II) sulphate, sodium hydroxide and potassium sodium tartrate—Rochelle salt—in water): on warming they give a red precipiate of copper(I) oxide (Cu_2O). This reaction is used to test for glucose in the urine of diabetic persons: they are unable to digest the usual foods without a supply of insulin.

29.19 The Carbon Cycle

The proportion of carbon dioxide in the atmosphere remains remarkably constant (0·03%) despite the vast quantities of coal and petroleum that are burnt, because it is soluble in water. The carbon cycle in nature is illustrated in figure 29.21.

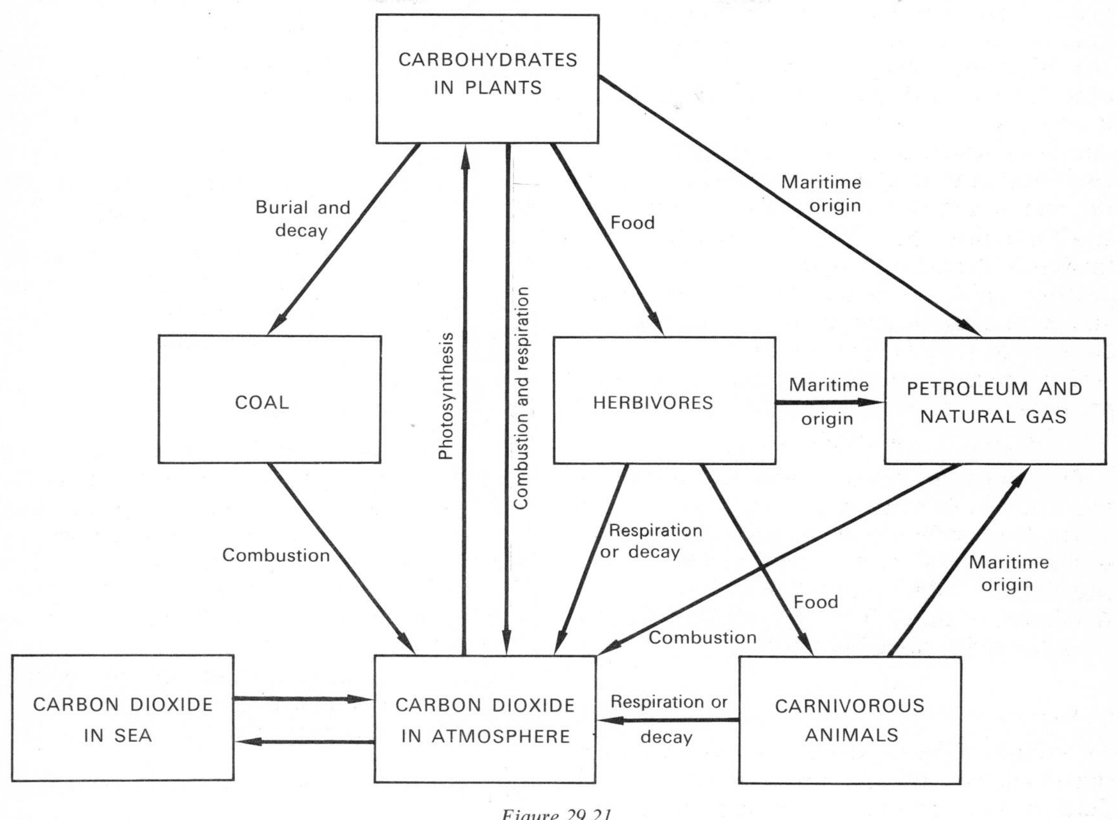

Figure 29.21
The carbon cycle

29.20 Plastics

A plastic material can be defined as one which at some stage in its manufacture has shown fluid properties. Plastics can be divided into two categories: thermoplastics which will soften on heating and harden on cooling again, and thermosetting which on heating become harder. The U.K. production in 1973 was 1.9 million tonnes of which thermoplastics constituted 77%. See section 2.1.

		Percentage
Thermoplastics	Poly(ethene)—low density	22
	—high density	4
	Poly(vinyl chloride)	22
	Poly(styrene)	9
	Poly(propene)	8
	Others	12
Thermosetting	Amino compounds	8
	Others	15

QUESTIONS 29

The Organic Chemistry of Carbon

1 Suppose that you were given a sample of ethene, contaminated with a small quantity of methane. Would this sample be less dense or more dense than air? Describe how you would: (*a*) isolate the methane from the sample; (*b*) convert the ethene into a liquid derivative of ethene; (*c*) prepare fairly pure ethene from ethanol. [OC]

2 When a solution of sucrose is *fermented* by yeast, ethanol is formed. This ethanol can be isolated from the liquid by *distillation*.
(*a*) Explain the meaning of the two italicized words.
(*b*) What is the purpose of the yeast?
(*c*) Write down the structural formula of ethanol and account for its name.
(*d*) Give three reactions of ethanol. [OC]

3 Indicate (without giving experimental details) how: (*a*) ethanol can be converted into ethene; (*b*) ethene can be reconverted to ethanol. State how ethene is obtained on a large scale and mention one important use to which ethene is put. [OC]

4 State one chemical and one physical property of ethane. Suppose that, in the molecule of ethane, (*a*) one H is replaced by OH, (*b*) one H is replaced by NH_2, (*c*) two H atoms are removed. State, in each case, the name, one chemical property and one physical property of the new compound formed. If two hydrogen atoms in the molecule of ethane are replaced by chlorine atoms, it is possible to obtain two different compounds as products. Explain this. [OC]

5 Outline one method of manufacturing ethanoic acid on a large scale. Write down its empirical formula, relative molecular mass, and structural formula. Give two physical properties of this compound. Give the name and formula, in each case, of the product formed when ethanoic acid reacts with (i) calcium oxide, (ii) ethanol. State two significant differences between these two products. [OC]

6 Answer briefly any four of the following:
(*a*) How is ethene obtained from petroleum?
(*b*) Describe one method by which products similar to petroleum can be obtained from coal.
(*c*) What are the advantages and disadvantages of preparing ethanol from ethene as compared with preparing it by the fermentation method?
(*d*) How can ethanol be converted to ethyl ethanoate?
(*e*) What reaction takes place when a fat is boiled with sodium hydroxide solution? [A]

7 Briefly indicate how ethanol (ethyl alcohol) is obtained from cane sugar ($C_{12}H_{22}O_{11}$). When ethanol is heated with concentrated sulphuric acid, the gas ethene is liberated: write an equation for this reaction. What are the products of the reaction of ethene with bromine? Give the equations and name the products. Compare this reaction with the behaviour of bromine towards ethane and state what adjectives are applied to ethene and to ethane because of these (and similar) reactions. [O]

8 (*a*) Write equations for reactions (one only in each instance) by which jars of (i) methane, (ii) ethyne can be prepared in the laboratory.
(*b*) Select one of these gases and draw a labelled diagram of the apparatus you would use to prepare and collect it.
(*c*) Write equations for the reaction of each of these gases with oxygen.
(*d*) Which of these hydrocarbons is said to be 'unsaturated'? Explain what this term means, and describe a test which would enable you to demonstrate that a hydrocarbon was unsaturated. [S]

9 What is meant by the word *polymer*? Here is a list of substances: Terylene, styrene, cellulose, ethane, Perspex, ammonia. From this list, name: (*a*) a naturally occurring polymer, (*b*) a man-made polymer, (*c*) a substance which is not a polymer but can readily be converted into one, (*d*) a substance which is not a polymer and which could not be polymerized. [Sc]

10 What are isomers? Write the structural formulae and names of the isomers of C_4H_{10}. 'Poly(ethene) is a polymer': carefully explain the meaning of this statement. Describe in outline the steps by which poly(ethene) is obtained from petroleum. Name the chemical processes involved. What is the chief chemical characteristic of poly(ethene) which makes it so useful? [J]

11 Use equations to illustrate one example of each of the following: (*a*) the preparation of an aldehyde from an alcohol; (*b*) the oxidation of an aldehyde; (*c*) the chlorination of a saturated hydrocarbon; (*d*) the addition of a halogen to an unsaturated hydrocarbon; (*e*) an addition reaction of an aldehyde; (*f*) the preparation of an unsaturated hydrocarbon from an alcohol. Mention the conditions under which each reaction takes place and name the products. Write the structural formula for your product in (*e*) above. [E]

12 This question concerns the following changes:
A The formation of ethanol (ethyl alcohol) from glucose.
B The formation of nylon from hexanedioyl chloride and hexane-1,6-diamine.
C The formation of alkenes (unsaturated hydrocarbons such as ethene) from medicinal paraffin.
D The formation of soap from olive oil.

In your answer you may refer to the changes by their letters.
(*a*) Which one of the changes can be achieved by fermentation?
(*b*) Which one of the changes can be achieved by a cracking and dehydrogenation process?
(*c*) Which one of the changes can be achieved by hydrolysis?
(*d*) In which one of the changes does the product consist of larger molecules than the reactants?
(*e*) All the reactants and products have two elements in common; what are they?
(*f*) Describe an experiment in which one of these changes can be brought about in the laboratory. You should describe briefly the materials and apparatus which were used, the conditions of the reaction, and how the products were isolated. It is not necessary to draw diagrams. [Nf]

13 What are carbohydrates? Why are they of importance to man? Explain carefully how you would, from a selected carbohydrate, obtain specimens of carbon, carbon dioxide and water. [O]

14 What is understood by the cracking of hydrocarbons? Write down the formulae of three hydrocarbons that might result from the cracking of propane (C_3H_8). If 10 cm^3 of propane are decomposed into the constituent elements, what volume of gas remains? [OC]

15 Two white powders A and B, each a pure compound, were tested as follows:
(1) When a dry sample of A or B was heated with dry copper(II) oxide, carbon dioxide and water were formed.
(2) On boiling with water a small quantity of each dissolved: a solution of A had no action on Fehling's solution but a solution of B gave a positive test with the Fehling's solution.
(3) On boiling solutions of A or B with dilute hydrochloric acid for two minutes, both gave a positive reaction with Fehling's solution.
(4) A chromatogram of these acid hydrolysates of A and B and also of three simple sugars was made with the result shown below.

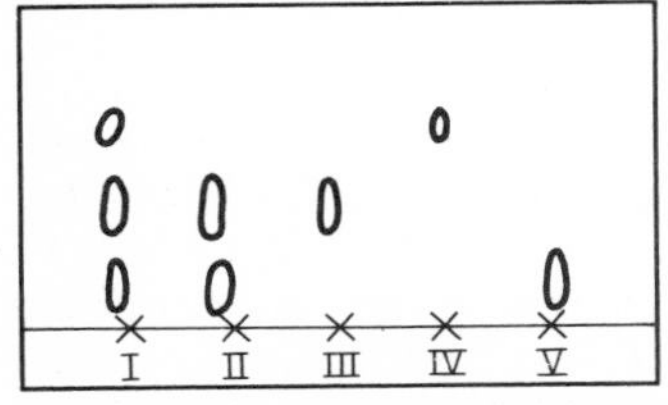

(*a*) What two elements must be present in powders A and B as shown by test (1)?
(*b*) Devise an experiment to show that these powders are carbohydrates and not hydrocarbons.
(*c*) What is observed in a positive result to Fehling's test?
(*d*) Explain why the results of tests (2) and (3) are different.
(*e*) How do you know that A and B are not simple sugars?
(*f*) Identify A and B and state which is the reducing sugar. It is known that: lactose hydrolyzes to glucose and galactose; maltose hydrolyzes to glucose; sucrose hydrolyzes to glucose and fructose; raffinose hydrolyzes to glucose, fructose and galactose. [Nf]

16 A pupil wrote the following account of an investigation into an unknown substance:
'When a sample of A was burned in a jar of oxygen, and calcium hydroxide solution was added, a cloudy white precipitate formed. This indicated the presence of carbon dioxide; therefore A contained carbon.

A colourless liquid was also formed when A burned in oxygen; this liquid boiled at 100°C and turned anhydrous copper sulphate blue. This indicated that the liquid was water; therefore A contained hydrogen.

These experiments indicate that A is a hydrocarbon.'

Comment on: (*a*) his reasoning in the first paragraph, (*b*) his reasoning in the second paragraph, (*c*) his conclusion.

Carbon forms a very large number of compounds. Here are the names of some of them: methane, ethanoic acid, starch, ethene (ethylene), sucrose, octane, glucose, tetrachloromethane, ethane-1,2-diol, butane. From this list of substances name (*d*) a substance found in petrol, (*e*) a substance used to put out fires, (*f*) a substance which is used as an antifreeze, (*g*) a substance that gives a red precipitate when warmed with Fehling's solution. The presence of two substances in the above list can be shown by their reaction with particular halogen elements. Name the substances and state which halogen is used to test for each.

What happens when concentrated sulphuric acid is added to sucrose, and what is a possible explanation of this?

When fuels containing carbon are burned, oxygen is used up and carbon dioxide is frequently produced, yet the respective amounts of these substances in the atmosphere as a whole remain constant. Why is this? [Sc]

17 Hot ashes from a coal fire sometimes smell of sulphur dioxide. Using this evidence alone, discuss whether each of the following hypotheses is likely to be true:
(*a*) Coal contains free sulphur trapped in its pores.
(*b*) Coal is a compound of sulphur.
(*c*) Coal contains compounds of sulphur.
(*d*) The observation tells us nothing about coal.
If (*a*) were correct, how would you separate some of the sulphur from the coal? [Sc]

30 Group V: Nitrogen and Phosphorus

30.1 Introduction

Of the five elements in Group V, nitrogen, phosphorus, arsenic, antimony and bismuth, only the first two are considered: they are both non-metals but they show many differences in their properties.

They attain octets of electrons when they exhibit a covalency of three: their hydrides, oxides, sulphides and halides are pyramidal in shape. Phosphorus can also be pentavalent: an oxide, a sulphide and halides exist. Frequently these elements are met in the form of anions together with oxygen, e.g. nitrite (NO_2^-), nitrate (NO_3^-) and phosphate (PO_4^{3-}). Nitrogen also forms the cation NH_4^+, ammonium, which in many respects behaves as a monovalent simple cation such as sodium (Na^+), except that it will often undergo thermal dissociation.

Whereas phosphorus shows allotropy nitrogen does not. The molecules of nitrogen and phosphorus differ in structure: nitrogen consists of diatomic molecules, white phosphorus of tetraatomic molecules and red phosphorus is a macromolecule. Phosphorus is very reactive towards the air and non-metals but forms an unreactive acid; nitrogen is unreactive as an element but forms a fairly reactive acid.

30.2 Occurrence

78% by volume of the air consists of nitrogen and this is the usual source of the element. It dilutes the oxygen of the air which would otherwise be too concentrated for human life. In a combined form it is found as Chile saltpetre (sodium nitrate, $NaNO_3$). It is an essential part of all plant and animal life but it has to be 'fixed' before it can be assimilitated by plant life which is then often eaten by animals (see section 30.23).

Phosphorus is only found in compound form: the chief source is calcium phosphate ($Ca_3(PO_4)_2$) found in minerals such as rock phosphate in the U.S.A. (39% of the world supply of 97 million tonnes of the mineral in 1973), the U.S.S.R. (22%) and Morocco (17%). The compound is a major constituent (80%) of animal bones. Phosphates are an essential part of our diet and are valuable fertilizers.

I THE ELEMENTS

30.3 The Manufacture of Nitrogen

Air is passed through sodium hydroxide solution to remove carbon dioxide and then through aluminium oxide or silica gel to remove water vapour because both these substances would otherwise solidify and block the apparatus. The air is then compressed to 150 atmospheres and cooled back to room temperature before being allowed to expand through a valve into a region where it is at 6 atmospheres pressure: this causes further cooling. The situation can be compared to a pump becoming hot when a tyre is blown up; if the valve of the tyre is suddenly removed the air rushing out feels cold. The cold air is used to cool the next batch of compressed air so that on expansion the air falls to a still lower temperature. This cycle is repeated until liquid air is obtained.

This liquid air is then fractionally distilled: the nitrogen boils off at $-196°C$ (77 K) and the oxygen, which has a boiling point of $-183°C$ (90 K), remains. If the distillation is carried out very carefully many of the noble gases can be obtained. The nitrogen and oxygen obtained are over 99% pure. The world produces about 34 m tonnes of nitrogen every year (U.K. about 1.5 m tonnes), most of it being 'fixed' in fertilizers.

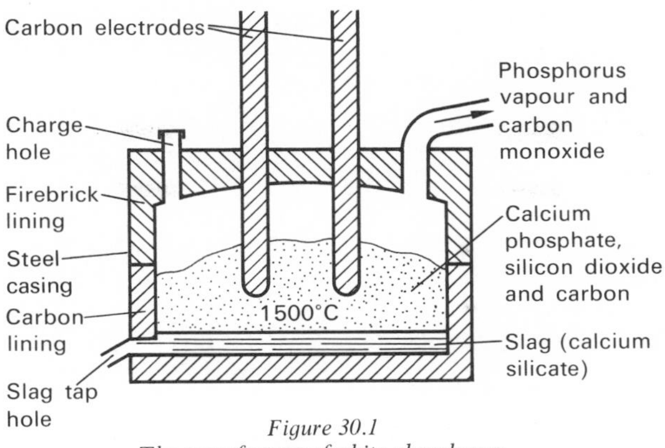

Figure 30.1
The manufacture of white phosphorus

30.4 The Manufacture of Phosphorus

The rock phosphate ($Ca_3(PO_4)_2$) is crushed and dried before being mixed with silica chippings (granite, SiO_2) and coke. The mixture is fed into an electric furnace (50 MW h) where at 1500°C a reaction occurs (see figure 30.1).

$$2Ca_3(PO_4)_2 + 6SiO_2 + 10C \rightarrow 6CaSiO_3 + P_4\uparrow + 10CO\uparrow$$

The reactions are electrothermal (**not** electrolytic). The phosphorus is separated from the carbon monoxide by condensing the former with warm (50°C) water; the liquid phosphorus is run off and may be cast into sticks of white phosphorus (m.p. 44°C). After solidification the calcium silicate slag is used in making roads.

If white phosphorus is heated to 400°C in the absence of air it changes into another form (allotrope) of the element: red phosphorus which is usually seen as a powder.

The total world production of phosphorus is about 1·1 million tonnes annually; the U.S.A. contributing 45%, the U.S.S.R. 25%, Holland 15%, Canada 7% and West Germany 6%. The U.K. imports its requirements.

30.5 Laboratory Preparation

There are two important ways of preparing **nitrogen** in the laboratory.

(*a*) The other constituents of the air are removed with the exception of the noble gases, whose inertness will not upset laboratory experiments. The carbon dioxide of the air is removed by bubbling the air through potassium (or sodium) hydroxide solution. Then the gas is passed very slowly over red-hot copper to remove the oxygen (see figure 30.2).

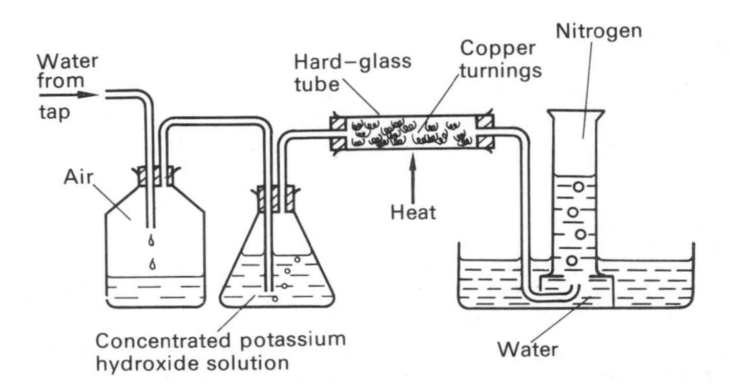

Figure 30.2
The preparation of nitrogen from air

$$NaOH + CO_2 \rightarrow NaHCO_3$$
$$2Cu + O_2 \rightarrow 2CuO$$

The nitrogen, containing about 1% by volume (and mass) of the noble gases, is collected over water.

(*b*) This method of making nitrogen yields the pure gas. A concentrated solution of ammonium chloride and sodium nitrate(III) (nitrite) is heated until the reaction starts and then the Bunsen burner is removed (see figure 30.3). Ammonium nitrite is not heated directly because it may be dangerously unstable.

$$NH_4Cl + NaNO_2 \rightarrow NH_4NO_2 + NaCl$$
$$NH_4NO_2 \rightarrow 2H_2O + N_2\uparrow$$

or $$NH_4^+ + NO_2^- \rightarrow 2H_2O + N_2\uparrow$$

A dry mixture of these two reagents is used in tennis balls to bring them to the correct internal pressure for a certain degree of bounce.

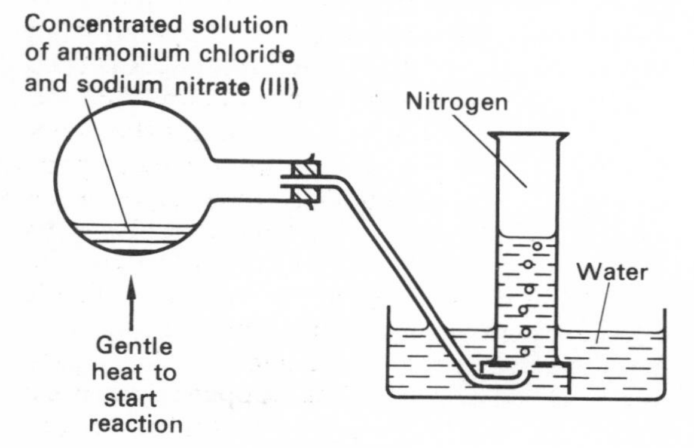

Figure 30.3
The preparation of nitrogen

(*c*) Three minor methods of preparing nitrogen are as follows:

(i) By passing ammonia over hot copper(II) oxide (see section 30.11(c)).

(ii) By igniting ammonium dichromate. In a spectacular reaction the orange crystals swell up leaving a green powder.

$$(NH_4)_2Cr_2O_7 \rightarrow Cr_2O_3 + 4H_2O\uparrow + N_2\uparrow$$

(iii) By adding urea to sodium bromate(I) and sodium hydroxide solutions.

$$CO(NH_2)_2 + 3NaOBr + 2NaOH \rightarrow 3NaBr + Na_2CO_3 + 3H_2O + N_2\uparrow$$

or

$$CO(NH_2)_2 + 3OBr^- + 2OH^- \rightarrow 3Br^- + CO_3^{2-} + 3H_2O + N_2\uparrow$$

White **phosphorus** cannot be prepared in the laboratory but the conversion of white to red can be accomplished on a small scale by warming a small piece of white phosphorus with a minute crystal of iodine (a catalyst, although it is not regained) in a test tube. Any traces of white phosphorus left are removed by boiling the red powder with dilute sodium hydroxide solution.

30.6 Physical Properties

Phosphorus shows allotropy: on cooling the vapour the white form is produced, which slowly changes on keeping to the red form, the change being accelerated by heating (see above). The red form is converted to the white form by distillation under reduced pressure. In this type of allotropy, known as monotropy, there is no transition temperature between the allotropes below the melting point.

White phosphorus consists of P_4 molecules with the atoms arranged in a tetrahedron (see figure 30.4); red phosphorus is usually seen as a powder, its structural nature being complex.

The physical properties of white and red

	Nitrogen	*Phosphorus (Red)*
Atomic number	7	15
Electronic structure	2,5	2,8,5
Relative atomic mass	14·01	30·97
Density (kg/m^3)	1·26 (almost the same as air)	2300
Melting point (°C)	−210	Sublimes at about 400
Boiling point (°C)	−196	
Structure	N_2 molecules	see above
Colour	None	Red
Physiological action	Non-poisonous, diluent for oxygen	Dangerous if impure
Smell	None	None
Solubility in water	Very small (neutral solution)	Very small

	White	*Red*
Melting point (°C)	44	Sublimes at
Boiling point (°C)	280	about 400
Density (kg/m^3)	1820	About 2300
Appearance	Wax-like, translucent	Powder, opaque
Smell	Like garlic	None
Physiological action	Poisonous	Poisonous if impure
Solvents	Carbon disulphide and benzene	None
Preserved	Under water	In jar with tight lid

phosphorus are tabulated above. Red phosphorus is the stable allotrope usually encountered, hence its properties are in the first table in this section. A black form also exists, see figure 30.5, but it is not usually seen in laboratories.

30.7 Chemical Properties

Nitrogen is a very unreactive gas; white phosphorus is dangerously reactive (unsafe to handle) but red phosphorus is only moderately reactive.

(a) With Air

Nitrogen does not burn nor support combustion. In the obsolete (in the United Kingdom) Birkeland-Eyde process for manufacturing nitric acid the first stage was to pass air through an electric arc in which, at 3000°C, about 5% of the air was converted into nitrogen oxide.

$$N_2 + O_2 \rightleftharpoons 2NO \qquad \Delta H = +184 \text{ kJ}$$

White phosphorus catches fire in moist air at 30°C but red phosphorus does not burn until about 260°C. In a plentiful supply of air phosphorus(V) oxide (phosphorus pentaoxide) is produced; in a limited supply of air, some phosphorus(III) oxide (phosphorus trioxide) is formed.

$$P_4 + 5O_2 \rightarrow \underset{\text{Phosphorus(V) oxide}}{P_4O_{10}}$$

$$P_4 + 3O_2 \rightarrow \underset{\text{Phosphorus(III) oxide}}{P_4O_6}$$

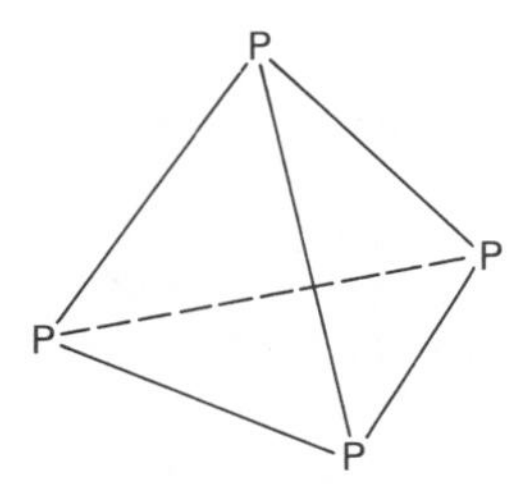

Figure 30.4
The structure of white phosphorus

White (or yellow) phosphorus should be kept under water but red phosphorus can be safely kept in an ordinary bottle, not under water.

The low ignition temperature of white phosphorus can be shown by dissolving a minute piece in the minimum quantity necessary of carbon disulphide and soaking a piece of filter paper with the solution. The filter paper is then held by long tongs in a fume cupboard until the solvent has evaporated. At this point the finely divided phosphorus on the filter paper spontaneously catches fire and the filter paper burns away. Even the heat of the hand is sufficient to ignite white phosphorus so it should never be touched directly. Phosphorus fires are difficult to put out and the element must be carefully handled with tongs. If phosphorus does come into contact with the skin, copper(II) sulphate solution should be applied so that it can be seen more clearly and removed.

The white form of phosphorus emits a yellow-green glow in the dark as it oxidizes. This may be shown by boiling some water in a flask containing a few small lumps of white phosphorus. The flask should be fitted with a jet: in a darkened room the issuing vapour can be seen to glow—this is due to oxidation of the phosphorus. The reaction evolves some of its energy as light and the phenomenon is called chemiluminescence.

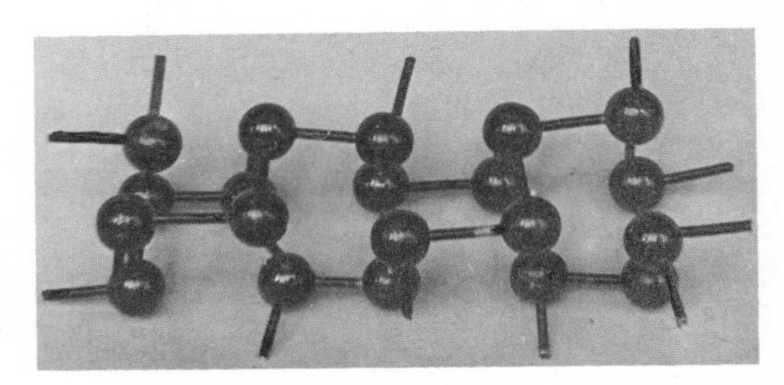

Figure 30.5
The structure of black phosphorus

(*b*) *With Dilute Acids and Alkalis*

Neither nitrogen nor phosphorus react with any of the common dilute acids; nitrogen does not react with alkalis.

White phosphorus reacts with sodium hydroxide solution, this being the way to eliminate any of it left when red phosphorus is made. The reaction is discussed in section 30.10.

(*c*) *With Metals*

A few very reactive metals will combine with nitrogen forming nitrides if they are continuously heated in a stream of the gas:

$$3Mg + N_2 \rightarrow Mg_3N_2$$
$$3Ca + N_2 \rightarrow Ca_3N_2$$

Most metals will react with phosphorus on heating yielding phosphides.

(*d*) *With Hydrogen*

Nitrogen under industrial conditions reacts with hydrogen yielding ammonia (see section 30.9).

(*e*) *With Non-metals*

Phosphorus reacts on heating with sulphur to yield a compound with the formula P_4S_3, used in matches. (P_4S_{10} is also made.) For the reactions of phosphorus with chlorine see section 30.22.

(*f*) *With Oxidizing Agents*

Two oxidizing agents not considered above which both react with phosphorus are concentrated nitric acid (see section 30.18) and concentrated sulphuric acid (see section 31.21).

30.8 Uses of Nitrogen and Phosphorus

Over 75% of the **nitrogen** made is converted into ammonia by the Haber process. 33% of the ammonia made is converted into nitric acid (most of the nitric acid and ammonia is used in fertilizers) and urea (see section 30.11 (f)). The second use of nitrogen is as an inert atmosphere when packaging food, making float glass, semi-conductors, vitamin A, nylon and in purging condensers in generating stations during shut-down. The third group of the uses of nitrogen is as a refrigerating agent: for food (e.g. strawberries) and medical (blood and corneas) storage and in precooling helium liquefiers.

The main use (over 90%) of **phosphorus** is in manufacturing phosphoric acid and phosphates, some of the latter being used as fertilizers. About 9% of phosphorus is converted into other phosphorus compounds, e.g. the chlorides. Some is made into the alloy phosphor bronze.

The original cause of manufacturing phosphorus, that of making matches, now accounts for less than 1% of the total amount produced. The head of a *safety match* is made of potassium chlorate(V), manganese(IV) oxide and sulphur; the side of the box is coated with red phosphorus and upon striking the match the phosphorus is oxidized. The head of a *strike-anywhere match* contains a phosphorus sulphide (P_4S_3) as well as potassium chlorate(V); the side of the box is merely a source of the friction necessary for ignition.

II THE HYDRIDES

30.9 The Manufacture of Ammonia

Hydrogen is manufactured by the steam reforming of naphtha (see section 23.3(a)) and nitrogen by the fractional distillation of liquid air (see section 30.3). In order that the catalyst for the synthesis of ammonia is not rapidly spoilt it is important to eliminate all the oxygen and sulphur compounds possibly present in the synthesis gas.

$$N_2 + 3H_2 \rightleftharpoons 2NH_3 \qquad \Delta H = -92\ kJ$$

The catalyst is iron, containing traces of potassium hydroxide and aluminium oxide. Gas or electric heating to 375°C is employed to initiate the reaction and thereafter the hot gases issuing from the catalyst chamber are used to heat the incoming gases. Ammonia is manufactured on a large scale (over 34 m tonnes each year), the U.S.A. contributing 21%, the U.S.S.R. 18%, France 8%, West Germany 7% and the U.K. 6%.

Although catalysts are only used in relatively small proportions, a typical ammonia convertor may contain 7·5 tonnes of catalyst, occupying about 4 m^3. It is renewed about every three years.

The reversible reaction is accompanied by a decrease in volume when proceeding from left to right in the equation. It is therefore advantageous to use reagents under high pressure: in the United Kingdom this is often 200–350 atmospheres, while elsewhere it may be up to

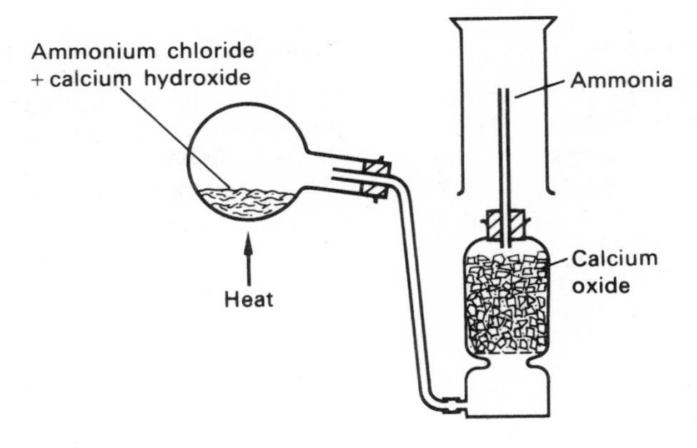

Figure 30.6
The preparation of ammonia

1000 atmospheres. A second important feature of the reaction is that it evolves heat: thus it is beneficial to carry out the reversible reaction at as low a temperature as possible, but this way of enhancing the yield is in conflict with the way of speeding up the reaction by carrying it out at as high temperature as possible. Hence the compromise temperature of about 390°C in the presence of the catalyst: the catalyst speeds up the attainment of the equilibrium but does not affect the yield of ammonia (see chapter 19).

The yield of ammonia for the single passage of the synthesis gas over the catalyst is about 15% conversion. The ammonia liquefies on cooling and can be drained out of the plant. The remaining nitrogen and hydrogen are replenished and recycled. Argon with other noble gases and a little methane (impurity in the hydrogen) accumulate in the plant and when their concentration reaches 5% they are blown out (purged) and processed to obtain the noble gases.

Phosphorus hydrides are not manufactured industrially.

30.10 Laboratory Preparation of the Hydrides

Ammonia is easily prepared in the laboratory by warming any ammonium salt (the chloride is the cheapest, the nitrate might be dangerously explosive) with any strong alkali (see figure 30.6). Sodium hydroxide solution is convenient in the laboratory but calcium hydroxide (suspension or solid) is cheaper.

$$2NaOH + (NH_4)_2SO_4 \rightarrow Na_2SO_4 + 2H_2O + 2NH_3\uparrow$$

$$Ca(OH)_2 + 2NH_4Cl \rightarrow CaCl_2 + 2H_2O + 2NH_3\uparrow$$

The second reaction is used in the Solvay process for sodium carbonate (see section 24.11).

The ionic equation in both cases is:

$$NH_4^+ + OH^- \rightarrow H_2O + NH_3\uparrow$$

The two commonest drying agents (concentrated sulphuric acid and anhydrous calcium

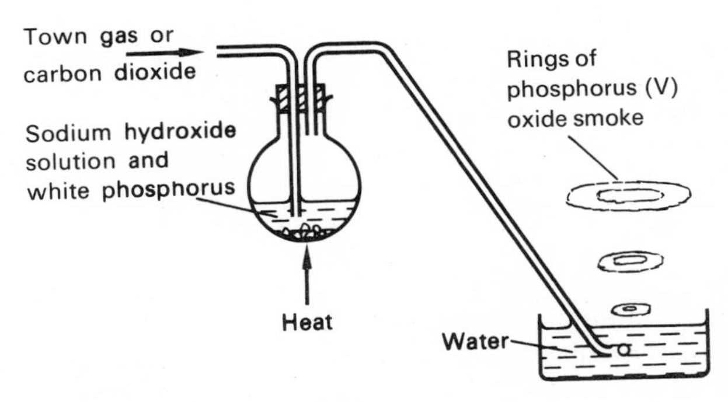

Figure 30.7
The preparation of phosphine

chloride) both react with ammonia, the former to give ammonium sulphate and the latter to yield a complex compound ($CaCl_2,6NH_3$). Calcium oxide is therefore used as the drying agent and the gas is collected by upward delivery because it is less dense than air. The gas cannot be collected over water because of its high solubility. If it is wanted in the form of a solution, however, the inverted funnel technique can be employed (see section 32.9).

Phosphine is prepared by boiling white phosphorus with concentrated sodium hydroxide solution (see figure 30.7). The product is usually spontaneously inflammable in air, due to impurities, and thus the air in the flask must be replaced by town gas or carbon dioxide. The phosphine can be collected over water. If the gas is allowed to escape into the air each bubble burns as it comes into contact with air giving smoke rings of phosphorus(V) oxide. The solution left in the flask contains sodium phosphinate (hypophosphite).

$$P_4 + 3NaOH + 3H_2O \rightarrow 3NaH_2PO_2 + PH_3\uparrow$$

or

$$P_4 + 3OH^- + 3H_2O \rightarrow 3H_2PO_2^- + PH_3\uparrow$$

30.11 Properties of the Hydrides

(a) Physical

Both ammonia and phosphine are colourless, pungent smelling, poisonous gases. Ammonia is given off in the decay of animal and vegetable matter. The smell of phosphine has been likened to that of rotten fish. Ammonia is very soluble in water and is a suitable gas for the fountain experiment (see section 32.10). Red litmus may be put in the water in the trough—it will turn blue as it emerges into the flask as ammonia in solution is alkaline. One volume of water will dissolve 770 volumes of ammonia at room temperature. Ammonia is the only common gas which turns moist red litmus paper blue. Phosphine on the other hand is almost insoluble in water, the solution being neutral.

Ammonia boils at $-34°C$; at $10°C$ it may be liquefied by the application of a pressure of 6 atmospheres.

(b) With Air or Oxygen

Ammonia will neither support combustion nor burn in air but it will burn in pure oxygen with a pale green-brown flame (see figure 30.8).

$$4NH_3 + 3O_2 \rightarrow 2N_2 + 6H_2O$$

In the presence of platinum as a catalyst the oxidation of ammonia yields nitrogen oxide, this reaction being the basis of the manufacture of nitric acid (see section 30.17).

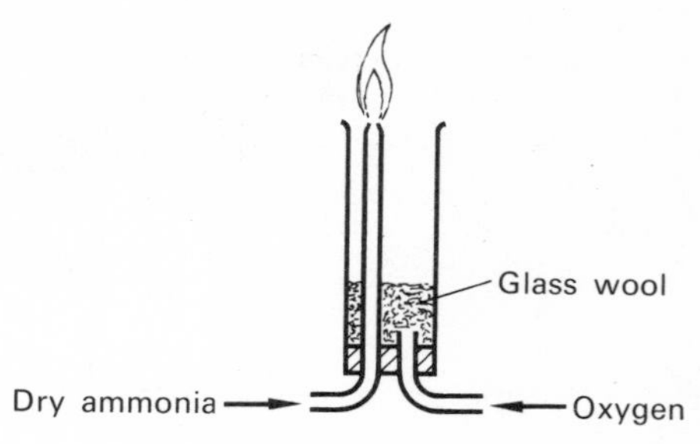

Figure 30.8
The combustion of ammonia

$$4NH_3 + 5O_2 \rightarrow 4NO + 6H_2O$$

Phosphine burns readily upon ignition, yielding phosphorus(V) oxide and steam.

$$4PH_3 + 8O_2 \rightarrow P_4O_{10} + 6H_2O$$

(c) With Metal Oxides

If ammonia or phosphine is passed over hot copper(II) oxide the latter is reduced to the metal (see figure 30.9).

$$\underset{\text{Black}}{2NH_3 + 3CuO} \rightarrow \underset{\text{Red}}{3Cu} + 3H_2O + N_2$$

(d) With Non-metals

Ammonia or phosphine react with chlorine—the ammonia should be in excess (see section 32.6(e)).

$$8NH_3 + 3Cl_2 \rightarrow 6NH_4Cl + N_2$$
$$(NH_4^+Cl^-)$$

Ammonia is the antidote for chlorine poisoning even though it is poisonous itself.

(e) With Hydrogen Chloride

An important test for the presence of ammonia is the addition of hydrogen chloride gas: the formation of a white cloud of ammonium chloride is obvious but not completely conclusive.

$$NH_3 + HCl \rightarrow NH_4Cl\downarrow$$
$$(NH_4^+Cl^-)$$

Phosphine behaves similarly with dry hydrogen iodide, forming phosphonium iodide.

(f) With Carbon Dioxide

Ammonia reacts with carbon dioxide at 200°C under a pressure of 200 atmospheres to yield urea which is an important substance in the manufacture of plastics and as a fertilizer.

$$2NH_3 + CO_2 \rightarrow CO(NH_2)_2 + H_2O$$

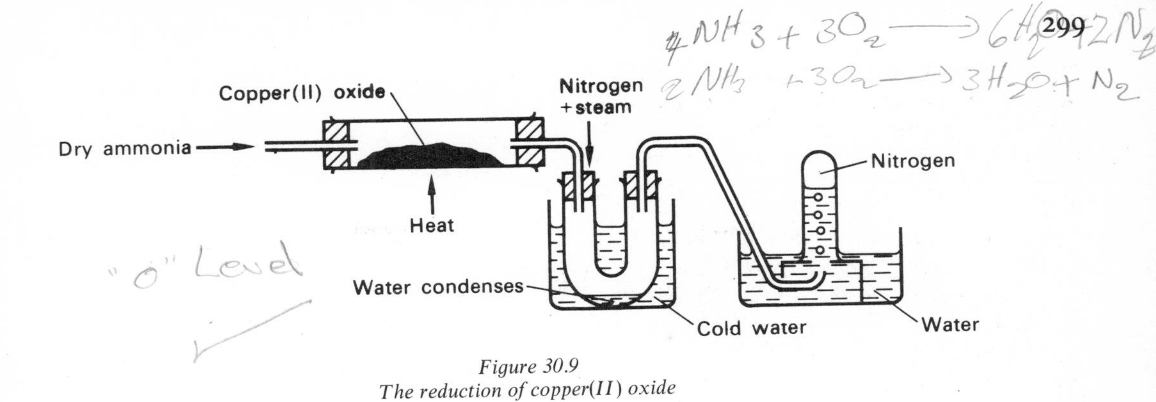

Figure 30.9
The reduction of copper(II) oxide

30.12 Ammonia in Solution

A saturated solution of ammonia in water has a density of 880 kg/m^3: it is often referred to as 880 ammonia. The saturated solution is approximately 20 M and contains 35% by mass of ammonia. The equilibrium in solution is best expressed by the equation

$$NH_3 + H_2O \rightleftharpoons NH_4^+ + OH^-$$

Ammonium hydroxide (NH_4OH) does not exist as an important individual substance and the solution is best referred to as aqueous ammonia. The concentration of free ammonia molecules is much higher than that of the ions—it is a weak electrolyte. The ammonium ion (NH_4^+) is a group of atoms which behaves in many respects as if it was a monovalent metal ion. Ammonia reacts with water to give a solution alkaline to the common indicators. This solution may be titrated with the common acids to yield salts, methyl orange being the most suitable indicator.

$$NH_3 + HCl \rightarrow NH_4Cl \quad (NH_4^+Cl^-)$$

$$NH_3 + HNO_3 \rightarrow NH_4NO_3 \quad (NH_4^+NO_3^-)$$

$$2NH_3 + H_2SO_4 \rightarrow (NH_4)_2SO_4 \quad [(NH_4^+)_2SO_4^{2-}]$$

All ammonium salts are soluble in water and are obtained from their solutions by crystallization.

Ammonia in solution, although a weak electrolyte, is sufficiently alkaline to precipitate the hydroxides of many substances. In some cases there is only one reaction:

$$MgCl_2 + 2NH_3 + 2H_2O \rightarrow \underset{\text{White}}{Mg(OH)_2\downarrow} + 2NH_4Cl$$

or $Mg^{2+} + 2OH^- \rightarrow Mg(OH)_2\downarrow$

$$FeSO_4 + 2NH_3 + 2H_2O \rightarrow \underset{\text{White or green}}{Fe(OH)_2\downarrow} + (NH_4)_2SO_4$$

or $Fe^{2+} + 2OH^- \rightarrow Fe(OH)_2\downarrow$

$$FeCl_3 + 3NH_3 + 3H_2O \rightarrow \underset{\text{Red-brown}}{Fe(OH)_3\downarrow} + 3NH_4Cl$$

or $Fe^{3+} + 3OH^- \rightarrow Fe(OH)_3\downarrow$

$$AlCl_3 + 3NH_3 + 3H_2O \rightarrow \underset{\text{White}}{Al(OH)_3\downarrow} + 3NH_4Cl$$

or $Al^{3+} + 3OH^- \rightarrow Al(OH)_3\downarrow$

$$Pb(NO_3)_2 + 2NH_3 + 2H_2O \rightarrow \underset{\text{White}}{Pb(OH)_2\downarrow} + 2NH_4NO_3$$

or $Pb^{2+} + 2OH^- \rightarrow Pb(OH)_2\downarrow$

In several other cases a precipitate is obtained which dissolves in excess ammonia because of the formation of a complex cation. Sodium hydroxide can be compared: in this case the dissolution of precipitates in excess alkali is due to the formation of a complex anion.

$$CuSO_4 + 2NH_3 + 2H_2O \rightarrow \underset{\text{Pale blue}}{Cu(OH)_2\downarrow} + (NH_4)_2SO_4$$

or $Cu^{2+} + 2OH^- \rightarrow Cu(OH)_2\downarrow$

Followed by

$$Cu(OH)_2 + 4NH_3 \rightarrow \underset{\text{Deep blue solution}}{[Cu(NH_3)_4]^{2+}(OH^-)_2}$$

$$ZnSO_4 + 2NH_3 + 2H_2O \rightarrow \underset{\text{White}}{Zn(OH)_2\downarrow} + (NH_4)_2SO_4$$

or $Zn^{2+} + 2OH^- \rightarrow Zn(OH)_2\downarrow$

Followed by

$$Zn(OH)_2 + 4NH_3 \rightarrow \underset{\text{Colourless solution}}{[Zn(NH_3)_4]^{2+}(OH^-)_2}$$

30.13 Uses of Ammonia and Its Salts

(*a*) Ammonium sulphate is the most important fertilizer used in agriculture as a source of combined nitrogen. Ammonium nitrate, ammonium dihydrogenphosphate and aqueous ammonia are also used. (80% used here.)
(*b*) Ammonia is used to manufacture nitric acid and urea, which are the basis for the manufacture of many other substances. (10%)
(*c*) Ammonia is used to manufacture nylon and rayon (cuprammonium rayon). (5%)
(*d*) Ammonia can be used to produce a suitable atmosphere of nitrogen for metallurgical operations: cracking the gas yields nitrogen and hydrogen which is a reducing atmosphere. This mixture may then be burnt in air to give steam (which is condensed out) leaving nitrogen which is an inert atmosphere.
(*e*) Ammonia was used in commercial refrigerators as a coolant but has been superseded in all new plants by organic fluids, e.g. Freon-22.
(*f*) Aqueous ammonia is used as a grease solvent, e.g. in window cleaning.
(*g*) Ammonium chloride (known as sal ammoniac) is used in dry cells and as a flux during soldering.
(*h*) Ammonium carbonate is the main constituent of smelling salts and is used in baking.

III THE OXIDES

30.14 The Preparation of the Oxides of Nitrogen

Dinitrogen oxide (nitrous oxide, laughing gas) is prepared in industry by heating ammonium nitrate. The alternative, to use a mixture of ammonium sulphate and sodium nitrate, is safer in the laboratory.

$$NH_4NO_3 \rightarrow 2H_2O + N_2O\uparrow$$

$$(NH_4)_2SO_4 + 2NaNO_3 \rightarrow Na_2SO_4 + 4H_2O + 2N_2O\uparrow$$

or

$$NH_4^+ + NO_3^- \rightarrow 2H_2O + N_2O\uparrow$$

The reagents should be heated in a small dry flask by a nearly luminous flame kept in motion so that the temperature does not rise above 250°C. The reagents should not be heated until they are completely decomposed. The gas may be purified by bubbling it through a little iron(II) sulphate solution to remove nitrogen oxide and can then be collected over hot water (see figure 30.10).

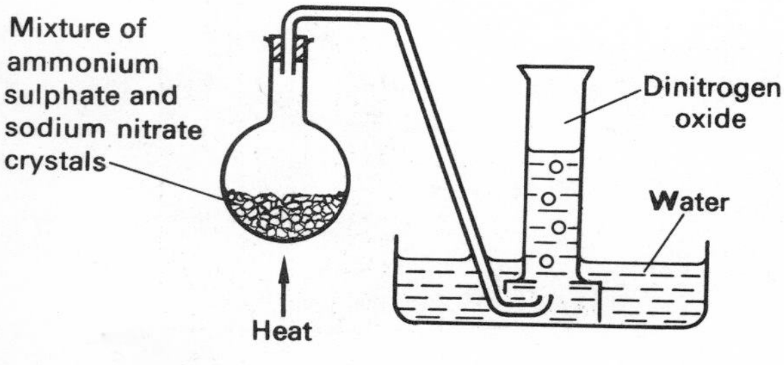

Mixture of ammonium sulphate and sodium nitrate crystals
Heat
Dinitrogen oxide
Water

Figure 30.10
The preparation of dinitrogen oxide

Nitrogen oxide (nitric oxide) is manufactured by passing ammonia and air over a platinum catalyst at 900°C.

$$4NH_3 + 5O_2 \rightarrow 4NO + 6H_2O$$

The nitrogen oxide is then converted to nitric acid (see section 30.17).

In the laboratory it is convenient to start with nitric acid and to reduce it: cold moderately concentrated nitric acid is poured on to copper turnings (see figure 30.11).

$$3Cu + 8HNO_3 \rightarrow 3Cu(NO_3)_2 + 4H_2O + 2NO\uparrow$$

or

$$3Cu + 8H^+ + 2NO_3^- \rightarrow 3Cu^{2+} + 4H_2O + 2NO\uparrow$$

The air in the flask initially reacts with some of the nitrogen oxide to give nitrogen dioxide but the latter is soluble in the water over which the former is collected.

The gas may be purified by absorbing it in iron(II) sulphate and then heating the dark solution obtained.

Nitrogen dioxide (and dinitrogen tetraoxide) is manufactured by allowing nitrogen oxide to react with air, a reaction that will only occur to an appreciable extent below 150°C.

$$2NO + O_2 \rightarrow 2NO_2$$

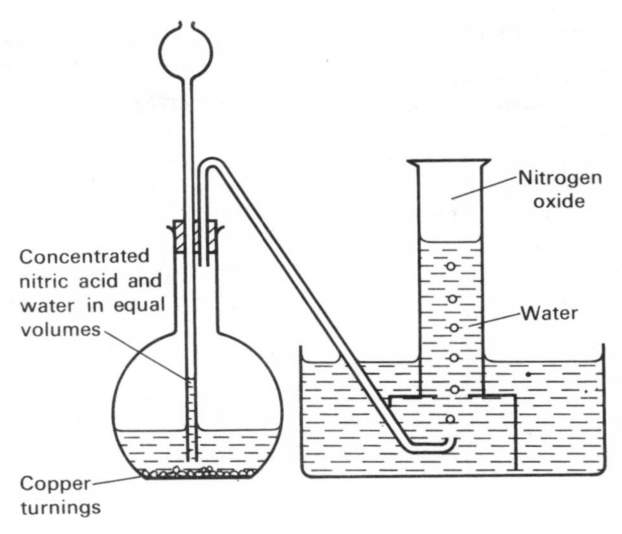

Figure 30.11
The preparation of nitrogen oxide

Nitrogen dioxide is often produced when nitric acid reacts with a metal, e.g. copper with concentrated nitric acid (see figure 30.12).

$$Cu + 4HNO_3 \rightarrow Cu(NO_3)_2 + 2H_2O + 2NO_2\uparrow$$

or

$$Cu + 4H^+ + 2NO_3^- \rightarrow Cu^{2+} + 2H_2O + 2NO_2\uparrow$$

This preparation, which must be carried out in a fume cupboard, yields brown poisonous fumes of nitrogen dioxide which are denser than air. When nitrogen dioxide is cooled the molecules dimerize and the product is a liquid, dinitrogen tetraoxide (N_2O_4). The best way of preparing dinitrogen tetraoxide in the laboratory is by heating lead nitrate (see figure 30.13). Lead nitrate is the most satisfactory heavy metal nitrate to use because it does not contain any water of crystallization.

$$\underset{\text{Colourless}}{2Pb(NO_3)_2} \rightarrow 2PbO + \underset{\text{Orange}}{4NO_2\uparrow} + O_2\uparrow$$

The crystals decrepitate as they decompose; the gases evolved are passed through a freezing mixture of ice and salt, the dinitrogen tetraoxide condenses to a yellow liquid (green if not dry or if the heating is too rapid) and the oxygen passes on.

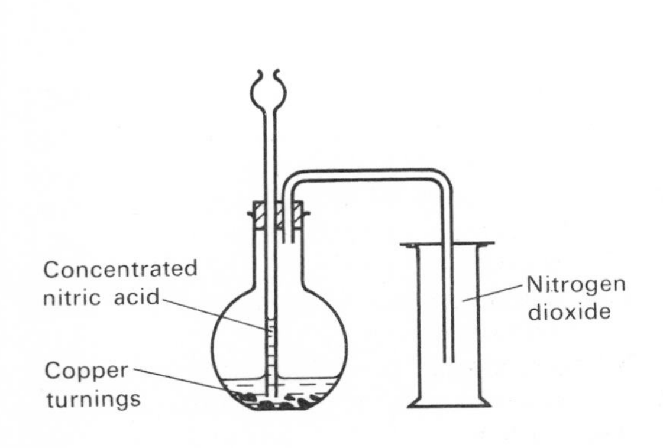

Figure 30.12
The preparation of nitrogen dioxide

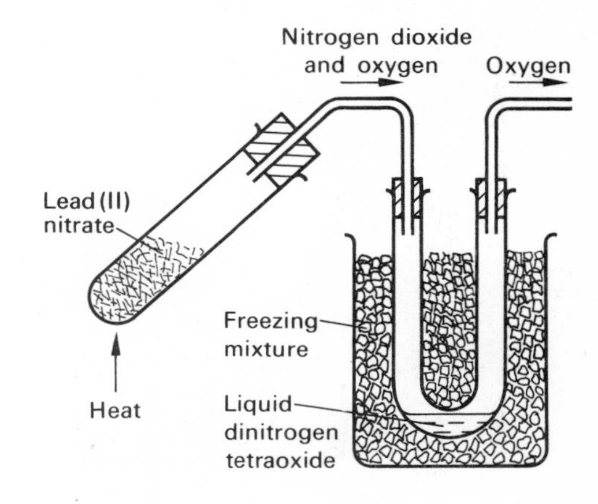

Figure 30.13
The preparation of dinitrogen tetraoxide

30.15 The Properties and Uses of the Oxides of Nitrogen

Property	*Dinitrogen oxide*	*Nitrogen oxide*	*Nitrogen dioxide (or dinitrogen tetraoxide)*
Appearance	Colourless gas	Colourless gas	Pale yellow liquid. Brown gas at moderate temperatures, colourless at high temperatures
Boiling point (°C)	-88	-152	$+21$
Formula	N_2O	NO	N_2O_4 at low temperatures, NO_2 mainly 50–500°C
Relative density	22	15	Variable—about 46 at 25°C
Solubility in water	Moderate in cold	Slight	High
Smell	Sweet	Cannot be smelt (it reacts with air)	Pungent
Physiological action	Causes laughter, then sleep	Dangerous, forms nitrogen dioxide	Poisonous
Action on heating	At 600°C $2N_2O \rightarrow 2N_2+O_2$	At 1000°C $2NO \rightarrow N_2+O_2$	Gas dissociates on heating in two stages: completely NO_2 at 140°C, completely $NO+O_2$ at 620°C
Support of combustion	Many substances are hot enough to decompose it and then they burn in an atmosphere which is 33% O_2, e.g. a strongly glowing splint, brightly burning sulphur, white phosphorus, carbon and a well-lit candle	Some substances are hot enough to decompose it and then they burn in an atmosphere which is 50% O_2, e.g. strongly burning white phosphorus and red-hot carbon	Few substances will burn in it.
Action on damp litmus	None	Usually forms nitrogen dioxide and hence acidic	Acidic $2NO_2+H_2O \rightarrow HNO_3+HNO_2$
Action with air	None	Instantly forms brown fumes of nitrogen dioxide. $2NO+O_2 \rightarrow 2NO_2$	None
Oxidation of hot copper	$Cu+N_2O \rightarrow CuO+N_2$	$2Cu+2NO \rightarrow 2CuO+N_2$	$4Cu+2NO_2 \rightarrow 4CuO+N_2$
	In conjunction with the relative densities these reactions are used in the determination of the formulae of the gases		
Action on acidified potassium manganate(VII) solution	None	Decolorized	Decolorized
Action of carbon disulphide	Burns on ignition	Vivid blue flash upon ignition, sulphur deposited	Burns on ignition
Action on iron(II) sulphate solution	None	Forms dark brown solution containing $FeSO_4,NO$—basis of brown ring test for nitrates	None

Dinitrogen oxide is used as an anaesthetic in dentistry and minor surgery. It is employed in testing for leaks in large containers and water mains. It is used as the solvent and propellant in cake icing aerosols (it is liquefied by the application of a pressure of 50 atm at 15°C).

Nitrogen oxide and nitrogen dioxide are important intermediates in the manufacture of nitric acid. A hypergolic mixture, i.e. one that spontaneously reacts on mixing, of dinitrogen tetraoxide and hydrazine (N_2H_4) is used in the control rocket engines of the service module of the Apollo spacecraft.

30.16 The Oxides of Phosphorus

The oxides of phosphorus that merit consideration are not comparable with the oxides of nitrogen discussed above because phosphorus shows valencies expected of an element in group five, i.e. three and five.

Phosphorus(III) oxide is obtained by burning phosphorus in a limited supply of air; phosphorus and phosphorus(V) oxide can be condensed out of the stream of gases at 50°C and the required oxide frozen out (yield 10%). It has the formula P_4O_6 and it readily oxidizes to phosphorus(V) oxide (P_4O_{10}).

$$P_4 + 3O_2 \rightarrow P_4O_6$$

Phosphorus(V) oxide is obtained by burning phosphorus in a plentiful supply of air.

$$P_4 + 5O_2 \rightarrow P_4O_{10} \qquad \Delta H = -3010 \text{ kJ}$$

It is a white solid like the lower oxide but its affinity for water is very much greater: it is one of the most powerful drying and dehydrating agents known. With a little cold water it gives complex products, but on standing or boiling with plenty of water, phosphoric(V) acid is produced.

$$P_4O_{10} + 6H_2O \rightarrow 4H_3PO_4$$

This oxide cannot be used to dry ammonia. Warm phosphorus(V) oxide will dehydrate concentrated sulphuric acid yielding sulphur(VI) oxide, and ethanol yielding ethene.

Figure 30.14
The structures of white phosphorus, phosphorus(III) oxide and phosphorus(V) oxide

IV THE ACIDS

30.17 The Manufacture of Nitric and Phosphoric Acids

Nitric acid is a very important substance: the United Kingdom manufactures about 2 million tonnes each year. Nitrogen from the distillation of liquid air and hydrogen obtained by the steam reforming of naphtha are converted into ammonia as described in section 30.9.

In the Ostwald process for nitric acid, the most important, the first stage is the catalytic oxidation of ammonia by air (1:9 by volume).

$$4NH_3 + 5O_2 \rightarrow 4NO + 6H_2O \qquad \Delta H = -904 \text{ kJ}$$

The catalyst, platinum (now), is heated electrically to start the reaction. The hot gases coming off the catalyst are cooled to below 150°C and compressed to 4 atmospheres: the nitrogen oxide combines with the excess oxygen giving nitrogen dioxide

$$2NO + O_2 \rightarrow 2NO_2$$

and this in turn dimerizes to dinitrogen tetraoxide (N_2O_4). The dinitrogen tetraoxide dissolves in the steam, most of these gases condensing at this stage.

$$N_2O_4 + H_2O \rightleftharpoons \underset{\text{Nitric(V) acid}}{HNO_3} + \underset{\text{Nitric(III) acid}}{HNO}$$

Figure 30.15
Platinum gauze catalyst being placed in an ammonia burner for the manufacture of nitric acid

The solution produced is one of nitric(V) acid containing some nitric(III) (nitrous) acid but the latter decomposes leaving only nitric(V) acid in solution. The gases are led into the base of a large tower down which this dilute nitric acid is sprayed from the half way position and water from the top. A secondary supply of air is also admitted at the base. The solution finally obtained contains about 70% nitric acid. Any unchanged oxides of nitrogen emerging from the final tower are absorbed in sodium carbonate solution forming sodium nitrate(V) and sodium nitrate(III) (nitrite).

$$Na_2CO_3 + N_2O_4 \rightarrow NaNO_3 + NaNO_2 + CO_2$$

If a 100% solution of nitric acid is required the 70% solution, which is the concentrated acid usually employed in laboratories, must be distilled with concentrated sulphuric acid.

Phosphoric(V) acid is manufactured by one of two processes. In the first white phosphorus is burnt in dry air and the phosphorus(V) oxide formed is condensed and dissolved in dilute phosphoric acid: the product is very pure. In the second process (cheaper), the naturally occurring calcium phosphate is dissolved in phosphoric acid. Then 77–98% sulphuric acid is added and calcium sulphate, in its dihydrate form, is precipitated leaving a greater quantity of phosphoric acid in solution then originally used. The acid is not so pure as by the first process.

30.18 The Laboratory Preparation and Physical Properties of the Acids

The preparation of **nitric acid** is a version of the old industrial process, utilizing a reaction discovered by Glauber in 1560: a nitrate is distilled with concentrated sulphuric acid in a glass retort (see figure 30.16).

$$NaNO_3 + H_2SO_4 \rightarrow NaHSO_4 + HNO_3\uparrow$$

This preparation yields the 100% acid which will attack rubber and cork. The fumes are condensed and then air is blown through the acid to eliminate nitrogen dioxide (formed by the thermal decomposition of the acid) which colours the acid yellow or even brown.

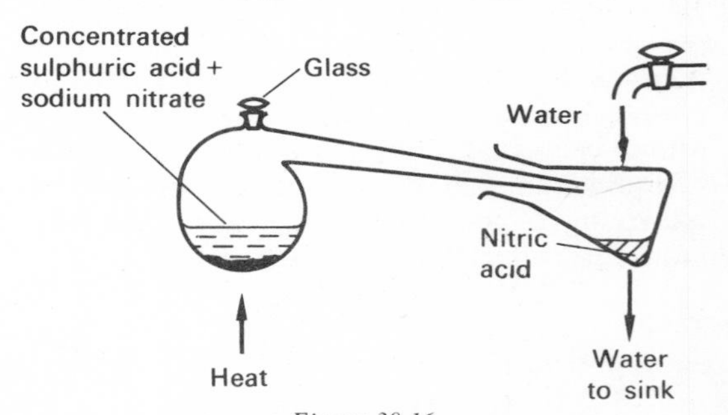

Figure 30.16
The preparation of nitric acid

Pure nitric acid (100%) is a colourless, fuming liquid with a boiling point of 86°C. The 70% acid is a constant boiling point mixture (b.p. 120°C). Nitric acid is a strong monobasic (monoprotic) acid; it is a violent, corrosive poison.

Phosphoric acid can be made by heating red phosphorus with moderately concentrated nitric acid. The reaction may be explosive and certainly must not be attempted with the white allotrope.

$$3P + 5HNO_3 + 2H_2O \rightarrow 3H_3PO_4 + 5NO\uparrow$$

The excess nitric acid and water can be distilled off in a fume cupboard and, by cooling in a freezing mixture, crystals of phosphoric acid can be obtained; as commonly supplied it is a syrupy liquid (85% acid).

Phosphoric acid (the ortho- prefix is not vital) is a deliquescent solid, melting point 42°C, which is readily soluble in water. It has a pleasant taste in dilute solution and is used in food and drink. The acid is tribasic (triprotic) and strong.

30.19 The Chemical Properties of Nitric Acid

Nitric acid can behave in three different ways:

(*a*) In dilute solution—as a strong acid.
(*b*) In concentrated, or sometimes dilute, solution—as a powerful oxidizing agent.
(*c*) In concentrated solution—as a nitrating agent (it behaves as a base).

(*a*) As an acid

(*i*) *With Metals.* Hydrogen is given off when cold dilute nitric acid reacts with magnesium but with most metals more complex reactions occur.

$$Mg + 2HNO_3 \rightarrow Mg(NO_3)_2 + H_2\uparrow$$

or

$$Mg + 2H^+ \rightarrow Mg^{2+} + H_2\uparrow$$

All nitrates are soluble in water and so they are prepared by the evaporation of the requisite solution to the point of crystallization.

(*ii*) *With Basic Oxides and Hydroxides.* These dissolve in dilute nitric acid to produce the corresponding salt and water.

$$\underset{\text{Black}}{CuO} + 2HNO_3 \rightarrow \underset{\text{Green-blue}}{Cu(NO_3)_2} + H_2O$$

or

$$CuO + 2H^+ \rightarrow Cu^{2+} + H_2O$$

$$NaOH + HNO_3 \rightarrow NaNO_3 + H_2O$$

or

$$OH^- + H^+ \rightarrow H_2O$$

$$\underset{\text{Red-brown}}{Fe(OH)_3} + 3HNO_3 \rightarrow \underset{\text{Pale yellow}}{Fe(NO_3)_3} + 3H_2O$$

or

$$Fe(OH)_3 + 3H^+ \rightarrow Fe^{3+} + 3H_2O$$

(*iii*) *With Carbonates.* There is an immediate effervescence as the carbon dioxide is evolved.

$$Na_2CO_3 + 2HNO_3 \rightarrow 2NaNO_3 + H_2O + CO_2\uparrow$$

or

$$CO_3^{2-} + 2H^+ \rightarrow H_2O + CO_2\uparrow$$

$$CuCO_3 + 2HNO_3 \rightarrow Cu(NO_3)_2 + H_2O + CO_2\uparrow$$

or

$$CuCO_3 + 2H^+ \rightarrow Cu^{2+} + H_2O + CO_2\uparrow$$

(*b*) As an Oxidizing Agent

(*i*) *Thermal Decomposition.* The ease with which nitric acid undergoes decomposition is a pointer to its ability to act as a powerful oxidizing agent. It is usually stored in a dark bottle because light as well as heat causes it to decompose and become discoloured. Using the apparatus shown in figure 30.17 the acid is heated at the bottom of the test tube and when it has vaporized the upper portions of the tube are heated. The nitrogen dioxide dissolves in the water in the beaker and the gas collected can be tested with a glowing splint.

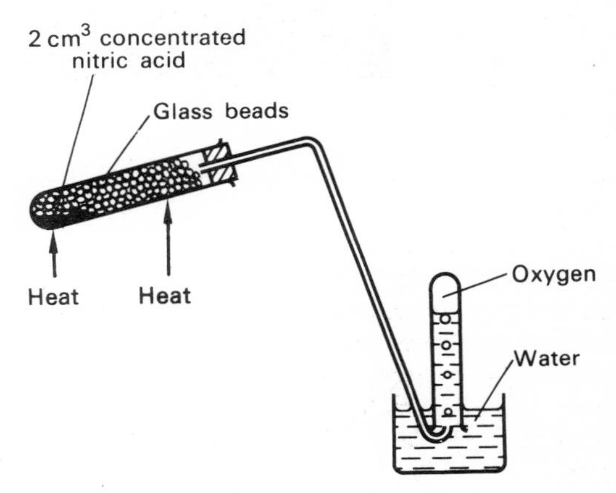

Figure 30.17
The thermal decomposition of nitric acid

$$4HNO_3 \rightarrow 2H_2O + 4NO_2 + O_2$$

(*ii*) *With Metals.* Although the reaction of magnesium is an example of a redox reaction the more far-reaching changes undergone when other metals react are better classified under this heading. The products range from nitrogen dioxide and nitrogen oxide to nitrogen and even ammonia depending on the concentration of the acid, the state of division of the metal and the temperature. Equations are sometimes quoted but should be regarded as approximations. When nitrogen oxide is produced it reacts with the air to give brown fumes of nitrogen dioxide.

$$Cu + 4HNO_3 \rightarrow Cu(NO_3)_2 + 2H_2O + 2NO_2\uparrow$$
Concentrated

or

$$Cu + 4H^+ + 2NO_3^- \rightarrow Cu^{2+} + 2H_2O + 2NO_2\uparrow$$

$$3Cu + 8HNO_3 \rightarrow 3Cu(NO_3)_2 + 4H_2O + 2NO\uparrow$$
Dilute

or

$$3Cu + 8H^+ + 2NO_3^- \rightarrow 3Cu^{2+} + 4H_2O + 2NO\uparrow$$

Devarda's alloy (copper, aluminium and zinc) is a very good reducing agent for nitrates in alkaline solution, converting them to ammonia which when it escapes from the reaction mixture is easily identified (see section 21.9(a)).

However, there are a few metals which appear not to react with the concentrated acid because they rapidly become coated with an oxide film rendering them 'passive': iron, chromium and aluminium are the commonest, the last named being used in constructing storage tanks for the acid.

(*iii*) *With Non-metals.* The oxidation of red phosphorus by warm concentrated nitric acid to give phosphoric acid has already been discussed in section 30.18. Other non-metals such as carbon, sulphur and iodine are likewise oxidized to the respective acids.

(*iv*) *With Inorganic Compounds.* Iron(II) salts (the sulphate is usually employed) are readily oxidized to iron(III) salts.

$$6FeSO_4 + 3H_2SO_4 + 2HNO_3 \rightarrow 3Fe_2(SO_4)_3 + 4H_2O + 2NO\uparrow$$

or $$3Fe^{2+} + 4H^+ + NO_3^- \rightarrow 3Fe^{3+} + 2H_2O + NO\uparrow$$

The solution is a pale green colour at the start and finishes as pale yellow-brown; the sulphuric acid should be dilute and the nitric acid concentrated; warming will hasten the conversion. When the reaction is employed in the brown ring test for a nitrate the suspected nitrate in **cold** solution should be acidified with a little dilute sulphuric acid and then fresh iron(II) sulphate solution added; finally concentrated sulphuric acid is carefully poured into the test-tube held at an angle of 45°. A brown ring forms at or near the junction of the two layers and then disappears: it is due to the transient formation of an addition compound having the formula $FeSO_4,NO$.

If hydrogen sulphide is bubbled into concentrated nitric acid yellow-white sulphur is precipitated.

$$3H_2S + 2HNO_3 \rightarrow 3S\downarrow + 4H_2O + 2NO\uparrow$$

When concentrated nitric acid is added to potassium iodide solution an immediate coloration or precipitate of iodine is observed.

$$6KI + 8HNO_3 \rightarrow 3I_2 + 4H_2O + 2NO\uparrow + 6KNO_3$$

or $$6I^- + 8H^+ + 2NO_3^- \rightarrow 3I_2 + 4H_2O + 2NO\uparrow$$

(*v*) *With Organic Compounds.* These are sometimes drastically and possibly dangerously oxidized. A laboratory demonstration is the conflagration caused when the 100% nitric acid is put on to dry sawdust.

(*c*) **As a Nitrating Agent**

Nitration is the substitution of the NO_2 radical for hydrogen in an organic compound, the reaction usually being performed upon the aromatic substances. The reagent is concentrated nitric acid in the presence of concentrated sulphuric acid. Toluene under these conditions gives trinitrotoluene (T.N.T.), a useful explosive for mining and quarrying.

30.20 The Chemical Properties of Phosphoric Acid

Orthophosphoric acid on heating to 300°C loses a molecule of water yielding metaphosphoric acid; the latter on standing or boiling with water reverts to the former.

$$H_3PO_4 \rightleftharpoons HPO_3 + H_2O$$

Most of the salts of phosphoric acid are insoluble in water; only the ammonium, sodium and potassium salts are soluble and are prepared by neutralization and then evaporation to the point of crystallization.

e.g. $NaH_2PO_4,2H_2O$, sodium dihydrogenphosphate, acid to litmus, neutral to methyl orange

$Na_2HPO_4,12H_2O$, disodium hydrogenphosphate, 'sodium phosphate', just alkaline to litmus, neutral to phenolphthalein

$Na_3PO_4,12H_2O$, trisodium phosphate, very alkaline to litmus and phenolphthalein

The first two salts can be prepared directly from the quantities of solutions found by titration.

There are two tests for a phosphate. The first is to warm a solution of the suspected phosphate with a little concentrated nitric acid and some ammonium molybdate: a yellow precipitate of a complex nature separates. The second test is the addition of silver nitrate to a solution of phosphate: a pale yellow precipitate of silver phosphate forms if the solution is neutral.

$$3AgNO_3 + Na_2HPO_4 \rightarrow Ag_3PO_4\downarrow + 2NaNO_3 + HNO_3$$

or $$3Ag^+ + PO_4^{3-} \rightarrow Ag_3PO_4\downarrow$$

30.21 The Uses of Nitric and Phosphoric Acids

Nitric acid is used for etching and for refining precious metals. It is used in the manufacture of many nitrates: fertilizers (80% to calcium, sodium, potassium and ammonium nitrates), photography (silver nitrate) and explosives (glyceryl trinitrate). It is used in the production of nitro-compounds, e.g. T.N.T. Lastly nitric acid is employed as an oxidizing agent in the manufacturing processes for nylon, Terylene and ethanedioic (oxalic) acid.

Phosphoric acid is used for rust-proofing steel, by forming a protective layer of iron phosphate, e.g. a car body is fashioned in presses and then dipped into the acid before painting. The acid is non-poisonous and is used to adjust the pH in soft drinks and jellies, and in penicillin and streptomycin manufacture. The acid is also used in polishing metals.

Phosphorus is essential for life and the phosphates are valuable fertilizers. Calcium phosphate, a naturally occurring material, is not very soluble and can only be used directly if it is very finely ground. Lawes, in the United Kingdom in 1842, devised the important process of treating it with moderately concentrated sulphuric acid to convert it to calcium dihydrogenphosphate and calcium sulphate.

$$Ca_3(PO_4)_2 + \underset{70\%}{2H_2SO_4} \rightarrow \underset{\text{'Superphosphate'}}{Ca(H_2PO_4)_2 + 2CaSO_4}$$

Over a quarter of the sulphuric acid manufactured in this country is used for making superphosphate: it is much more soluble than the natural phosphate and so can be utilized by the plants. There is sufficient calcium phosphate in blast furnace slag to make the latter useful as a fertilizer.

Trisodium phosphate is used in detergents, e.g. 'Tide'. Sodium metaphosphate ($NaPO_3$) in the form of a polymer is used for descaling boilers (known as 'Calgon') and softening water (see section 25.17). Other forms of sodium phosphate are used in baking powder, in self-raising flour and in 'Instant' puddings.

V THE CHLORIDES

30.22 The Chlorides of Phosphorus

Only the chlorides of phosphorus merit discussion at this stage.

Phosphorus trichloride is prepared by heating white phosphorus in a stream of dry chlorine in a fume cupboard. The whole apparatus must be perfectly dry and the air must be displaced from the apparatus by carbon dioxide (see figure 30.18).

$$P_4 + 6Cl_2 \rightarrow 4PCl_3$$

The trichloride, which is a colourless liquid (b.p. 76°C), distils into the cooled receiver. In moist air the trichloride fumes vigorously; the reaction with water is instantaneous, hydrolysis yielding phosphonic acid.

$$PCl_3 + 3H_2O \rightarrow H_3PO_3 + 3HCl\uparrow$$

Phosphorus pentachloride is prepared by dripping the trichloride into a dry vessel, e.g. a Woulfe bottle, through which a stream of dry chlorine is passing (see figure 30.19). The

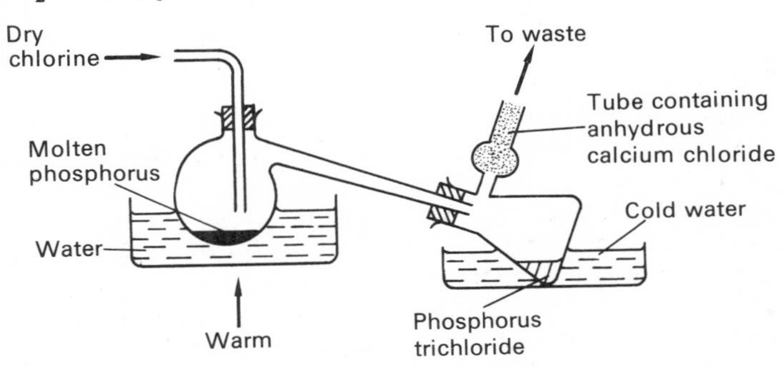

Figure 30.18
The preparation of phosphorus trichloride

experiment must be performed in a fume cupboard. The vessel must be cooled in a freezing mixture.

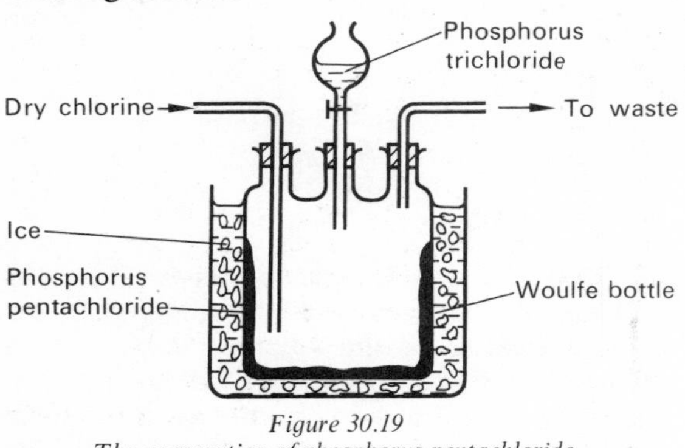

Figure 30.19
The preparation of phosphorus pentachloride

$$PCl_3 + Cl_2 \rightarrow PCl_5$$

The pentachloride is a pale yellow solid which on heating reforms the trichloride and chlorine. The pentachloride also reacts vigorously with water: there is an immediate expulsion of hydrogen chloride, a gas which fumes in air.

$$PCl_5 + H_2O \rightarrow \underset{\text{Phosphorus trichloride oxide}}{POCl_3} + 2HCl\uparrow$$

Both the chlorides can be used to chlorinate organic compounds but the pentachloride is the more useful. It is the standard test for an —OH radical in a compound:

$$C_2H_5OH + PCl_5 \rightarrow C_2H_5Cl + POCl_3 + HCl\uparrow$$

$$CH_3COOH + PCl_5 \rightarrow CH_3COCl + POCl_3 + HCl$$

VI NITROGEN IN NATURE

30.23 The Nitrogen Cycle

The nodules on the roots of leguminous plants, e.g. peas, clover, contain colonies of bacteria which are capable of 'fixing' atmospheric nitrogen in the form of proteins, these plants thus enriching the soil. The plant proteins are taken in by animals during feeding, broken down and

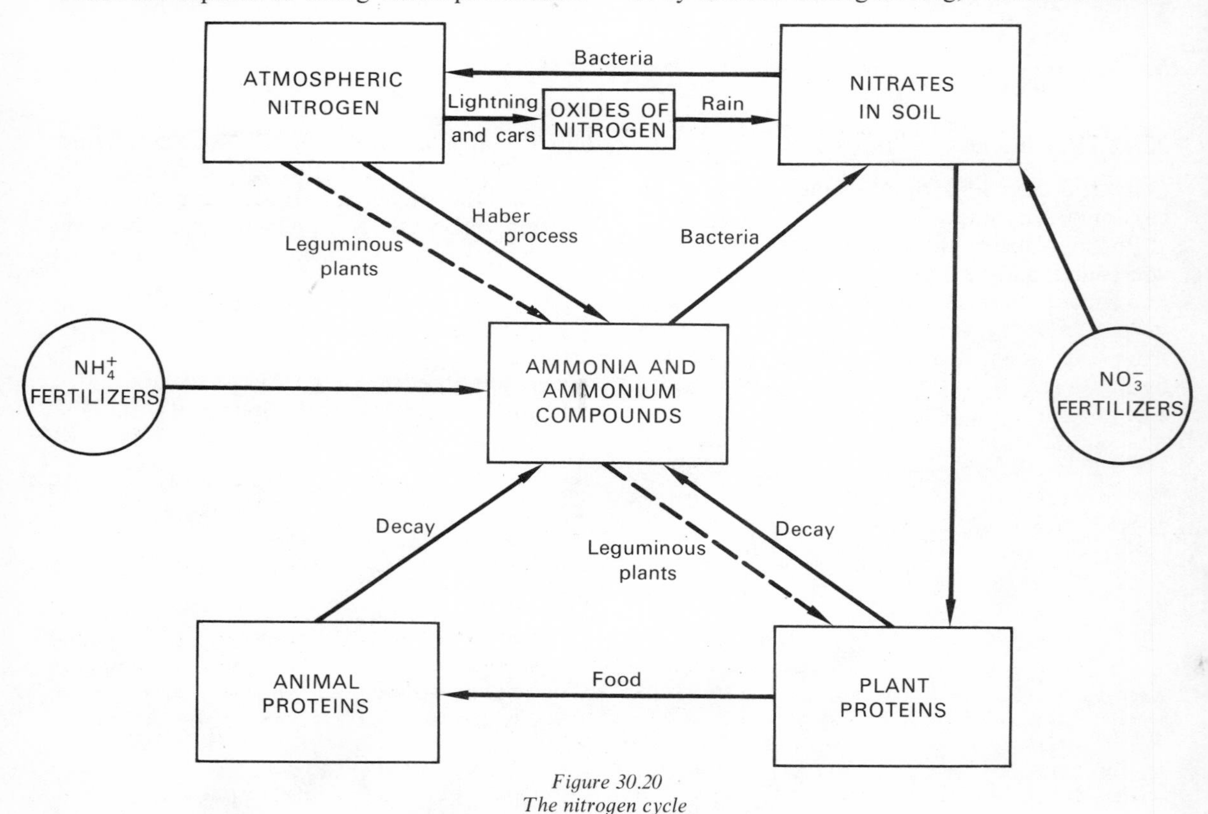

Figure 30.20
The nitrogen cycle

then rebuilt to give animal proteins: animals cannot fix atmospheric nitrogen and are dependent upon plant life (see figure 30.20).

Lightning causes nitrogen and oxygen to combine to give nitrogen oxide, eventually producing nitric acid which is washed out of the atmosphere by the rain into the soil.

$$N_2 + O_2 \rightarrow 2NO$$
$$2NO + O_2 \rightarrow 2NO_2$$
$$2H_2O + 4NO_2 + O_2 \rightarrow 4HNO_3$$

The petrol engines of cars also emit oxides of nitrogen and thus there are two 'natural' sources of nitrates in the soil. The nitrates are absorbed by plants which are in turn eaten by animals. When plants and animals die and decay, bacteria in the soil convert the proteins back to ammonia.

Intensive agriculture and the traditional methods of sewage disposal upset the rhythm of nature so that fertilizers are required to maintain the production of crops.

30.24 The Soil and Fertilizers

The major nutrients of the soil are potassium, magnesium, calcium, nitrogen, phosphorus, sulphur, carbon, hydrogen and oxygen. The commonest minor nutrients are boron, chlorine, manganese, iron, cobalt, copper and zinc.

The world applies over 60 million tonnes of fertilizers every year to the soil but one quarter of the land receives three quarters of the fertilizer. In the United Kingdom over 3 million tonnes of fertilizer are applied, about 15 times as much as 40 years ago; thus food production has risen from 30% to 60% of our requirements. Ammonium sulphate is the most widely used nitrogen (and sulphur) containing fertilizer and calcium dihydrogenphosphate is the most important source of phosphorus. Only a small proportion of the phosphorus in fertilizers is taken up by the plants but, unlike nitrate fertilizers, little is lost by leaching. Calcium hydroxide or carbonate is employed to reduce the acidity of the soil.

It is possible to convert sewage into an acceptable fertilizer, the operating costs being less than the normal disposal costs in most towns. The other advantage of adopting this process is that the rivers and seas would no longer be repulsively polluted.

QUESTIONS 30

Group V: Nitrogen and Phosphorus

1 Write equations for reactions by which nitrogen can be obtained (*a*) from ammonia, (*b*) from nitrogen dioxide, (*c*) from a mixture of salts. [S]

2 State the conditions under which ammonia is manufactured from nitrogen and hydrogen. Calculate the mass of ammonia that can be made from one kilogram of nitrogen. [OC]

3 Describe, with a sketch of the apparatus, how you would prepare several gas jars of dry ammonia. Starting from ammonia how would you show that it contains nitrogen and hydrogen? Give one large scale use, different in each case, for ammonium chloride, ammonium sulphate, ammonium nitrate. [OC]

4 Ammonia gas dissolves in water to form an alkaline solution.
(*a*) Describe how, starting with ammonium chloride and calcium hydroxide, you would prepare such a solution.
(*b*) Write down the names and formulae of two ions and two molecules that could be detected in the solution.
(*c*) Give two examples of the covalent bond to be found in the substances in the solution.
(*d*) State how each of the ions, quoted in (*b*), has been formed.
(*e*) Suggest a method by which, without applying heat, such a solution could be made to evolve part of the dissolved gas. [OC]

5 Ammonia is obtained industrially from nitrogen and hydrogen by the Haber process. The equation for the reaction is

$$N_2 + 3H_2 \rightleftharpoons 2NH_3$$

(*a*) What does the symbol $\rightleftharpoons$ mean?
(*b*) Approximately what temperature and pressure are used?
(*c*) Explain why the conditions you mention in (*b*) are used.
(*d*) What catalyst is used to speed up the reaction?
(*e*) What happens to the nitrogen and hydrogen which remain unconverted?
(*f*) If it were possible to convert 100 cm^3 of nitrogen and 300 cm^3 of hydrogen completely to ammonia, what volume of it would be obtained (all volumes being measured at the same temperature and pressure)? State the chemical law you make use of in your calculation.
Describe an experiment by which you could obtain a specimen of nitrogen from ammonia. [A]

6 (*a*) How is dry ammonia prepared and collected in the laboratory? Illustrate your answer by means of a diagram and an equation.
(*b*) Describe what is seen and give equations, where appropriate, for the reactions between ammonia and (i) heated copper(II) oxide, (ii) hydrogen chloride, (iii) damp neutral litmus paper.
(*c*) Describe an experiment to show that nitric acid can be broken down to yield oxygen as one of the products of decomposition. [Sc]

7 (*a*) What is the effect of ammonia solution on (i) iron(III) chloride solution, (ii) copper sulphate solution, (iii) precipitated silver chloride?
(*b*) Ammonia is oxidized to nitrogen when in contact with hot copper(II) oxide; write an equation for this reaction and calculate what volume of nitrogen (at s.t.p.) could be produced by oxidation of 6 g ammonia. [O]

8 Describe, with the aid of a labelled diagram, how you would use solid ammonium sulphate to obtain a moderately concentrated solution of ammonia. Write the equation for the reaction concerned. State and explain, with equations, what would be observed when this solution (*a*) is added to iron(III) chloride solution, (*b*) has chlorine passed through it. What volume of gaseous ammonia, at s.t.p., should be obtainable from 6·6 g of solid ammonium sulphate? [W]

9 Describe, by means of labelled diagrams and brief explanations, how you would prove that ammonium chloride contains nitrogen by preparing from it a sample of the gas. State why none of the other reagents you employ in your experiments should contain nitrogen. Contrast what is seen in each of the following pairs of experiments: (*a*) Solid samples of sodium chloride and ammonium chloride are heated in separate test tubes.
(*b*) An aqueous solution of ammonia is added in turn to separate solutions of iron(II) and iron(III) salts. [W]

10 How would you prepare a small quantity of dry ammonium chloride from ammonia gas? Draw diagrams to show (*a*) how the bonds are formed in ammonia, (*b*) the shape of an ammonia molecule, and (*c*) the types of bonding in ammonium chloride. [J]

11 Nitric acid is manufactured by catalytic oxidation of ammonia to oxides of nitrogen and conversion of these to nitric acid.
(*a*) Name the catalyst used.
(*b*) Give the approximate temperature at which the oxidation is carried out.
(*c*) Name the oxidation products and give the equation.
(*d*) How is the first product further oxidized and then converted to nitric acid?
(*e*) Outline how the dilute acid so produced can be concentrated. Describe two reactions in which nitric acid acts as an acid and two reactions in which it acts as an oxidizing agent, stating your reasons for regarding the reactions as oxidations. [A]

12 Nitric acid may be prepared in the laboratory by the action of concentrated sulphuric acid on a suitable nitrate and distilling off the nitric acid. Describe in detail such a preparation and answer the following questions concerning this preparation.
(*a*) Why is an apparatus consisting of glass only desirable for the preparation?
(*b*) Nitric acid and sulphuric acid are both liquids. Why does only nitric acid distil off and not sulphuric acid?
(*c*) Pure nitric acid is colourless but the product in this preparation is usually pale yellow or even brown. Why is this? What precaution would you take in the preparation to reduce this coloration to a minimum?
(*d*) If your product were yellow or brown how would you obtain a colourless sample of nitric acid from it? [L]

13 Describe the laboratory preparation of nitric acid. Outline the stages by which this substance is manufactured from atmospheric nitrogen via ammonia. Describe the preparation of a crystalline specimen of lead nitrate from nitric acid.

14 Describe briefly the manufacture of nitric acid from ammonia, giving the essential chemistry but no technical details. Give two uses for nitric acid and one physical property. Describe two experiments that illustrate the oxidizing power of concentrated nitric acid. Include in your accounts evidence that oxidation has occurred. [C]

15 Describe, in outline, the stages of the manufacture of nitric acid from atmospheric nitrogen. State two important uses of this acid, or derivatives of it. How would you demonstrate that nitric acid contains oxygen? [OC]

16 The two gases, nitrogen oxide and oxygen, react immediately at room temperature to form another gas, nitrogen dioxide. The two reactants are said to react in the ratio 2 moles of nitrogen oxide: 1 mole of oxygen. Nitrogen dioxide dissolves readily in 2M sodium hydroxide, but pure nitrogen oxide and oxygen do not. Devise an apparatus and procedure by which you could check whether nitrogen oxide and oxygen react in the given ratio. Indicate what results you would expect and how these results lead to your conclusions. [Nf]

17 A brown gas *A* is readily condensed by a freezing mixture of ice and salt. The gas *A* is absorbed by sodium hydroxide to give a colourless solution. This solution is made just acid with hydrochloric acid.
(*a*) When this acid solution is added to iron(II) sulphate solution the whole solution turns brown.
(*b*) When the acid solution is warmed the original gas *A* is evolved and a solution is left which gives a positive result for the *brown ring* test. Identify the compound *A* and trace the course of reactions mentioned, giving equations. [L]

18 Briefly describe and explain what happens when oxygen is bubbled into a jar of nitrogen oxide standing over water containing blue litmus. [S]

19 Describe in detail how you would prepare specimens of the following in the laboratory: (*a*) dinitrogen tetraoxide; (*b*) nitrogen oxide; (*c*) dinitrogen oxide. Describe briefly (i) how nitrogen may be obtained from nitrogen oxide; (ii) the reaction of nitrogen oxide with the air.

20 Making use of concentrated nitric acid, magnesium, copper and a supply of water, but no other chemicals, describe briefly experiments by which you could show the formation of three different gases. [C]

21 Describe laboratory methods, one in each case, by which you could obtain from the air samples of (*a*) oxygen, and (*b*) nitrogen. Outline the method by which these gases are obtained on an industrial scale from the atmosphere. (A diagram is not required.) Under what conditions do these two gases react together, and what is the product first formed? State whether this reaction is exothermic or endothermic. [J]

22 Describe and explain what happens when lead nitrate is heated in a dry test tube. How can you separate the two gases which are evolved? One of these gases is an oxide of nitrogen. Describe briefly how you would prepare a sample of another oxide of nitrogen. Which oxide of nitrogen is nicknamed *laughing gas*, and for what is it used? Which oxide of nitrogen is easily converted into one of the others, and how is the conversion brought about? [S]

23 State what would happen if the following tests were carried out on nitrogen gas: (*a*) litmus solution was added to the gas, (*b*) a lighted taper was placed in the gas, (*c*) calcium hydroxide solution was added to the gas, (*d*) oxygen was added to the gas and a spark passed through the mixture for some time.

One substance which is reported to release nitrogen compounds into soil is dust from a leather factory. Describe how you would try to find out if a sample of such dust actually did contain compounds of nitrogen. [Sc]

24 Write brief statements, without diagrams, explaining how (*a*) nitrogen oxide, (*b*) nitrogen dioxide, are commonly prepared and collected in the laboratory. Describe, and account for, what is seen when (i) a freshly filled bottle of concentrated nitric acid is left exposed to light, (ii) a loosely stoppered flask of nitrogen dioxide is heated in a Bunsen flame, (iii) a piece of copper foil is added to some sodium nitrate crystals, and the mixture is warmed with a little concentrated sulphuric acid. [W]

25 Describe and explain how you would use a supply of ammonia, and any other essential reagents, to prepare (*a*) a crystalline sample of ammonium sulphate, (*b*) jars of nitrogen. Describe one experiment which illustrates the oxidizing property of nitric acid, showing clearly why you classify the reaction concerned as oxidation. [W]

26 Red and white phosphorus are said to be *allotropes*. Explain what this means. How can white phosphorus be converted into the red allotrope? How can the red form be changed back to the white? Describe in outline: (*a*) the preparation of orthophosphoric acid from white phosphorus; (*b*) the production of *superphosphate* from calcium phosphate. What advantage has superphosphate over calcium phosphate as a fertilizer?

27 Describe the manufacture of white phosphorus from calcium phosphate. State in tabular form three differences, other than colour, between the common allotropes. State briefly the importance of phosphorus to the human body.

28 How and under what conditions does phosphorus react with (*a*) oxygen; (*b*) nitric acid; (*c*) chlorine; (*d*) sodium hydroxide?

Given some phosphorus pentachloride, describe how you would obtain from it: (*a*) pure dry hydrogen chloride; (*b*) orthophosphoric acid free from hydrogen chloride. [O]

29 Give an account of the importance of phosphorus and its compounds: (*a*) in agriculture (*b*) in the manufacture of matches.

Describe with necessary detail how, starting from red phosphorus, you would prepare a well-crystallized specimen of sodium phosphate, $Na_2HPO_4,12H_2O$. Describe how you would show that your preparation *was* of sodium phosphate. [O]

30 How can phosphine be prepared? Compare the physical and chemical properties of phosphine with those of ammonia.

31 Group VI: Oxygen and Sulphur

31.1 Introduction

Oxygen and sulphur are both non-metals but they show several differences in their properties.

Octets of electrons are attained by a transfer of electrons to oxygen and sulphur as in metal oxides and sulphides respectively (O^{2-} and S^{2-}). Octets of electrons are also attained by covalency and there are many compounds in which oxygen and sulphur have a valency of two. There are a few compounds in which sulphur shows a greater valency: four in sulphur dioxide and six in sulphur trioxide.

Both elements show allotropy. Oxygen (O_2) and ozone (trioxygen, O_3) are usually encountered as gases so differences in crystal structure are not apparent. Sulphur shows allotropy in the solid state and in the liquid state.

31.2 Occurrence

Oxygen was first isolated by Priestley and by Scheele separately in 1774 (see section 5.4). Sulphur has been known since Greek times.

Oxygen is the most abundant element in the earth's crust (see figure 1.1) but it is not readily available from rocks. It occurs in air to an extent of 23% by mass or 21% by volume. There is a layer of ozone 25 kilometres up in the atmosphere which cuts off the ultra violet radiation from the sun which would be dangerous to life. There is no ozone at the seaside, only the smell of rotting seaweed. Oxygen is the main constituent by mass of water—89%.

Sulphur occurs to a large extent in underground deposits in the U.S.A. and in volcanic regions, e.g. Sicily. There are small proportions of hydrogen sulphide in many samples of natural gas and it must be removed before the gas can be further utilized. Many elements occur in the form of their sulphides, e.g. lead, zinc and mercury which are important as sources of the metal contained and of sulphur dioxide; iron disulphide (iron pyrites), which also occurs naturally, is important only as a source of sulphur dioxide. Sulphur is a constituent of all living matter, being found in certain proteins, e.g. hair and eggs. As the world uses more and more petroleum, which contains a small proportion of sulphur compounds, refineries produce increasing quantities of sulphur as a by-product because it must be removed. Several sulphates occur in nature but with the minor exception of calcium sulphate (anhydrite) they are not important as sources of sulphur or its compounds.

I THE ELEMENTS

31.3 The Manufacture of Oxygen and Sulphur

Oxygen is manufactured by the liquefaction and subsequent fractional distillation of liquid air. The oxygen is 99·5% pure. The U.K. produces 2 m tonnes. See section 30.3.

Sulphur is extracted from the underground deposits in the U.S.A. by the Frasch process. The geological nature of the ground and the low melting point of sulphur led to a novel way of obtaining the element directly in a state of high purity (over 99·5%). A pipe, about 20 cm in diameter, is sunk to the 'cap rock' of the sulphur dome and pipes of smaller diameters are lowered inside it as shown in figure 31.1. Superheated water (at 180°C and under 10 atmospheres pressure) is sent down the outside tube to melt the sulphur. Hot compressed air is sent down the central tube and this forces a foam of sulphur, water and air back to the surface; this is led into vast wooden vats. The air escapes back into the atmosphere and the water drains off the surface of the molten sulphur as the latter solidifies.

The sulphur in petroleum is converted into hydrogen sulphide; in natural gas this sub-

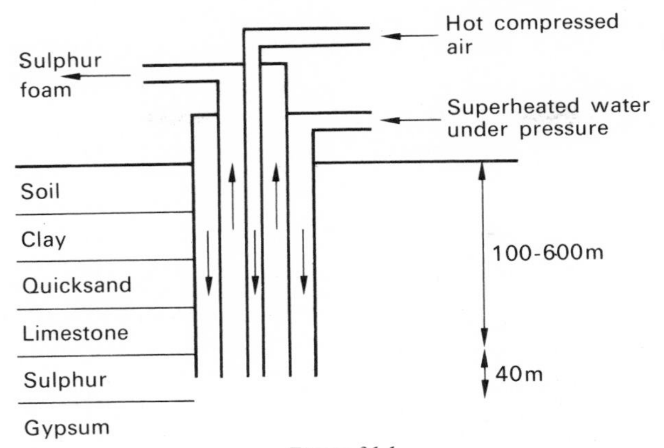

Figure 31.1
The Frasch process for sulphur

stance is already present. The hydrogen sulphide is then extracted and converted to sulphur. Where gas is still made from coal, e.g. in a coke oven serving an iron works, the hydrogen sulphide is removed using the oxide of zinc, aluminium or iron(III). The 'spent oxide' is then roasted in air to yield sulphur dioxide.

The world production of sulphur in 1973 was 29·8 m tonnes, to which can be added 19·3 m tonnes in the form of other sulphur compounds. The U.S.A. contributed 34% of the 29·8 m tonnes (26% by the Frasch process, 8% recovered from petroleum and natural gas), Canada 24% (recovered sulphur) and France 6% (recovered sulphur).

31.4 The Laboratory Preparation of Oxygen and Ozone (Trioxygen)

The easiest and safest way to prepare **oxygen** in the laboratory is to pour hydrogen peroxide (20- or 50-volume solution) on to manganese(IV) oxide and a little water (see figure 31.2). The manganese(IV) oxide is a catalyst for the reaction, which is very slow in its absence.

$$2H_2O_2 \rightarrow 2H_2O + O_2\uparrow$$

Another way is to heat potassium chlorate(V) crystals in a hard-glass test tube: again manganese(IV) oxide is a catalyst. The chlorate will decompose in the required manner on heating alone but the temperature required is much higher. It is sufficient to mix 5—10% of manganese(IV) oxide into the potassium chlorate but higher proportions are often recommended.

$$2KClO_3 \rightarrow 2KCl + 3O_2\uparrow$$

The potassium chloride can be extracted after the test tube has cooled by shaking the contents with water, filtering and evaporating the solution; the catalyst is insoluble in water and remains on the filter paper. Accidents have

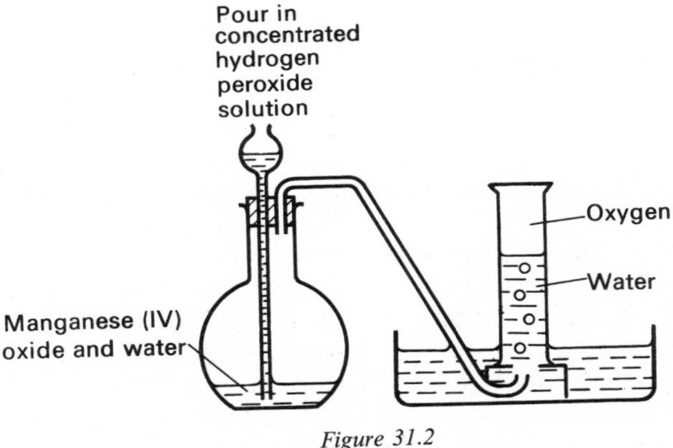

Figure 31.2
The preparation of oxygen

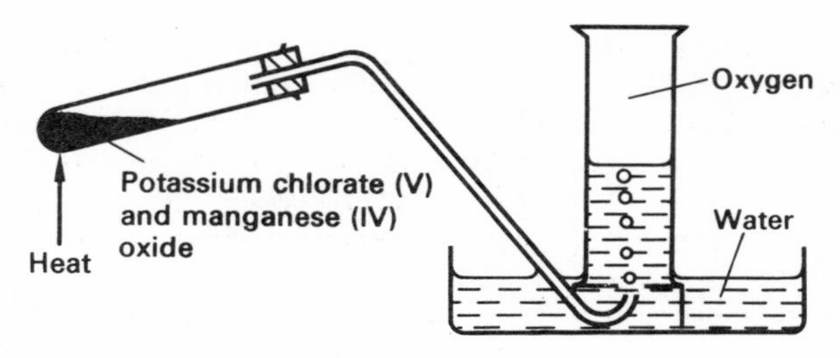

Figure 31.3
The preparation of oxygen

happened in the past when the wrong substances have been heated and it is always best to check that the 'oxygen mixture' is as stated by heating a small portion in an ignition tube before heating a large quantity. The gas is conveniently collected over water, as shown in figure 31.3. Many other oxygen containing substances release the gas on heating but the reactions are not sufficiently important to repeat the details here.

Oxygen can be partially converted to **ozone** by supplying energy in the form of a silent electrical discharge. An electric spark is liable to decompose the gas because it produces heat. However, when two metal plates are connected to an induction coil and kept at such a distance that sparking cannot occur, it is found that a peculiar smell is produced. A similar smell is often noticed near electrical machinery, and is due to ozone (trioxygen).

The best way of producing ozonized oxygen is to use an ozonizer, a laboratory form of which, using a Liebig condenser, is shown in figure 31.4. The outside of the outer tube and the inside of the inner tube are covered with aluminium foil and these are connected to the induction coil. When dry oxygen is passed through the tube with the induction coil switched on, about 5% of the gas issuing is ozone (trioxygen).

$$3O_2 \rightleftharpoons 2O_3 \qquad \Delta H = +289 \text{ kJ}$$

31.5 The Laboratory Preparation of Sulphur

(a) Rhombic or Octahedral Sulphur

If some roll sulphur is ground into a powder and then shaken with carbon disulphide in a fume cupboard, (the solvent is very toxic and inflammable), the solution so obtained can be left to evaporate and it will soon yield crystals. The rate of evaporation can be reduced, so that the crystals obtained are larger, by covering the vessel with a filter paper which has a few holes pierced in it. The use of dimethylbenzene (xylene) or methylbenzene (toluene), which are safer solvents, usually yields smaller crystals.

(b) Monoclinic or Prismatic Sulphur

Some crushed roll sulphur is warmed in a test-tube over a small Bunsen flame and the sulphur tipped out the moment it is all molten into a

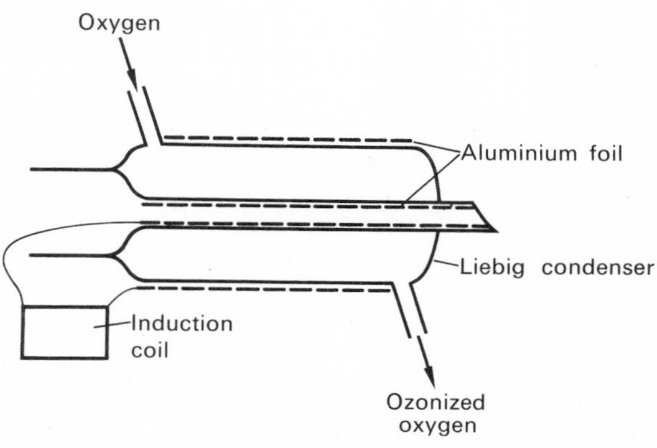

Figure 31.4
The preparation of ozonized oxygen

piece of filter paper in a funnel. The sulphur should be allowed to solidify and then the crust gently broken revealing crystals of monoclinic sulphur. Monoclinic sulphur is only stable between 96–119°C.

Larger but more fragile monoclinic crystals can be obtained by warming a spatula load of sulphur in 25 cm^3 of dimethylbenzene in a boiling tube by means of a beaker full of boiling water. On cooling, needle shaped crystals grow upwards from the excess sulphur in the bottom of the tube. On examination under a microscope a few hours later, the crystals are found to have reverted to rhombic sulphur.

There are other forms of sulphur which are not crystalline and cannot therefore be said to be allotropes. Sulphur is not often prepared deliberately in reactions because it is readily available as an element.

(c) Amorphous (or Colloidal) Sulphur

When a dilute acid is added to sodium thiosulphate, sulphur is precipitated, the reaction often being studied from the point of view of rates of reactions (see section 18.9).

(d) Liquid Sulphur

Sulphur undergoes some peculiar changes as it is heated from its melting point to its boiling point. When just molten it is a pale yellow mobile liquid but on further heating it soon darkens. At about 160°C it suddenly becomes very viscous and cannot be poured. On further heating it remains dark but slowly becomes more mobile again.

An easy way of showing the difference between the forms of liquid sulphur is to pour the just-molten liquid into a beaker of cold water: this yields small beads of a brittle solid. On the other hand, if hot molten sulphur (250–350°C) is poured into cold water an elastic form is given, known as plastic sulphur: it is a super-cooled liquid (like glass) but it is not very stable and soon reverts to a hard yellow solid.

31.6 Physical Properties and Allotropy

Although oxygen and ozone are not obtained in the laboratory as solids so that their crystalline forms can be compared, they are none-the-less allotropes: different crystalline forms of the same element (monotropes, see section 8.4).

The structure of oxygen given in section 14.3 is an approximation because its properties cannot be fully explained in terms of a double bond between the two oxygen atoms.

Property	*Oxygen*	*Sulphur* (α)
Atomic number	8	16
Electronic structure	2,6	2,8,6
Relative atomic mass	16·00	32·06
Density (kg/m^3)	1·44	2070
Melting point (°C)	−219	114
Boiling point (°C)	−183	445
Colour	Gas—none. Liquid—pale blue	Yellow
Physiological action	Essential to life	Constituent of proteins but some compounds poisonous
Solubility	In water—slight. Fish and water plants depend on it	In water—none. More soluble in carbon disulphide, methylbenzene, dimethylbenzene etc.

Property	*Oxygen*	*Ozone*
Usual state	Colourless gas	Colourless gas
Molecular state	O_2	O_3
Odour	None	Like dilute chlorine
Physiological action	Essential to life	Irritant 100 p.p.m., large quantities poisonous
Colour of liquid	Pale blue	Dark blue
Boiling point (°C)	−183	−112
Solubility in water	Slight	Moderate

There are very few differences between rhombic (α) and monoclinic (β) sulphur which are the only two solid allotropes (enantiotropic, see section 8.4). The differences in crystal structure are shown in figure 31.5: they arise because below 96°C, the transition temperature, the rings of S_8 molecules can be packed more densely than above 96°C (densities 2070 and 2000 kg/m^3 respectively). Rhombic sulphur is transparent and is a pale amber-yellow while monoclinic sulphur is translucent and is a pastel yellow colour.

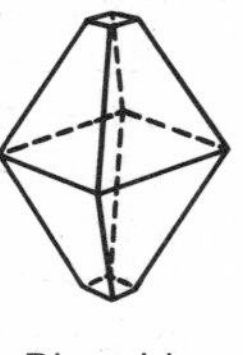

Figure 31.5
The crystal forms of sulphur

The changes in liquid sulphur can also be explained in terms of structures. At low temperatures the liquid (like the rhombic and monoclinic forms) consists of S_8 molecules (see figure 31.6). On heating, the vibrations within the molecules reach a critical level at about 160°C, the rings break and form long chains: this is the very viscous form of sulphur. On further heating, the chains decrease in length, and so the molecules are less tangled and the liquid is more mobile. It is a case of dynamic allotropy (see section 8.4). The zig-zag chain molecule in plastic sulphur is similar to the molecules found in rubber and fibres.

31.7 The Chemical Properties of Oxygen

Oxygen is a very reactive gas which is involved in burning, breathing, rusting and photosynthesis. It reacts directly with nearly all the elements except the noble gases and the halogens.

(a) With a Glowing Splint
A splint is lit and the flame then blown out so that only a small part remains red-hot. On plunging it into a jar of oxygen the splint relights. This is the test for oxygen: the only other gas which will give this result is dinitrogen oxide which has a characteristic smell.

(b) With Hydrogen
Hydrogen will burn in oxygen (see section 23.6(d)).

(c) With Sodium
After warming on a deflagrating spoon over a Bunsen burner, a small piece of clean sodium will burn brightly in a gas jar of dry oxygen with a brilliant yellow flame. The main product is sodium peroxide, a yellowish solid, which when moistened releases oxygen.

$$2Na + O_2 \rightarrow Na_2O_2 \quad [(Na^+)_2O_2^{2-}]$$

$$2Na_2O_2 + 2H_2O \rightarrow 4NaOH + O_2\uparrow \quad (Na^+OH^-)$$

(d) With Magnesium
A piece of magnesium ribbon is hung on a deflagrating spoon. After ignition by a Bunsen burner, it is placed in a jar of oxygen where it

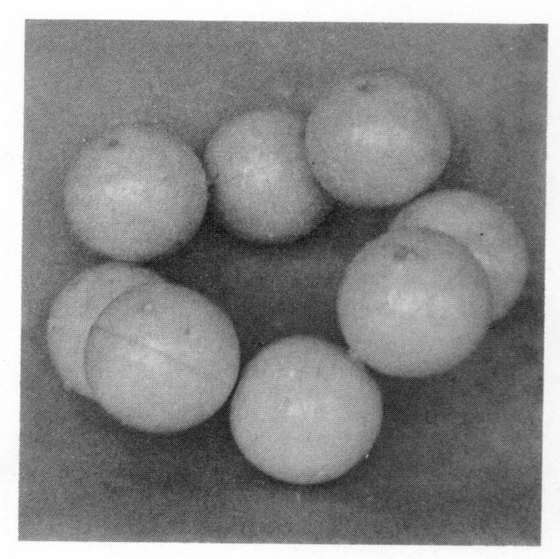

Figure 31.6
The sulphur molecule, S_8, viewed at 45°

piece of filter paper in a funnel. The sulphur should be allowed to solidify and then the crust gently broken revealing crystals of monoclinic sulphur. Monoclinic sulphur is only stable between 96–119°C.

Larger but more fragile monoclinic crystals can be obtained by warming a spatula load of sulphur in 25 cm^3 of dimethylbenzene in a boiling tube by means of a beaker full of boiling water. On cooling, needle shaped crystals grow upwards from the excess sulphur in the bottom of the tube. On examination under a microscope a few hours later, the crystals are found to have reverted to rhombic sulphur.

There are other forms of sulphur which are not crystalline and cannot therefore be said to be allotropes. Sulphur is not often prepared deliberately in reactions because it is readily available as an element.

(c) Amorphous (or Colloidal) Sulphur

When a dilute acid is added to sodium thiosulphate, sulphur is precipitated, the reaction often being studied from the point of view of rates of reactions (see section 18.9).

(d) Liquid Sulphur

Sulphur undergoes some peculiar changes as it is heated from its melting point to its boiling point. When just molten it is a pale yellow mobile liquid but on further heating it soon darkens. At about 160°C it suddenly becomes very viscous and cannot be poured. On further heating it remains dark but slowly becomes more mobile again.

An easy way of showing the difference between the forms of liquid sulphur is to pour the just-molten liquid into a beaker of cold water: this yields small beads of a brittle solid. On the other hand, if hot molten sulphur (250–350°C) is poured into cold water an elastic form is given, known as plastic sulphur: it is a supercooled liquid (like glass) but it is not very stable and soon reverts to a hard yellow solid.

31.6 Physical Properties and Allotropy

Although oxygen and ozone are not obtained in the laboratory as solids so that their crystalline forms can be compared, they are none-the-less allotropes: different crystalline forms of the same element (monotropes, see section 8.4).

The structure of oxygen given in section 14.3 is an approximation because its properties cannot be fully explained in terms of a double bond between the two oxygen atoms.

Property	*Oxygen*	*Sulphur* (α)
Atomic number	8	16
Electronic structure	2,6	2,8,6
Relative atomic mass	16·00	32·06
Density (kg/m^3)	1·44	2070
Melting point (°C)	−219	114
Boiling point (°C)	−183	445
Colour	Gas—none. Liquid—pale blue	Yellow
Physiological action	Essential to life	Constituent of proteins but some compounds poisonous
Solubility	In water—slight. Fish and water plants depend on it	In water—none. More soluble in carbon disulphide, methylbenzene, dimethylbenzene etc.

Property	*Oxygen*	*Ozone*
Usual state	Colourless gas	Colourless gas
Molecular state	O_2	O_3
Odour	None	Like dilute chlorine
Physiological action	Essential to life	Irritant 100 p.p.m., large quantities poisonous
Colour of liquid	Pale blue	Dark blue
Boiling point (°C)	−183	−112
Solubility in water	Slight	Moderate

There are very few differences between rhombic (α) and monoclinic (β) sulphur which are the only two solid allotropes (enantiotropic, see section 8.4). The differences in crystal structure are shown in figure 31.5: they arise because below 96°C, the transition temperature, the rings of S_8 molecules can be packed more densely than above 96°C (densities 2070 and 2000 kg/m^3 respectively). Rhombic sulphur is transparent and is a pale amber-yellow while monoclinic sulphur is translucent and is a pastel yellow colour.

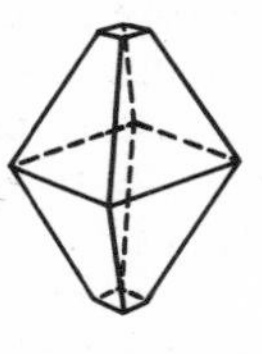
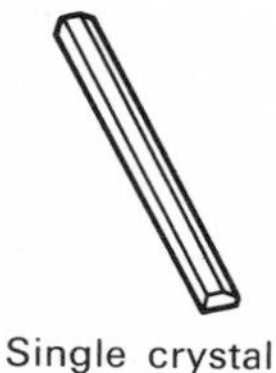

Figure 31.5
The crystal forms of sulphur

The changes in liquid sulphur can also be explained in terms of structures. At low temperatures the liquid (like the rhombic and monoclinic forms) consists of S_8 molecules (see figure 31.6). On heating, the vibrations within the molecules reach a critical level at about 160°C, the rings break and form long chains: this is the very viscous form of sulphur. On further heating, the chains decrease in length, and so the molecules are less tangled and the liquid is more mobile. It is a case of dynamic allotropy (see section 8.4). The zig-zag chain molecule in plastic sulphur is similar to the molecules found in rubber and fibres.

31.7 The Chemical Properties of Oxygen

Oxygen is a very reactive gas which is involved in burning, breathing, rusting and photosynthesis. It reacts directly with nearly all the elements except the noble gases and the halogens.

(a) With a Glowing Splint
A splint is lit and the flame then blown out so that only a small part remains red-hot. On plunging it into a jar of oxygen the splint relights. This is the test for oxygen: the only other gas which will give this result is dinitrogen oxide which has a characteristic smell.

(b) With Hydrogen
Hydrogen will burn in oxygen (see section 23.6(d)).

(c) With Sodium
After warming on a deflagrating spoon over a Bunsen burner, a small piece of clean sodium will burn brightly in a gas jar of dry oxygen with a brilliant yellow flame. The main product is sodium peroxide, a yellowish solid, which when moistened releases oxygen.

$$2Na + O_2 \rightarrow Na_2O_2$$
$$[(Na^+)_2O_2^{2-}]$$
$$2Na_2O_2 + 2H_2O \rightarrow 4NaOH + O_2\uparrow$$
$$(Na^+OH^-)$$

(d) With Magnesium
A piece of magnesium ribbon is hung on a deflagrating spoon. After ignition by a Bunsen burner, it is placed in a jar of oxygen where it

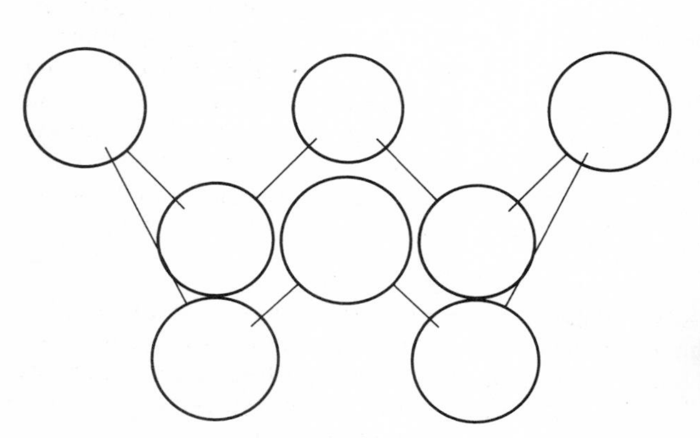
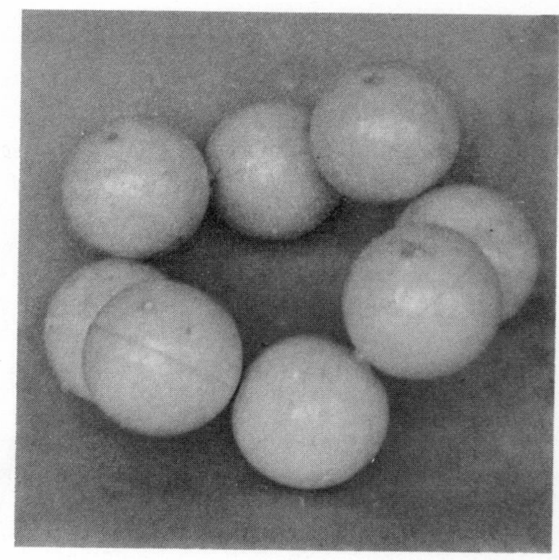

Figure 31.6
The sulphur molecule, S_8, viewed at 45°

burns with a blinding white light, dangerous to the eyes if viewed directly. A white powder, magnesium oxide, is produced.

$$2Mg + O_2 \rightarrow 2MgO \qquad (Mg^{2+}O^{2-})$$

(e) With Calcium
When a piece of calcium is strongly heated in oxygen or air it suddenly, almost explosively, bursts into flame.

$$2Ca + O_2 \rightarrow 2CaO \qquad (Ca^{2+}O^{2-})$$

(f) With Iron
Steel wool (iron wire) can be heated similarly and will burn vigorously in a jar of oxygen. Hot iron filings, when sprinkled in a Bunsen flame or into a jar of oxygen, react likewise. There should be some water in the bottom of the jar to prevent it being cracked by hot particles.

$$3Fe + 2O_2 \rightarrow \underset{\text{Triiron tetraoxide (black)}}{Fe_3O_4}$$

(g) With Carbon
If a small piece of charcoal is heated on a deflagrating spoon until it glows, and then put into a jar of oxygen, it will glow more brightly and burst into flame. Carbon dioxide will form, which can be shown by shaking the jar after adding some calcium hydroxide solution.

$$C + O_2 \rightarrow CO_2$$

(h) With Carbon Compounds
Coal, gas, wood and petroleum (paraffin, naphtha, naphthalene, benzene etc.) all contain compounds of hydrogen and carbon so that on combustion steam and carbon dioxide are produced.

(i) With Phosphorus
White phosphorus (P_4) ignites spontaneously in oxygen; red phosphorus (P) requires warming. In both gases the main product of the vigorous reaction is phosphorus(V) oxide.

$$P_4 + 5O_2 \rightarrow P_4O_{10}$$

(j) With Sulphur
Sulphur requires warming on a deflagrating spoon before it will burn: it does so with a blue flame, producing a poisonous gas with a choking smell, sulphur dioxide.

$$S + O_2 \rightarrow SO_2$$

Most of the elements (and many compounds) burn to give oxides. Materials that can be made to burn in air will burn much better in oxygen. The reactions involve the oxidation of the material. Oxygen is a very effective oxidizing agent at high temperatures, the only important reactions occurring at ordinary temperatures being in the body, in photosynthesis and in rusting.

31.8 The Classification of Oxides

The oxides made as above can be shaken with litmus solution: litmus is a vegetable dye which turns red with acids and blue with alkalis. The oxides of the metals (if soluble), such as sodium, turn litmus blue but the oxides of the non-metals such as carbon, phosphorus and sulphur, turn litmus red. Triiron tetraoxide is insoluble in water and magnesium oxide is only slightly soluble. The basic nature of metallic oxides, which are mostly insoluble in water, is more evident from the fact that they react with acids to give salts and water (see section 17.5).

There is a chemical reaction between the water and the oxide before the litmus is affected. Thus:

$$2Na_2O_2 + 2H_2O \rightarrow \underset{\text{Sodium hydroxide}}{4NaOH} + O_2\uparrow$$

or

$$2O_2^{2-} + 2H_2O \rightarrow 4OH^- + O_2\uparrow$$

$$MgO + H_2O \rightarrow \underset{\text{Magnesium hydroxide}}{Mg(OH)_2} \qquad \text{(slow)}$$

or

$$O^{2-} + H_2O \rightarrow 2OH^-$$

$$CO_2 + H_2O \rightarrow \underset{\text{Carbonic acid (a few ions)}}{H_2CO_3}$$

$$P_4O_{10} + 2H_2O \rightarrow \underset{\text{Metaphosphoric acid (many ions)}}{4HPO_3}$$

$$SO_2 + H_2O \rightarrow \underset{\text{Sulphuric(IV) acid (many ions)}}{H_2SO_3}$$

Acidic and basic oxides are the most important types but a full classification is as follows:

(a) Basic Oxides
These are the normal oxides of metals; the valency exerted is usually the group number or two below this. They neutralize acids, forming salts and water. They are usually ionic substances and the element is the cation. Some of them are soluble in water (e.g. sodium and calcium oxides) but most of them are insoluble in water (e.g. copper(II) oxide and iron(III) oxide).

(b) Acidic Oxides
These are the normal oxides of non-metals; the valency exerted may be the group number, two or four or even six below this. They neutralize bases forming salts and water. As these oxides dissolve in water yielding the respective acids, they are often called acid anhydrides. They

are usually covalent substances but when they dissolve in water or react with a base the element becomes the anion (or part thereof), e.g. the oxides of carbon, phosphorus and sulphur.

(c) Amphoteric Oxides

These are the oxides of some metals, particularly ones near the middle of the periodic table, which can react with acids and bases, e.g. zinc and aluminium oxides.

(d) Neutral Oxides

These are the oxides of some non-metals, often ones near the dividing line between metals and non-metals, which react neither with acids nor with alkalis, e.g. carbon monoxide (under laboratory conditions), nitrogen oxide (NO), dinitrogen oxide (N_2O) and can be said to include water itself.

(e) Mixed or Salt Oxides

These are metallic oxides which react with acids to give either two salts or a salt and the other oxide, e.g. trilead tetraoxide (red lead oxide, Pb_3O_4) which behaves as PbO_2+2PbO and tri-iron tetraoxide (magnetite, Fe_3O_4) which behaves as $FeO+Fe_2O_3$, although in neither case are the constituents apparent in the overall structure.

When the classification of oxides is based upon structural rather than acid-base considerations, two further categories are often distinguished.

(f) Peroxides

In these the valency of the metal is apparently higher than usual if oxygen is taken as having a valency of two. They are electrovalent oxides of metals which contain the ion O_2^{2-}. On treatment with cold dilute acids they yield hydrogen peroxide. The best-known examples are sodium peroxide (Na_2O_2) and barium peroxide (BaO_2). They are salts of hydrogen peroxide.

e.g. $$Na_2O_2+H_2SO_4 \rightarrow Na_2SO_4+H_2O_2$$

or $$O_2^{2-}+2H^+ \rightarrow H_2O_2$$

(A very weak acid)

(g) Dioxides

They have formulae similar to the peroxides but are covalent. They do not yield hydrogen peroxide with acids. Two similarities are that they yield oxygen on heating and are strong oxidizing agents.

e.g. $$MnO_2+4HCl \rightarrow MnCl_2+2H_2O+Cl_2\uparrow$$

Hot, concentrated acid

31.9 The Chemical Properties of Ozone (Trioxygen)

Ozone is a much more powerful oxidizing agent than oxygen: the course of the reaction is usually the same but the time shorter than in the case of oxygen.

(a) Stability

Ozone decomposes to give oxygen at 300°C. The decomposition is catalyzed by manganese(IV) oxide, platinum etc. The oxygen molecule is very stable.

$$2O_3 \rightarrow 3O_2$$

(b) With Mercury

Ozone makes mercury 'tail', i.e. causes it to cease running freely on a watch-glass. Oxygen does not react until the temperature reaches 350°C and at a slightly higher temperature, decomposition of the red oxide occurs.

(c) With Potassium Iodide

Potassium iodide solution, especially if acidified, rapidly discolours in the presence of ozone.

$$2KI+H_2SO_4+O_3 \rightarrow I_2\downarrow+H_2O+K_2SO_4+O_2\uparrow$$

or $$2I^-+2H^++O_3 \rightarrow I_2\downarrow+H_2O+O_2\uparrow$$

The iodine (a black precipitate) will dissolve if there is any excess of potassium iodide giving a brown solution; it may also be detected by starch with which it gives a dark blue coloration. This is a standard test for an oxidizing agent.

Ozone will also react in the usual manner with many other reducing agents, e.g. iron(II) salts, hydrogen sulphide, lead sulphide etc.

(d) With Alkenes

Substances which contain alkenic double bonds, e.g. rubber and turpentine, and aromatic substances, e.g. benzene, are attacked by ozone.

31.10 The Chemical Properties of Sulphur

These properties were not considered at the same time as those of oxygen because the properties of oxygen are usually considered at a very early stage in the study of chemistry when a comparison with sulphur would not be required. The chemical properties of oxygen and sulphur are very similar, the reactions of the latter taking a similar course but needing more heat to initiate them.

(a) With Oxygen

When heated, sulphur burns with a blue flame in air or oxygen, forming sulphur dioxide.

$$S+O_2 \rightarrow SO_2$$

This reaction can be utilized to show that the allotropes of sulphur are identical substances burns with a blinding white light, dangerous to the eyes if viewed directly. A white powder, magnesium oxide, is produced.

$$2Mg+O_2 \rightarrow 2MgO \qquad (Mg^{2+}O^{2-})$$

(e) With Calcium
When a piece of calcium is strongly heated in oxygen or air it suddenly, almost explosively, bursts into flame.

$$2Ca+O_2 \rightarrow 2CaO \qquad (Ca^{2+}O^{2-})$$

(f) With Iron
Steel wool (iron wire) can be heated similarly and will burn vigorously in a jar of oxygen. Hot iron filings, when sprinkled in a Bunsen flame or into a jar of oxygen, react likewise. There should be some water in the bottom of the jar to prevent it being cracked by hot particles.

$$3Fe+2O_2 \rightarrow \underset{\text{Triiron tetraoxide (black)}}{Fe_3O_4}$$

(g) With Carbon
If a small piece of charcoal is heated on a deflagrating spoon until it glows, and then put into a jar of oxygen, it will glow more brightly and burst into flame. Carbon dioxide will form, which can be shown by shaking the jar after adding some calcium hydroxide solution.

$$C+O_2 \rightarrow CO_2$$

(h) With Carbon Compounds
Coal, gas, wood and petroleum (paraffin, naphtha, naphthalene, benzene etc.) all contain compounds of hydrogen and carbon so that on combustion steam and carbon dioxide are produced.

(i) With Phosphorus
White phosphorus (P_4) ignites spontaneously in oxygen; red phosphorus (P) requires warming. In both gases the main product of the vigorous reaction is phosphorus(V) oxide.

$$P_4+5O_2 \rightarrow P_4O_{10}$$

(j) With Sulphur
Sulphur requires warming on a deflagrating spoon before it will burn: it does so with a blue flame, producing a poisonous gas with a choking smell, sulphur dioxide.

$$S+O_2 \rightarrow SO_2$$

Most of the elements (and many compounds) burn to give oxides. Materials that can be made to burn in air will burn much better in oxygen. The reactions involve the oxidation of the material. Oxygen is a very effective oxidizing agent at high temperatures, the only important reactions occurring at ordinary temperatures being in the body, in photosynthesis and in rusting.

31.8 The Classification of Oxides

The oxides made as above can be shaken with litmus solution: litmus is a vegetable dye which turns red with acids and blue with alkalis. The oxides of the metals (if soluble), such as sodium, turn litmus blue but the oxides of the non-metals such as carbon, phosphorus and sulphur, turn litmus red. Triiron tetraoxide is insoluble in water and magnesium oxide is only slightly soluble. The basic nature of metallic oxides, which are mostly insoluble in water, is more evident from the fact that they react with acids to give salts and water (see section 17.5).

There is a chemical reaction between the water and the oxide before the litmus is affected. Thus:

$$2Na_2O_2+2H_2O \rightarrow \underset{\text{Sodium hydroxide}}{4NaOH}+O_2\uparrow$$

or

$$2O_2^{2-}+2H_2O \rightarrow 4OH^-+O_2\uparrow$$

$$MgO+H_2O \rightarrow \underset{\text{Magnesium hydroxide}}{Mg(OH)_2} \quad \text{(slow)}$$

or

$$O^{2-}+H_2O \rightarrow 2OH^-$$

$$CO_2+H_2O \rightarrow \underset{\text{Carbonic acid (a few ions)}}{H_2CO_3}$$

$$P_4O_{10}+2H_2O \rightarrow \underset{\text{Metaphosphoric acid (many ions)}}{4HPO_3}$$

$$SO_2+H_2O \rightarrow \underset{\text{Sulphuric(IV) acid (many ions)}}{H_2SO_3}$$

Acidic and basic oxides are the most important types but a full classification is as follows:

(a) Basic Oxides
These are the normal oxides of metals; the valency exerted is usually the group number or two below this. They neutralize acids, forming salts and water. They are usually ionic substances and the element is the cation. Some of them are soluble in water (e.g. sodium and calcium oxides) but most of them are insoluble in water (e.g. copper(II) oxide and iron(III) oxide).

(b) Acidic Oxides
These are the normal oxides of non-metals; the valency exerted may be the group number, two or four or even six below this. They neutralize bases forming salts and water. As these oxides dissolve in water yielding the respective acids, they are often called acid anhydrides. They

are usually covalent substances but when they dissolve in water or react with a base the element becomes the anion (or part thereof), e.g. the oxides of carbon, phosphorus and sulphur.

(c) *Amphoteric Oxides*

These are the oxides of some metals, particularly ones near the middle of the periodic table, which can react with acids and bases, e.g. zinc and aluminium oxides.

(d) *Neutral Oxides*

These are the oxides of some non-metals, often ones near the dividing line between metals and non-metals, which react neither with acids nor with alkalis, e.g. carbon monoxide (under laboratory conditions), nitrogen oxide (NO), dinitrogen oxide (N_2O) and can be said to include water itself.

(e) *Mixed or Salt Oxides*

These are metallic oxides which react with acids to give either two salts or a salt and the other oxide, e.g. trilead tetraoxide (red lead oxide, Pb_3O_4) which behaves as PbO_2+2PbO and tri-iron tetraoxide (magnetite, Fe_3O_4) which behaves as $FeO+Fe_2O_3$, although in neither case are the constituents apparent in the overall structure.

When the classification of oxides is based upon structural rather than acid-base considerations, two further categories are often distinguished.

(f) *Peroxides*

In these the valency of the metal is apparently higher than usual if oxygen is taken as having a valency of two. They are electrovalent oxides of metals which contain the ion O_2^{2-}. On treatment with cold dilute acids they yield hydrogen peroxide. The best-known examples are sodium peroxide (Na_2O_2) and barium peroxide (BaO_2). They are salts of hydrogen peroxide.

e.g. $Na_2O_2+H_2SO_4 \rightarrow Na_2SO_4+H_2O_2$

or $O_2^{2-}+2H^+ \rightarrow H_2O_2$

(A very weak acid)

(g) *Dioxides*

They have formulae similar to the peroxides but are covalent. They do not yield hydrogen peroxide with acids. Two similarities are that they yield oxygen on heating and are strong oxidizing agents.

e.g. $MnO_2+4HCl \rightarrow MnCl_2+2H_2O+Cl_2\uparrow$

Hot, concentrated acid

31.9 The Chemical Properties of Ozone (Trioxygen)

Ozone is a much more powerful oxidizing agent than oxygen: the course of the reaction is usually the same but the time shorter than in the case of oxygen.

(a) *Stability*

Ozone decomposes to give oxygen at 300°C. The decomposition is catalyzed by manganese(IV) oxide, platinum etc. The oxygen molecule is very stable.

$$2O_3 \rightarrow 3O_2$$

(b) *With Mercury*

Ozone makes mercury 'tail', i.e. causes it to cease running freely on a watch-glass. Oxygen does not react until the temperature reaches 350°C and at a slightly higher temperature, decomposition of the red oxide occurs.

(c) *With Potassium Iodide*

Potassium iodide solution, especially if acidified, rapidly discolours in the presence of ozone.

$$2KI+H_2SO_4+O_3 \rightarrow I_2\downarrow+H_2O+K_2SO_4+O_2\uparrow$$

or $2I^-+2H^++O_3 \rightarrow I_2\downarrow+H_2O+O_2\uparrow$

The iodine (a black precipitate) will dissolve if there is any excess of potassium iodide giving a brown solution; it may also be detected by starch with which it gives a dark blue coloration. This is a standard test for an oxidizing agent.

Ozone will also react in the usual manner with many other reducing agents, e.g. iron(II) salts, hydrogen sulphide, lead sulphide etc.

(d) *With Alkenes*

Substances which contain alkenic double bonds, e.g. rubber and turpentine, and aromatic substances, e.g. benzene, are attacked by ozone.

31.10 The Chemical Properties of Sulphur

These properties were not considered at the same time as those of oxygen because the properties of oxygen are usually considered at a very early stage in the study of chemistry when a comparison with sulphur would not be required. The chemical properties of oxygen and sulphur are very similar, the reactions of the latter taking a similar course but needing more heat to initiate them.

(a) *With Oxygen*

When heated, sulphur burns with a blue flame in air or oxygen, forming sulphur dioxide.

$$S+O_2 \rightarrow SO_2$$

This reaction can be utilized to show that the allotropes of sulphur are identical substances from the chemical point of view (see section 9.5). The same mass of sulphur yields double that mass of sulphur dioxide, (within the limits of experimental error).

(b) With Hydrogen

The reaction only occurs to an appreciable extent if hydrogen and gaseous sulphur are passed over a nickel catalyst at 450°C.

$$H_2 + S \rightarrow H_2S$$

(c) With Metals

Most metals will combine directly with sulphur. The reaction with metals at the top of the electrochemical series may be dangerously vigorous. The reaction with iron is frequently used to demonstrate the differences between elements, mixtures and compounds and between physical and chemical changes.

$$Fe + S \rightarrow FeS$$

Even mercury, which is near the bottom of the series, reacts on rubbing with sulphur in the cold.

$$Hg + S \rightarrow HgS$$

(d) With Non-metals

For the reaction of carbon with sulphur, see section 28.16. A phosphorus sulphide is mentioned in section 30.7.

31.11 The Uses of Oxygen, Ozone and Sulphur

The use of **oxygen** in the iron and steel industry has revolutionized production there and led to a great upsurge in the manufacture of oxygen.

(a) Iron and Steel Industry. About 80% of the oxygen manufactured is used directly to enrich the air blown into blast furnaces and to convert the impure iron produced there into steel.

(b) Cutting Metals. Ethyne (C_2H_2) or propane (C_3H_8) burning in a forced supply of oxygen gives a very hot flame which can be used to cut steel up to 30 cm thick, the temperature being over 3000°C.

(c) Rockets. Stage 1 of the Saturn V rocket for the Apollo spacecraft burns kerosene in oxygen; stages 2 and 3 burn hydrogen in oxygen.

(d) Low Temperature Work. The boiling point of oxygen (−183°C, 90 K) is a temperature at which many phenomena have been investigated. The cooling property of oxygen is made use of in the shrink fitting of linings to cylinders.

(e) Support for Life. Patients undergoing surgical and dental operations are given an anaesthetic and oxygen. Cylinders of compressed oxygen are also needed for patients with breathing difficulties, for rescue work in mines, for fire-fighting, for flying and climbing at high altitude, in submarines and for sub-aqua exploration by swimmers.

(f) For Combustion. Vast quantities of air are needed for the combustion of coal, coal gas, petroleum and the conversion of ammonia to nitrogen oxide, of sulphur dioxide to sulphur trioxide and in many other important reactions.

Ozone is not a very widely used substance because of the difficulty and expense of making it.

(a) Sterilizing Water. A five minute dose of ozone (1 p.p.m.) is more efficient at rendering water fit for drinking than an hour's treatment with chlorine. It degrades detergents, chlorinated phenols and pesticides, all of which give rise to increasing concern nowadays. It is used in sterilizing Manchester's water supply.

(b) Purifying Air. Ozone is a strong germicide because of its oxidizing powers and it is used in underground railway systems to keep the air fresh.

(c) Industrial Wastes and Sewage. Ozone can be used to destroy cyanides which industry often uses in electroplating solutions, and its use to render sewage innocuous or, even better, useful, is now being investigated.

Sulphur like oxygen, has one use which is far more important than all the others put together.

(a) Sulphuric Acid. The main use of sulphur is as a starting point in the manufacturing sequence that involves sulphur dioxide, sulphur trioxide and sulphuric acid. Some sulphites (e.g. of calcium) and sulphates (e.g. ammonium sulphate) are very important. Calcium hydrogensulphite is used in the manufacture of paper from wood.

(b) Sulphides. Carbon disulphide and phosphorus sulphide are dealt with in sections 28.16 and 30.7(e) respectively.

(c) Rubber. Raw rubber as obtained from the natural latex is too soft and is heated with sulphur to make it a useful substance: it is vulcanized.

(d) Medicine. Many of the modern drugs, e.g. the sulphonamides such as M and B 693, and the modern antibiotics, e.g. penicillin, contain sulphur in their molecules. Sulphur ointments are also used for dermatological purposes.

(e) Pest Control. Insects and fungi which attack fruit trees are destroyed by spraying with sulphur and sulphur derivatives.

II THE HYDRIDES

31.12 Water

An earlier chapter (6) was devoted to a simple discussion of water and its use as a solvent. The facts that it is the most common liquid and the most widely used solvent disguise the fact that, in comparison with the hydrides of the neighbours of oxygen in the periodic table, it is one of the oddest substances known.

Water melts at 0°C, a temperature at which most of its neighbours (ammonia, phosphine, hydrogen sulphide and hydrogen chloride) have already boiled and become gases. One structure of ice is based on that of diamond: it is tetrahedral with oxygen in the place of carbon and with two hydrogen atoms associated with every oxygen atom, there being one hydrogen for every oxygen-oxygen axis (see figure 31.7). The water molecules in ice thus form cages in which other molecules may be trapped. The molecules are relatively widely spaced and when ice melts overlapping of the molecules can take place. (Analogy: fingers held tip to tip with palms of hands apart; when fingers are interwoven the palms are closer.) Most substances expand on heating because the individual particles take up a greater effective volume by virtue of their increased vibrations on heating: this effect does not outweigh the former one until above 4°C. Thus ice has a density of 920 kg/m^3 and water shows a maximum density of 1000 kg/m^3 at 4°C.

The ability of hydrogen to come between and to bridge two oxygen atoms as shown in figure 31.7 is unexpected and very important: it is known as hydrogen bonding. A consequence of this bonding is that the effective molecular weight of water is much greater than 18 and its enthalpies of fusion and evaporation, its surface tension and its viscosity, are much higher than expected. The non-identical twin molecules in RNA and DNA (see section 29.16) are held together by hydrogen bonding into a double spiral form. Hydrogen bonds are not very stable to heat, e.g. when meat is cooked the bonds are broken and other changes take place making the meat edible.

Water is the standard of neutrality (pH = 7) but its neighbours are alkaline or acidic in their reactions. Water is essential to life and is odourless but its neighbours are poisonous, pungent smelling substances. Water is too often taken for granted: as industrialization increases the demand increases also. In the United Kingdom some 15 million tonnes of water are used every day, only one quarter of which is recycled, the remainder being sent back to the sea. The choice of water for drinking and industrial purposes has been discussed in section 25.13. Highly purified water is needed for a few purposes, e.g. in laboratories. Industries that use large quantities of water include electricity generation, steel manufacturing, paper manufacturing, sugar beet processing, rayon manufacturing and oil refining.

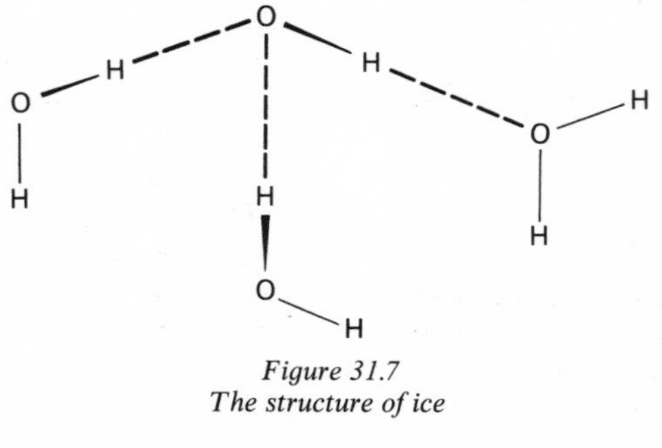

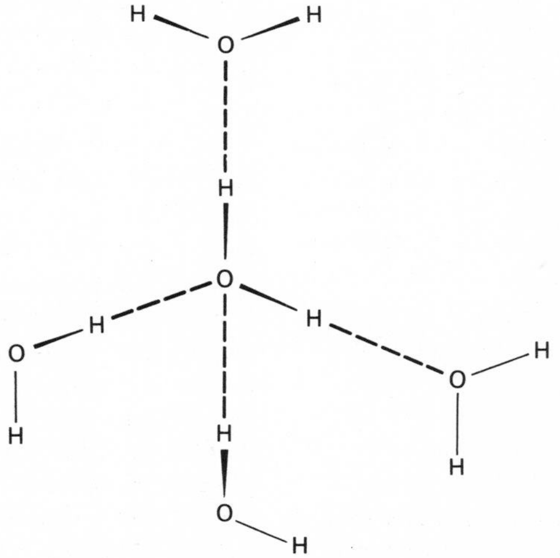

Figure 31.7
The structure of ice

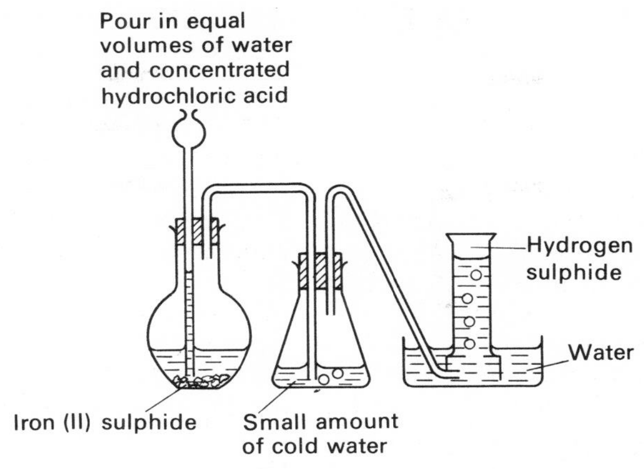

Figure 31.8
The preparation of hydrogen sulphide

31.13 The Preparation and Physical Properties of Hydrogen Sulphide

Hydrogen sulphide is formed when many animal and vegetable substances decompose (e.g. bad eggs), because sulphur is a constituent of the proteins they contain. It is present in volcanic gases, spring waters and in many samples of natural gas, e.g. at Lacq in France. Synthesis of the gas in the laboratory is difficult (see section 31.10(b)).

The laboratory preparation may start with iron and sulphur which are heated together in the correct proportion by mass (56:32) or with commercial iron sulphide. The iron sulphide is treated with moderately concentrated hydrochloric acid either in a Kipp's apparatus or as shown in figure 31.8.

$$Fe + S \rightarrow FeS$$

and $$FeS + 2HCl \rightarrow FeCl_2 + H_2S\uparrow$$

or $$FeS + 2H^+ \rightarrow Fe^{2+} + H_2S\uparrow$$

The gas should be rinsed with a little water to remove hydrogen chloride that may have boiled off and then collected over water or by downward delivery. If the latter method of collection is employed the preparation should be done in a fume cupboard because the gas is poisonous. As free iron is usually present in iron sulphide some hydrogen will contaminate the gas, but not adversely so.

$$Fe + 2HCl \rightarrow FeCl_2 + H_2\uparrow$$

or $$Fe + 2H^+ \rightarrow Fe^{2+} + H_2\uparrow$$

Hydrogen sulphide is a colourless gas with a smell of bad eggs. It is very dangerous because er a while people become insensitive to its The gas is denser than air (R.D. 17) and moderately soluble in cold water—it is better to lose some by solution in water when collecting it than to attempt to collect the gas over warm water.

31.14 The Chemical Properties of Hydrogen Sulphide

The properties can be divided into three categories: it is an acid, a reducing agent and an oxidizing agent.

(a) As an Acid

It is a weak acid in aqueous solution, turning litmus a claret colour (like carbon dioxide); its salts are the sulphides and hydrogensulphides because it is a dibasic acid.

(*i*) *Soluble sulphides.* Sodium and ammonium hydrogensulphides are first prepared by saturating the respective alkaline solutions with hydrogen sulphide.

e.g. $$NaOH + H_2S \rightarrow NaHS + H_2O$$

Then a portion of alkali equal to that used originally is added to convert the acid salt to a normal salt

e.g. $$NaHS + NaOH \rightarrow Na_2S + H_2O$$

(*ii*) *Insoluble sulphides.* These are prepared by precipitation. There are many insoluble sulphides and so the test may be performed under acidic or alkaline conditions so that some distinctions can be made.

The sulphides of copper (black), lead (black), silver (black-brown), mercury (black), tin(II) (brown) and tin(IV) (yellow) are precipitated in acidic solutions—a little dilute hydrochloric acid is added to a solution of the substance under test.

e.g. $Pb(NO_3)_2 + H_2S \rightarrow PbS\downarrow + 2HNO_3$
or $Pb^{2+} + H_2S \rightarrow PbS\downarrow + 2H^+$

This particular reaction is used to characterize hydrogen sulphide in gaseous mixtures (the alternative being lead ethanoate solution) and to detect lead ions in drinking water. Lead paints tarnish in town air because of the formation of lead sulphide. The formation of a film of the sulphide on silver articles is the explanation of their tarnishing in town air and with some foods, such as egg yolk.

The sulphides of iron(II) (black), zinc (off white) and manganese (buff) are only precipitated in neutral or alkaline solution—a spatula load of ammonium chloride and then some ammonia solution are added to the solution of the substance under test.

e.g. $FeSO_4 + H_2S \rightarrow FeS\downarrow + H_2SO_4$
or $Fe^{2+} + H_2S \rightarrow FeS\downarrow + 2H^+$

The hydrogen sulphide test, which should be performed on a small quantity of a substance, is most useful when analyzing substances and is done whenever flame tests, sodium hydroxide and ammonia solution tests do not yield a clear result. Hydrogen sulphide has no uses as an individual substance except in analysis; the industrialist converts all that he obtains to sulphur.

(b) As a Reducing Agent

Hydrogen sulphide is readily oxidized and the appearance of the product colloidal sulphur sometimes makes analysis difficult.

$$H_2S \rightarrow 2H^+ + 2e^- + S\downarrow$$

(*i*) *With Oxygen.* If there is a plentiful supply of air the products of combustion are both gaseous initially.

$$2H_2S + 3O_2 \rightarrow 2H_2O + 2SO_2\uparrow$$

If the air supply is limited or the blue flame is chilled by holding a cold porcelain surface over it then sulphur is deposited (compare methane and ammonia, sections 29.8 and 30.11(b)).

$$2H_2S + O_2 \rightarrow 2S\downarrow + 2H_2O$$

(*ii*) *With Potassium Manganate(VII).* The purple colour of the solution, which should be acidified with a little dilute sulphuric acid, rapidly disappears and a milky yellow suspension of sulphur remains (compare sulphur dioxide).

(*iii*) *With Potassium Dichromate* (*or Chromate*). Again the test proceeds readily if the solution (orange or yellow respectively) is slightly acidified: a green solution with a suspension of sulphur is produced (compare sulphur dioxide).

(*iv*) *With Iron(III) Chloride.* Reduction of the yellow iron(III) salt to a pale green iron(II) salt occurs, with the liberation of sulphur.

$$2FeCl_3 + H_2S \rightarrow 2FeCl_2 + 2HCl + S\downarrow$$
or $$2Fe^{3+} + H_2S \rightarrow 2Fe^{2+} + 2H^+ + S\downarrow$$

If ammonium sulphide is used instead of hydrogen sulphide a black precipitate of iron(II) sulphide and a white one of sulphur are deposited.

(*v*) *With Sulphur Dioxide.* If jars of the moist gases are brought mouth-to-mouth sulphur is deposited; water is a catalyst for the reaction.

$$2H_2S + SO_2 \rightarrow 3S\downarrow + 2H_2O$$

(*vi*) *Other Oxidizing Agents.* Concentrated nitric and sulphuric acids and the halogens react in a similar manner producing sulphur.

(c) As an Oxidizing Agent

If hydrogen sulphide is passed over a heated metal the metal sulphide is produced.

e.g. $Sn + H_2S \rightarrow SnS + H_2$

31.15 The Preparation of Hydrogen Peroxide

The existence of two compounds of hydrogen and oxygen is an illustration of the Law of Multiple Proportions (see section 9.6).

Sodium peroxide is easily made by burning sodium in air free from carbon dioxide and moisture (see section 24.5(a)); when it is added to dilute sulphuric acid hydrogen peroxide and sodium sulphate, which may be crystallized out, are formed.

$$Na_2O_2 + H_2SO_4 \rightarrow Na_2SO_4 + H_2O_2$$

Alternatively, powdered hydrated barium peroxide can be added to dilute sulphuric acid. In both cases the acid should be ice-cold and acid should be left in slight excess because hydrogen peroxide is not a very stable substance.

$$BaO_2 + H_2SO_4 \rightarrow BaSO_4\downarrow + H_2O_2$$

These reactions yield a dilute (about 3%) solution of hydrogen peroxide. In industry a more complex process involving an organic substance (X) is first reduced and then oxidized; a more concentrated solution, which can be further concentrated by distillation, is obtained.

$$X + H_2 \rightarrow XH_2$$
$$XH_2 + O_2 \rightarrow X + H_2O_2$$

31.16 The Reactions of Hydrogen Peroxide

The reactions can be divided into three classes: its decomposition, as an oxidizing agent and a reducing agent.

(a) Thermal and Catalytic Decomposition

The solution will decompose on standing, even in the dark at ordinary temperatures, and on heating.

$$2H_2O_2 \rightarrow 2H_2O + O_2\uparrow$$

Catalysts that promote its decomposition include manganese(IV) oxide, alkalis and powdered metals. Inhibitors include sulphuric and phosphoric acids; when it is used for household bleaching the acid should be neutralized with a little ammonia.

Hydrogen peroxide is usually sold by its volume strength, e.g. '10-volume' peroxide: this means that 1 dm^3 of the solution will yield 10 dm^3 of oxygen (at s.t.p.) on decomposition. From the equation quoted above

$$68\text{ g } H_2O_2 \rightarrow 22{\cdot}4\text{ dm}^3\ O_2 \text{ (at s.t.p.)}$$

i.e. 10 dm^3 O_2 would be produced by

$$\frac{68 \times 10}{22{\cdot}4}\text{ g } H_2O_2 = 30\text{ g } H_2O_2$$

Thus if the 30 g of H_2O_2 is in 1 dm^3 of solution it is a 10 volume solution, or 3% solution in terms of the mass of pure hydrogen peroxide.

(b) As an Oxidizing Agent

$$H_2O_2 + 2H^+ + 2e^- \rightarrow 2H_2O$$

(*i*) *With Potassium Iodide*. The use of acidified potassium iodide is a standard test for an oxidizing agent: iodine is readily formed here.

$$2KI + H_2SO_4 + H_2O_2 \rightarrow I_2\downarrow + K_2SO_4 + 2H_2O$$
or $$2I^- + 2H^+ + H_2O_2 \rightarrow I_2\downarrow + 2H_2O$$

(*ii*) *With Iron(II) Sulphate*. The pale green solution of the iron(II) salt will change to a yellow colour and the oxidation can be readily confirmed by adding an alkali.

$$2FeSO_4 + H_2SO_4 + H_2O_2 \rightarrow Fe_2(SO_4)_3 + 2H_2O$$
or $$2Fe^{2+} + 2H^+ + H_2O_2 \rightarrow 2Fe^{3+} + 2H_2O$$

(*iii*) *With Lead Sulphide*. When lead compounds are used for painting and the pictures are then hung in town air, black lead sulphide forms and gradually discolours them. The paintings are restored to their original hues by careful treatment using a solution of hydrogen peroxide in ethoxyethane.

$$PbS + 4H_2O_2 \rightarrow PbSO_4 + 4H_2O$$

(*iv*) *In Bleaching*. Hydrogen peroxide is used as a bleach for wool, silk, cotton, straw and hair (peroxide blondes used a 0·5% solution containing a little sodium silicate etc.). It destroys the dyes by oxidation but does not harm the fibre, as the more powerful bleach chlorine sometimes does.

(*v*) *As an Antiseptic*. Hydrogen peroxide can be used in moderation for cleaning infected wounds: too much will damage the tissues (blood catalyzes its rapid decomposition).

(c) As a Reducing Agent

$$H_2O_2 \rightarrow 2H^+ + 2e^- + O_2\uparrow$$

(*i*) *With Potassium Manganate(VII)*. The purple solution, acidified with a little dilute sulphuric acid, is rapidly decolorized and oxygen is released.

(*ii*) *With Potassium Dichromate* (*or Chromate*). The solution (orange or yellow respectively) is acidified with dilute sulphuric acid and then hydrogen peroxide is added. The solution becomes blue due to the formation of perchromic acid, which will dissolve in ether. The blue ethereal layer is fairly stable but the aqueous layer soon becomes green because of the formation of a chromium(III) salt. Oxygen is evolved. This is the best test for hydrogen peroxide.

(*iii*) *After Bleaching*. If chlorine is used as the bleaching agent it is essential to destroy the chloric(I) acid formed when the chlorine dissolves in water.

$$Cl_2 + H_2O \rightarrow HOCl + HCl$$

Hydrogen peroxide will do this and cause the formation of less reactive substances which can easily be washed out of the fabric: this use accounts for about two thirds of the hydrogen peroxide manufactured.

$$H_2O_2 + HOCl \rightarrow HCl + H_2O + O_2\uparrow$$

III THE OXIDES

31.17 The Preparation of Sulphur Dioxide

This gas has been used since ancient times as a disinfectant but it was first described in scientific terms by Priestley (1772). It occurs in the atmosphere to an increasing extent as more and more sulphur-containing fuels are burnt and the waste gases escape, polluting the atmosphere.

The simplest way of manufacturing the gas is to burn sulphur in air.

$$S + O_2 \rightarrow SO_2\uparrow$$

There are many metals which occur as their

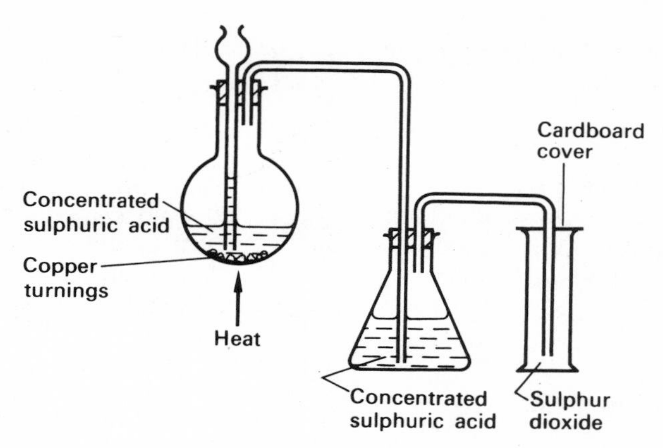

Figure 31.9
The preparation of sulphur dioxide

sulphides and the first step in extracting the metal is to burn the ore in air to obtain the metal oxide and sulphur dioxide.

e.g. $$2ZnS + 3O_2 \rightarrow 2ZnO + 2SO_2\uparrow$$

Iron(II) disulphide, iron pyrites, is treated similarly although it is not a source of iron.

$$4FeS_2 + 11O_2 \rightarrow 2Fe_2O_3 + 8SO_2\uparrow$$

The 'spent oxide' from coke ovens and gasworks is also burnt in air when the concentration of sulphur reaches 50%.

In the United Kingdom some sulphur dioxide is manufactured from calcium sulphate (anhydrite). It is powdered and mixed with siliceous materials (clay, ash, etc.) and coke before being fed into a rotary kiln similar to that used in the manufacture of cement and of aluminium oxide. At a temperature of 1400°C attained by the combustion of a fuel, sulphur dioxide (easily separated from the carbon monoxide also formed) and a slag of calcium silicate (used for cement) are formed.

$$CaSO_4 + SiO_2 + C \rightarrow CaSiO_3 + SO_2\uparrow + CO\uparrow$$

The usual laboratory preparation of sulphur dioxide, that of heating concentrated sulphuric acid with copper turnings (see figure 31.9) is economic nonsense because the sulphuric acid has just been made from sulphur dioxide. It is better to use a siphon of liquid sulphur dioxide. The reaction is only of interest as an example of a redox reaction and the equation

$$Cu + 2H_2SO_4 \rightarrow CuSO_4 + 2H_2O + SO_2\uparrow$$

is only an approximation because black sulphides of copper (Cu_2S, CuS) discolour the residue. The copper sulphate can be extracted by pouring off the excess acid, adding water, filtering and evaporating the solution to the point of crystallization.

Sulphur dioxide is more easily obtained in the laboratory by the action of a dilute acid on a sulphite [sulphate(IV)].

e.g. $$Na_2SO_3 + 2HCl \rightarrow 2NaCl + H_2O + SO_2\uparrow$$

or $$SO_3^{2-} + 2H^+ \rightarrow H_2O + SO_2\uparrow$$

31.18 The Properties of Sulphur Dioxide

Sulphur dioxide is a colourless, dense (R.D. 32) gas with a poisonous and pungent smell. The boiling point of the liquid is -10°C but it is easy to compress the gas at room temperature (about 2·5 atmospheres suffices) and obtain the liquid in siphons or to use it as a refrigerant. The gas is very soluble in water and forms a moderately strong, dibasic acid.

$$H_2O + SO_2 \rightleftharpoons \underset{\text{Sulphuric(IV) (sulphurous) acid}}{H_2SO_3}$$

The solution is not stable, decomposing by evaporation and oxidation.

The chemical properties of sulphur dioxide can be divided into three classes: as an acid, as a reducing agent and as an oxidizing agent.

(a) As an Acid

If sulphur dioxide is bubbled into sodium hydroxide solution until no more will dissolve, sodium hydrogensulphite is produced.

$$NaOH + SO_2 \rightarrow NaHSO_3$$

If a second portion of sodium hydroxide equal to the first is then added the acid salt is converted to the normal salt (compare hydrogen sulphide, section 31.14).

$$NaOH + NaHSO_3 \rightarrow Na_2SO_3 + H_2O$$

Sulphurous [sulphuric(IV)] acid is sufficiently strong to liberate carbon dioxide from a carbonate in solution.

e.g. $$Na_2CO_3 + SO_2 \rightarrow Na_2SO_3 + CO_2\uparrow$$

Sulphites in solution will perform many of the reactions discussed in sections (b) and (c).

(b) As a Reducing Agent

$$2H_2O+SO_2 \rightarrow 4H^+ + SO_4^{2-} + 2e^-$$

(i) With Potassium Manganate(VII)

The purple solution is decolorized, no additional acid being necessary (compare hydrogen sulphide). The absence of a precipitate of sulphur means that sulphur dioxide is preferred as a reducing agent.

(ii) With Potassium Dichromate (or Chromate)

The orange (or yellow) solution becomes a clear bright green colour (compare hydrogen sulphide). This is a good test for the gas.

(iii) With Air

A solution of sulphur dioxide or a sulphite slowly oxidizes in the air to one of a sulphate.

$$2H_2O+2SO_2+O_2 \rightarrow 2H_2SO_4$$

This reaction is the basis of the lead chamber process for manufacturing sulphuric acid, nitrogen dioxide being used to speed production.

(iv) As a Bleaching Agent.

Straw hats can be bleached with sulphur dioxide but, in the light, atmospheric oxidation slowly restores the yellow colour. Sulphur dioxide is still used for bleaching wood pulp for newspapers.

(v) After Bleaching

Sulphur dioxide, like hydrogen peroxide, can be used after bleaching by chlorine: it is another antichlor.

$$SO_2+2H_2O+Cl_2 \rightarrow H_2SO_4+2HCl$$

(vi) With Iron(III) Chloride

If sulphur dioxide is bubbled into warm yellow iron (III) chloride solution, reduction occurs via orange substances to pale green iron(II) chloride.

$$2FeCl_3+SO_2+2H_2O \rightarrow 2FeCl_2+2HCl+H_2SO_4$$

or

$$2Fe^{3+}+SO_2+2H_2O \rightarrow 2Fe^{2+}+4H^+ + SO_4^{2-}$$

(vii) With Hydrogen Peroxide

Oxidation of a sulphite to a sulphate occurs.

e.g. $H_2O_2+SO_2 \rightarrow H_2SO_4$

(viii) With Oxygen

Unlike the above reactions which occur in solution the dry gas can be oxidized rapidly in a catalytic reaction yielding sulphur(VI) oxide (see next section). This is by far the most important use of sulphur dioxide.

(c) As an Oxidizing Agent

(i) With Magnesium

Although the gas does not burn, it will support the combustion of magnesium (compare carbon dioxide).

$$2Mg+SO_2 \rightarrow \underset{\text{White}}{2MgO}+\underset{\text{Yellow}}{S}$$

(ii) With Hydrogen Sulphide

Sulphur is produced if the two gases are moist or in solution.

$$2H_2S+SO_2 \rightarrow 3S\downarrow+2H_2O$$

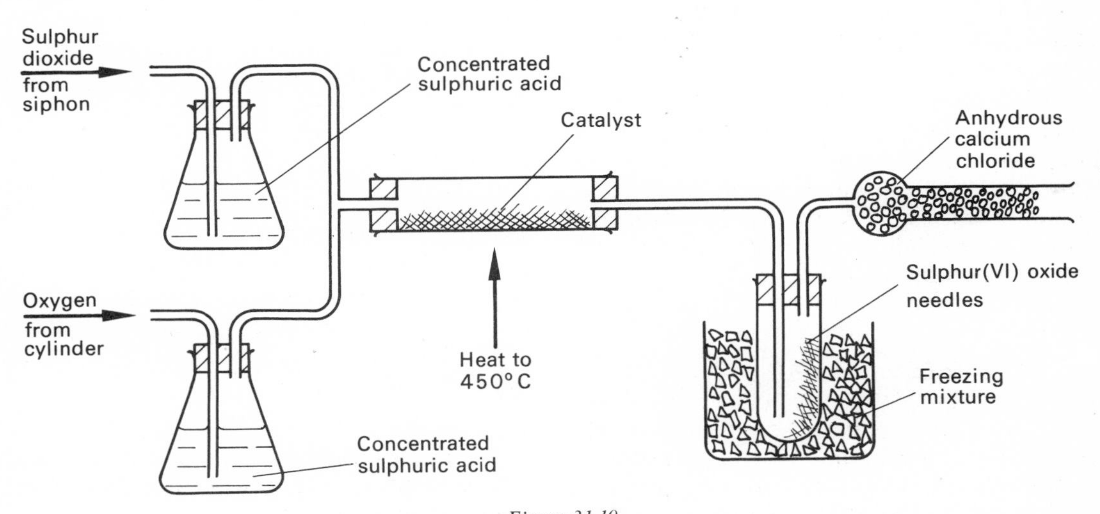

Figure 31.10
The preparation of sulphur(VI) oxide

31.19 Uses

The main use for sulphur dioxide is in the manufacture of sulphuric acid. Minor uses include acting as a bleaching agent, being a preservative for fruit and jam, preventing the liquid effluent from chrome-plating works being poisonous, and as a liquid solvent for refining petroleum.

31.20 Sulphur(VI) Oxide

In 1777 Scheele recognized sulphur(VI) oxide to be the anhydride of sulphuric acid. It is liberated when sulphates, e.g. iron(II) sulphate, decompose on gentle heating. The laboratory preparation that bears most similarity to the industrial process is illustrated in figure 31.10. The gases must be carefully dried before being passed over the catalyst which is gently warmed; vanadium(V) oxide or platinized asbestos can be used.

$$2SO_2 + O_2 \rightleftharpoons 2SO_3(g) \qquad \Delta H = -188 \text{ kJ}$$

This reaction is the basis of the contact process of manufacturing sulphuric acid. The sulphur(VI) oxide is condensed to a mass of silky needles if the apparatus is thoroughly dried beforehand.

The direct reaction of sulphur(VI) oxide with water is noisy and is inefficient on a commercial scale: the factory would be filled with a mist of corrosive vapours.

$$H_2O + SO_3 \rightarrow H_2SO_4$$

IV THE ACIDS

31.21 The Manufacture of Sulphuric Acid

In 1843 Liebig said 'It is no exaggeration to say that we may fairly judge of the commercial prosperity of a country from the amount of sulphuric acid it consumes' and his claim was true for well over a century. Even today when chlorine is a better economic barometer, sulphuric acid is still one of the most important substances manufactured. In the last century, whilst the population of the United Kingdom has doubled, the use of sulphuric acid has increased tenfold. The first stage is to obtain sulphur dioxide (see section 31.17); the relative importance of the starting materials for making this gas in the United Kingdom in 1973 was:

	%
Sulphur (or hydrogen sulphide)	88·5
Anhydrite	5·4
Sulphide minerals	5·1
Spent oxide	1·0

The second stage is the catalytic oxidation of sulphur dioxide to sulphur(VI) oxide (see section 31.20). The method devised by Phillips in the United Kingdom in 1831, but not used commercially until 1901, is known as the Contact process; it accounts for 98% of the sulphuric acid manufactured in the United Kingdom (total 3·9 million tonnes in 1973). The alternative conversion, known as the lead chamber process, is cheaper to run and can run on a dirtier gas supply but it does not yield an acid of a very high concentration (78 as opposed to 98–100%) nor is the acid so pure. The commercial use of platinum as a catalyst has almost ceased, because although it is more efficient than vanadium(V) oxide, it is dearer and more liable to poisoning.

The final stage is to dissolve the sulphur(VI) oxide in concentrated sulphuric acid (98%) and then to add this solution of fuming sulphuric acid or oleum to water, producing yet more 98% acid. The world production of sulphuric acid is about 85 million tonnes a year but the average person does not see it in use except in his motor car battery.

31.22 The Properties of Sulphuric Acid

The concentrated acid is a colourless, oily liquid which causes serious burns if it comes into contact with the skin. Its density is 1·84 g/cm^3 and it is a constant boiling point mixture, like concentrated hydrochloric and nitric acids, boiling at 338°C. The concentrated acid has a strong affinity for water and its dilution must be carried out carefully with stirring by **pouring the acid into the water**, this being accompanied by the evolution of a large amount of heat. It is a strong acid but it is not as strong (i.e. having such a great proportion of ions) as hydrochloric and nitric acid even in dilute solution: there is a large proportion of hydrogensulphate ions.

The chemical properties of sulphuric acid can be divided into four categories: as an acid, as a drying and dehydrating agent, as a non-volatile substance and as an oxidizing agent.

(*a*) As an Acid

It is a dibasic (diprotic) acid and its properties as an acid are best seen when it is employed as a dilute solution.

(i) With Metals

With zinc, iron, magnesium and other moderately reactive metals it gives hydrogen.

e.g. $$Mg + H_2SO_4 \rightarrow MgSO_4 + H_2\uparrow$$
or $$Mg + 2H^+ \rightarrow Mg^{2+} + H_2\uparrow$$

Figure 31.11
A sulphuric acid plant in Ireland

In this reaction it is also behaving as an oxidizing agent.

(*ii*) *With Metal Oxides*

Metal oxides dissolve, if the sulphate is soluble, to give a solution of a salt and water.

e.g. $CuO + H_2SO_4 \rightarrow CuSO_4 + H_2O$

or $CuO + 2H^+ \rightarrow Cu^{2+} + H_2O$

(*iii*) *With Metal Hydroxides*

The reactions are very similar to those above where the metal hydroxide is insoluble in water.

e.g. $Cu(OH)_2 + H_2SO_4 \rightarrow CuSO_4 + 2H_2O$

or $Cu(OH)_2 + 2H^+ \rightarrow Cu^{2+} + 2H_2O$

If, however, the metal hydroxide is soluble in water a titration can be performed (see section 17.11(c)).

e.g. $2NaOH + H_2SO_4 \rightarrow Na_2SO_4 + 2H_2O$

or $OH^- + H^+ \rightarrow H_2O$

(*iv*) *The Insoluble Sulphates*

These must be prepared by precipitation reactions using a solution of the metal salt and of a sulphate, not necessarily the acid. Lead, calcium and barium sulphates are usually prepared this way; the formation of white barium sulphate, a precipitate insoluble in dilute hydrochloric acid, is the general test for a sulphate.

e.g. $BaCl_2 + H_2SO_4 \rightarrow BaSO_4\downarrow + 2HCl$

or $Ba^{2+} + SO_4^{2-} \rightarrow BaSO_4\downarrow$

(*v*) *With Carbonates*

Unless the respective carbonate is insoluble in dilute sulphuric acid will release carbon dioxide when added to the metal carbonate.

e.g. $Na_2CO_3 + H_2SO_4 \rightarrow Na_2SO_4 + H_2O + CO_2\uparrow$

or $CO_3^{2-} + 2H^+ \rightarrow H_2O + CO_2\uparrow$

(*b*) **As a Drying and Dehydrating Agent**

Concentrated sulphuric acid has such an affinity for water that it can be used to dry many gases such as hydrogen, nitrogen, oxygen, chlorine, carbon dioxide, hydrogen chloride, etc., but it cannot be used for alkaline gases, e.g. ammonia, nor for gases which are strong reducing agents, e.g. hydrogen sulphide. The concentrated acid is also able to extract water from substances containing hydrogen and oxygen.

(*i*) *With Hydrates*

The water of crystallization is removed and there may be further reactions.

$$\underset{\text{Blue}}{CuSO_4{,}5H_2O} \rightarrow \underset{\text{White}}{CuSO_4} + 5H_2O$$

(*ii*) *With Carbohydrates*

If moist sugar (sucrose) is treated with concentrated sulphuric acid a black mass of carbon is produced in a vigorous reaction.

$$C_{12}H_{22}O_{11} \rightarrow 12C + 11H_2O$$

Similarly, a piece of paper on which the concentrated acid has been dripped will, if held in warm air over a Bunsen burner, be converted into carbon.

$$C_6H_{10}O_5 \rightarrow 6C + 5H_2O$$

(Cellulose is a polymer of this)

(iii) With Methanoates and Ethanedioates

The acids or their sodium salts are often treated with concentrated sulphuric acid when a supply of carbon monoxide is required.

$$\underset{\text{Sodium methanoate}}{HCOONa} + H_2SO_4 \rightarrow NaHSO_4 + H_2O + CO\uparrow$$

$$\underset{\text{Sodium ethanedioate}}{Na_2C_2O_4} + 2H_2SO_4 \rightarrow 2NaHSO_4 + H_2O + CO\uparrow + CO_2\uparrow$$

(iv) With Ethanol

See section 29.10.

(c) **As a Non-volatile Substance**

The boiling point of concentrated sulphuric acid is 338°C and it will displace acids of lower boiling points from their salts.

e.g. (i) $NaCl + H_2SO_4 \rightarrow NaHSO_4 + \underset{\text{B.p. } -85°C}{HCl\uparrow}$

(ii) $NaNO_3 + H_2SO_4 \rightarrow NaHSO_4 + \underset{\text{B.p. } 86°C}{HNO_3\uparrow}$

The second reaction proceeds on gentle warming.

(d) **As an Oxidizing Agent**

The concentrated acid is a vigorous oxidizing agent, especially when hot.

(i) With Metals

Moderately reactive metals such as copper, zinc and aluminium reduce concentrated sulphuric acid to water and sulphur dioxide.

e.g. $$Cu + 2H_2SO_4 \rightarrow CuSO_4 + 2H_2O + SO_2\uparrow$$

(ii) With Non-metals

With carbon, sulphur, etc., reduction of the acid again occurs.

e.g. $$C + 2H_2SO_4 \rightarrow 2H_2O + CO_2\uparrow + 2SO_2\uparrow$$

(iii) With Hydrogen Sulphide

In this case a reaction occurs with the cold concentrated acid.

$$H_2S + H_2SO_4 \rightarrow S\downarrow + 2H_2O + SO_2\uparrow$$

Hydrogen bromide and iodide behave similarly.

(iv) With Iron(II) Sulphate

When heated with concentrated sulphuric acid, iron(II) sulphate is slowly oxidized to iron(III) sulphate.

$$2FeSO_4 + 2H_2SO_4 \rightarrow Fe_2(SO_4)_3 + 2H_2O + SO_2\uparrow$$

31.23 The Uses of Sulphuric Acid

In 1973 the United Kingdom used 4·3 million tonnes of sulphuric acid (assessed as 100% acid) divided as shown.

%	
25·7	Calcium hydrogenphosphate, superphosphate } Fertilizers
5·4	Ammonium sulphate } Fertilizers
14·6	Paints and pigments, especially titanium(IV) oxide
15·1	Natural and man-made fibres and transparent cellulose film
13·1	Other chemicals—Plastics Magnesium, barium, aluminium, copper and zinc sulphates Hydrogen fluoride (and acid) Hydrogen chloride (and acid)
9·6	Detergents and soap
2·9	Metallurgy, including 'pickling' steel before coating with tin or zinc
2·1	Dyestuffs and intermediates
1·1	Oil and petrol
10·4	Miscellaneous.

QUESTIONS 31

Group VI: Oxygen and Sulphur

1 *(a)* In an experiment it was found that 100 volumes of oxygen united with 200 volumes of hydrogen to form 200 volumes of steam, measurements being made throughout at the same temperature and pressure. *(b)* In another experiment it was found that 100 volumes of oxygen formed 96 volumes of ozonized oxygen and that when all the ozone was absorbed in turpentine there remained 88 volumes of oxygen, measurements being made throughout at the same temperature and pressure. From these two experiments show that (i) the molecule of oxygen is probably diatomic; (ii) the molecule of ozone is triatomic. [J]

2 Oxygen is now prepared in the laboratory by catalytic decomposition of hydrogen peroxide.
(a) What is meant by a catalyst?
(b) Name the catalyst usually used.
(c) State how the oxygen is collected.
(d) Give the equation for the reaction.
(e) Give a diagram of an apparatus you could use to prepare and collect oxygen by this method.
(f) Name the chemicals used and give the equation for any other reaction by which oxygen can be prepared.

Describe the following experiments: (i) to show ammonia burning in oxygen, (ii) to determine fairly accurately the proportion of oxygen present in air. [A]

3 (*a*) Describe how oxygen is manufactured from air commercially. How is such oxygen conveyed to the consumer? Give three uses of oxygen.
(*b*) Name three compounds which when heated alone give off fairly pure oxygen. Write equations for their decomposition.
(*c*) How is oxygen made by electrolysis?

4 (*a*) Describe in outline the commercial preparation of oxygen from the air. Mention three large scale uses of oxygen.
(*b*) How do the processes of respiration and the burning of fuels (i) resemble one another, (ii) differ from one another? [C]

5 How is ozonized oxygen prepared? Mention two reactions by which it can be distinguished from ordinary oxygen. Explain one important use of ozonized oxygen.

6 'Water is so often regarded solely as a solvent that its capacity to take part in a chemical change is sometimes overlooked.' Classify the behaviour of water into the following types of reaction. Summarize one experiment to illustrate each type of reaction.
(*a*) The formation of hydrates.
(*b*) The action of water on elements.
(*c*) The action of water with substances to form acidic solutions.
(*d*) Hydrolysis.
(*e*) The formation of hydrogen and oxygen from water. [Nf]

7 How can hydrogen and oxygen be made to combine? Describe briefly how you could decompose water to give hydrogen and oxygen. (The explanation of the method is not required.)

8 Describe two different ways of producing a specimen of water from hydrogen. Draw diagrams to show the apparatus you would use and state four different tests you would perform on the product to prove that it was, in fact, pure water. [O]

9 Hydrogen peroxide is a weak dibasic acid, miscible with water.
(*a*) Explain the terms *weak* and *dibasic*.
(*b*) Write down the empirical formula of this compound.
(*c*) Suggest the formula of calcium peroxide, and give the equation for its reaction with hydrochloric acid.
(*d*) Explain why hydrogen peroxide bleaches hair.
(*e*) Describe an experiment for obtaining and collecting a sample of oxygen from hydrogen peroxide. [OC]

10 Give an account of the chemistry involved in the industrial preparation of sulphuric acid by the Contact Process. Why is it necessary to use a catalyst in this process? Describe, mentioning any points of interest, the action of sulphuric acid on (*a*) copper metal, (*b*) ethanedioic acid, (*c*) copper sulphate crystals, (*d*) ethanol (excess of acid being used at 180°C). [S]

11 Sulphuric acid is manufactured by converting sulphur dioxide to sulphur(VI) oxide and dissolving this in 95–98% sulphuric acid, whilst adding water to the appropriate extent.
(*a*) How is the sulphur dioxide obtained?
(*b*) State the approximate temperature and relative proportions of gases used in the conversion.
(*c*) Name one of the catalysts commonly used.
(*d*) Explain why one uses a moderate temperature and a catalyst.
(*e*) Why is the sulphur(VI) oxide not dissolved in water directly?
(*f*) State two important commercial uses of sulphuric acid.
Describe one reaction in each case in which sulphuric acid acts as (i) an acid, (ii) a dehydrating agent, (iii) an oxidizing agent. [A]

12 Write an essay on *sulphuric acid*. (In your essay you are recommended to consider the essential raw materials from which it is manufactured, the method of manufacture on a large scale, the physical and chemical properties of the acid both when pure and in aqueous solution and the economic importance of the substance.)

13 Describe and explain the reactions between (*a*) hydrogen sulphide gas and concentrated sulphuric acid, (*b*) warm concentrated sulphuric acid and either ethanedioic acid or methanoic acid, (*c*) iron(III) sulphide solution, sulphuric acid and zinc. [C]

14 Give the equation for one chemical reaction for the production of hydrogen sulphide naming all substances concerned and mentioning the conditions normally employed. State three physical properties of hydrogen sulphide. What is the chemical behaviour of hydrogen sulphide towards (*a*) chlorine water, (*b*) copper sulphate solution, (*c*) water, (*d*) iron(III) chloride solution? To what classification or type of chemical change does reaction (*a*) belong, and which (if any) of reactions (*b*), (*c*) and (*d*) are similar to (*a*) in type? [O]

15 (*a*) Describe how you would prepare and collect some gas jars of hydrogen sulphide.
(*b*) Explain the reaction by which sulphur dioxide bleaches wood pulp. Why does the colour slowly return to paper made from the pulp when it is exposed to the atmosphere? [C]

16 Starting from flowers of sulphur describe how you would prepare (*a*) crystals of rhombic sulphur, (*b*) crystals of monoclinic (or prismatic) sulphur, (*c*) plastic sulphur. Name the phenomenon that the occurrence of these substances illustrates. Describe two experiments that you can carry out with sulphur which show that its chemical reactions are those of a typical non-metal. [S]

	α-Sulphur	β-Sulphur
Appearance	Rhombic shaped amber-yellow crystals	Needle-shaped pastel yellow crystals
Melting point (°C)	114	119
Volume of one mole	15·8 cm^3	16·3 cm^3
Structure	Ring of eight atoms	Ring of eight atoms
Molecular formula	S_8	S_8

17 The information given in the table above concerns two crystalline allotropes of sulphur.

A group of pupils discovered from reference books that a specimen of α-sulphur crystals could be made by dissolving sulphur in carbon disulphide, filtering if necessary, and then allowing the solution to evaporate slowly to form crystals. Warning was given about carrying out the experiment away from all flames e.g. lit Bunsen burners. Some pupils used sulphur from a bottle labelled 'Powdered Roll Sulphur' and obtained specimens of the crystals by following these instructions. It was found to be unnecessary to filter the solution. Other pupils used sulphur from a bottle labelled 'Flowers of Sulphur' and could not get the sulphur to dissolve completely in carbon disulphide. After filtering, a yellow solution was collected from which a specimen of α-sulphur appeared on slow evaporation. A quantity of pale yellow powder was left on the filter paper.

(*a*) Why was the warning given to carry out the experiment away from all flames?

(*b*) How would you show by a simple chemical method that the yellow powder left on the filter paper was a form of sulphur?

(*c*) When some powdered roll sulphur was melted, a clear mobile amber liquid was formed. When some flowers of sulphur was melted, a cloudy mobile liquid was obtained. What could be the cause of the cloudiness?

(*d*) When powdered roll sulphur is melted and then heated to about 160°C, the liquid becomes very viscous. How do you account for this great increase in viscosity?

(*e*) Some α-sulphur was dissolved in dimethylbenzene (b.p. 139°C) and this solution was placed in a distillation apparatus. Some solvent was distilled off and after a time the liquid became cloudy. Distillation was stopped and a pool of orange liquid was seen to collect at the bottom of the solution in the flask. (i) What is the orange liquid at the bottom of the flask? (ii) What conclusion can you draw about the dimethylbenzene solution also in the flask? (iii) The flask and contents were then allowed to cool in an oil bath maintained at 115°C. Crystals were seen to form from solution. What shape are these crystals likely to be? State your reasoning.

(*f*) The change from α-sulphur to β-sulphur is accompanied by an absorption of heat. Suggest an explanation from the information given above. [Nf]

18 Name three different chemical sources of sulphur. Give two reasons for regarding sulphur as a non-metal. Describe how you could obtain a sample of pure sulphur from a mixture of sulphur and sand. Calculate the mass of sulphuric acid that could be made from one kilogram of sulphur. [OC]

19 Quote two chemical differences which support the statement that sulphur and iron are respectively non-metallic and metallic elements. How would you use these two elements in the preparation and collection of jars of dry hydrogen sulphide? What impurity would you expect to be present in your product, and why? Explain and contrast what would be observed in each of the following pairs of reactions:
(*a*) Hydrogen sulphide is ignited in a gas jar, and also at a jet with free access of air.
(*b*) Hydrogen sulphide is passed through separate solutions of lead nitrate and iron(III) chloride.
(*c*) Moist sulphur dioxide is mixed separately with hydrogen sulphide and with nitrogen dioxide. [W]

20 Describe, with a diagram, a laboratory experiment to prepare sulphur(VI) oxide. Describe the action of water on sulphur(VI) oxide. How and under what conditions does sulphuric acid react with (*a*) copper, (*b*) crystalline copper sulphate, (*c*) copper(II) oxide, (*d*) potassium nitrate? [S]

21 Sulphur dioxide can be prepared and collected by the reaction of hydrochloric acid on sodium sulphite.
(*a*) State the conditions under which the reaction is carried out.
(*b*) State how the sulphur dioxide is collected.
(*c*) Give the equation for the reaction.
(*d*) Give a diagram of an apparatus you could use.
(*e*) Give the equation for any other reaction in which sulphur dioxide is liberated. Name the chemicals used.
(*f*) State two reactions of sulphur dioxide, in the first of which it acts as a reducing agent and in the second of which it acts as an oxidizing agent. Give equations. [A]

22 How may (*a*) sulphur(VI) oxide be prepared from sulphur dioxide, (*b*) hydrogen sulphide from sulphur and (*c*) crystals of zinc sulphate from concentrated sulphuric acid? Write an equation in which (i) sulphur dioxide, (ii) sulphuric acid, behaves as an oxidizing agent, pointing out clearly why the reaction you give is an oxidation. [OC]

32 Group VII: Chlorine, Bromine and Iodine

32.1 Introduction

The Group VII elements (fluorine, chlorine, bromine, iodine and astatine) are known as the halogens, i.e. the salt formers. The three members here constitute a good example of a 'family' of elements. Fluorine is too reactive and astatine too radioactive and unstable to be the subjects of experiments in school.

The elements are non-metallic in most of their physical and chemical properties but iodine has a density greater than the important metal, titanium. The elements are very electronegative: they are powerful oxidizing agents.

Most of the compounds studied contain either the ion E^- formed by an atom gaining one electron i.e. being a vigorous oxidizing agent, or the atom sharing two electrons with another atom, i.e. less vigorous oxidation. The halogens also form compounds in which the halogen and oxygen constitute the anion, e.g. OCl^- hypochlorite and ClO_3^- chlorate. For a given metal the chloride is usually the most electrovalent compound, chlorine having a greater affinity for electrons than bromine and iodine: the electron gained is closest to the nucleus in the case of chlorine.

32.2 Occurrence

The principal source of **chlorine** is sodium chloride (see figure 24.1) found in sea water (2·7% by mass on crystallization), but usually extracted from dried up sea beds. **Bromine** ions also occur in sea water (0·0067%). Sodium and magnesium bromides can be obtained by crystallization—in the U.K. there is a bromine factory in Anglesey. One of the chief sources of **iodine** is the solution left after the crystallization of sodium nitrate (caliche, Chile saltpetre): the iodine is originally present as calcium iodate ($Ca(IO_3)_2$). Some iodine is extracted from natural and volcanic spring water in Japan. Bromine and iodine occur in seaweed. Iodine (usually in the form of potassium iodide), sodium chloride and potassium chloride are essential articles of diet. The iodine in the body is concentrated in the thyroid gland in the neck.

I THE ELEMENTS

32.3 The Manufacture of the Halogens

The manufacture of **chlorine** is performed in electrolytic cells which have already been described:

	Section	*% output*
Diaphragm cells, e.g. Gibbs	24.8	60
Mercury cathode cell	24.8	35
Downs' cell	24.3	5

The world production (1972) of chlorine was about 20 million tonnes, the U.S.A contributing 50%, Japan 10%, West Germany 9%, and the United Kingdom 6%. In the United Kingdom the diaphragm cells are less important than the mercury cathode brine cells. Whereas formerly it was said that the production of sulphuric acid by a country was a measure of its commercial prosperity, today the production of chlorine is a better guide. The cells which used to be run primarily to produce sodium hydroxide are now run to produce chlorine, surplus sodium hydroxide being treated with carbon dioxide to give sodium carbonate. See also section 32.8.

Bromine is produced from sea water: huge quantities have to be handled in order to extract the minute proportion present. The sea water is acidified with hydrochloric and sulphuric acids to prevent hydrolysis. Then chlorine is passed in to displace the bromine.

$$Cl_2 + 2NaBr \rightarrow 2NaCl + Br_2$$

or

$$Cl_2 + 2Br^- + 2Cl^- + Br_2$$

The bromine is displaced by air and then sulphur

Figure 32.1(a)
The plant for the extraction of bromine from sea water at Amlwch

dioxide is added: this converts the bromine back into a compound.

$$SO_2 + 2H_2O + Br_2 \rightarrow H_2SO_4 + 2HBr$$

These acid mists are condensed to liquids by passing them through a tower packed with glass fibre. Chlorine and steam are passed into the acids displacing bromine as before, the remaining hydrochloric and sulphuric acid being used to acidify the sea water. The bromine distils out and is easily separated from the water, in which it is not very soluble.

The world production of bromine in 1973 was over 300000 tonnes of which the U.S.A. produced 67% and the U.K. about 3%.

Half of the **iodine** manufactured comes from caliche (impure sodium nitrate). The caliche is crushed and hot (40°C) water added and on cooling sodium nitrate crystallizes out. The remaining solution (the mother liquor) is treated with sulphur dioxide (which in solution is sulphurous acid) in an absorption tower.

$$2NaIO_3 + 6H_2SO_3 \rightarrow 6H_2SO_4 + 2NaI$$

or $$2IO_3^- + 6SO_3^{2-} \rightarrow 6SO_4^{2-} + 2I^-$$

The solution containing sodium iodide is added to some fresh mother liquor.

$$5NaI + NaIO_3 + 3H_2SO_4 \rightarrow 3I_2\downarrow + 3H_2O + 3Na_2SO_4$$

or $$5I^- + IO_3^- + 6H^+ \rightarrow 3I_2\downarrow + 3H_2O$$

The suspended iodine flows to flotation cells which are kept alkaline with sodium carbonate. The slurry or froth from the flotation cells with free iodine in suspension flows into special reactors where it is heated to 125°C at above atmospheric pressure. The iodine melts and forms a dense layer at the bottom of the reactor

Figure 32.1(b)
Top part of the counter current conversion tower for the manufacture of iodine at Coya Sur (Chile)

and is tapped off on to a rotating water-cooled steel drum where it solidifies into thin flakes.

The world production of iodine in 1973 was about 9000 tonnes, of which Japan produced 67% and Chile 33%. The United Kingdom uses about 1000 tonnes each year.

32.4 The Laboratory Preparation

Chlorine was first made by the Swedish chemist Scheele in 1774. He heated manganese-(IV) oxide with concentrated hydrochloric acid, a redox reaction which can be carried out in the apparatus illustrated in figure 32.2.

$$MnO_2 + 4HCl \rightarrow MnCl_2 + 2H_2O + Cl_2\uparrow$$

or $$MnO_2 + 4H^+ + 2Cl^- \rightarrow Mn^{2+} + 2H_2O + Cl_2\uparrow$$

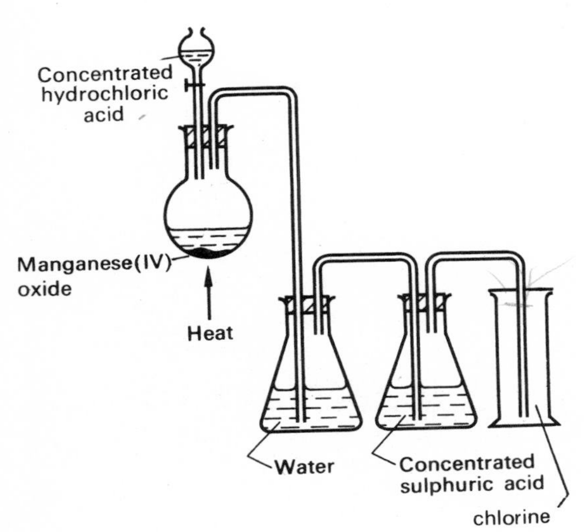

Figure 32.2
The preparation of chlorine

Alternatively lead(IV) oxide can be used or potassium manganate(VII) crystals—the latter react so readily that the reaction proceeds in the cold.

$$2KMnO_4 + 16HCl \rightarrow 2KCl + 2MnCl_2 + 8H_2O + 5Cl_2\uparrow$$

or $$2MnO_4^- + 16H^+ + 10Cl^- \rightarrow 2Mn^{2+} + 8H_2O + 5Cl_2\uparrow$$

The safest place to carry out any of the above reactions is in the fume cupboard. The chlorine is washed with water to remove any hydrogen chloride fumes and then dried with concentrated sulphuric acid before being collected by downward delivery (upward displacement of air). Chlorine may also be collected over concentrated sodium chloride solution in which it is not very soluble.

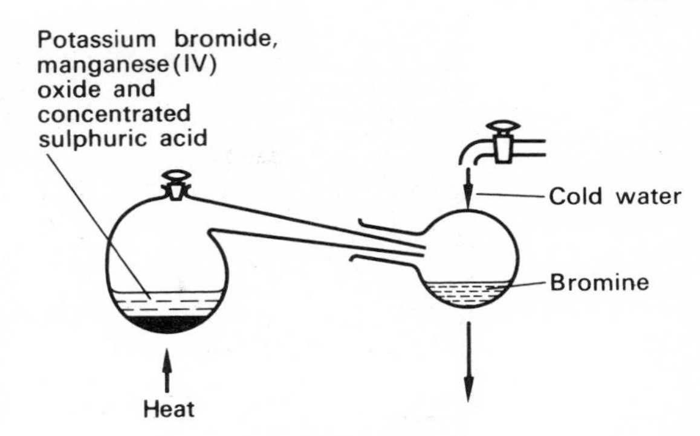

Figure 32.3
The preparation of bromine

Likewise **bromine** is made by the oxidation of hydrobromic acid: usually this is prepared at the time required by the action of concentrated sulphuric acid on a bromide, manganese(IV) oxide is a suitable oxidizing agent:

$$KBr + H_2SO_4 \rightarrow KHSO_4 + HBr$$

then $$MnO_2 + 2HBr + H_2SO_4 \rightarrow MnSO_4 + 2H_2O + Br_2\uparrow$$

or $$MnO_2 + 4H^+ + 2Br^- \rightarrow Mn^{2+} + 2H_2O + Br_2\uparrow$$

The apparatus (figure 32.3) must be completely made of glass because rubber and cork are attacked by bromine. The three reagents are mixed and heated and the bromine condenses as a dark red liquid in the cooled receiver. Bromine is very poisonous and the preparation is best performed in a fume cupboard.

Iodine is prepared in a similar manner to bromine (see figure 32.4). Iodine is a solid but, on warming, the black crystals sublime giving a puple vapour:

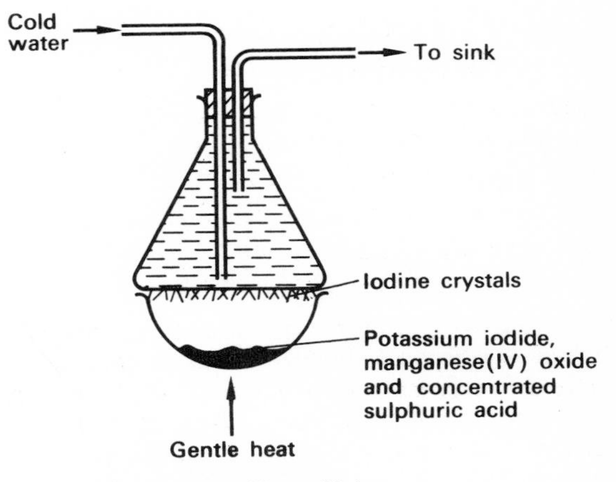

Figure 32.4
The preparation of iodine

$$KI + H_2SO_4 \rightarrow KHSO_4 + HI$$

then $$MnO_2 + 2HI + H_2SO_4 \rightarrow MnSO_4 + 2H_2O + I_2\uparrow$$

or $$MnO_2 + 4H^+ + 2I^- \rightarrow Mn^{2+} + 2H_2O + I_2\uparrow$$

The iodine can be further purified by resublimation.

32.5 Physical Properties

	Chlorine	*Bromine*	*Iodine*
Atomic number	17	35	53
Electronic structure	2,8,7	2,8,18,7	2,8,18,18,7
Relative atomic mass (equal to relative density)	35·45	79·90	126·9
Density (kg/m^3)	3·2	3100	4900
Melting point (°C)	−101	−7	113
Boiling point (°C)	−34	59	184
Colour	Pale yellow-green	Red-brown	Purple vapour Black solid
State at s.t.p.	Gas	Liquid	Solid
Odour	Pungent	Pungent	Faintly irritating
Physiological action	Poisonous vapour: irritates mucous membrane	Poisonous suffocating vapour: irritates eyes, burns flesh severely	Dilute solutions in ethanol used as mild antiseptic
Solubility in water	2·4 vol./1 vol. H_2O	33 g/dm^3	0·3 g/dm^3

The halogens all exist as diatomic molecules. Chlorine is easily liquefied by pressure (3·5 atm at 15°C) and so is stored and transported in the liquid state in steel tanks or cylinders. When chlorine and bromine dissolve in water there is some chemical reaction as well (see section 32.6). The halogens are also soluble in organic solvents such as tetrachloromethane. Iodine dissolves in carbon disulphide and tetrachloromethane giving purple solutions but in water, ethanol and potassium iodide solution (in which it is very soluble) it gives brown solutions.

32.6 Chemical Properties

All the halogens are oxidizing agents but the power diminishes from chlorine to iodine.

(*a*) *With Water*

When chlorine dissolves in water it gives a greenish solution called chlorine water: some reaction occurs between solute and solvent giving an acidic solution.

$$H_2O + Cl_2 \rightarrow HOCl + HCl$$

Chloric(I) (or hypochlorous) acid

Chloric(I) acid is responsible for the fact that moist chlorine, unlike dry chlorine, is a powerful bleaching agent. It will bleach vegetable dyes, like litmus, and the colouring matter of many flowers; it does this by oxidation, converting the dye to colourless compounds.

$$\text{Coloured Substance} + HOCl \rightarrow \text{Colourless Substance} + HCl$$

If two pieces of paper, one printed and the other freshly written in ordinary ink, are placed in a jar of chlorine, the ordinary ink is rapidly bleached. The printer's ink, because it consists of carbon, is unaffected.

Chloric(I) acid is not very stable and the solution readily decomposes, especially when exposed to sunlight.

$$2HOCl \rightarrow 2HCl + O_2\uparrow$$

If a long tube is filled with a saturated solution of chlorine in water, inverted and left exposed to sunlight, a gas (oxygen) collects at the top of the tube and can be tested by its ability to ignite a glowing splint.

The ease with which it loses oxygen makes chloric(I) acid a powerful antiseptic and disinfectant—it kills bacteria by oxidation.

Bromine reacts with water in a manner similar to chlorine but iodine is almost insoluble and there is little reaction. Thus bromine is a poor bleaching agent and iodine has no such properties. Bromine water is useful as a test for unsaturation in organic compounds (see section 29.10).

(*b*) *With Hydrogen*

If hydrogen issuing from a jet is carefully ignited

and lowered into a gas jar containing chlorine it will continue to burn forming hydrogen chloride which fumes in the air. If a mixture of hydrogen and chlorine is sparked or exposed to a bright light an explosive reaction takes place.

$$H_2 + Cl_2 \rightarrow 2HCl$$

Bromine reacts with hydrogen on heating but the yield of hydrogen bromide is poor, even in the presence of a platinum catalyst.

If iodine is heated with hydrogen there is hardly any reaction.

(c) With Metals

The vigour with which chlorine reacts depends on the state of division of the metal: fine powders often react spontaneously but coarse materials require heating. If the metal exhibits more than one valency the highest one is shown in the product, e.g. when chlorine is passed over heated iron the product is iron(III) chloride (see section 26.13). If hydrogen chloride is used iron(II) chloride is produced (see section 26.12). This method of synthesis is used whenever the chloride reacts with its water of crystallization on heating, thus preventing successful dehydration of the hydrated crystals. Sodium, aluminium and tin(IV) chlorides can be prepared in the same manner as iron(III) chloride (see figure 26.10).

Dutch metal (an alloy of copper and zinc supplied in powder form or thin sheets) inflames spontaneously when dropped into a gas jar of chlorine giving brown copper(II) chloride and white zinc chloride.

$$Cu + Cl_2 \rightarrow CuCl_2$$
$$Zn + Cl_2 \rightarrow ZnCl_2$$

Bromine and iodine react similarly but less vigorously.

(d) With Non-metals

White phosphorus reacts instantaneously with chlorine giving mainly phosphorus trichloride (see section 30.22). Red phosphorus requires warming to start the reaction. Molten sulphur reacts with chlorine to give sulphur monochloride (S_2Cl_2) but carbon, nitrogen and oxygen do not have a direct reaction.

(e) With Hydrogen-containing Substances

A little ammonia is the best antidote for chlorine in the atmosphere, although both gases are poisonous.

$$2NH_3 + 3Cl_2 \rightarrow N_2 + 6HCl$$

This may be followed by

$$NH_3 + HCl \rightarrow NH_4Cl$$

if there is still an excess of ammonia. The opposite situation, in which there is an excess of chlorine, is dangerous.

If hydrogen sulphide as a gas or in solution is mixed with a halogen, sulphur is deposited and the respective halogen hydride is formed in solution:

$$H_2S + Cl_2 \rightarrow S\downarrow + 2HCl$$

When a lighted wax-paper (or candle) is put into a gas jar of chlorine it continues burning: the flame is coloured red and black clouds of soot together with hydrogen chloride are produced. The reaction with turpentine ($C_{10}H_{16}$) is similar because both substances are hydrocarbons. The turpentine can be warmed and a piece of filter paper moistened with it before dropping into the jar.

$$C_{10}H_{16} + 8Cl_2 \rightarrow 10C\downarrow + 16HCl$$

(f) With Alkalis

Chlorine will react with a cold dilute solution of an alkali to give a mixture of a chloride and a chlorate(I) (or hypochlorite).

$$Cl_2 + 2NaOH \rightarrow NaCl + NaOCl + H_2O$$

or $$Cl_2 + 2OH^- \rightarrow Cl^- + OCl^- + H_2O$$

The chlorate(I)'s are salts of chloric(I) acid (see above) and, like the acid, are strong oxidizing and bleaching agents. They are more stable than the free acid and so are extensively used as bleaches and antiseptics. Commercial products such as 'Milton' and 'Domestos' are dilute solutions of sodium chlorate(I). On treatment with dilute acids chlorate(I)'s give chlorine.

$$NaCl + NaOCl + H_2SO_4 \rightarrow Na_2SO_4 + H_2O + Cl_2\uparrow$$

or $$Cl^- + OCl^- + 2H^+ \rightarrow H_2O + Cl_2\uparrow$$

If, however, chlorine is passed into a hot concentrated solution of an alkali (or if a solution of sodium chlorate(I) is boiled) the chlorate(V) is produced.

$$3Cl_2 + 6NaOH \rightarrow NaClO_3 + 5NaCl + 3H_2O$$

or $$3Cl_2 + 6OH^- \rightarrow ClO_3^- + 5Cl^- + 3H_2O$$

and $$3OCl^- \rightarrow ClO_3^- + 2Cl^-$$

Sodium and potassium chlorate(V) are powerful (possibly dangerous) oxidizing agents but they do not cause bleaching nor do they give chlorine with dilute acids. The former is deliquescent so the latter is used in explosives, matches, fireworks and in making oxygen (see section 31.4). Sodium chlorate(V) is used as a weed killer and in the extraction of uranium. Both chlorate(V)'s are used in throat lozenges because they destroy bacteria.

If chlorine is passed over moist calcium hydroxide, bleaching powder is produced (Bachmann's process). Bleaching powder is a complex substance containing calcium chlorate(I) ($Ca(OCl)_2$); on treatment with a dilute acid it yields chlorine—up to 35% by mass is 'available' chlorine. Bleaching powder is declining in importance as liquid chlorine is transported to bleaching works; canisters of it are sold in shops under the misleading name of 'chloride of lime'. Even the very weak carbonic acid in the atmosphere will release chlorine so the powder smells and deteriorates rapidly once the container is opened. When the powder is used for bleaching, the cloth is first steeped in a suspension of the powder and then dipped in a bath of dilute acid. The moist chlorine liberated oxidizes the dye destroying the colour. To remove the excess chlorine the cloth is rinsed in a solution of an antichlor:

(i) $H_2SO_3 + Cl_2 + H_2O \rightarrow 2HCl + H_2SO_4$

or $SO_3^{2-} + Cl_2 + H_2O \rightarrow 2H^+ + 2Cl^- + SO_4^{2-}$

(ii) $\underset{\text{Hydrogen peroxide}}{H_2O_2} + Cl_2 \rightarrow 2HCl + O_2$

or $H_2O_2 + Cl_2 \rightarrow 2H^+ + 2Cl^- + O_2$

(iii) $\underset{\text{Sodium thiosulphate}}{Na_2S_2O_3} + 4Cl_2 + 5H_2O \rightarrow 2NaCl + 6HCl + 2H_2SO_4$

or $S_2O_3^{2-} + 4Cl_2 + 5H_2O \rightarrow 10H^+ + 8Cl^- + 2SO_4^{2-}$

The fabric must then be thoroughly washed with water.

Bromine and iodine behave similarly towards alkalis but sodium iodate(I) is more unstable than sodium bromate(I) which in turn is more unstable than sodium chlorate(I).

(g) With other Substances

Each halogen will displace those below it in the periodic table from a solution of the appropriate salt:

(i) $Cl_2 + 2KBr \rightarrow Br_2\downarrow + 2KCl$

or $Cl_2 + 2Br^- \rightarrow Br_2\downarrow + 2Cl^-$

and $Cl_2 + 2KI \rightarrow I_2\downarrow + 2KCl$

or $Cl_2 + 2I^- \rightarrow I_2\downarrow + 2Cl^-$

(ii) $Br_2 + 2KI \rightarrow I_2\downarrow + 2KBr$

or $Br_2 + 2I^- \rightarrow I_2\downarrow + 2Br^-$

These reactions are a direct illustration of the decrease in oxidizing power from chlorine to iodine.

Chlorine will oxidize iron(II) salts to iron(III) salts either in the solid state or in solution:

$$2FeCl_2 + Cl_2 \rightarrow 2FeCl_3$$

or $2Fe^{2+} + Cl_2 \rightarrow 2Fe^{3+} + 2Cl^-$

(h) Reactions peculiar to Iodine

Iodine forms potassium triiodide when dissolved in potassium iodide solution.

$$KI + I_2 \rightleftharpoons KI_3$$

or $I^- + I_2 \rightleftharpoons I_3^-$

The reaction is easily reversible so that such a solution behaves as free iodine.

Iodine with cold starch solution gives a dark blue coloration. This is a very sensitive test for either substance.

32.7 Uses of the Halogens

Chlorine is used:

(*a*) to manufacture inorganic chlorides (aluminium chloride, iron(III) chloride, tetrachloromethane, silicon tetrachloride, phosphorus tri- and penta-chlorides, sulphur monochloride);
(*b*) to manufacture chlorate(I)'s and chlorate(V)'s the uses of which have been mentioned above;
(*c*) to manufacture organic chlorine compounds (often indirectly), weedkillers such as 'Gammexane' (benzene hexachloride), insecticides such as D.D.T. (**d**ichloro**d**iphenyl**t**richloroethane), in chlorinated rubber paints and in plastics such as P.V.C. (**p**oly**v**inyl**c**hloride): in the U.S.A. uses under this heading account for 80% of the chlorine manufactured;
(*d*) for bleaching cotton, linen and wood pulp: animal fibres such as silk and wool are too delicate and would be damaged;
(*e*) for the sterilization of water: it kills harmful bacteria.

Bromine is used:

(*a*) mostly to manufacture 1,2-dibromoethane (ethylene dibromide) which is added to petrol as well as tetraethyllead. The tetraethyllead prevents preignition (knocking) of the petrol vapour when the piston compresses the fuel and air in the cylinder. The 1,2-dibromoethane causes the formation of volatile lead(II) bromide which is carried away in the exhaust gases. The drawback to these additives is the effect of lead compounds on human beings and their environment. 1,2-dibromoethane is also used in the synthesis of dyes, pharmaceuticals and fumigation compounds and as a solvent for gums;
(*b*) to manufacture other organic bromine compounds:
(i) bromoethane for organic syntheses and

the manufacture of pharmaceuticals, especially barbiturates;
(ii) chlorobromomethane for a non-corrosive fire extinguishing fluid;

(*c*) to manufacture silver bromide which is the light sensitive constituent of photographic films;

(*d*) to manufacture other inorganic bromine compounds:
(i) potassium bromide as a nerve sedative;
(ii) lithium bromide in air conditioners;
(iii) zinc bromide for windows in radioactivity experiments;

(*e*) as a bleaching and disinfecting agent.

Iodine is used:

(*a*) 39% in human medicine and pharmacy: X-ray contrast media, tincture (a solution in ethanol and water) for antiseptics, drinking water purification and salt iodization (potassium iodide is added to supply iodine for the thyroid gland in the neck: a shortage of iodine causes the swelling of the neck known as 'goitre');
(*b*) 34% in technical industry, laboratory research and teaching: quartz-iodine bulbs, ammonium iodide for polaroid, in lubricants, and silver iodide for photographic films;
(*c*) 27% in agriculture and veterinary medicine: animal feed additives, selective weedkillers, dairy sanitizers and antiseptic solutions.

II THE HYDRIDES

32.8 The Manufacture of Halogen Hydrides

Other than hydrogen fluoride which is not relevant here, hydrogen chloride is the only halogen hydride to be made on a large scale—about 200 000 tonnes each year in the United Kingdom.
(*a*) Hydrogen and chlorine are both products of the electrolysis of sodium chloride solution. The hydrogen may be burnt in an atmosphere of chlorine to yield hydrogen chloride.
(*b*) Hydrogen chloride is a by-product of the chlorination of organic compounds: this is often a substitution reaction in which one chlorine atom takes the place of one hydrogen atom in the organic compound and the second chlorine atom combines with the hydrogen displaced.

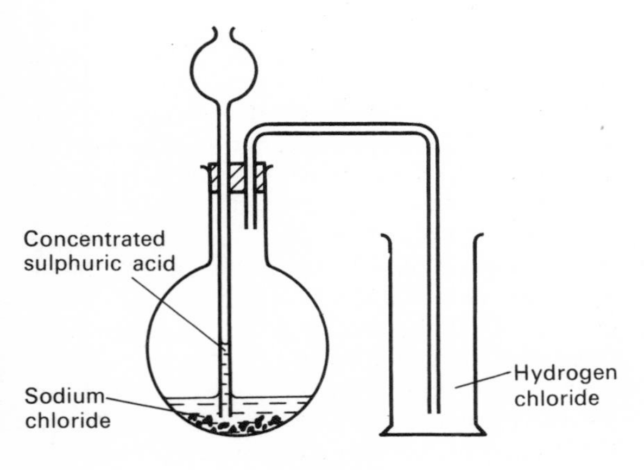

Figure 32.5
The preparation of hydrogen chloride

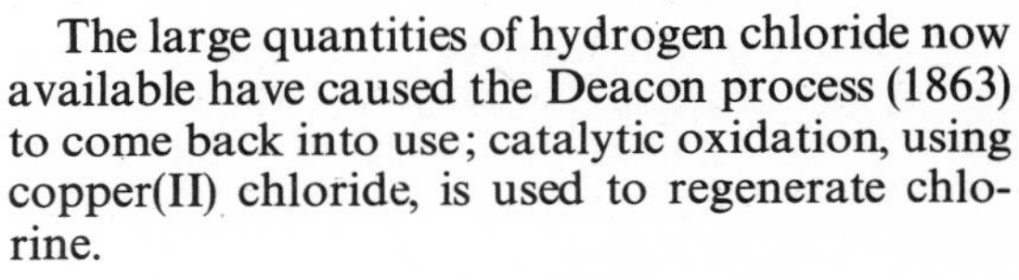

The large quantities of hydrogen chloride now available have caused the Deacon process (1863) to come back into use; catalytic oxidation, using copper(II) chloride, is used to regenerate chlorine.

$$4HCl + O_2 \rightleftharpoons 2H_2O + 2Cl_2$$

32.9 The Laboratory Preparation of Halogen Hydrides

A steady stream of **hydrogen chloride** can be obtained by dripping concentrated sulphuric acid on to sodium chloride crystals (see figure 32.5). Small pieces of rock salt may be used: heating is not vital but will speed up the reaction. The gas is denser than air and thus can be easily collected by the upward displacement of air (downward delivery).

$$NaCl + H_2SO_4 \rightarrow NaHSO_4 + HCl\uparrow$$

The above method is not suitable for hydrogen bromide and even less so for hydrogen iodide because they are more powerful reducing agents than hydrogen chloride and will react with the concentrated sulphuric acid (an oxidizing agent) releasing the appropriate halogen.

$$2HBr + H_2SO_4 \rightarrow 2H_2O + SO_2\uparrow + Br_2\uparrow$$
$$8HI + H_2SO_4 \rightarrow 4I_2\downarrow + 4H_2O + H_2S\uparrow$$

Hydrogen bromide is usually made by dripping bromine on to moist red phosphorus (see figure 32.6). The reaction is vigorous so that no heating is necessary. Some bromine may vaporize and so the U-tube is fitted with glass beads (or wool) smeared with moist red phosphorus—the reaction is the same in both vessels.

$$2P + 3Br_2 + 6H_2O \rightarrow \underset{\text{Phosphonic acid}}{2H_3PO_3} + 6HBr\uparrow$$

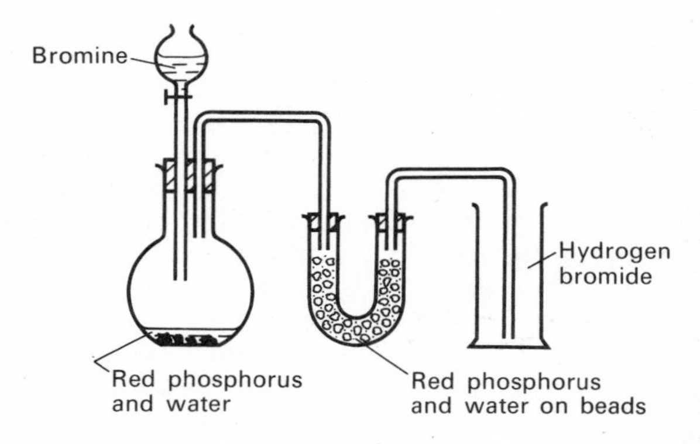

Figure 32.6
The preparation of hydrogen bromide

Hydrogen iodide is prepared in the same apparatus as that used for hydrogen bromide, the only variations in procedure being that water is dripped on to a mixture of red phosphorus, iodine and a little potassium iodide and that the U-tube can be eliminated because iodine is not as volatile as bromine.

Solutions of these three gases are easily made: they are known as hydrochloric acid, hydrobromic acid and hydriodic acid. Whereas the gases are covalent the acids are electrovalent. The gases are very soluble in water: it is highly likely that if a tube were used to deliver the gas into water the water would 'suck back' up the tube and enter the flask containing the reagents for the preparation. The safest technique is the inverted funnel method in which the tube from the reagent vessel is connected to a funnel (see figure 32.7). The rim of the funnel must be just below the surface of the water in the beaker. If sucking back starts to occur the water level in the beaker falls below the rim and a bubble of air can get round to the inside of the funnel to equalize the pressure inside and outside the apparatus.

32.10 Physical Properties of the Halogen Hydrides

They are all colourless, poisonous, pungent smelling gases. The gases fume in moist air because the solution formed with water has a lower vapour pressure than water at the same temperature and thus the minute droplets grow.

The high solubility in water can be illustrated by inverting a gas jar in a bowl of water or by the fountain experiment (see figure 32.8):

	Hydrogen chloride HCl	*Hydrogen bromide* HBr	*Hydrogen iodide* HI
Boiling point (°C)	−85	−66	−36
Solubility in water:			
(vol. gas in 1 vol. H_2O at 15°C)	458	600	425
% mass in H_2O	42	69	70
Relative density	18	40	64

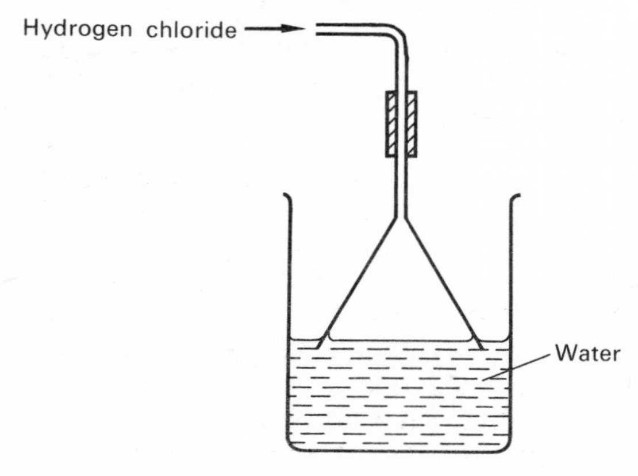

Figure 32.7
The inverted funnel technique.

A tube with a jet at one end and a piece of rubber tubing at the other (which can be closed with a clip) is filled with water and connected into a round-bottomed flask filled with dry hydrogen chloride gas by means of a rubber stopper. A trough is filled with water coloured with litmus made blue by a trace of ammonia. A drop of water is then allowed to fall from the jet into the flask: it dissolves a large volume of hydrogen chloride so that the pressure inside the flask becomes less than that of the atmosphere. Thus when the clip is opened under the water in the trough the atmospheric pressure acting on the water forces water up the tube into the flask, creating a fountain. The hydrochloric acid formed changes the colour of the litmus to red.

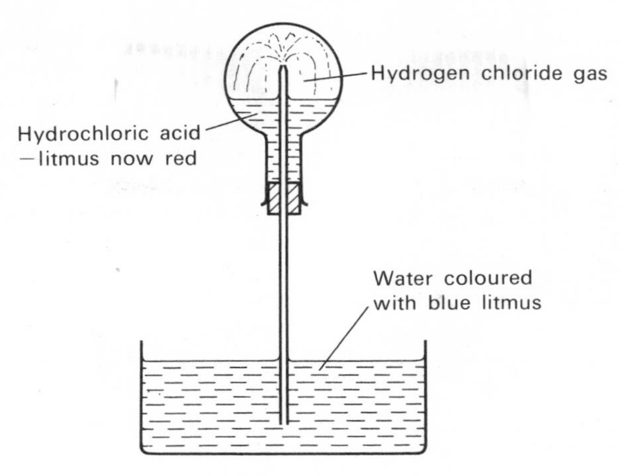

Figure 32.8
The fountain experiment.

32.11 Physical Properties of the Acids

They are all strong monobasic (monoprotic) acids.

When a dilute solution of hydrogen chloride in water is boiled the vapour is mostly water; when a concentrated solution (which fumes in air) is boiled the vapour is mostly hydrogen chloride. Separation of the two components of the mixture cannot be achieved and eventually a mixture distils which has a constant boiling point. The same is true of the other two acids.

	Hydrogen chloride	*Hydrogen bromide*	*Hydrogen iodide*
Boiling point of constant b.p. mixture (°C)	110	126	127
% halogen hydride in mixture	20	48	57

32.12 Chemical Properties of Halogen Hydrides

(*a*) The dry gases are not acidic.
(*b*) The gases do not burn nor do they support the combustion of splints, tapers, sulphur, etc.
(*c*) If a gas jar of hydrogen chloride is brought mouth-to-mouth with one of ammonia and the cover slips removed, white clouds of the solid ammonium chloride are formed. If the gas jars are carefully chosen a partial vacuum is obtained and the bottom one will adhere to the top one.

$$NH_3 + HCl \rightarrow NH_4Cl\downarrow$$

or $$NH_3\,(g) + HCl\,(g) \rightarrow NH_4^+Cl^-\,(s)$$

Hydrogen bromide and iodide react likewise.
(*d*) The gases will react with some metals if the latter are warmed, e.g. aluminium, zinc. If a metal exhibits two valencies the halide corresponding to the lower valency is produced, e.g. iron(II) chloride (see section 26.12). The iron filings are packed into a combustion tube and dry hydrogen chloride is passed over the heated metal; the hydrogen escaping from the far end of the tube may be ignited.

32.13 Chemical Properties of the Acids

Hydrochloric acid is considered as a representative (and the cheapest) acid—a dilute solution is used.
(*a*) The acid properties are only present when the gases are dissolved in water. Thus damp litmus may be used as a test for the gases: it is not a conclusive test (see section 17.3). The reaction between the gas and water may be represented by the equation

$$HCl \rightleftharpoons H^+ + Cl^-$$

but it is often shown by

$$HCl + H_2O \rightleftharpoons \underset{\text{Hydroxonium ion}}{H_3O^+} + \underset{\text{Chloride ion}}{Cl^-}$$

or by $$HCl + xH_2O \rightleftharpoons H^+\,(aq) + Cl^-\,(aq)$$

(*b*) Reactive metals, e.g. magnesium, aluminium, zinc, iron, dissolve in the acids, hydrogen being evolved.

$$Mg + 2HCl \rightarrow MgCl_2 + H_2\uparrow$$

or

$$Mg + 2H^+ \rightarrow Mg^{2+} + H_2\uparrow$$

The acids are thus oxidizing agents.

(*c*) The acids dissolve many oxides yielding a solution of the appropriate chloride.

$$\underset{\text{Black}}{CuO} + 2HCl \rightarrow \underset{\text{Green}}{CuCl_2} + H_2O$$

or

$$CuO + 2H^+ \rightarrow Cu^{2+} + H_2O$$

See sections 7.10–7.12 for details of the preparation of salts.

(*d*) Many metal hydroxides behave in a similar manner to the oxides.

$$Cu(OH)_2 + 2HCl \rightarrow CuCl_2 + 2H_2O$$

or

$$Cu(OH)_2 + 2H^+ \rightarrow Cu^{2+} + 2H_2O$$

If the metal hydroxide is soluble, i.e. an alkali, then a titration is performed to find appropriate quantities.

$$NaOH + HCl \rightarrow NaCl + H_2O$$

or

$$OH^- + H^+ \rightarrow H_2O$$

$$NH_3 + HCl \rightarrow NH_4Cl$$

In solution

or

$$NH_3 + H^+ \rightarrow NH_4^+$$

(*e*) Metal carbonates dissolve in hydrochloric acid with the evolution of carbon dioxide:

$$CaCO_3 + 2HCl \rightarrow CaCl_2 + H_2O + CO_2\uparrow$$

or

$$CaCO_3 + 2H^+ \rightarrow Ca^{2+} + H_2O + CO_2\uparrow$$

(*f*) The insoluble salts of metals are prepared by precipitation: the commonest insoluble halides are those of lead(II) and silver. The lead(II) halides are only precipitated in cold: lead(II) chloride and bromide are white but lead(II) iodide is a beautiful yellow colour and on recrystallization from hot water yields iridescent crystals.

$$Pb(NO_3)_2 + 2HCl \rightarrow PbCl_2\downarrow + 2HNO_3$$

or

$$Pb^{2+} + 2Cl^- \rightarrow PbCl_2\downarrow$$

The bromide and iodide are formed likewise.

The precipitation of the silver halide is the standard test for the halide in solution: silver chloride is white, silver bromide a very pale yellow and silver iodide a pale yellow. The precipitates are all insoluble in dilute nitric acid but they can be distinguished by adding a dilute solution of ammonia to the precipitates (after pouring off excess acid): silver chloride dissolves, silver bromide dissolves to some extent whereas silver iodide is insoluble. The reason for the dissolution of silver chloride is the formation of the soluble substance diamminesilver chloride.

$$AgNO_3 + HCl \rightarrow AgCl\downarrow + HNO_3$$

or

$$Ag^+ + Cl^- \rightarrow AgCl\downarrow$$

then

$$AgCl + 2NH_3 \rightarrow [Ag(NH_3)_2^+]Cl^-$$

(*g*) The halogen hydrides and the acids occasionally act as reducing agents: hydrogen iodide is the most powerful one because it decomposes the most readily into its elements. Hydriodic acid soon becomes coloured because atmospheric oxygen releases iodine.

$$4HI + O_2 \rightarrow 2I_2 + 2H_2O$$

Oxidizing agents such as dilute nitric acid, iron(III) salts and concentrated sulphuric acid readily cause the liberation of iodine; with the last named reagent the reaction may be most complex. The reactions of concentrated hydrochloric acid with manganese(IV) oxide and potassium manganate(VII) have already been discussed (see section 32.4).

32.14 Uses of the Acids

Hydrochloric acid is used for preparing

(*a*) metal surfaces before galvanizing (or electroplating, e.g. rust is removed from iron by 'pickling');

(*b*) chlorine (again) after organic chlorinations (see section 32.7);

(*c*) vinyl chloride (chloroethene) from ethyne: polymerization yields poly(vinylchloride) (P.V.C.), a useful plastic;

(*d*) glucose from starch (see section 29.18(d));

(*e*) glue from bones: it dissolves the calcium phosphate;

(*f*) many substances which are formed under acidic conditions: drugs, dyes and photographic chemicals.

Hydrobromic acid is used as an intermediate in other chemical processes. **Hydriodic acid** is a useful reducing agent.

QUESTIONS 32

Group VII: Chlorine, Bromine and Iodine

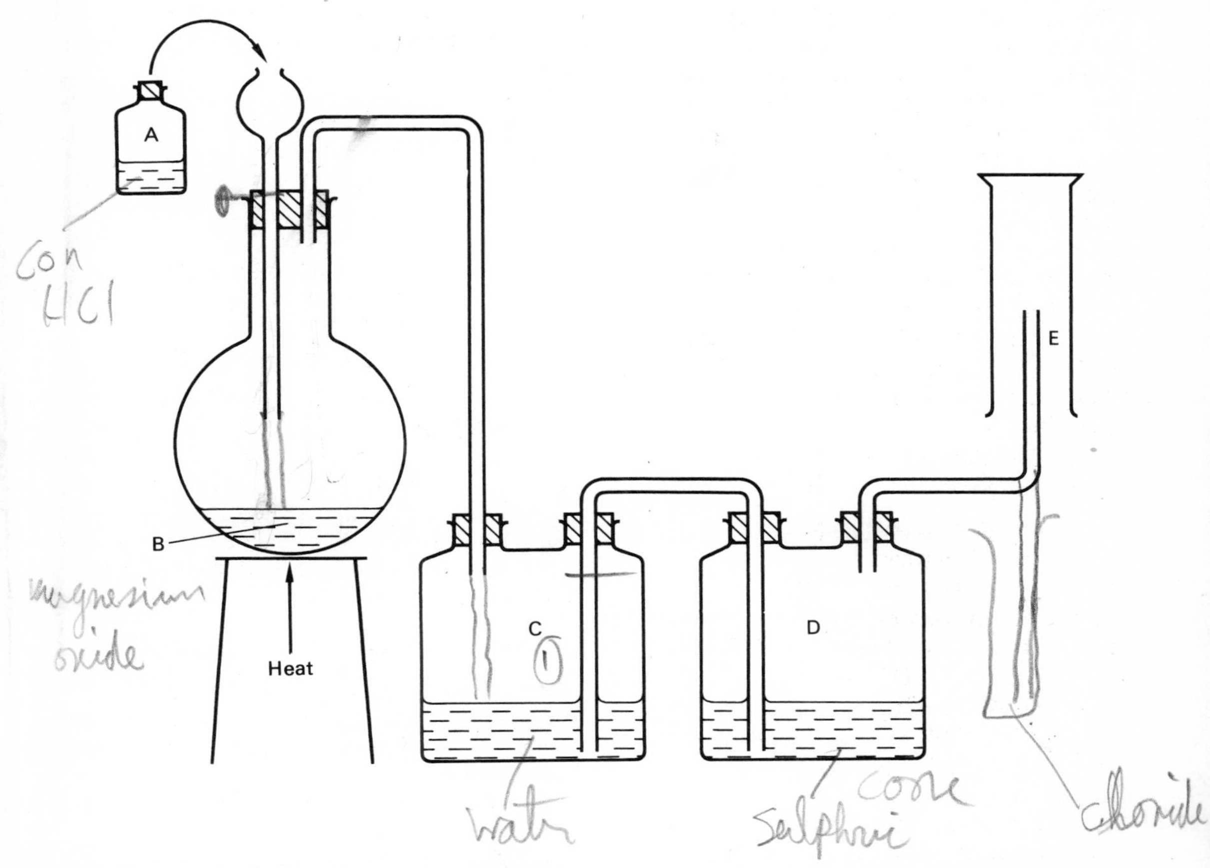

1 A student assembled the apparatus shown above in order to prepare and collect some dry chlorine gas.

(*a*) Before he was allowed to start he was told to make four alterations in the apparatus. What were they?
(*b*) What two substances *A* and *B* would be suitable?
(*c*) What liquid would be in bottle *C* and what would be its purpose?
(*d*) What liquid would be in bottle *D* and what would be its purpose?
(*e*) Write the equation for the reaction.

When several jars of chlorine had been collected, the bottle *D* and the apparatus to the right of it were removed and replaced by apparatus for passing the gas through cold dilute sodium hydroxide solution, the apparatus being designed to prevent *sucking back*.
(*f*) What is meant by *sucking back*? Sketch a piece of apparatus that would prevent it.
(*g*) What two sodium salts would be formed in this last experiment? [L]

2 Sketch and name the apparatus and materials you would use in the laboratory preparation of dry chlorine. Write the equation for your preparation. How, and under what conditions, does chlorine react with (*a*) iron filings, (*b*) calcium hydroxide, (*c*) a solution of sulphur dioxide? [S]

3 Describe the preparation of dry chlorine in the laboratory without using electrolysis. Sketch the apparatus required. Describe an experiment to show the volume composition of hydrogen chloride. [OC]

4 Describe the preparation of a pure sample of chlorine from common salt (do not use electrolysis). Sketch the apparatus required. How does chlorine react with (*a*) hot concentrated sodium hydroxide solution, (*b*) hydrogen sulphide? How would you prove that a given soluble salt was a chloride? How would you convert gaseous chlorine into chloride ions? [OC]

5 Give reasons for the items which are italicized in the following directions for the laboratory preparation of chlorine:
'Some manganese(IV) oxide is placed in a flask fitted with a stopper carrying a thistle funnel and a delivery tube. Concentrated hydrochloric acid is poured in through the thistle funnel *which reaches almost to the bottom of the flask.* The mixture is heated and the chlorine evolved is bubbled through a *little water* and then through *concentrated sulphuric acid* before being collected by *downward delivery* through a tube *which reaches almost to the bottom of the glass jar.*' Write an equation for the reaction. What part is played by the manganese(IV) oxide? Name one other compound which could be used instead of manganese(IV) oxide. Describe and explain the effect of chlorine on (*a*) a starch-potassium iodide paper, (*b*) iron(II) chloride solution. [J]

6 Chlorine may be generated by gently warming a mixture of concentrated hydrochloric acid with manganese(IV) oxide. Sketch an apparatus in which you could carry out this reaction, dry the gas, and collect samples. How would you identify the gas as chlorine? What are the effects of (*a*) chlorine upon heated iron, (*b*) sunlight upon a solution of chlorine in water, and (*c*) chlorine upon potassium bromide solution? [O]

7 State what is seen and explain the reaction which occurs when the following substances are introduced into jars containing chlorine: (*a*) a lighted taper; (*b*) a paper soaked in litmus solution; (*c*) a piece of white phosphorus. Give two commercial uses for chlorine. [J]

8 Group VII of the periodic table includes the elements chlorine, bromine and iodine; the compounds of these elements with potassium are crystalline solids whose aqueous solutions are electrolytes. State in terms of electrons how the particles in these solids are held together. What are the results of bubbling chlorine into (*a*) cold dilute sodium hydroxide solution, (*b*) hot concentrated potassium hydroxide solution, (*c*) potassium iodide solution? How would you release (i) hydrogen, (ii) chlorine from an aqueous solution of hydrogen chloride? A different method should be used for each element. [O]

9 Describe an experiment to show that hydrogen chloride is a very soluble gas. How and under what conditions does hydrochloric acid react with (*a*) calcium carbonate, (*b*) lead(IV) oxide, (*c*) silver nitrate?

10 In the laboratory, hydrogen chloride is usually prepared from sodium chloride and sulphuric acid.
(*a*) State the conditions necessary to obtain a steady supply of the gas.
(*b*) Give the equation for the reaction under these conditions.
(*c*) Give a diagram of the apparatus you would use.
(*d*) How is hydrogen chloride usually collected?
(*e*) With the aid of a diagram, explain the precautions you would take when dissolving hydrogen chloride in water.
(*f*) How could you show when the solution is saturated and dissolving no more gas?
Describe and give equations for the reaction between (i) hydrogen chloride and ammonia gas, (ii) hydrochloric acid and calcium carbonate, (iii) hydrochloric acid and zinc sulphide. [A]

11 Describe the manufacture either of a solution of sodium chlorate(I) or of solid bleaching powder. [C]

12 Bromine (A_r = 80) is a dark red-brown corrosive liquid boiling at 59°C. It is a member of the halogen family of elements and closely resembles chlorine in its chemical properties. 1 g of bromine vapour occupies a volume of 140 cm^3, the volume being corrected to s.t.p.
(*a*) Name one member of the halogen family other than chlorine and bromine.
(*b*) Calculate the relative molecular mass and the atomicity of bromine vapour.
(*c*) Bromine can be prepared by heating a mixture of sodium bromide and manganese(IV) oxide with concentrated sulphuric acid. Draw a diagram of the apparatus you would use for this preparation and explain why manganese(IV) oxide and concentrated sulphuric acid are used to obtain bromine from sodium bromide.
(*d*) Describe one other reaction in which bromine is formed from sodium bromide.
(*e*) What particles are present in a crystal of sodium bromide? [C]

13 Bromine (symbol Br) resembles chlorine in many of its chemical properties, and the corresponding compounds of the two elements also have similar properties.
(*a*) How is the chemical similarity of the two elements explained by the modern theory of atomic structure?
(*b*) Give an equation for the reaction between bromine and hydrogen.
(*c*) What compound would be formed when bromine reacts with sodium? Give the equation for the reaction.
(*d*) What products would you expect to get when hydrobromic acid reacts with manganese(IV) oxide?
(*e*) What is the reaction between bromine and hydrogen sulphide? Give the equation.
(*f*) How would you test for bromide ions in solution?
(*g*) Calculate the mass of zinc bromide which could be obtained by the action of excess hydrobromic acid on 6·5 g of zinc. [A]

14 Describe, with the aid of a labelled diagram, how you would prepare and collect a sample of bromine. Describe briefly three reactions in which bromine and chlorine behave in a similar manner. Describe and explain a test by which you could distinguish between samples of sodium chloride and sodium bromide. [S]

15 Describe a laboratory method for the preparation of bromine from potassium bromide. Outline the commercial preparation of the element from sea water and explain its importance.

16 How could you prepare pure crystals of iodine from an impure specimen of the element? Describe the preparation of an aqueous solution of hydriodic acid. In what ways does this solution differ from dilute hydrochloric acid?

17 Compare the reactions of sodium chloride, sodium bromide and sodium iodide with (*a*) concentrated sulphuric acid, (*b*) silver nitrate solution.

18 Construct a table showing, in as much detail as possible, the differences between the halogens and their principal compounds.

19 Outline experiments by which you could prepare (*a*) hydrogen chloride from chlorine, (*b*) chlorine from hydrogen chloride. Explain the bleaching action of chlorine on a dye such as litmus. Describe what happens in two reactions (other than bleaching) in which chlorine and bromine behave similarly. Write the equations for these reactions.

20 Starting with sodium chloride describe how you would prepare a solution of hydrochloric acid. What do you observe and what reaction takes place, if you add a sample of your solution of hydrochloric acid to (*a*) iron filings, (*b*) sodium sulphite crystals, (*c*) silver nitrate solution, (*d*) bleaching powder? [S]

33 Revision

33.1 How to Revise

At the end of a month, a term or a year it is advisable or necessary to review the progress made and to consolidate the work done. In your practical book you have a record of the laboratory work accomplished and that is a most valuable asset to use in conjunction with this book. In looking through the practicals try to see what the object of the experiment was and remember the principles of the method—the details are not important.

The revision of this book is more difficult. To begin with there may be paragraphs which are not required in your syllabus: these could be lightly crossed out diagonally in pencil. Then comes the task of assessing what should be known and here also principles are more important than facts. Chemistry can be compared to a language: the grammar is equated with the principles and the vocabulary with the facts. It is better to know the grammar and have a reasonable vocabulary in a language than to know all the vocabulary and be unable to construct a sentence. So too in chemistry it is better to have a thorough knowledge of the principles and know a reasonable number of facts rather than to know many isolated facts and be unable to relate them.

Thus chapters 1–22, especially chapters 12–17, and to a slightly lesser extent 18–20, are more important than the factual side of chemistry dealt with in chapters 23–32. The **methods** of analysis discussed in chapters 21–22 are more important than the individual experiments. The process must not be carried to extremes, but as you progress through school and university or technical college more and more time is spent on physical (or general) chemistry than on inorganic or organic chemistry. Most full-time chemists have studied physics and mathematics to a high level but many of the exciting developments of the subject are being made by people interested in the chemical aspects of biology.

The method by which one revises is partly applicable to oneself only but some general courses of action are set out below.

(*a*) Try answering a selection of questions (including calculations) from the ends of the chapters and if you are unable to answer the questions read the appropriate sections.
(*b*) Read the chapters and then try answering questions.
(*c*) Read the chapters slowly making sure you understand the principles at each stage.
(*d*) Write rough notes on what you consider are the key features of each chapter: it should not be necessary to have a book of permanent notes unless there are omissions due to syllabuses having changed.

To help you, the other sections of this chapter pinpoint some key facts and give a summary of all the definitions, laws and important terms used in this book.

33.2 Some Important Metals

Group		*Manufacture*	*M.P.* (°C)	*Electronic Structure*	*Crystal Structure*	*S.E.P.* (*volts*)
1	Sodium Na	Electrolysis at 600°C of fused sodium chloride plus calcium chloride	98	2,8,1	b.c.c.	−2·71
2	Magnesium Mg	Electrolysis of fused chloride or by carbon or silicon reduction	650	2,8,2	h.c.p.	−2·37
2	Calcium Ca	Electrolysis of fused chloride	850	2,8,8,2	f.c.c.	−2·87
T.E.	Iron Fe	Blast furnace—reduction of oxide by carbon monoxide	1535	2,8,14,2	b.c.c.	−0·44
T.E.	Copper Cu	Complex, usually includes electrolysis of sulphate for purification	1080	2,8,18,1	f.c.c.	+0·34
T.E.	Zinc Zn	Sulphide converted to oxide, then either add sulphuric acid and electrolyze or reduce oxide with carbon	419	2,8,18,2	h.c.p.	−0·76
3	Aluminium Al	Careful purification and conversion of bauxite to aluminium oxide; electrolysis of oxide dissolved in fused sodium hexafluoroaluminate(III) at 1000°C	660	2,8,3	f.c.c.	−1·66
4	Tin Sn	Carbon reduction of oxide	232	2,8,18,18,4	tetragonal	−0·14
4	Lead Pb	Lead sulphide and silicon dioxide heated in air; silicate reduced by carbon	327	2,8,18,32,18,4	f.c.c.	−0·13

33.3 Hydrogen and Some Important Non-metals

Group		*Manufacture*	*Laboratory Preparation*	*Identification*
–	Hydrogen H_2	Steam reforming of natural gas or naphtha	Zinc + dilute sulphuric acid	Explodes on ignition
4	Carbon C	Diamonds mostly natural, graphite by electrothermal processing of coke	Sugar charcoal by adding concentrated sulphuric acid to sugar and then washing with water	On burning yields carbon dioxide
5	Nitrogen N_2	Fractional distillation of liquid air	Heat ammonium chloride and sodium nitrite solution carefully	Negative result to all simple tests: splint, litmus, smell, calcium hydroxide solution
5	Phosphorus P_4	Electrothermal processing of calcium phosphate + carbon + silicon dioxide	Red allotrope from white by gentle heating	White crystals burn spontaneously in air, red crystals on ignition: giving phosphorus(V) oxide
6	Oxygen O_2	Fractional distillation of liquid air	Catalytic decomposition by manganese(IV) oxide of hydrogen peroxide solution or of fused potassium chlorate	Reignites glowing splint, no smell
6	Sulphur S_8	Frasch process and by-product of petroleum refineries	Allotropes by crystallization	Yellow powder, on burning yields sulphur dioxide
7	Chlorine Cl_2	Electrolysis of sodium chloride solution	Concentrated hydrochloric acid and potassium manganate(VII) crystals	Bleaches damp litmus, pale yellow-green, pungent smell
7	Bromine Br_2	From sea water by displacement using chlorine	Warm potassium bromide crystals, manganese(IV) oxide and concentrated sulphuric acid	Red-brown gas or liquid, pungent smell, slowly bleaches damp litmus, liberates iodine from potassium iodide
7	Iodine I_2	Sulphur dioxide reduction of sodium iodate(V)	Warm potassium iodide crystals, manganese(IV) oxide and concentrated sulphuric acid	Black solid, purple vapour, dissolves in potassium iodide solution and then turns starch dark blue

33.4 Some Important Solid and Liquid Compounds

Group	*Name*	*Formula*	*Reason*
1	Sodium hydroxide	$NaOH$	The most important alkali (caustic)
	Sodium chloride	$NaCl$	The source of most sodium compounds
	Sodium carbonate	Na_2CO_3	The important mild alkali
	Sodium hydrogencarbonate	$NaHCO_3$	Used in baking powder and health salts
2	Magnesium sulphate	$MgSO_4$	For demonstrating the reactions of Mg^{2+}
	Calcium oxide	CaO	For drying ammonia
	Calcium hydroxide	$Ca(OH)_2$	Solution used to test for carbon dioxide, a cheap alkali but not very soluble in water
	Calcium chloride	$CaCl_2$	For demonstrating the reactions of Ca^{2+}; a drying agent
	Calcium carbonate	$CaCO_3$	For the production of many compounds
T.E.	Potassium manganate(VII)	$KMnO_4$	A powerful oxidizing agent
	Iron(II) sulphate	$FeSO_4$	For demonstrating the reactions of Fe^{2+}; a reducing agent
	Iron(III) chloride	$FeCl_3$	For demonstrating the reactions of Fe^{3+}
	Copper(II) sulphate	$CuSO_4$	For demonstrating the reactions of Cu^{2+}; electrolyte in the purification of copper
	Zinc sulphate	$ZnSO_4$	For demonstrating the reactions of Zn^{2+}
3	Aluminium chloride	$AlCl_3$	For demonstrating the reactions of Al^{3+}; a useful catalyst
4	Tin(II) chloride	$SnCl_2$	For demonstrating the reactions of Sn^{2+}; a reducing agent
	Lead nitrate	$Pb(NO_3)_2$	For demonstrating the reactions of Pb^{2+}; one of the few soluble lead compounds
Org.	Ethanol	C_2H_5OH	The alcohol, a useful solvent
	Ethanoic (acetic) acid	CH_3COOH	A weak acid
5	Nitric acid	HNO_3	A strong acid and a vigorous oxidizing agent
6	Hydrogen peroxide	H_2O_2	An oxidizing agent and a reducing agent
	Sulphuric acid	H_2SO_4	A strong acid, an oxidizing agent, a dehydrating agent, has a high boiling point
	Sodium thiosulphate	$Na_2S_2O_3$	Photography
7	Hydrochloric acid	HCl	A strong acid, can be an oxidizing agent and a reducing agent

33.5 Some Important Gaseous Compounds

Group		*Laboratory Preparation*	*Identification*
Org.	Methane CH_4 Ethene (Ethylene) C_2H_4	Heat sodium ethanoate and soda-lime Heat ethanol and concentrated sulphuric acid to 180°C	Saturated (see page 275), no smell, inflammable Unsaturated (see page 278), slight smell, inflammable
4	Carbon monoxide CO	Concentrated sulphuric acid added to sodium methanoate or methanoic acid	Does not affect calcium hydroxide solution. Burns without explosion upon ignition, blue flame, producing carbon dioxide
4	Carbon dioxide CO_2	Dilute acid on a carbonate; heating most carbonates and all hydrogencarbonates	Turns calcium hydroxide solution milky
5	Ammonia NH_3	Heat an ammonium compound with an alkali	Pungent smell, turns damp litmus paper blue, white cloud with hydrogen chloride
5	Dinitrogen oxide N_2O	Heat ammonium nitrate crystals carefully	Sweetish smell, relights strongly glowing splint, but no action on nitrogen oxide
5	Nitrogen oxide NO	Moderately concentrated nitric acid on many metals	Turns brown in air becoming nitrogen dioxide
5	Nitrogen dioxide NO_2	Heat crystals of many nitrates; concentrated nitric acid on many metals	Brown gas, pungent smell, very acidic to damp litmus
6	Hydrogen sulphide H_2S	Moderately concentrated hydrochloric or sulphuric acid on a sulphide	Pungent smell, turns lead ethanoate or nitrate solution black. Burns with blue flame upon ignition, without explosion
6	Sulphur dioxide SO_2	Acid on a sulphite, concentrated sulphuric acid on many metals	Pungent smell, acidic to damp litmus, decolorizes potassium manganate(VII) solution
6	Sulphur(VI) oxide SO_3	Catalytic oxidation of sulphur dioxide	Pungent smell, fumes in air, very acidic to damp litmus, usually contaminated with sulphur dioxide
7	Hydrogen chloride HCl	Concentrated sulphuric acid on a chloride	Colourless but fumes in air, fumes intensified by ammonia, very acidic to damp litmus

33.6 Definitions, Laws and Statements

An Acid is a solution of a compound which contains hydrogen (hydroxonium) ions as the only positive ions. (Section 17.2)

An Acid Anhydride is a non-metallic oxide, which will react with water to form an acid. (Section 17.4)

The Acidity of a base is the number of hydrogen atoms (of an acid) which will react with one molecule of the base. (Section 17.15)

An Acid Salt is one in which only part of the replaceable hydrogen of an acid has been replaced by a metal and which upon dissolving in water yields both the hydrogen and the metal as cations. (Section 17.14)

An Addition Reaction is one in which an unsaturated organic compound becomes a more saturated one by adding on new atoms. (Sections 29.10 and 29.11)

An Alkali is a base which is soluble in water. (Section 17.6)

Allotropy is the property possessed by some elements of being able to exist in two or more different forms, all in the same physical state. (Section 8.4)

An Alloy is a mixture or compound of metals. (Sections 25.6, 27.6 and 28.7)

An Amalgam is an alloy of a metal with mercury. (Section 15.2)

An Amorphous substance is one which has no recognizable crystalline shape. (Section 8.1)

An Amphoteric Oxide or Hydroxide is one which exhibits both basic and acidic properties. (Section 31.8)

Analysis is the identification and estimation of the elements in a compound or compounds in a mixture. (Qualitative, Chapter 21; Quantitative, Chapter 22)

An Anhydrous substance is one which does not contain any water of crystallization. (Section 6.4)

The Anode of any cell is the electrode to which the negatively charged ions (anions) migrate during electrolysis, or the electrode at which the conventional current enters the cell. (Section 20.2)

An Atom is the smallest possible particle of an element that can exist. (Section 3.2)

The Atomic Heat of a solid element is the product of its relative atomic mass and its specific heat capacity. It is equal to approximately $26 \text{ J mol}^{-1} \text{ °C}^{-1}$ (Dulong and Petit's Law). (Section 10.3)

The Atomicity of a substance, whether an element or a compound, is the number of atoms in one molecule. (Section 3.4)

The Atomic Number of an element is the number of positive charges on the nucleus of one atom of the element (i.e. number of protons). (Section 12.2)

Atomic Theory (Dalton's):
(*a*) Matter is composed of a large number of atoms which are minute particles that cannot be split up, destroyed or created.
(*b*) All the atoms of the same element are identical in all respects; in particular they all have exactly the same mass. Different elements have atoms differing in mass.
(*c*) Chemical combination takes place between small whole numbers of atoms of the elements concerned forming molecules; all the molecules of a compound are identical in all respects. (Section 9.8)

The Relative Atomic Mass of an element is the average mass of an atom of the substance compared with the mass of one atom of the nuclide ^{12}C taken as 12 (or, as an approximation, H = 1). (Section 10.1)

Avogadro's Constant is 6×10^{23} particles per mole. (Section 7.11)

Avogadro's Law: Equal volumes of all gases under the same conditions of temperature and pressure contain equal numbers of molecules. (Section 7.11)

A Base is a compound which will react with an acid to give a salt and water only. (Section 17.5)

The Basicity of an acid is the number of replaceable hydrogen atoms in one molecule of the acid. (Section 17.14.) Or monoprotic, diprotic, etc.

A Basic Salt is one formed by combination of the normal salt with the oxide or hydroxide of the metal (Section 17.15)

The Boiling Point of a liquid is the temperature at which its saturated vapour pressure is equal to the external pressure. (Sections 2.1, 6.2 and 6.9)

Boyle's Law: The volume occupied by a fixed mass of gas is inversely proportional to the pressure, provided that the temperature remains constant. (Section 7.2)

A Catalyst is a substance which increases the rate of a chemical reaction but remains itself unchanged in mass and chemical composition at the end of the reaction. (Section 18.7)

The Cathode of any cell is the electrode to which the positively charged ions (cations) migrate during electrolysis, or is the electrode at which the conventional current leaves the cell. (Section 20.2)

A Cell is a device from which a chemical reaction gives an electrical current in an external circuit. (Sections 20.2–20.5 and 20.15)

Centrifuging is the process of separating a suspension into a solid and a solution by means of revolving a tube containing the suspension: this is done rapidly in order that the denser solid particles travel to the bottom of the tube after which the solution may be poured off. (Section 4.9)

Charles' Law: The volume of a fixed mass of gas is directly proportional to the thermodynamic temperature, provided that the pressure is constant. (Section 7.3)

A Chemical Change is a change in which a new substance is formed. (Sections 2.2–2.7)

Chromatography is the analysis of a mixture by splitting it into its constituents using a bound solvent on a stationary phase and a mobile solvent. (Section 6.17)

Combining Mass—see Equivalent Mass.

The Combustion (or burning) of a substance is an oxidation reaction which is accompanied by the evolution of energy (heat or light or both). (Sections 5.1–5.5)

A Compound is a substance which contains two or more elements combined in such a way that their properties are changed. (Section 3.3)

Conservation of Mass (Law of): In a chemical reaction the total mass of the reacting substances is equal to the total mass of the substances formed, i.e. matter can neither be created nor destroyed. (Section 9.2)

Constant Composition (Law of): All pure samples of the same chemical compound contain the same elements united in the same proportions by mass. (Section 9.3)

Co-ordinate (Dative) Valency is the combination of atoms by the sharing of electrons, both electrons for a bond being supplied by one of the atoms. (Section 14.4)

Covalency is the combination of atoms by the sharing of electrons, one (or more) from each atom. (Section 14.1)

Cracking is the splitting up of large molecules into small molecules; it may be done by heat alone (thermal cracking) or by hot catalysts (catalytic cracking). The term is usually only applied to petroleum. (Section 29.2)

A Crystal has a definite geometric shape, is bounded by plane faces and straight edges, has constant angles between its faces, and can be split into smaller crystals. (Section 8.3)

Deliquescence is the absorbing of moisture from the atmosphere by certain substances, forming a saturated solution. (Section 6.5)

Destructive Distillation (otherwise known as pyrolysis or thermal degradation) is the decomposition of a substance by heating so that the solid or liquid gives gaseous products which may or may not condense. (Sections 29.4 and 28.3)

Diffusion is the process by which, independently of gravity, a gas passes through a porous material. (Sections 7.8 and 23.5)

Diffusion (Graham's Law of): The rate of diffusion of a gas at constant pressure is inversely proportional to the square root of its density. (Section 7.9)

Displacement (Replacement) occurs when one atom (or radical) takes the place of another in a compound. (Sections 2.5 and 15.6)

Distillation is the process of heating a liquid to convert it to the gaseous state. It is frequently performed to purify liquids and to separate mixtures. (Sections 4.11, 29.2 and 30.3)

Double Decomposition—see Ionic Association.

Dulong and Petit's Law—see Atomic Heat.

Dynamic Allotropy is shown by substances which exist in several forms, the properties of which change as the temperature changes. (Sections 8.4 and 31.5)

Efflorescence is the giving up of water of crystallization by a hydrated crystal to the atmosphere. (Section 6.5)

The Electrochemical Equivalent of an element is the mass in grams liberated by the passage of 1 coulomb of electricity through the solution of an electrolyte. (Section 20.9)

The Electrodes are the wires or plates which carry the current in and out of a solution. (Section 20.2)

The Electrode Potential of a substance is measured by comparing the tendency of a metal atom to form ions in solution relative to the tendency of hydrogen atoms to form ions in solution. (Section 15.5)

Electrolysis is the passage of an electric current through a fused substance or a solution accompanied by chemical reactions at the electrodes. (Section 20.6)

Electrolysis (Faraday's Laws of):
(*a*) The mass of a substance liberated in electrolysis is proportional to the quantity of electricity passed.
(*b*) The masses of different substances liberated by the same quantity of electricity are proportional to their relative atomic masses divided by the valencies of their ions. (Section 20.9)

An Electrolyte is a substance which in the fused state or in solution, usually aqueous, conducts an electric current accompanied by chemical reactions at the electrodes. (Sections 20.2 and 20.6)

Electrolytic Dissociation occurs when ions separate, e.g. on melting a substance or dissolving it in water. (Sections 14.2 and 20.7)

An Electrothermal Reaction is one accomplished using electricity to supply heat only. (Sections 25.7, 26.4 and 30.4)

Electrovalency is the combination of atoms by the transfer of electrons. (Section 14.1)

An Element is a substance which cannot be split up into two or more substances by chemical means. (Section 3.1)

The Empirical Formula of a substance is the simplest formula of that substance: it expresses the relative numbers of the atoms of the elements. (Section 11.5)

Enantiotropy is exhibited by allotropes which have a transition temperature below the melting point of the solid. (Section 8.4)

An Endothermic Reaction is one in which energy is taken in. (Section 18.1)

An Enzyme is a very specific catalyst found associated with living cells but able to act away from them. (Sections 29.13 and 29.18)

The Equivalent (Combining) Mass of an element is the number of parts by mass of it that will combine with or displace one part by mass of hydrogen or 8 parts by mass of oxygen or 35·5 parts by mass of chlorine. (Section 9.4)

An Eudiometer is a vessel in which volumes of gases are measured before and after a reaction. (Sections 6.7 and 7.15)

Evaporation is the process of conversion of a liquid into a gas which occurs at all temperatures (not just at the boiling point). (Section 4.10)

An Exothermic Reaction is one in which energy is given out. (Section 18.1)

An Explosion is an extremely rapid reaction in which a large amount of energy is liberated very quickly and carried away by gases. (Section 18.6)

The Faraday Constant is the quantity of electricity required to liberate in electrolysis a mole of monovalent ions, i.e. it is 96 500 C/mol. (Section 20.9)

Fermentation is the slow decomposition process of organic substances catalyzed by enzymes and usually accompanied by the evolution of a gas and heat. (Sections 29.13 and 29.18)

Filtration is the process of separating a suspension into the solid (the residue, precipitate) and a solution (the filtrate) by means of a selective barrier (the filter paper). (Section 4.8)

A Flame is a reaction occurring at the boundary of two substances which gives out heat and light. (Section 5.14)

The Formula of a substance expresses the composition of a substance in terms of the relative numbers of atoms that have combined. (Section 11.2)

A Functional Group is an atom or a set of atoms which are responsible for the typical properties of a homologous series of organic compounds. (Section 29.7)

The Fundamental Particles of Matter of importance to a chemist are the electron, the proton and the neutron. (Section 12.1)

A Gas is the state of matter in which a substance has no definite shape and completely fills the containing vessel. (Section 2.1 and Chapter 7)

The Gram Atomic (Equivalent, Formula, Molecular) Mass of any substance is the mass in grams numerically equal to its relative atomic (equivalent, formula, molecular) mass.

Gram Molecular Volume—see Molar Volume.

A Group is a vertical division of the periodic table containing elements of closely similar physical and chemical properties. (Section 13.7)

The Half-Life of an isotope is the time taken for half a given mass of the substance to disintegrate. (Section 12.5)

A Hard Water is one which will not lather readily with soap. (Section 25.13)

A Heavy Metal is a metal other than an alkali metal, e.g. iron, copper and lead; it is a dense substance. (Sections 15.1 and 15.6)

A Homologous Series contains members which have the same general formula, are made by similar methods, have physical properties which may show a steady gradation, have chemical properties which show a considerable measure of similarity, but have a first member which may be slightly different from the other members. (Section 29.7)

A Hydrate is a compound which contains water combined in the molecular state. (Section 6.4)

Hydrolysis is a reaction in which a compound is decomposed by water. (Section 26.13)

A Hygroscopic Compound is one which absorbs moisture from the air without change of state. (Section 6.5)

An Indicator is a compound which, by changing colour, shows whether a solution is acidic or alkaline, and can be used to show the end point of a reaction. (Section 17.3)

An Ion is an electrically charged atom or radical. Electrolytes contain ions which, during electrolysis, move through the solution or fused substance, the flow of ions constituting the electric current in the liquid. (Section 20.7)

Ionic Association is the process in which oppositely charged ions are attracted to one another and remain in contact, often forming a precipitate. (Section 17.12)

Ionization is the process of forming ions: it occurs whenever atoms or molecules combine by the transfer of electrons. (Section 20.7)

Isomerism is the existence of two or more compounds having the same molecular formula but different structural formulae. (Section 29.9)

Isomorphism is the existence of several compounds which have the same crystalline shape. (Section 8.4)

Isotopes (nuclides) are varieties of atoms of the same element differing in the neutron content of their nuclei, and therefore in their masses. (Section 12.2)

Le Chatelier's Principle. If any of the factors affecting a reversible reaction is changed the system will react to diminish the change. (Section 19.4)

A Liquid is the state of matter in which a substance has no definite shape but which flows so that it takes up the shape of the bottom of the container. (Section 2.1 and Chapter 6)

Local Action is the reaction of an electrode with the electrolyte in a cell when it is not supplying an external current. (Section 20.2)

A Lone Pair of electrons is a non-bonding pair of electrons in the outermost shell of the atom of an element. It may give rise to dative covalency. (Section 14.4)

A Macromolecule (giant molecule) is a large molecule usually composed of small easily recognizable units from which it has been formed or into which it is decomposed. (Sections 14.5, 28.5, 28.12, 28.18, 29.10, 29.17, 29.18 and 30.6)

Mass Action (Law of): The rate of a chemical reaction is proportional to the molecular concentrations of the reacting substances (but the precise dependence must be found by experiment). (Section 18.6)

The Mass Spectrometer is an instrument in which the vapour of a substance is sorted by magnetic and electrical fields into components having the same ratio of mass to charge. (Section 10.6)

The Melting Point of a pure element or compound is that definite temperature at which solid and liquid are in equilibrium at a given pressure. (Sections 2.1, 6.2 and 6.9)

A Metal is an element whose atoms tend to lose electrons. It is distinguished by the majority of its physical and its chemical properties to be one of those on the left hand side or centre of the periodic table. (Section 15.2)

The Mineral Acids. This collective term is used for hydrochloric, nitric and sulphuric acids which are prepared from minerals (salts), as contrasted with organic acids. (Sections 17.1 and 17.3)

A Mixture contains two or more different substances, either elements or compounds, which are not chemically joined together. (Section 3.5)

The Molar Mass of a substance is its relative molecular mass considered in grams. (Page 362)

A 1 M Solution contains one mole of a substance in 1000 cm^3 (1 dm^3 or litre) of solution. (Section 22.2)

The Molar Volume of any gas is the volume occupied at s.t.p. by 1 mole and is equal to 22·4 dm^3. (Section 11.14)

The Mole is the amount of substance of a system which contains as many elementary units as there are carbon atoms in 0·012 kilogrammes of carbon-12. The elementary unit must be specified and may be an atom, a molecule, an ion, an electron, etc., or a specified group of such entities. (Section 7.11 and page 362)

The Molecular Formula of a substance records the number of each kind of atom in one molecule of the substance. (Section 11.7)

The Relative Molecular Mass of an element or compound is the average mass of one molecule of the substance compared with the mass of one atom of the nuclide ${}^{12}_{6}C$ taken as 12 (or, as an approximation, H = 1). (Section 10.1)

A Molecule is the smallest particle of a substance, whether it is an element or a compound, which can exist in a free state. (Section 3.4)

Monotropy is exhibited by allotropes which do not have a transition temperature below the melting point of the solid. (Section 8.4)

The Mother Liquor is the cold solution above crystals which have formed as the hot solution cooled. (Sections 17.11 and 32.3)

Multiple Proportions (Law of): If two elements combine to form more than one compound then the masses of one element which combine with a fixed mass of the other element are in a simple ratio. (Section 9.6)

Neutralization is a reaction between an acid and a base producing a salt and water only. (Section 17.8)

The Noble Gases (Inert or Rare Gases) constitute Group O of the periodic table. They are chiefly of use for their physical properties and their structures are very important in valency. (Section 5.12 and Chapter 14)

A Non-metal is an element whose atoms tend to gain electrons. It has more electrons in the outermost shell than its period number. (Section 15.2)

A Normal Salt is a compound formed when all the replaceable hydrogen of an acid has been replaced by a metal. (Section 17.14)

A Normal Solution contains the gram combining (equivalent) mass of the compound dissolved in 1000 cm^3 (1 dm^3 or litre) of solution.

Oxidation is (*a*) the addition of oxygen or any other non-metal, (*b*) an increase in the proportion of the electronegative constituent, or decrease in the proportion of the electropositive constituent, (*c*) the removal of hydrogen or a metal, (*d*) the loss of electrons. (Section 16.2)

Oxidation Number (See pages 7 and 362)

An Oxidizing Agent is a substance which can bring about oxidation. (Section 16.3)

Partial Pressures (Dalton's Law of): The total pressures of a mixture of gases, which do not react with each other, is the sum of the pressures that each gas would exert if it alone occupied the volume of the mixture at the same temperature. (Section 7.6)

A Period or Series is a horizontal division of the periodic table. (Section 13.7)

The Periodic Law (Mendeléeff): The properties of elements (and of their corresponding compounds) are periodic functions of their relative atomic (and molecular) masses. The Law is now emended from relative atomic masses to atomic numbers. (Sections 13.5 and 13.7)

The Periodic Table is drawn up by placing the elements (*a*) in order of increasing atomic number, (*b*) starting a new line every time a new outer shell of electrons is started, and (*c*) putting the elements into columns according to the number of electrons in the outermost shell (or if these are the same for successive elements into columns according to the number of electrons in the penultimate shell. (Section 13.7)

The pH of a solution is the negative logarithm of the hydrogen ion concentration in mol/dm^3, i.e. $pH = -\log_{10}[H^+]$. (Section 17.16)

Photosynthesis is a reaction in which plants make sugars and starches from carbon dioxide and water catalyzed by chlorophyll. (Section 5.11)

A Physical Change is a change which a substance undergoes without its properties being vastly altered. (Sections 2.2–2.7)

A Plastic is a substance which at some stage has been a liquid but which now may have more properties akin to those of a solid. Plastics can be divided into thermoplastics (ones which soften on heating) and thermosetting (ones which harden on heating). (Sections 2.1, 29.10, 29.17 and 29.20)

Polarization is the formation of a film of a gas on an electrode which hinders further electrolysis. (Section 20.2)

Polymerization is the process in which simple molecules link together to form a large molecule of the same empirical formula (addition polymerization, Section 29.10). In condensation polymerization (Section 29.17) a small molecule is eliminated when each linkage in the large molecule is formed.

Polymorphism is the existence of a compound in two or more different forms without changing its physical state. (Section 8.4)

Precipitation occurs either when a solid separates out from a gas or liquid or when a liquid separates out from a gas. It is frequently employed as a method of preparing an insoluble substance by mixing two solutions containing the necessary components. (Section 17.12)

Pressures (Law of): The pressure exerted by a fixed mass of gas is directly proportional to the thermodynamic temperature, provided that the volume is constant. (Section 7.4)

Pyrolysis—see Destructive Distillation.

Qualitative Analysis is an examination of a substance which discovers the components of a mixture or a compound. (Chapter 21)

Quantitative Analysis is an examination of a substance which determines the quantities or proportions of the components of a mixture or a compound (Chapter 22)

A Radical is a group of atoms found as a unit in many compounds: it often does not have a separate existence. (Section 11.3)

Radioactivity is the spontaneous disintegration of the nucleus of an atom of an element emitting helium nuclei (α particles), electrons (β) or electromagnetic radiation (γ). (Section 12.5)

Reciprocal Proportions (Law of): The proportion by mass in which two substances each combine with or displace a fixed amount of a third substance is a simple ratio of the proportion in which they combine with or displace one another. (Section 9.7)

Redox Reactions are those in which reduction and oxidation occur. (Chapter 16)

A Reducing Agent is a substance which can bring about reduction. (Section 16.4)

Reduction is the opposite of oxidation: it is (*a*) the removal of oxygen or any other non-metal, (*b*) a decrease in the proportion of the electronegative constituent, or increase in the proportion of the electropositive constituent, (*c*) the addition of hydrogen or a metal, (*d*) the gain of electrons. (Section 16.2)

The Relative (Vapour) Density of a gas is the

$$\frac{\text{Mass of one volume of gas}}{\text{Mass of one volume of hydrogen}}$$

under the same conditions of temperature and pressure. (Sections 7.14 and 10.4)

Replacement—see Displacement.

A Reversible Reaction is one which can be made to go in either direction by changing the conditions under which it is carried out. (Section 19.2)

The R_f Value of a solute dissolved in a bound solvent when moved by a mobile solvent is $\frac{\text{Distance flowed by solute}}{\text{Distance flowed by solvent}}$. (Section 6.17)

Rusting is the oxidation of iron caused by air and a film of moisture on its surface. Rust is hydrated iron(III) oxide. (Sections 15.6 and 26.7)

A Salt is a compound formed when part or all of the replaceable hydrogen of an acid is replaced by a metal. (Section 17.8)

Saponification is the process of making soap. (Section 29.15)

A Saturated Compound is one in which all the covalency bonds are single ones so that it can only react by disintegrating and then adding on new atoms: it performs substitution rather than addition reactions. (Section 29.8)

A Saturated Solution of a solid at a given temperature is one which can exist in equilibrium with undissolved solid at that temperature. (Section 6.11)

The Selective Discharge Theory states that the lower a metal is in the electrochemical series the more likely its ions are to be discharged upon electrolysis. (Sections 15.5 and 15.6)

A Solid is the state of matter in which a substance has a definite shape. (Section 2.1 and Chapter 8).

The Solubility of a solid in a solvent, at a given temperature, is the maximum number of grammes that will dissolve in 1 kg of the solvent, in the presence of excess of the solid, at this temperature. (Section 6.11)

A Solute is the solid which dissolves in a solvent to form a solution. (Section 6.9)

A Solution is a completely perfect (homogeneous) mixture of two or more substances. (Section 6.9)

A Solvent is the liquid in which a solid is dissolved to form a solution. (Section 6.9)

A Standard Solution is one of which the concentration is known. (Section 22.1)

The Standard Temperature and Pressure (s.t.p.) are 0°C and 101 325 Pa. (Section 7.5)

The Structural Formula shows the arrangement in three dimensions of the atoms in one molecule of the substance. (Section 14.3)

Sublimation is the process in which a solid evaporates without melting and the vapours recondense directly to crystalline solids if cooled (Section 4.12)

Substitution is a type of reaction undergone by saturated compounds in which an atom is lost by the original molecule and then another one takes its place. (Section 29.8)

A Supersaturated Solution is one in which the solvent contains more solute than is necessary to form a saturated solution at that temperature. (Section 6.15)

A Suspension is a mixture of a liquid and a finely divided but insoluble solid. (Section 6.9)

The Symbol for an element represents one atom of the element and is shown by one or two letters, the first of which only is a capital. (Section 11.1)

Synthesis is the building up of a compound from its elements or from simple substances the synthesis of which is known. (Section 6.7, 2.5, 3.6 and 17.13)

Thermal Decomposition is a non-reversible reaction brought about by heating the compound alone. (Section 2.5)

Thermal Degradation—see Destructive Distillation.

Thermal Dissociation is the splitting up of a compound by heat; if the products are cooled they will reform the original compound, i.e. the reaction is reversible. (Sections 19.3, 30.15 and 30.22)

A Titration is the process of adding an acid to an alkali using an indicator to find the end point. Later extended to other pairs of reactants. (Sections 17.11, 22.5, 20.14 and 18.2)

Titres are the volumes of acid (etc.) found to react precisely with a given volume of an alkali (etc.). (Section 22.5)

The Transition Elements (Metals) are ones in which the increase in electron content of successive members (according to atomic number) occurs in the penultimate shell. (Section 13.7 and Chapter 26)

An Unsaturated Compound is one in which the covalent bonds are double or triple ones and atoms may be added on by the multiple bonds becoming single bonds: it performs addition reactions. (Sections 14.3, 29.10 and 29.11)

The Valency of an element is the number of hydrogen atoms which will combine with or displace one atom of the element. (Section 11.3)

Van der Waals' Forces are the weak forces of attraction between the molecules of a covalent substance. (Section 14.3)

A Voltameter is a vessel in which electrolysis occurs. (Sections 6.7, 20.6 and 20.15)

Water of Crystallization is that definite amount of water with which some compounds are combined on crystallizing out from aqueous solutions. (Sections 6.4 and 11.5)

Answers to Numerical Questions

Chapter 6 (page 51)

1 (*a*) 25°C (*b*) dissolves at 58°C (*c*) 68·5g crystallizes (*d*) 70°C 11·5 g A left; 10°C 17·5 g A + 19·0 g B crystallize. 48% A, 52% B.
3 36·3.
4 12 g NaCl; 1 g NaCl, 25 g KNO_3; 37 g NaCl, 25 g KNO_3
5 6
10 11·1% H_2, 88·9% O_2
12 117, saturated
13 (*a*) 155 (*b*) 213, 25 g
16 (*a*) 225, 470 (*b*) 29 g
22 56
23 7
25 (*a*) 58 (*b*) 144 g, 8

Chapter 7 (page 61)

1 1 dm^3 H_2 : 0·5 dm^3 O_2; 1 dm^3 H_2 : 1 dm^3 Cl_2
2 20 cm^3 N_2; 10 cm^3 O_2; 12 cm^3 CO_2
3 (*a*) 34 (*b*) 5 g
4 10 dm^3 N_2; 35·7 g CuO
5 (*a*) (i) 40 cm^3 O_2; 40 cm^3 SO_2 (ii) 33% O_2; 33% SO_2; 33% H_2O (g) (*b*) 3·0; 12·0
6 3% CO_2; 30% O_2; 67% N_2
8 1 500 m^3
9 (*a*) 1 370 cm^3 (*b*) −194°C (*c*) 90 mmHg (*d*) 249 cm^3 (*e*) −83·8°C (*f*) 608 mmHg (*g*) 500 cm^3 (*h*) 10 000 cm^3
10 88 seconds
11 0·5 atm O_2; 0·25 atm N_2; total 0·75 atm
12 44·9 cm^3
13 18 seconds
14 (*a*) 1 120 dm^3 (*b*) 100 g
16 (*a*) 100 cm^3 (*b*) 22
17 5 cm^3, 10 cm^3; 15 cm^3, 10 cm^3
18 CH_4
19 H_2O
20 (*a*) 75 cm^3 O_2 (*b*) 75 cm^3 O_2 (*c*) 500 cm^3 SO_2
21 33%
23 (*a*) 22; 44 (*b*) 12

Chapter 9 (page 73)

1 3:4
2 (*a*) 56 (*b*) 2; 3
3 1:2
5 3:2
6 3 g
7 7·45 g
8 4:1:2
10 64
11 2:3
12 118
14 2:1
16 4·44:1
17 64
18 32
19 32:64
20 14; 7

Chapter 10 (page 81)

1 (*a*) 2; 3 (*b*) 48; 32 (*d*) 0·28 $Jg^{-1}\,°C^{-1}$
2 18·7; 56
3 96
4 (*a*) 14·5 (*b*) 40
5 112
6 (*a*) 81 (*b*) 1 (*c*) 80
7 59·1; 71·1 g
8 64
9 15; 7·58 g; 3
10 20·2
11 75% ^{63}Cu

Chapter 11 (page 91)

1 55·9
2 2·9 g
3 50
5 $FeCl_2,4H_2O$
6 Fe_2S_3
7 (*a*) 62·9 (*b*) 5·6 dm^3
8 (*a*) 75% C; 25% H (*b*) 80% Cu; 20% O (*c*) 52·2% C; 13·0% H; 34·8% O (*d*) 36·5% Na; 25·4% S; 38·1% O
9 (*a*) 20 (*b*) $SnCl_4$
10 14·7
11 (*a*) 25·6 (*b*) 5·6 dm^3 (*c*) 20 cm^3 (*d*) 10 cm^3
12 (*a*) 48·8 (*b*) 11·2 dm^3 (*c*) 11·2 dm^3
13 CH_3; C_2H_6
14 $Na_2SO_3,7H_2O$
15 $Na_2S_2O_3,5H_2O$
16 (*d*)
17 $Na_2CO_3,NaHCO_3,2H_2O$
18 $CuSO_4,H_2O$
19 H_2SO_5
20 (*a*) CH_2 (*b*) C_4H_{10}
21 CO_2H; $H_2C_2O_4$
22 C_2H_4O
23 Na_2SO_3
24 TiO_2; $TiCl_4$
25 $C_6H_8O_7$
26 EO_2; EO_3
27 (*b*) SO_3
28 $C_2H_4Br_2$

29 7
30 (*b*) $ZnCO_3$
31 (*a*) and (*b*) C_2H_5Br
32 40
33 560 cm^3; 2·92 g
34 (*a*) 373 dm^3 (*b*) 2 570 dm^3
35 (*a*) 34·9 g (*b*) 36·8 g
36 (*a*) 896 cm^3 (*b*) 11·1 g
37 (*a*) 89·6 dm^3 (*b*) 168 g
38 2·24 dm^3, 10 g; (*a*) 1·2 g (*b*) 2·24 dm^3
39 (*a*) 12·6 g (*b*) 2·24 dm^3
41 6·50 g
42 (*a*) and (*b*) CH_3Br
44 5
45 (*a*) 23 mm (*b*) 0·003 (*c*) 6 cm^3 (*d*) 0·006
53 (*b*) 208 g (*d*) 6 cm^3 (*e*) 3 dm^3 (*f*) 3
(*g*) $M(SO_4)_3$ (*h*) 54
56 (*a*) 0·125 (*b*) 56; 84 (*c*) C_4H_8 (*d*) 1
(*h*) 60 cm^3; 40 cm^3
57 (*a*) 140 cm^3 (*b*) 280 cm^3
58 58
59 (*a*) 14 (*b*) 28 (*c*) 5·6 dm^3
60 30
61 34
62 32

Chapter 14 (page 116)
6 (*a*) (ii) 0·8 g

Chapter 17 (page 142)
19 $C_2H_3O_3$ (*a*) $C_4H_6O_6$ (*b*) 355

Chapter 18 (page 153)
13 (*a*) 173 kJ, exothermic (*c*) −160 kJ
15 (*a*) −3·85 kJ (*b*) 50 cm^3 (*c*) −5·45 kJ (*d*) 0·1
(*e*) −54·5 kJ (*f*) −54·5 kJ/dm^3 acid
16 (*a*) B, E, G (*c*) 330 seconds
17 (*b*) 0·01 (*c*) 0·02 (*d*) 2
18 (*c*) 50 cm^3 (*d*) 2 dm^3 (*e*) 72 cm^3 (*f*) $FeCl_2$
19 (*d*) 3:1 (*e*) 3·4:1

Chapter 20 (page 170)
4 0·166 g
5 520 cm^3
6 1·4 dm^3; 0·7 dm^3; 1·12 g
8 13·5 g; 145 000
9 (i) 4·78 g each (ii) 1·67 dm^3 H_2; 0·84 dm^3 O_2
15 (*a*) 0·2 A h or 720 C (*b*) 66 g (*c*) 290 000
(*d*) 3+
18 195 kC

Chapter 22 (page 187)
1 (*a*) 200 cm^3 (*b*) 1·27 dm^3
2 2·5 g; 560 cm^3
3 (*a*) 6·9 g (*b*) (i) 0·06 (ii) 90
4 10
5 (*a*) 2·45 g (*b*) (i) 25 cm^3 (ii) 50 cm^3
6 27
7 4·48 dm^3; 400 cm^3; 28·4 g
8 0·6; 24
9 (*b*) 0·5 (*c*) 560 cm^3
10 (*a*) 2·8 g (*b*) 0·1 (*c*) 500 cm^3 (*d*) 6·8 g $KHSO_4$
12 (*a*) 5·3 g (*c*) (i) 100 cm^3 (ii) 200 cm^3
13 49 g; 160 cm^3; 11·4 g Na_2SO_4
14 (*a*) 2 g (*b*) 0·5 (*c*) (i) 1·12 dm^3 (ii) 50 cm^3
15 (*b*) 10 cm^3 (*c*) 2·65 g
16 (*a*) 5·6 dm^3 (*b*) 60 g $NaHSO_4$
17 (*a*) 63 g (*b*) 56 g; 250 cm^3 KOH
18 (*a*) and (*b*) 0·167
19 49 g (i) 11·2 dm^3 (ii) 500 cm^3
20 40

Chapter 24 (page 207)
5 (*b*) 530 tonnes

Chapter 25 (page 220)
7 2·44 g
25 40

Chapter 26 (page 238)
7 2 g; 1·4 dm^3
12 2·38
15 (*a*) 0·005 (*b*) 0·0025 (*c*) 2 (*d*) 4+ (*h*) 0·05
22 25 g
24 39

Chapter 28 (page 262)
14 (*a*) 75 cm^3 (*b*) 25 cm^3 O_2
18 10 cm^3 CO_2; 20 cm^3 N_2

Chapter 29 (page 290)
14 40 cm^3 H_2

Chapter 30 (page 309)
2 1·21 kg
5 200 cm^3
7 (*b*) 89·6 dm^3
8 2·24 dm^3

Chapter 31 (page 328)
17 3·06 kg

Chapter 32 (page 341)
13 (*g*) 22·5 g

Logarithms

Proportional Parts

	0	1	2	3	4	5	6	7	8	9	1	2	3	4	5	6	7	8	9
10	0000	0043	0086	0128	0170	0212	0253	0294	0334	0374	4	8	12	17	21	25	29	33	37
11	0414	0453	0492	0531	0569	0607	0645	0682	0719	0755	4	8	11	15	19	23	26	30	34
12	0792	0828	0864	0899	0934	0969	1004	1038	1072	1106	3	7	10	14	17	21	24	28	31
13	1139	1173	1206	1239	1271	1303	1335	1367	1399	1430	3	6	10	13	16	19	23	26	29
14	1461	1492	1523	1553	1584	1614	1644	1673	1703	1732	3	6	9	12	15	18	21	24	27
15	1761	1790	1818	1847	1875	1903	1931	1959	1987	2014	3	6	8	11	14	17	20	22	25
16	2041	2068	2095	2122	2148	2175	2201	2227	2253	2279	3	5	8	11	13	16	18	21	24
17	2304	2330	2355	2380	2405	2430	2455	2480	2504	2529	2	5	7	10	12	15	17	20	22
18	2553	2577	2601	2625	2648	2672	2695	2718	2742	2765	2	5	7	9	12	14	16	19	21
19	2788	2810	2833	2856	2878	2900	2923	2945	2967	2989	2	4	7	9	11	13	16	18	20
20	3010	3032	3054	3075	3096	3118	3139	3160	3181	3201	2	4	6	8	11	13	15	17	19
21	3222	3243	3263	3284	3304	3324	3345	3365	3385	3404	2	4	6	8	10	12	14	16	18
22	3424	3444	3464	3483	3502	3522	3541	3560	3579	3598	2	4	6	8	10	12	14	15	17
23	3617	3636	3655	3674	3692	3711	3729	3747	3766	3784	2	4	6	7	9	11	13	15	17
24	3802	3820	3838	3856	3874	3892	3909	3927	3945	3962	2	4	5	7	9	11	12	14	16
25	3979	3997	4014	4031	4048	4065	4082	4099	4116	4133	2	3	5	7	9	10	12	14	15
26	4150	4166	4183	4200	4216	4232	4249	4265	4281	4298	2	3	5	7	8	10	11	13	15
27	4314	4330	4346	4362	4378	4393	4409	4425	4440	4456	2	3	5	6	8	9	11	13	14
28	4472	4487	4502	4518	4533	4548	4564	4579	4594	4609	2	3	5	6	8	9	11	12	14
29	4624	4639	4654	4669	4683	4698	4713	4728	4742	4757	1	3	4	6	7	9	10	12	13
30	4771	4786	4800	4814	4829	4843	4857	4871	4886	4900	1	3	4	6	7	9	10	11	13
31	4914	4928	4942	4955	4969	4983	4997	5011	5024	5038	1	3	4	5	7	8	10	11	12
32	5051	5065	5079	5092	5105	5119	5132	5145	5159	5172	1	3	4	5	7	8	9	11	12
33	5185	5198	5211	5224	5237	5250	5263	5276	5289	5302	1	3	4	5	6	8	9	10	12
34	5315	5328	5340	5353	5366	5378	5391	5403	5416	5428	1	3	4	5	6	8	9	10	11
35	5441	5453	5465	5478	5490	5502	5514	5527	5539	5551	1	2	4	5	6	7	9	10	11
36	5563	5575	5587	5599	5611	5623	5635	5647	5658	5670	1	2	4	5	6	7	8	10	11
37	5682	5694	5705	5717	5729	5740	5752	5763	5775	5786	1	2	3	5	6	7	8	9	10
38	5798	5809	5821	5832	5843	5855	5866	5877	5888	5899	1	2	3	5	6	7	8	9	10
39	5911	5922	5933	5944	5955	5966	5977	5988	5999	6010	1	2	3	4	5	7	8	9	10
40	6021	6031	6042	6053	6064	6075	6085	6096	6107	6117	1	2	3	4	5	6	7	9	10
41	6128	6138	6149	6160	6170	6180	6191	6201	6212	6222	1	2	3	4	5	6	7	8	9
42	6232	6243	6253	6263	6274	6284	6294	6304	6314	6325	1	2	3	4	5	6	7	8	9
43	6335	6345	6355	6365	6375	6385	6395	6405	6415	6425	1	2	3	4	5	6	7	8	9
44	6435	6444	6454	6464	6474	6484	6493	6503	6513	6522	1	2	3	4	5	6	7	8	9
45	6532	6542	6551	6561	6571	6580	6590	6599	6609	6618	1	2	3	4	5	6	7	8	9
46	6628	6637	6646	6656	6665	6675	6684	6693	6702	6712	1	2	3	4	5	6	7	7	8
47	6721	6730	6739	6749	6758	6767	6776	6785	6794	6803	1	2	3	4	5	5	6	7	8
48	6812	6821	6830	6839	6848	6857	6866	6875	6884	6893	1	2	3	4	4	5	6	7	8
49	6902	6911	6920	6928	6937	6946	6955	6964	6972	6981	1	2	3	4	4	5	6	7	8
50	6990	6998	7007	7016	7024	7033	7042	7050	7059	7067	1	2	3	3	4	5	6	7	8
51	7076	7084	7093	7101	7110	7118	7126	7135	7143	7152	1	2	3	3	4	5	6	7	8
52	7160	7168	7177	7185	7193	7202	7210	7218	7226	7235	1	2	2	3	4	5	6	7	7
53	7243	7251	7259	7267	7275	7284	7292	7300	7308	7316	1	2	2	3	4	5	6	6	7
54	7324	7332	7340	7348	7356	7364	7372	7380	7388	7396	1	2	2	3	4	5	6	6	7

Proportional Parts

	0	1	2	3	4	5	6	7	8	9	1	2	3	4	5	6	7	8	9
55	7404	7412	7419	7427	7435	7443	7451	7459	7466	7474	1	2	2	3	4	5	5	6	7
56	7482	7490	7497	7505	7513	7520	7528	7536	7543	7551	1	2	2	3	4	5	5	6	7
57	7559	7566	7574	7582	7589	7597	7604	7612	7619	7627	1	2	2	3	4	5	5	6	7
58	7634	7642	7649	7657	7664	7672	7679	7686	7694	7701	1	1	2	3	4	4	5	6	7
59	7709	7716	7723	7731	7738	7745	7752	7760	7767	7774	1	1	2	3	4	4	5	6	7
60	7782	7789	7796	7803	7810	7818	7825	7832	7839	7846	1	1	2	3	4	4	5	6	6
61	7853	7860	7868	7875	7882	7889	7896	7903	7910	7917	1	1	2	3	4	4	5	6	6
62	7924	7931	7938	7945	7952	7959	7966	7973	7980	7987	1	1	2	3	3	4	5	6	6
63	7993	8000	8007	8014	8021	8028	8035	8041	8048	8055	1	1	2	3	3	4	5	6	6
64	8062	8069	8075	8082	8089	8096	8102	8109	8116	8122	1	1	2	3	3	4	5	5	6
65	8129	8136	8142	8149	8156	8162	8169	8176	8182	8189	1	1	2	3	3	4	5	5	6
66	8195	8202	8209	8215	8222	8228	8235	8241	8248	8254	1	1	2	3	3	4	5	5	6
67	8261	8267	8274	8280	8287	8293	8299	8306	8312	8319	1	1	2	3	3	4	4	5	6
68	8325	8331	8338	8344	8351	8357	8363	8370	8376	8382	1	1	2	3	3	4	4	5	6
69	8388	8395	8401	8407	8414	8420	8426	8432	8439	8445	1	1	2	3	3	4	4	5	6
70	8451	8457	8463	8470	8476	8482	8488	8494	8500	8506	1	1	2	2	3	4	4	5	6
71	8513	8519	8525	8531	8537	8543	8549	8555	8561	8567	1	1	2	2	3	4	4	5	5
72	8573	8579	8585	8591	8597	8603	8609	8615	8621	8627	1	1	2	2	3	4	4	5	5
73	8633	8639	8645	8651	8657	8663	8669	8675	8681	8686	1	1	2	2	3	4	4	5	5
74	8692	8698	8704	8710	8716	8722	8727	8733	8739	8745	1	1	2	2	3	4	4	5	5
75	8751	8756	8762	8768	8774	8779	8785	8791	8797	8802	1	1	2	2	3	3	4	5	5
76	8808	8814	8820	8825	8831	8837	8842	8848	8854	8859	1	1	2	2	3	3	4	5	5
77	8865	8871	8876	8882	8887	8893	8899	8904	8910	8915	1	1	2	2	3	3	4	4	5
78	8921	8927	8932	8938	8943	8949	8954	8960	8965	8971	1	1	2	2	3	3	4	4	5
79	8976	8982	8987	8993	8998	9004	9009	9015	9020	9025	1	1	2	2	3	3	4	4	5
80	9031	9036	9042	9047	9053	9058	9063	9069	9074	9079	1	1	2	2	3	3	4	4	5
81	9085	9090	9096	9101	9106	9112	9117	9122	9128	9133	1	1	2	2	3	3	4	4	5
82	9138	9143	9149	9154	9159	9165	9170	9175	9180	9186	1	1	2	2	3	3	4	4	5
83	9191	9196	9201	9206	9212	9217	9222	9227	9232	9238	1	1	2	2	3	3	4	4	5
84	9243	9248	9253	9258	9263	9269	9274	9279	9284	9289	1	1	2	2	3	3	4	4	5
85	9294	9299	9304	9309	9315	9320	9325	9330	9335	9340	1	1	2	2	3	3	4	4	5
86	9345	9350	9355	9360	9365	9370	9375	9380	9385	9390	1	1	2	2	3	3	4	4	5
87	9395	9400	9405	9410	9415	9420	9425	9430	9435	9440	0	1	1	2	2	3	3	4	4
88	9445	9450	9455	9460	9465	9469	9474	9479	9484	9489	0	1	1	2	2	3	3	4	4
89	9494	9499	9504	9509	9513	9518	9523	9528	9533	9538	0	1	1	2	2	3	3	4	4
90	9542	9547	9552	9557	9562	9566	9571	9576	9581	9586	0	1	1	2	2	3	3	4	4
91	9590	9595	9600	9605	9609	9614	9619	9624	9628	9633	0	1	1	2	2	3	3	4	4
92	9638	9643	9647	9652	9657	9661	9666	9671	9675	9680	0	1	1	2	2	3	3	4	4
93	9685	9689	9694	9699	9703	9708	9713	9717	9722	9727	0	1	1	2	2	3	3	4	4
94	9731	9736	9741	9745	9750	9754	9759	9764	9768	9773	0	1	1	2	2	3	3	4	4
95	9777	9782	9786	9791	9795	9800	9805	9809	9814	9818	0	1	1	2	2	3	3	4	4
96	9823	9827	9832	9836	9841	9845	9850	9854	9859	9863	0	1	1	2	2	3	3	4	4
97	9868	9872	9877	9881	9886	9890	9894	9899	9903	9908	0	1	1	2	2	3	3	4	4
98	9912	9917	9921	9926	9930	9934	9939	9943	9948	9952	0	1	1	2	2	3	3	4	4
99	9956	9961	9965	9969	9974	9978	9983	9987	9991	9996	0	1	1	2	2	3	3	4	4

	0	1	2	3	4	5	6	7	8	9	1	2	3	4	5	6	7	8	9
											Proportional Parts								
·00	1000	1002	1005	1007	1009	1012	1014	1016	1019	1021	0	0	1	1	1	1	2	2	2
·01	1023	1026	1028	1030	1033	1035	1038	1040	1042	1045	0	0	1	1	1	1	2	2	2
·02	1047	1050	1052	1054	1057	1059	1062	1064	1067	1069	0	0	1	1	1	1	2	2	2
·03	1072	1074	1076	1079	1081	1084	1086	1089	1091	1094	0	0	1	1	1	1	2	2	2
·04	1096	1099	1102	1104	1107	1109	1112	1114	1117	1119	0	1	1	1	1	2	2	2	2
·05	1122	1125	1127	1130	1132	1135	1138	1140	1143	1146	0	1	1	1	1	2	2	2	2
·06	1148	1151	1153	1156	1159	1161	1164	1167	1169	1172	0	1	1	1	1	2	2	2	2
·07	1175	1178	1180	1183	1186	1189	1191	1194	1197	1199	0	1	1	1	1	2	2	2	2
·08	1202	1205	1208	1211	1213	1216	1219	1222	1225	1227	0	1	1	1	1	2	2	2	3
·09	1230	1233	1236	1239	1242	1245	1247	1250	1253	1256	0	1	1	1	1	2	2	2	3
·10	1259	1262	1265	1268	1271	1274	1276	1279	1282	1285	0	1	1	1	1	2	2	2	3
·11	1288	1291	1294	1297	1300	1303	1306	1309	1312	1315	0	1	1	1	2	2	2	2	3
·12	1318	1321	1324	1327	1330	1334	1337	1340	1343	1346	0	1	1	1	2	2	2	2	3
·13	1349	1352	1355	1358	1361	1365	1368	1371	1374	1377	0	1	1	1	2	2	2	2	3
·14	1380	1384	1387	1390	1393	1396	1400	1403	1406	1409	0	1	1	1	2	2	2	3	3
·15	1413	1416	1419	1422	1426	1429	1432	1435	1439	1442	0	1	1	1	2	2	2	3	3
·16	1445	1449	1452	1455	1459	1462	1466	1469	1472	1476	0	1	1	1	2	2	2	3	3
·17	1479	1483	1486	1489	1493	1496	1500	1503	1507	1510	0	1	1	1	2	2	2	3	3
·18	1514	1517	1521	1524	1528	1531	1535	1538	1542	1545	0	1	1	1	2	2	2	3	3
·19	1549	1552	1556	1560	1563	1567	1570	1574	1578	1581	0	1	1	1	2	2	3	3	3
·20	1585	1589	1592	1596	1600	1603	1607	1611	1614	1618	0	1	1	1	2	2	3	3	3
·21	1622	1626	1629	1633	1637	1641	1644	1648	1652	1656	0	1	1	2	2	2	3	3	3
·22	1660	1663	1667	1671	1675	1679	1683	1687	1690	1694	0	1	1	2	2	2	3	3	3
·23	1698	1702	1706	1710	1714	1718	1722	1726	1730	1734	0	1	1	2	2	2	3	3	4
·24	1738	1742	1746	1750	1754	1758	1762	1766	1770	1774	0	1	1	2	2	2	3	3	4
·25	1778	1782	1786	1791	1795	1799	1803	1807	1811	1816	0	1	1	2	2	3	3	3	4
·26	1820	1824	1828	1832	1837	1841	1845	1849	1854	1858	0	1	1	2	2	3	3	3	4
·27	1862	1866	1871	1875	1879	1884	1888	1892	1897	1901	0	1	1	2	2	3	3	3	4
·28	1905	1910	1914	1919	1923	1928	1932	1936	1941	1945	0	1	1	2	2	3	3	4	4
·29	1950	1954	1959	1963	1968	1972	1977	1982	1986	1991	0	1	1	2	2	3	3	4	4
·30	1995	2000	2004	2009	2014	2018	2023	2028	2032	2037	0	1	1	2	2	3	3	4	4
·31	2042	2046	2051	2056	2061	2065	2070	2075	2080	2084	0	1	1	2	2	3	3	4	4
·32	2089	2094	2099	2104	2109	2113	2118	2123	2128	2133	0	1	1	2	2	3	3	4	4
·33	2138	2143	2148	2153	2158	2163	2168	2173	2178	2183	0	1	1	2	2	3	3	4	4
·34	2188	2193	2198	2203	2208	2213	2218	2223	2228	2234	1	1	2	2	3	3	4	4	5
·35	2239	2244	2249	2254	2259	2265	2270	2275	2280	2286	1	1	2	2	3	3	4	4	5
·36	2291	2296	2301	2307	2312	2317	2323	2328	2333	2339	1	1	2	2	3	3	4	4	5
·37	2344	2350	2355	2360	2366	2371	2377	2382	2388	2393	1	1	2	2	3	3	4	4	5
·38	2399	2404	2410	2415	2421	2427	2432	2438	2443	2449	1	1	2	2	3	3	4	4	5
·39	2455	2460	2466	2472	2477	2483	2489	2495	2500	2506	1	1	2	2	3	3	4	5	5
·40	2512	2518	2523	2529	2535	2541	2547	2553	2559	2564	1	1	2	2	3	3	4	5	5
·41	2570	2576	2582	2588	2594	2600	2606	2612	2618	2624	1	1	2	2	3	4	4	5	5
·42	2630	2636	2642	2648	2655	2661	2667	2673	2679	2685	1	1	2	2	3	4	4	5	6
·43	2692	2698	2704	2710	2716	2723	2729	2735	2742	2748	1	1	2	2	3	4	4	5	6
·44	2754	2761	2767	2773	2780	2786	2793	2799	2805	2812	1	1	2	3	3	4	4	5	6
·45	2818	2825	2831	2838	2844	2851	2858	2864	2871	2877	1	1	2	3	3	4	5	5	6
·46	2884	2891	2897	2904	2911	2917	2924	2931	2938	2944	1	1	2	3	3	4	5	5	6
·47	2951	2958	2965	2972	2979	2985	2992	2999	3006	3013	1	1	2	3	3	4	5	6	6
·48	3020	3027	3034	3041	3048	3055	3062	3069	3076	3083	1	1	2	3	4	4	5	6	6
·49	3090	3097	3105	3112	3119	3126	3133	3141	3148	3155	1	1	2	3	4	4	5	6	7
	0	**1**	**2**	**3**	**4**	**5**	**6**	**7**	**8**	**9**	**1**	**2**	**3**	**4**	**5**	**6**	**7**	**8**	**9**

	0	1	2	3	4	5	6	7	8	9	Proportional Parts 1	2	3	4	5	6	7	8	9
·50	3162	3170	3177	3184	3192	3199	3206	3214	3221	3228	1	1	2	3	4	4	5	6	7
·51	3236	3243	3251	3258	3266	3273	3281	3289	3296	3304	1	2	2	3	4	5	5	6	7
·52	3311	3319	3327	3334	3342	3350	3357	3365	3373	3381	1	2	2	3	4	5	5	6	7
·53	3388	3396	3404	3412	3420	3428	3436	3443	3451	3459	1	2	2	3	4	5	6	6	7
·54	3467	3475	3483	3491	3499	3508	3516	3524	3532	3540	1	2	2	3	4	5	6	6	7
·55	3548	3556	3565	3573	3581	3589	3597	3606	3614	3622	1	2	2	3	4	5	6	7	7
·56	3631	3639	3648	3656	3664	3673	3681	3690	3698	3707	1	2	3	3	4	5	6	7	8
·57	3715	3724	3733	3741	3750	3758	3767	3776	3784	3793	1	2	3	3	4	5	6	7	8
·58	3802	3811	3819	3828	3837	3846	3855	3864	3873	3882	1	2	3	4	4	5	6	7	8
·59	3890	3899	3908	3917	3926	3936	3945	3954	3963	3972	1	2	3	4	5	5	6	7	8
·60	3981	3990	3999	4009	4018	4027	4036	4046	4055	4064	1	2	3	4	5	6	7	7	8
·61	4074	4083	4093	4102	4111	4121	4130	4140	4150	4159	1	2	3	4	5	6	7	8	9
·62	4169	4178	4188	4198	4207	4217	4227	4236	4246	4256	1	2	3	4	5	6	7	8	9
·63	4266	4276	4285	4295	4305	4315	4325	4335	4345	4355	1	2	3	4	5	6	7	8	9
·64	4365	4375	4385	4395	4406	4416	4426	4436	4446	4457	1	2	3	4	5	6	7	8	9
·65	4467	4477	4487	4498	4508	4519	4529	4539	4550	4560	1	2	3	4	5	6	7	8	9
·66	4571	4581	4592	4603	4613	4624	4634	4645	4656	4667	1	2	3	4	5	6	7	8	10
·67	4677	4688	4699	4710	4721	4732	4742	4753	4764	4775	1	2	3	4	5	7	8	9	10
·68	4786	4797	4808	4819	4831	4842	4853	4864	4875	4887	1	2	3	4	6	7	8	9	10
·69	4898	4909	4920	4932	4943	4955	4966	4977	4989	5000	1	2	3	5	6	7	8	9	10
·70	5012	5023	5035	5047	5058	5070	5082	5093	5105	5117	1	2	4	5	6	7	8	9	11
·71	5129	5140	5152	5164	5176	5188	5200	5212	5224	5236	1	2	4	5	6	7	8	10	11
·72	5248	5260	5272	5284	5297	5309	5321	5333	5346	5358	1	2	4	5	6	7	9	10	11
·73	5370	5383	5395	5408	5420	5433	5445	5458	5470	5483	1	3	4	5	6	8	9	10	11
·74	5495	5508	5521	5534	5546	5559	5572	5585	5598	5610	1	3	4	5	6	8	9	10	12
·75	5623	5636	5649	5662	5675	5689	5702	5715	5728	5741	1	3	4	5	7	8	9	10	12
·76	5754	5768	5781	5794	5808	5821	5834	5848	5861	5875	1	3	4	5	7	8	9	11	12
·77	5888	5902	5916	5929	5943	5957	5970	5984	5998	6012	1	3	4	6	7	8	10	11	12
·78	6026	6039	6053	6067	6081	6095	6109	6124	6138	6152	1	3	4	6	7	8	10	11	13
·79	6166	6180	6194	6209	6223	6237	6252	6266	6281	6295	1	3	4	6	7	9	10	12	13
·80	6310	6324	6339	6353	6368	6383	6397	6412	6427	6442	1	3	4	6	7	9	10	12	13
·81	6457	6471	6486	6501	6516	6531	6546	6561	6577	6592	2	3	5	6	8	9	11	12	14
·82	6607	6622	6637	6653	6668	6683	6699	6714	6730	6745	2	3	5	6	8	9	11	12	14
·83	6761	6776	6792	6808	6823	6839	6855	6871	6887	6902	2	3	5	6	8	9	11	13	14
·84	6918	6934	6950	6966	6982	6998	7015	7031	7047	7063	2	3	5	6	8	10	11	13	14
·85	7079	7096	7112	7129	7145	7161	7178	7194	7211	7228	2	3	5	7	8	10	12	13	15
·86	7244	7261	7278	7295	7311	7328	7345	7362	7379	7396	2	3	5	7	8	10	12	14	15
·87	7413	7430	7447	7464	7482	7499	7516	7534	7551	7568	2	3	5	7	9	10	12	14	16
·88	7586	7603	7621	7638	7656	7674	7691	7709	7727	7745	2	4	5	7	9	11	12	14	16
·89	7762	7780	7798	7816	7834	7852	7870	7889	7907	7925	2	4	5	7	9	11	13	14	16
·90	7943	7962	7980	7998	8017	8035	8054	8072	8091	8110	2	4	6	7	9	11	13	15	17
·91	8128	8147	8166	8185	8204	8222	8241	8260	8279	8299	2	4	6	8	10	11	13	15	17
·92	8318	8337	8356	8375	8395	8414	8433	8453	8472	8492	2	4	6	8	10	12	14	15	17
·93	8511	8531	8551	8570	8590	8610	8630	8650	8670	8690	2	4	6	8	10	12	14	16	18
·94	8710	8730	8750	8770	8790	8810	8831	8851	8872	8892	2	4	6	8	10	12	14	16	18
·95	8913	8933	8954	8974	8995	9016	9036	9057	9078	9099	2	4	6	8	10	12	14	17	19
·96	9120	9141	9162	9183	9204	9226	9247	9268	9290	9311	2	4	6	9	11	13	15	17	19
·97	9333	9354	9376	9397	9419	9441	9462	9484	9506	9528	2	4	7	9	11	13	15	17	20
·98	9550	9572	9594	9616	9638	9661	9683	9705	9727	9750	2	4	7	9	11	13	16	18	20
·99	9772	9795	9817	9840	9863	9886	9908	9931	9954	9977	2	5	7	9	11	14	16	18	21
	0	1	2	3	4	5	6	7	8	9	1	2	3	4	5	6	7	8	9

MOLES AND FORMULA UNITS

Dry solids (or pure liquids or pure gases)

$$\text{Number of moles} = \frac{\text{Mass of solid in grams}}{\text{Molar mass}}$$

e.g. 1·06 g Na_2CO_3, $M_r = 2\times 23+12+3\times 16$
$= 106$

Molar mass $= 106$ g

Number of moles $= \frac{1{\cdot}06}{106} = 0{\cdot}01$

Solutions

A 1 M solution contains 1 mole in 1 dm^3 (litre) or 1000 cm^3 (ml). M stands for mole/dm^3.

$$\text{Number of moles} = \frac{\text{Volume of solution in cm}^3}{1000} \times \text{Multiplier (multiplicand) of M.}$$

e.g. 10 cm^3 3 M HCl.

$$\text{Number of moles} = \frac{10}{1000}\times 3 = 0{\cdot}03$$

Gases

1 mole of a gas at s.t.p. (0°C and 1 atm or 760 mm Hg or 101 kPa or 101 kN/m^2) occupies 22·4 dm^3 or 22 400 cm^3. At room temperature, say 15°C, the volume is about 24 000 cm^3 and this value is accurate enough for many purposes.

$$\text{Number of moles} = \frac{\text{Volume of gas in cm}^3}{24\,000}$$

e.g. 96 cm^3 CO_2.

$$\text{Number of moles} = \frac{96}{24\,000} = 0{\cdot}004$$

Formula units

To find the number of formula units (molecules only for covalent substances), multiply the number of moles by 6×10^{23} (the Avogadro constant).

e.g. 1·06 g Na_2CO_3 = 0·01 moles
$= 0{\cdot}01\times 6\times 10^{23}$ formula units
$= 6\times 10^{21}$ formula units.

OXIDATION NUMBER

The oxidation number (O.N.) approach to naming substances is being applied more and more. It can be used whether the substance is electrovalent or covalent. Hydrogen has an O.N. of plus one and oxygen of minus two. All elements and compounds have a total oxidation number of zero.

Thus in water, H_2O, the O.N. equals

$$(2\times(+1))+(-2) = 0$$

In sodium hydroxide, NaOH, the O.N. of sodium is $0-(-2+1) = +1$. In hydrogen chloride, the O.N. of chlorine is $0-(+1) = -1$. This crosschecks because in sodium chloride $+1\ -1$ does equal zero. In a carbonate radical, CO_3, we have one carbon (+4) and three oxygen (−6) so the overall O.N. is −2. This is the charge on a carbonate ion and it is confirmed by the formula of sodium carbonate being Na_2CO_3.

The idea can be extended to finding the O.N. or valency of a metal in a complex ion such as MnO_4^- which is present in potassium permanganate, $KMnO_4$. The total O.N. is zero which equals $+1\ +x+4(-2)$, so the O.N. of manganese (x) is +7 and a systematic name is potassium manganate(VII). O.N.'s are frequently omitted when substances exhibit a constant valency, e.g. the elements in groups I, II and III. Some old and new names are:

carbonate	carbonate(IV)*
nitrite	nitrate(III)
nitrate	nitrate(V)*
phosphate	phosphate(V)*
sulphite	sulphate(IV)
sulphate	sulphate(VI)*
hypochlorite	chlorate(I)
chlorate	chlorate(V)
dichromate	dichromate(VI)*
chromate	chromate(VI)*

(*) In these cases the key (central) element is exhibiting its group valency and the oxidation number is implied if it is omitted.

Index

A page number in **bold** type guides you to a section dealing with that topic